Controlled Release Systems

Controlled Release Systems

Advances in Nanobottles and Active Nanoparticles

edited by

Alexander van Herk
Jacqueline Forcada
Giorgia Pastorin

Pan Stanford Publishing

Published by

Pan Stanford Publishing Pte. Ltd.
Penthouse Level, Suntec Tower 3
8 Temasek Boulevard
Singapore 038988

Email: editorial@panstanford.com
Web: www.panstanford.com

British Library Cataloguing-in-Publication Data
A catalogue record for this book is available from the British Library.

Controlled Release Systems: Advances in Nanobottles and Active Nanoparticles

ISBN 978-981-4613-21-7 (Hardcover)
ISBN 978-981-4613-22-4 (eBook)

Cover image courtesy: Dr. Younan Xia, Brock Family Chair and GRA Eminent Scholar in Nanomedicine, Georgia Institute of Technology and Emory University, USA

Printed in the USA

Contents

Preface

This book focuses on active nanoparticles, bringing together two interesting areas, nanoparticles and responsive polymers. This is due to the fact that nanoparticles are increasingly popular as carriers of therapeutic agents and as diagnostic tools.

The book consist of two parts: an introductory text targeting the fundamentals of the synthesis and characterization of active nanoparticles for controlled release of drugs and other molecules as well as nanoparticles containing inorganic moieties for therapeutic purposes. This part also includes a chapter on how to select particles for certain applications. The second part of the book contains contributions from experts in the field giving a state-of-the-art overview on the most intriguing biomedical applications of functionalized nanoparticles, with both contributions focusing on a class of particles as well as nanoparticulate formulations targeting certain diseases.

The book defines the new field of nanobottles along with the first scarce examples of such structures. It would be useful for students and researchers in the field of nanoparticle preparation and controlled release systems and bio-pharmaceutical scientists with an interest in the development of new therapeutic strategies.

Alexander van Herk
Jacqueline Forcada
Giorgia Pastorin
Autumn 2015

Chapter 1

General Introduction and Definitions

Alexander van Herk
Institute for Chemical and Engineering Sciences, 1 Pesek Road, Jurong Island, 627833 Singapore
alexander_herk@ices.a-star.edu.sg

1.1 Introduction to Controlled Release Systems: Advances in Nanobottles and Active Nanoparticles

In the area of controlled release of active substances, like drugs, a strong interest in nanoparticles as carriers of active ingredients has arisen. Some of the active components are extremely hydrophobic, without cellular permeability, and susceptible to metabolic degradation; due to this, their use is limited. This kind of agents can be transported without any problem through physiological media by using nanoparticles. The size of the particles is an important parameter because it governs the efficiency of the delivery system. For this type of application, particles having a diameter smaller than 1 μm are especially useful. Micelles and

Controlled Release Systems: Advances in Nanobottles and Active Nanoparticles
Edited by Alexander van Herk, Jacqueline Forcada, and Giorgia Pastorin

ISBN 978-981-4613-21-7 (Hardcover), 978-981-4613-22-4 (eBook)
www.panstanford.com

liposomes are the most common vehicles for this purpose but polymeric nanoparticles are a strongly upcoming field. Polymeric nanoparticles having diameters in the colloidal range are produced by means of polymerization processes in dispersed media. Drugs are taken up into the nanoparticles by adsorption, absorption or "entrapment," or covalent bonding and they are delivered (released) by desorption, diffusion, polymer degradation, physical or chemical polymer transformation (e.g., lower critical solution temperature (LCST), protonation/deprotonation), or a combination of these mechanisms.

New trends are indicating that the objectives to reach require a high degree of control of their properties. Among them are an excellent stability for circulation in blood, new functionalities for subsequent bioconjugation/biovectorization, narrow nanoparticle-size distributions, biocompatibility and/or biodegradability, and sustained drug delivery. Normally the release profiles follow a steady decay curve over a certain period. Nanoparticles that release their contents by external triggering open up new possibilities for therapeutic strategies. External triggering by, for example, light, heat, change in pH, application of ultrasound, etc., open up the possibility to release material on demand. Linking the external triggering to probes that monitor the possible demand for the active substance can lead to a new generation of nanotherapeutics. Such nanostructures that can be opened and closed by external triggers could be seen as nanobottles. In general the walls of the nanocapsules are made out of a responsive polymer. If the system is triggered the complete wall changes its properties in order to release the contents. In the true sense of the word, "nanobottles" would be nanocontainers with the active substance and a lid (only part of the wall) on the container that can be "opened" and "closed" by external triggering. In this book, the concept of the nanocapsules and nanobottles is introduced; the different means of preparing these structures, loading, targeting, and release of drugs is discussed.

1.2 Concept of the Book

In contrast to other books on nanoparticles applied in, for example, therapeutics, the book is focusing on responsive nanoparticles, bringing together two interesting areas, nanoparticles and responsive polymers. The concept of the book is that in Part I we introduce the reader to the concepts of responsive polymers, responsive nanoparticles, the different ways of making these nanostructures, and selection criteria for the use of these nanoparticles. Although most of the applications mentioned in this book are in the area of controlled drug release, many of these particles also find applications in the areas of personal care products, consumer care products, and specialty coatings, among others.

Part I is targeting at the novice level and no particular background in polymer chemistry and nanoparticle synthesis is assumed. The second part of the book consists of contributions from experts in the field, giving a state-of-the-art overview on the field. Besides the experts in the field, the book is also aiming at novices and students in the field of nanoparticle preparation for controlled release and pharmaceutists with an interest in development in new therapeutic strategies.

1.3 Relevant Definitions

In this section we will first introduce the reader to some relevant definitions used in the field of nanoparticles on the basis of the International Union of Pure and Applied Chemistry (IUPAC) document published in 2011 (Slomkowski et al., 2011).

1.3.1 Particle-Related Definitions

Polymer microparticle (often a microsphere): A polymer particle of any shape with an equivalent diameter approximately from 0.1 to 100 μm.

Polymer nanoparticle (often a polymeric nanosphere): A polymer particle of any shape with an equivalent diameter from a few to a hundred nanometers. In the literature this definition is not strictly held and sometimes particles up to 1000 nm are still called nanoparticles (see Chapter 5).

Polymer gel: A gel consisting of a polymer solution in a liquid solvent in which a three-dimensional structure has been formed.

This structure may arise because the polymer is slightly crosslinked or because bridging has occurred by hydrogen bonds, ionic or hydrophobic interactions between lyophobic blocks in the polymer chains, or lyophobic moieties pendant on the polymer chains. Other more complex, three-dimensional structures are possible. Commonly gels have low solid content. However, in spite of that they usually retain their shape and behave as elastic solids with a low yield stress.

Hydrogel: A gel that swells in water, which is the sustaining liquid phase.

Aerogel: A microporous solid in which the disperse phase is a gas. Microporous silica, glass, carbon, and zeolites are common examples, many of which are of extremely low density.

Gel particle: A particle composed of a three-dimensional polymer network; this could be nano- or microparticles. Often (reversibly) responsive polymers are slightly crosslinked to keep the polymer chains together in the expanded state. A gel can still absorb considerable amounts of solvents or molecules. A **microgel** is a gel in the micrometer range. A **nanogel** is a gel in the nanometer range (see Chapter 8).

Latex: A colloidal dispersion of polymer particles in a liquid medium. The polymer may be organic or inorganic.

Latex particle: A polymer particle that is present in a latex.

Synthetic latex: A latex obtained as a product of a heterogeneous polymerization like emulsion polymerization (see Chapter 5).

Artificial latex: A latex obtained by emulsification of polymer/solvent or oligomer/solvent mixtures in a liquid medium.

Hybrid latex:

a. A latex comprising a relatively high-molecular-mass polymer and an oligomer, such as an alkyd, in which there is usually chemical bonding between the two either during synthesis or subsequently after film formation.
b. A latex comprising multicomponent particles, which contain both organic and inorganic material domains (see Chapter 11).

Natural latex: A latex, obtained from a natural source, the disperse phase of which is often polyisoprene obtained from various plants, such as the rubber tree, *Hevea brasiliensis*. Many plants when wounded produce a milky, sticky, sap which is referred to as *latex*.

Inverse latex: A nonaqueous latex in which the disperse phase comprises an aqueous solution of a polymer. These inverse latexes are, for example, produced in inverse emulsion polymerizations.

Core–shell particle: A multicomponent particle comprising at least two domains, one of which (the core) lies within the other(s) forming the polymeric outer layer(s), the shell(s).

Multilayered particle: A multicomponent particle made of at least two different polymers, with an inner core of one polymer and with at least two concentric layers of different polymers.

Occluded particle: A multicomponent particle in which the second-stage polymer forms more than one domain within a matrix of the seed polymer. The number and size of these domains can vary greatly, and their spatial distribution within the particles is often not uniform.

Macroporous particle: A particle containing pores with widths exceeding about 50 nm.

Mesoporous particle: A particle containing pores with widths between 2 nm and 50 nm.

Microporous particle: A particle containing pores with widths not exceeding 2 nm.

Micelle: An aggregate of colloidal dimensions composed of surface-active agents in solution. Micelles are in equilibrium with the molecules or ions from which they are formed (see Chapter 4). The **c**ritical **m**icelle **c**oncentration (CMC) is the concentration of the surfactant above which micelles are formed.

Hemimicelle (also admicelle): Local zone of compact organized surfactants on a surface, usually a monolayer of surfactant on a surface. In the case of a charged surface the surfactants are oriented with their charged head groups toward the solid surface. In case of further addition of surfactant a surface bilayer may form, which is sometimes called admicelle (**ad**sorbed **micelle**).

Vesicle: A bilayer of surfactant molecules in the form of a closed structure that contains solvent (usually water) inside (see Chapter 4). If there is only one bilayer, they are called unilamellar vesicles; otherwise they are called multilamellar vesicles.

Liposome: An artificially prepared spherical vesicle composed of a lamellar-phase lipid bilayer (see Chapter 4). Liposomes are often composed of phosphatidylcholine-enriched phospholipids and may also contain mixed lipid chains with surfactant properties, such as egg phosphatidylethanolamine. If the liposome contains a gas it is called a bubble liposome.

Microcapsule, nanocapsule: A core–shell particle of the appropriate dimensions wherein the core is a liquid, a solid, or a gas (microbubble) that is to be isolated from the exterior environment and that may subsequently be released (see Chapter 10).

Carbon nanotube: An allotrope of carbon with a cylindrical structure (see Chapter 12). Nanotubes are categorized as single-walled nanotubes (SWNTs) and multiwalled nanotubes (MWNTs).

Dendrimer: A repetitively branched molecules (see Chapter 13).

1.3.2 Polymerization Processes Related to Nanoparticle Formation

Emulsion polymerization: A process whereby monomer(s), an initiator, the dispersion medium (and possibly colloid stabilizer) are mixed, forming initially an inhomogeneous system conducive to the polymerization of the monomer(s) resulting in particles of colloidal dimensions. The term "emulsion polymerization" is a misnomer because historically it was thought that polymerization occurred in the droplets of a monomer emulsion. It is now well understood that this is generally not the case and the polymer is formed in the micelles (see Chapter 5).

Inverse emulsion polymerization: An emulsion polymerization in a nonaqueous medium in which the disperse phase is usually an aqueous solution, initially of monomer(s), and finally of a polymer.

Emulsifier-free emulsion polymerization: An emulsion polymerization without the addition of a colloid stabilizer. In emulsifier-free emulsion polymerization, a colloid stabilizer is produced in situ (e.g., the polymerization of styrene initiated with potassium persulfate yields macromolecules with anionic end groups, providing ionic stabilization of the colloidal polystyrene particles).

Seeded emulsion polymerization: Emulsion polymerization with preformed particles (seeds) formed in situ or added from the very beginning to the polymerizing mixture.

Vesicle polymerization: Polymerization inside a vesicle bilayer structure, leading to the formation of a polymer inside this bilayer (see Chapter 5).

Vesicle-templating polymerization: Polymerization onto a vesicle bilayer structure, leading to the formation of a polymer surrounding this bilayer (see Chapter 5).

Encapsulating emulsion polymerization: Polymerization leading to the encapsulation of the solid particle or liquid droplet and often performed inside a surface adsorbed bilayer (admicelle) (see Chapter 5).

Microemulsion polymerization: Emulsion polymerization in which the starting system is a microemulsion and the final synthetic latex comprises polymer colloidal particles dispersed in an aqueous medium (see Chapter 5).

Micellar polymerization: Polymerization in solution of a surfactant bearing a polymerizable functional group, above its critical micelle concentration, that is, in the micellar state (see Chapter 5).

Miniemulsion polymerization: Polymerization of a monomer miniemulsion. All of the polymerization occurs within preexisting monomer droplets without the formation of new particles (see Chapter 5).

Precipitation polymerization: A polymerization process in which the continuous phase is a solvent for the monomer and initiator and a nonsolvent for the formed polymer (see Chapter 5).

Dispersion polymerization: A precipitation polymerization whereby monomer(s), an initiator, and a colloid stabilizer are mixed in a solvent, forming initially a homogeneous system conducive to the polymerization of the monomer(s) and producing a polymer that is insoluble in the solvent, resulting in the formation of polymer microparticles (see Chapter 5).

Suspension (macroemulsion) polymerization: A polymerization process in which a polymer is formed in monomer (or monomer-solvent) droplets in a continuous phase that is a nonsolvent for the polymer. Polymer particles with diameters usually in the range from 50 μm to 2000 μm and with a broad diameter nonuniformity factor are obtained in this process.

Microsuspension polymerization: A suspension polymerization process in which the size of the monomer droplets is close to 1 μm.

1.4 Conclusions

In Part I of the book we introduce the reader to all the necessary concepts to better understand the state-of-the-art overviews in Part II of the book. In Chapter 2 we give an overview of responsive polymers, and in Chapter 3 we show how these polymers can be used to produce responsive nanoparticles. In Chapter 4 the formation of micelles, vesicles, and liposomes is discussed. In Chapter 5 all the polymerization techniques that can produce polymeric nanoparticles and nanocapsules are shown. In Chapter 6 some examples of the new nanobottle structures will be shown and in Chapter 7 selection criteria for the use of nanoparticles will be given.

In Part II of the book we give the state of the art in applying nanoparticles in controlled release using nanogels (Chapter 8), magnetic latex (Chapter 9), nanocapsules, liposomes and micelles (Chapter 10), inorganic nanoparticles (Chapter 11), carbon nanotubes (Chapter 12), and dendrimers (Chapter 13). In Chapter 14 nanoparticles applied in cancer therapy are discussed. In Chapter 15 nanoparticles used for treating neurodegenerative diseases like Alzheimer's and Parkinson's are discussed. In Chapter 16 nanoparticles in nanocardiology and lung and blood diseases are discussed. In Chapter 17 the application of nanoparticles in immunotherapy and vaccines is discussed.

Chapter 2

Responsive Polymers: Types and Properties

Alexander van Herk
Institute for Chemical and Engineering Sciences, 1 Pesek Road, Jurong Island, 627833 Singapore
alexander_herk@ices.a-star.edu.sg

Responsive polymers are polymers that show a drastic and discontinuous change in their physical properties under the effect of an environmental stimulus.

In this chapter, after giving some general principles of polymerization reactions, the behavior of stimuli-responsive polymer solutions are discussed, and in Chapter 3, stimuli-responsive nanoparticles produced by using some of these stimuli-sensitive polymers are presented.

2.1 General Principles of Polymerization Reactions

Polymers can in principle be formed by two classes of polymerization reactions, step-growth polymerizations

Controlled Release Systems: Advances in Nanobottles and Active Nanoparticles
Edited by Alexander van Herk, Jacqueline Forcada, and Giorgia Pastorin

ISBN 978-981-4613-21-7 (Hardcover), 978-981-4613-22-4 (eBook)
www.panstanford.com

(of which polyester formation is an example) and chain reactions (of which free-radical polymerization is an example).

In polycondensation reactions the monomers have two or more functional groups and each functional group is reacting with a complementary group to link up monomer units. The initial oligomers thus formed can by itself be regarded as (macro) monomers again because each oligomer again contains two or more functional groups. These step-growth reactions build up the molecular weight slowly and only at high conversion of the functional groups long polymer chains are formed. An alternative name for these step-growth polymerization reactions is polycondensation reactions because in many (but not all) cases a small molecule like water is produced in the reaction (which condensates at some location in the reactor, hence polycondensation). In a typical polyester polymerization a dialcohol is reacting with a diacid to form ester bonds and water molecules. In case we would like to form the polymer (to synthesize particles) in water this is a very unfavorable situation because the esterification reaction is an equilibrium reaction:

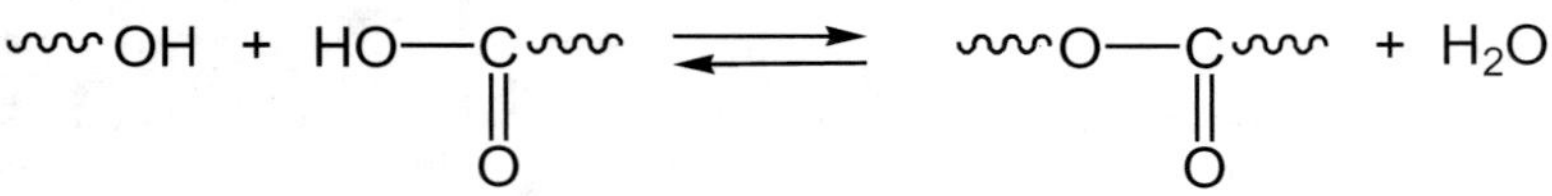

So many of the polymers used in aqueous responsive systems and for biomedical applications are not formed by step-growth polymerizations in water itself. One area where preformed step-growth polymers can easily be used is in the layer-by-layer approach (see Chapter 10). In that case the step-growth polymer is formed separately and subsequently used in aqueous systems to form capsules. Hydrolytical stability of polyester and polyamines remains a point of attention, although it can be brought to use in those cases where (bio) degradable polymers are desired.

The other group of polymerization reactions are the chain reactions where we have, for example, anionic, cationic, Ziegler–Natta, and radical polymerizations. The first three chain polymerization reactions are not easy to perform in water, so the most commonly used polymerization mechanism in preparing responsive polymers in water is radical polymerization, and we will go a bit more in depth into this mechanism.

The mechanism of free-radical polymerizations belongs to the class of so-called chain reactions. Chain reactions are characterized by the fast subsequent addition of monomers to an active center at the chain end. The activity of the growing chain is transferred to the adding unit. The active centers are present in very low concentrations (10^{-5}–10^{-8} mol/L). The rate of addition is very high (10^3–10^4 units per second) and the time of growth of a chain (time between initiation and termination of a chain) is quite short (0.1–10 seconds) relative to the total reaction time, which can be in the order of several hours.

This means that composition of the chain and chain length is determined in seconds. Terminated chains, in principle, do not take part in further reactions (except when transfer-to-polymer events occur). The final chemical composition distribution and molecular mass distribution is determined by the accumulation of rapidly produced dead chains (chains without an active center). In free-radical polymerization the active center is a free radical.

During the formation of a polymer chain a number of subsequent kinetic events take place: (1) radical formation, (2) initiation, (3) propagation, and (4) termination. Transfer of the radical activity to another molecule is a complication that will often take place also. Sometimes a chain transfer agent is added deliberately to control and limit the molecular weight.

2.1.1 Radical Formation in Free-Radical Polymerization

The formation of free radicals can take place in a number of ways. Radicals can be produced by photoinitiation and radiation (γ-radiation or electron beams), electrochemically, and by thermal initiation. Well-known examples of the thermal decomposition of initiators are:

$$\varphi - \mathrm{C} - \mathrm{O} - \mathrm{O} - \mathrm{C} - \varphi \xrightarrow{80^\circ\text{–}90^\circ\mathrm{C}} 2\varphi - \mathrm{C} - \mathrm{O}\bullet \rightarrow 2\varphi\bullet + \mathrm{CO}_2$$

benzoyl peroxide (BPO)

$$\mathrm{NC} - \mathrm{C(CH_3)_2} - \mathrm{N} = \mathrm{N} - \mathrm{C(CH_3)_2} - \mathrm{CN} \xrightarrow{60\text{–}70^\circ\mathrm{C}} 2\mathrm{NC} - \mathrm{C(CH_3)_2}\bullet + \mathrm{N_2}$$

$\alpha\alpha'$ azobis(isobutyronitrile) (AIBN)

A schematic representation of the decomposition of the initiator (I) into two radicals (R•), with a decomposition rate coefficient (k_d), and an expression for the rate of decomposition of the initiator (v_d) is given below:

$$I \xrightarrow{k_d} 2R\bullet \qquad \text{with rate } v_d = \frac{d[R\bullet]}{dt} = 2k_d[I] \tag{2.1}$$

This reaction has a high activation energy (140–160 kJ/mol, depending on the initiator), so k_d depends strongly on temperature.

In fact not all radicals will initiate a polymeric chain; some of the radicals are lost in side reactions (like recombination of the initiator fragments). For this reason the efficiency factor f is introduced. f is the fraction of radicals that actually initiate a polymeric chain, the rate of radical production (leading to an actual initiation step) ρ_i equals then:

$$\rho_i = 2k_d f[I] \tag{2.2}$$

2.1.2 Initiation

This is the addition of the first monomeric unit to the initially formed free radical:

$$R\bullet + M \xrightarrow{k_i} RM\bullet \quad \text{with rate } R_i = k_i[R\bullet][M] \tag{2.3}$$

2.1.3 Propagation

This is the process for the growth of the chains:

$$RM\bullet + M \xrightarrow{k_p} -M_2\bullet$$

$$-M_2\bullet + M \xrightarrow{k_p} -M_3\bullet$$

$$-M_i\bullet + M \xrightarrow{k_p} -M_{i+1}\bullet$$

$$\text{with rate} \quad R_p = k_p[M\bullet][M] \tag{2.4}$$

2.1.4 Termination

Termination takes place via two types of bimolecular free-radical reactions:

In combination the two radicals form a new bond, connecting the two growing chains to form one dead chain with the combined length of the two growing chains.

$$\text{-combination: } -M_i\bullet + -M_j\bullet \xrightarrow{ktc} -M_{i+j}-$$

In disproportionation a radical abstracts a proton from the chain end of another growing chain, leading to two dead chains, one with a double bond and one with a saturated chain end.

$$\text{-disproportionation: } -M_i\bullet + -M_j\bullet \xrightarrow{ktd} -M_i = + -M_j$$

at a rate $R_t = 2k_t[M\bullet]^2$ (2.5)

where k_t is a combination of k_{tc} and k_{td}. The rate of termination is usually determined by the rate of diffusion of the polymer chains. Because rates of diffusion are dependent on the viscosity of the medium and the size of the diffusing species this means that the rate of termination is dependent on conversion and on the chain length of the polymer chains.

An important feature of a polymer chain is its chain length and chain length distribution.

First we introduce the kinetic chain length (ν), which is defined independent of the mode of termination, and later we will link the kinetic chain length to the degree of polymerization.

A simple way of identifying the kinetic chain length is to multiply the rate of monomer addition at the growing chain end ($k_p[M]$) with the average growth time of a chain radical (τ_g):

$$\nu = k_p[M]\tau_g = \frac{k_p[M]}{2k_t[M\bullet]} \quad (2.6)$$

Equation 2.6 can easily be used to predict the effect of changes in recipe on the kinetic chain length:

- with increasing conversion the monomer concentration decreases (in a batch reaction) and therefore the kinetic chain length decreases approximately exponentially;

$$\nu \cong \nu_0 e^{-k't} \quad (2.7)$$

- when increasing the initiator concentration the rate of termination (Eq. 2.5) increases and therefore the average growth time of a chain decreases, which in turn will decrease the kinetic chain length;
- when comparing monomers with different k_p values it is obvious that, under otherwise similar conditions, a monomer

with a higher k_p value will produce chains with a higher kinetic chain length;

- another process that will affect the average growth time of a chain is a transfer reaction, terminating a growing chain sooner than without transfer, decreasing τ_g and thus creating a shorter chain;
- with increasing conversion the viscosity of the medium increases and therefore the rate of termination decreases, which will increase the average growth time of the chain and increasing the chain length.

For the number averaged degree of polymerization $\overline{P}_n$ of the chains that are formed at a particular point in time, the following holds: $\overline{P}_n$ = number of monomer units added in the time interval dt over the number of dead chains formed in dt. Therefore, $\overline{P}_n = 2\nu$ for termination by combination, because two growing chains produce one dead chain and $P_n = \nu$ for termination by disproportionation, because two growing chains will produce two dead chains.

Because of the very short lifetime of a growing polymer chain (less than a second) it is impossible to produce block copolymers (one simply cannot change from one monomer to another monomer within the growth time of a polymer chain). To change monomer that will result in other monomer units added within the same chain one needs much longer single-chain growth times. This is the strength of controlled radical polymerization (CRP, also living radical polymerization). In controlled or living radical polymerization the radical is protected against termination and continues to grow during the complete reaction time. Examples of CRP are reversible addition-fragmentation transfer (RAFT) polymerization, atom transfer radical polymerization (ATRP) and nitroxide-mediated polymerization (NMP). With these special forms of radical polymerization one is able to produce many interesting chain structures (diblock copolymers, gradient copolymers, etc.) and especially RAFT and ATRP are widely used in the field of the synthesis of responsive polymers, particles, and micellar structures. For further reading refer to Moad and Solomon (2005) and Davis (2006).

In Fig. 2.1 some typical polymers are shown that are responsive. All of these polymers are formed in radical polymerization.

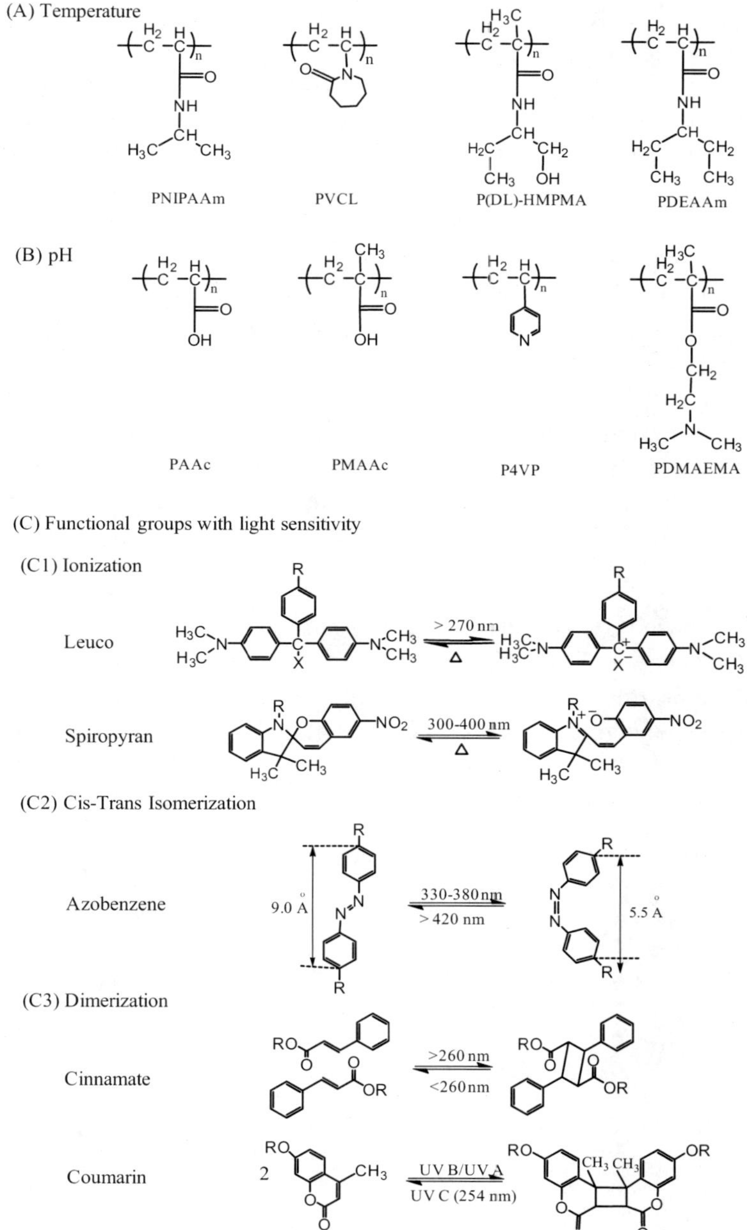

Figure 2.1 Typical polymers (A, B) and functional units (C) applied in (A) temperature-, (B) pH-, and (C) light-responsive systems.

2.2 Temperature-Responsive Polymers

When a polymer dissolves in a solvent it takes the form of a coil. The size of the coil is related to, amongst others, the molecular weight of the polymer, the structure of the polymer, and the interactions between the polymer and the solvent. A transition from dissolved polymer to polymer in the form of globules happens at the so-called lower critical solution temperature (LCST). Below the LCST there are favorable interactions between the polymer and the solvent (water in most of the examples in this book). The most studied synthetic responsive polymer is poly(*N*-isopropylacrylamide) (PNIPAm). The favorable polymer–solvent interactions are the hydrogen bonds between the amide groups and the water molecules (Fig. 2.2) and contribute to a negative enthalpy term ($\Delta H < 0$) in the Gibbs equation and an overall negative Gibbs free energy ($\Delta G < 0$).

$$\Delta G = \Delta H - T\Delta S \tag{2.8}$$

Above the LCST the entropic term dominates the overall Gibbs free energy unfavorably ($\Delta G > 0$), water molecules are released into the bulk aqueous phase, and the internal hydrophobic interactions in the polymer chain lead to compact globule formation.

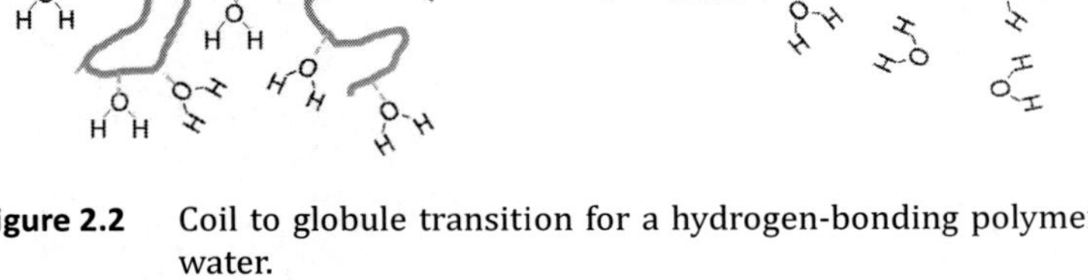

Figure 2.2 Coil to globule transition for a hydrogen-bonding polymer in water.

The LCST phenomenon is widespread for polymers containing hydrogen-bonding sites for water molecules. Other examples are poly(*N*-vinylcaprolactam) (PVCL), poly(ethylene oxide) (PEO, also PEG), and poly(propylene oxide) (PPO).

Copolymers of the hydrogen bond–forming monomer unit with other monomer units can shift the LCST value.

The collapse of the coil structure can lead to gelation giving a reversible thermal gelation (RTG) behavior.

Some polymers show an upper critical solution temperature (UCST), which means they render soluble at a higher temperature, for example, zwitterionic polymers that have a lot of intrachain interactions (Vasanatha, 2014) and nonionic polymers that contain both H bond donor and acceptor amide sites like poly(*N*-acryloylasparaginamide) (Glatzel, 2011).

Introducing these thermoresponsive polymers as blocks into block copolymer structures opens up the possibility of creating responsive micelles/particles (see Chapter 3).

2.3 pH-Responsive Polymers

pH-responsive polymer solutions represent another well-known group of responsive polymers. pH-responsive polymers contain ionisable functional groups capable of donating or accepting protons upon environmental pH changes.

Polyacids such as poly(acrylic acid) (PAAc) and poly(methacrylic acid) (PMAAc) can be deprotonated at higher pH values (the pKa values are in the range of 5). The polyacids become more water soluble and the coils expand. In contrast to low-molecular-weight acids the pKa values change with the degree of deprotonation (α) of the polyacid. This can be understood by the negative charge building up on the polyacid, making it more difficult for the next proton to be dissociated. This pKa value and also the changes in the pKa values with α are depending on the other ions present in solution. pH-responsive polybases accept protons when the pH is lowered and expand under acidic conditions. Both amino and amine functional groups in poly(*N,N′*-dimethyl aminoethyl methacrylate) (PDMAEMA) and poly(vinyl pyridine) can be protonated to give these transitions. The same remarks about changes in pK as a function of degree of protonation and other ions can be made for the polybases.

2.4 Salt-Responsive Polymers

Although not so likely to be used in biomedical applications, salt-responsive polymers are of general interest. Poly{[2-(methacryloyloxy)ethyl] trimethylammonium iodide} (PMETAI), a highly charged polymer, collapses at high salt concentrations. In many cases the LCST of thermoresponsive polymers is dependent on the type of salt and the salt concentration. This is not a surprise because the LCST is based on a balance between water that is hydrogen-bonded to functional groups in the polymer and the entropy gain when this water is released. When at one temperature, one can shift the LCST in such a way with the introduction of salt that the polymer dissolves or precipitates (Fig. 2.3).

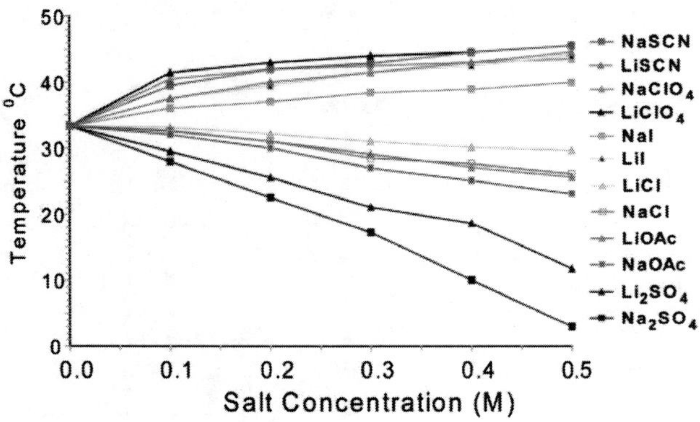

Figure 2.3 Effect of salt type and concentration on the LCST of PEGMA-EE 246/PEGMA-ME 475 copolymers. Adapted with permission from Magnusson et al. Copyright 2008, American Chemical Society.

Zwitterionic polymers are sometimes insoluble in pure water but dissolve in high salt concentrations. An example of such a zwitterionic polymer is based on the zwitterionic monomer depicted in Fig. 2.4.

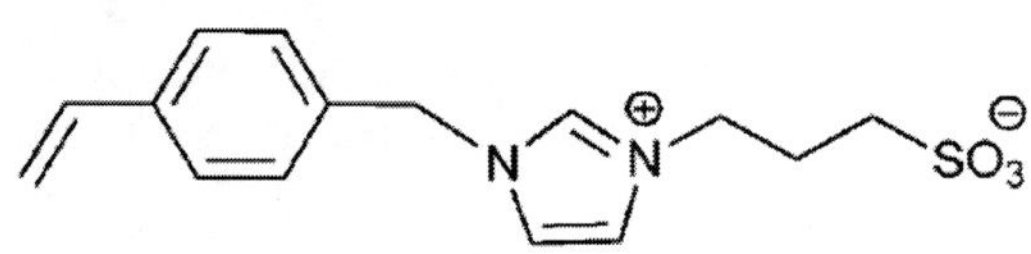

Figure 2.4 Imidazole-based sulfobetaine.

2.5 Electromagnetic Radiation-Responsive Polymers

The presence of photochromic groups in the polymer chains can lead to polymers that respond to electromagnetic radiation. The responses can be based on *cis-trans* isomers, ionization of monomer units, and dimerization of monomer units (Liu and Urban, 2010). Typically this concept works with UV radiation but also IR radiation.

2.6 Multiresponsive Systems

From the above it is clear that one and the same polymer can respond to different stimuli, for example, PNIPAm can respond to changes in temperature but also changes in salt concentration. Polyacids respond to changes in pH and also ionic strength. Combining more than one responsive monomer unit in a copolymer can lead to multiresponsive switchable copolymers like, for example, the copolymer of NIPAm and spirobenzopyran, which is temperature-, pH-, and photoresponsive (Sumaru, 2004). Furthermore see a review on multiresponsive polymers and gels (Pasparakis, 2011).

Chapter 3

Responsive Polymeric Nanoparticles

Jacqueline Forcada
POLYMAT and Applied Chemistry Department, University of the Basque Country UPV/EHU, Ado. 1072, 20080 Donostia-San Sebastián, Spain
jacqueline.forcada@ehu.es

3.1 Introduction

As described in Chapter 2, responsive polymers are available in a variety of forms and structures, from macroscopic or bulk materials to constituting nanoparticles. Responsive polymers capable of undergoing structural changes in response to an external stimulus have attracted much attention due to the possibility of producing materials able to give measurable responses to environmental changes. Great interest has been directed in recent years to those environmentally responsive polymers forming responsive polymeric nanoparticles (RPNPs) due to the following aspects.

The introduction of therapeutic agents into the body by means of the different administration routes that can be employed must overcome a set of efficient biological barriers, which constitute the system defense of the body. Since many drugs potencies and

Controlled Release Systems: Advances in Nanobottles and Active Nanoparticles
Edited by Alexander van Herk, Jacqueline Forcada, and Giorgia Pastorin

ISBN 978-981-4613-21-7 (Hardcover), 978-981-4613-22-4 (eBook)
www.panstanford.com

therapeutic effects are limited or otherwise reduced due to the partial degradation that occurs before they reach the desired target in the body, the design of a suitable carrier is necessary. For the administration of therapeutics the use of stimuli responsive systems at the nanoscale (nanocarriers) able to overcome these impediments and bypass such barriers is considered a safe, selective, and efficient way to achieve this objective.

The use of stimuli RPNPs in biomedical applications is an emerging field that has to intensify studies on biodistribution, biodisposability, and toxicity to provide real solutions to the challenges presented by their use under biological conditions. The interest on RPNPs has to be highlighted by comparison with other type of responsive materials as macroscopic polymeric responsive bulk assemblies, as, for example, flat-surface grafted responsive polymer brushes. RPNPs allow more quick response and efficient delivery of therapeutics to pathological regions due to their smaller size compared to that of the macroscopic homologues (Ramos et al., 2014).

There are some considerations in the design of RPNPs, such as the different factors contributing to their properties, which give the desirable and appropriate responses for the targeted application (Gibson and O'Reilly, 2013). As an example, if the responsivity required for the bioapplication should be given by a responsive corona or by the whole nanoparticle being responsive (in the case of a uniformous composition of the particle). The changes in chemical and/or physical properties occurring in response to stimuli endow the RPNPs the ability to retain cargoes during circulation, target the pathological region, and release the cargoes in the desired location. RPNPs have been produced based on different nanostructures such as micelles, vesicles, crosslinked nanoparticles, hybrid nanoparticles, and nanocapsules (see Chapters 4–6 for the synthesis of these nanostructures).

Different authors are indicating that the objectives to reach in the future design and development of RPNPs as delivery systems for in vivo applications require a high degree of control of their properties. Among them, an excellent stability during circulation in blood, new functionalities for subsequent bioconjugation/biovectorization, nanometric sizes, biocompatibility and/or biodegradability, and specific groups for selective absorption in specific cells in the case of using RPNPs in gene delivery.

RPNPs to be used as vehicles for delivering and releasing molecules or biologically active material must remain chemically unaltered and be pharmacologically stable during their transit from the administration site to the target where they are going to exert their effect. Likewise, the characteristics of the vehicle must be such that it has to be not only compatible with the path that it has to go through but also compatible with the dosing site, that is, it has to be ready to release the active ingredient at the precise moment when it has reached the target and also allowing bioavailability. On the other hand, the carrier or vehicle must not only maintain its characteristics and properties in an aqueous medium but must also be capable of showing its properties again if it needs to be stored in a dry (lyophilized) state for the application and/or transport. In other words, once redispersed, it must again show its properties and behavior with respect to changes in external stimuli.

Considering that one of the most interesting advances in nanomedicine (the use of nanotechnology in medicine), and more specifically in drug delivery, are multifunctional and responsive nanoparticulate systems for simultaneous imaging of tumors and drug delivery, taking into account the variety of features required to use these systems for in vivo applications, and the inherent difficulties in the transfer of new responsive nanoparticulate systems to the clinical practice, this chapter is devoted to the most promising RPNPs reported in the literature, which can be actually or potentially used in biomedical applications.

3.2 Design and Production of RPNPs

An ideal drug delivery carrier should have a few common features including a smaller particle size (10–200 nm), biodegradability and/or biocompatibility, prolonged blood circulation time, high drug or enzyme loading, and/or entrapment and protection of molecules from the body immune system. The multifunctional properties of RPNPs can be achieved by chemical functional group and by using surface-active and stimulus-responsive constituents (Vinogradov et al., 2002; Yallapu et al., 2011). Polymerization in dispersed media of the adequate monomers under specific reaction conditions will give as a result new RPNPs that could be useful in biomedical applications due to their biocompatibility (being even biodegradable) and could be used as delivery systems for biologically active agents or as vehicles/carriers of other nanoparticles (e.g., magnetic nanoparticles; see Chapter 9).

Table 3.1 summarizes the different types of RPNPs that meet the requirements previously commented for the nanoparticles to be used in bioapplications.

Among the different RPNPs, **micro-/nanogels** are the most interesting ones in terms of their use as drug delivery nanocarriers (see Chapter 8). From the biotechnological application's point of view, the interest in nanogels comes from their stimuli-responsive nature, that is, from their ability to suffer reversible phase transitions in response to stimuli or changes in the medium (Pelton, 2000). Moreover, nanogel particles can respond to changes in the medium more quickly than macroscopic gels due to their nanometric small size. Stimuli-responsive nanogels are crosslinked colloidal particles, which can swell by absorption of large amounts of solvent, but they do not dissolve due to the constituting structure of the polymer network, physically or chemically crosslinked. Nanogels have sizes comprised in the colloidal range, possessing the ability to vary their volume in response to external stimuli (Ramos et al., 2014). Despite the International Union of Pure and Applied Chemistry (IUPAC) recommendations of 2011 (Slomkowski et al., 2011) on defining a gel nanoparticle or nanogel (see Chapter 8), until recently, there was no worldwide consensus on the definition of nanogels; however, on the basis of the results from various research groups, gel nanoparticles smaller than 200 nm in size are considered to be nanogels. Some authors call them "smart" gels, but their behavior is governed by the solution thermodynamics of the crosslinked polymeric chains. In this way, their "smartness" is relative and in general their behavior (swelling–de-swelling) depends on the constituting components and interactions. Nanogels exhibit a behavior that goes from a polymeric solution (swell form) to a hard particle (collapsed form) (Bradley et al., 2011). Swelling and shrinking are caused by conformational changes of the subchains between two nearby crosslinking points inside the gel network. This behavior is governed by the balance and result of the competition between repulsive intermolecular forces acting to expand the

Table 3.1 Types of responsive polymeric nanoparticles

Responsive polymeric nanoparticle type	Characteristics	Comments/Possible uses	References
Micro-/Nanogels (NGs)	-Thermoresponsive, pH-responsive, light-sensitive, ionic strength, enzyme, redox potential, ultrasound, magnetic field–sensitive NGs -Biocompatible and/or degradable NGs	-PVCL-based, PNIPAM-based, PDEAEMA-based NGs -Poly(ethylene glycol)-, polyethylenimine-, poly(MeODEGM), poly(AEMA)–based NGs -Cargo can be conjugated to or entrapped inside the polymeric network.	Pelton 2000; Ramos et al., 2011, 2012, 2014; Imaz et al., 2008, 2011; Pikabea et al., 2014; Vihola et al., 2005; Bradley et al., 2011; Klinger and Landfester 2012; Loomis et al., 2011; Bhuchar et al., 2012; Chapter 8
	-Bacteria-responsive NGs	-They can differentially deliver antibiotics to bacterial infection sites.	Liu and An 2014
	-Multifunctional self-crosslinked NGs	-In some cases, NGs should be able to load both hydrophilic and hydrophobic drugs to obtain a synergistic effect.	Liu and An 2014
	-Based on cleavable crosslinking points	-Hydrolytically cleavable crosslinkers -Enzymatically cleavable crosslinkers -Photodegradable crosslinkers	Klinger and Landfester 2012

(*Continued*)

Table 3.1 (*Continued*)

Responsive polymeric nanoparticle type	Characteristics	Comments/Possible uses	References
	-Enzymatically crosslinked NGs	-Enzyme-catalyzed formation of NGs under mild conditions similar to native biological conditions to avoid the use of potentially toxic chemicals or detrimental polymerization conditions for delivery of biomacromolecules such as protein for treatment of disease -Enzymatically degradable NGs	Liu and An, 2014 Aguirre et al., 2013
Micelles (responsive polymer-based)	- Poly(lactic acid) (PLA)-, PLG- (polyglycolide), and poly(lactic-co-glycolic acid) (PLGA) based -PDEAEMA based -Poly(caprolactone, poly(histidine), poly(aspartic acid), poly(L-glutamic acid) -Dual pH- and temperature-responsive micelles	-For encapsulation of Pt(IV) prodrugs -For fast release of drugs in acidic conditions -Can carry cargo chemically conjugated to polymers or inside interior -The slight acidic environment and the body temperature found near tumors result in the specific release of drugs from the mixed micelles.	Callari et al., 2014 Loomis et al., 2011 Chen et al., 2012f Lo et al., 2009

(*Continued*)

Table 3.1 (*Continued*)

Responsive polymeric nanoparticle type	Characteristics	Comments/Possible uses	References
Responsive polymeric nanoconjugates (drug nanoconjugates and complexes)	-A highly innovative technology in nanomedicine -Nanoconjugate concept for cancer therapy to avoid multiple-drug resistance in chemotherapy	-Nanoconjugates that carry more than one functional group (–OH. –COOH, –NH_2) provide the capability to simultaneously inhibit several tumor pathways, deliver optimal drug concentrations to the site of treatment, and reduce adverse effects on healthy tissue -Proteins and peptides can form NPs by complexation -Doxorubicin chemically conjugated to polypeptides using thiol chemistry, which drives NP assembly, increases plasma circulation of drug and cellular internationalization	Kateb et al., 2011 Janib et al., 2010
Responsive nanocapsules	-By layer-by-layer assembly: poly(L-lysine) with poly(glutamic acid) on colloidal silica templates -Responsive nanocapsules	-For cisplatin and hydrophobic drugs encapsulation	Callari et al., 2014 Aguirre et al., 2014

(*Continued*)

Table 3.1 (*Continued*)

Responsive polymeric nanoparticle type	Characteristics	Comments/Possible uses	References
Responsive vesicles	-PLGA, PLA, polycaprolactone), chitosan	-Polymeric vesicles have the advantage of encapsulating hydrophobic and hydrophilic therapeutic agents: low-molecular-weight drugs, proteins, enzymes, and genes. -For encapsulation of functionalized proteins -Drugs can be located in the core or can be electrostatically complexed or covalently attached to polymers.	Janib et al., 2010
Prodrug*-responsive nanocarriers	-Drug–polymer conjugate-based colloids for sustained anticancer drug release	-"Conjugation through" strategy consists of grafting a drug on a monomer prior to its polymerization. -"Conjugation to" is the conjugation of a drug to a polymer scaffold to achieve drug-polymer NPs.	Delplace et al., 2014

*A **prodrug** is a drug that is administered in an inactive or less than fully active form. It becomes converted to its active form through a normal metabolic process, such as hydrolysis of an ester form of the drug.

tridimensional polymer network of polymeric chains loosely crosslinked forming the nanogel and attractive forces that act to shrink it: swelling occurs when the ionic repulsion and the osmotic forces are higher than the attractive forces, that is, hydrogen bonds, van der Waals, and hydrophobic interactions.

Different polymerization methods or techniques in dispersed media (see Chapters 5 and 8) are being used for the preparation of nanogels, among them, emulsion polymerization, inverse microemulsion polymerization, anionic copolymerization, crosslinking between neighboring chains, and others. At this point it is necessary to comment that some authors prefer to apply the terminology of polymers accepted in the IUPAC recommendations of 2011, and they use "precipitation polymerization" for the polymerization process used to produce nanogels instead of using "emulsion polymerization" due to the high monomer solubility in water compared to that of the produced polymer (Slomkowski et al., 2011). This IUPAC recommendation proposes the use of "precipitation polymerization" for a polymerization in which monomer(s), initiator(s), and colloidal stabilizer(s) are dissolved in a solvent, and this continuous phase is a nonsolvent for the formed polymer beyond a critical molecular weight. Nevertheless, a number of authors continue using the term "emulsion polymerization" for the polymerization technique used to synthesize nanogel particles.

The ability the nanogels have to undergo large reversible changes in volume makes them interesting and suitable materials to be used as carriers for the uptake and release of molecules, compounds, or other materials. Nanogels can respond to physical stimuli (temperature, ionic strength, magnetic, or electric fields), chemical stimuli (pH, ions, specific molecules, etc.), and biochemical stimuli (enzymatic substrates, affinity ligands, etc.) (Ramos et al., 2011). Among them, temperature is the most studied one because it is an effective stimulus in a number of applications. Nanogels that are able to undergo a volumetric phase change by changing the temperature of the dispersion medium are very interesting in biotechnological applications needing the delivery of an active compound or biomolecule in media in which the main variable to consider is temperature. The design and controlled production of thermoresponsive nanogels has received considerable interest due to their unique feature to swell at low temperatures and being collapsed at high ones in aqueous solutions, showing a volume phase transition temperature (VPTT) (see Fig. 3.1). At this point it is important to note that when polymers having a lower critical solution temperature (LCST) (see Chapter 2) are in the form of crosslinked nanoparticles, the phase transition is observed as a VPTT. This unique behavior makes those nanogels with VPTTs near physiological temperature and based on a biocompatible polymer very attractive for many potential bioapplications, such as drug delivery.

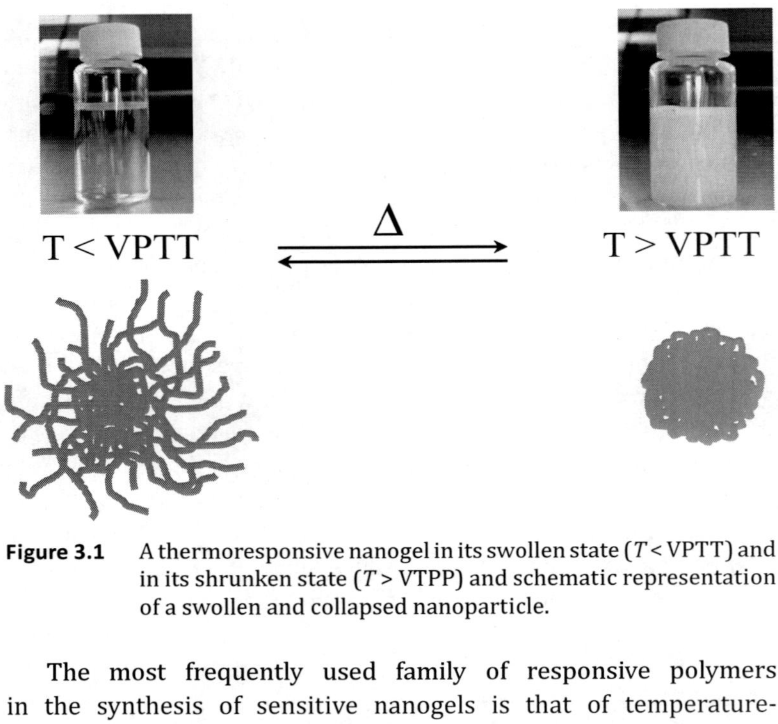

Figure 3.1 A thermoresponsive nanogel in its swollen state (T < VPTT) and in its shrunken state (T > VTPP) and schematic representation of a swollen and collapsed nanoparticle.

The most frequently used family of responsive polymers in the synthesis of sensitive nanogels is that of temperature-sensitive poly(alkylacrylamides), more specifically poly(*N*-isopropylacrylamide (PNIPAM). However, its toxicity prevents its use in biomedical applications (Ramos et al., 2012). Nevertheless, during the past few years a number of articles and patents have appeared on this type of nanogels. Among biocompatible and temperature-sensitive monomers there is *N*-vinylcaprolactam (VCL) (Vihola et al., 2005), which is a water-soluble monomer. The corresponding polymer, poly(*N*-vinylcaprolactam) (PVCL), combines useful and important properties because, together with its biocompatibility, it has a phase transition in the physiological temperature region (32°C–38°C). This combination of properties allows it to be considered as an adequate material for the design of biomedical devices and useful in drug delivery systems (Ramos et al., 2014).

Another type of sensitivity that is relevant to biomedical applications is the response to pH changes. This is the case of pH-sensitive nanogels: they swell when pH approximates the pKa of the ionic monomer incorporated by copolymerization in the crosslinked chains constituting the particles. These nanogels are useful in the case of releasing a biologically active compound in a physiological medium in which the main characteristic is the change in pH. pH-sensitive nanogels are composed of crosslinked polyelectrolytes with weakly acidic (i.e., carboxylic) or weakly basic (i.e., amino) groups that can be used either as proton donors or receptors or through a combination of both. The choice of polymer depends on the physiological conditions of the target in which the delivery is needed. pH-responsive nanogels are able to swell in response to small pH variations, showing a volume phase transition pH (VPTpH). Below this transition pH, nanogel particles are swollen, and above it, they are shrunken. The volume change is ascribed to the enhanced electrostatic repulsion among charges within the polymer network that appears due to the ionization of ionizable groups varying the pH. Considering this interesting property, they can be used at different levels: at the organ level, as the gastrointestinal tract is characterized by pH gradients; at the tissue level, taking advantage of the characteristic acidic extracellular environment of unhealthy tissues different from healthy tissues; and at the cellular level, since endolysosomes are more acidic in comparison to the cytoplasm (Gao et al., 2010; Zha et al., 2011).

As commented above, RPNPs for biomedical applications are complex nanomaterials demanding multifunctional properties. In this sense, apart from thermosensitivity other responsiveness is usually required. The design of multisensitive nanogels is well documented in the literature and basically it can be obtained by incorporating different stimulus responsive polymers into their network via (i) copolymerization of several monomers or (ii) "seed and feed" process.

The preparation of dually pH- and temperature-responsive nanogels by emulsion copolymerization of a thermoresponsive monomer (i.e., NIPAM or VCL) with an ionizable comonomer is well reported in the literature (see Chapter 8). In the case of the preparation of acid containing PVCL-based nanogels, the real challenge is the elimination or minimization of the hydrolysis of VCL (Imaz et al., 2008). Most of the works found in literature did not pay attention to the hydrolysis of VCL; however, with ionizable comonomers, such as AA, the synthesis strategies have an important role if one wants the hydrolysis to be minimized (Imaz and Forcada, 2011).

Recently, the synthesis of poly(2-(diethylamino)ethyl) methacrylate (PDEAEMA)-based nanogels with dual pH and T-sensitivities has been reported (Pikabea et al., 2014). It is well known that PDEAEMA is a pH-, temperature-, and ionic strength–responsive polymer with interesting and potential uses in biomedical applications. When this cationic polyelectrolyte is crosslinked, dual and tunable responsive nanogels are obtained. It is remarkable that their tunable and dual sensitivity, when dispersed in phosphate-buffered saline (PBS), is given at physiological conditions (38°C and pH 7.1), demonstrating the potential use of these RPNPs in drug delivery as they can be tailored to suit the needs of this particular bioapplication.

Apart from biocompatibility, which is a characteristic to fulfill if the RPNPs are designed for in vivo applications, biodegradability is in some cases compulsory. Ideally, nanoparticles would be capable of clearance following drug administration, reducing toxicity caused by particle accumulation. For a number of drug delivery devices, the renal clearance pathway is considered the most efficient means of elimination in comparison with others (e.g., uptake by hepatocytes and biliary excretion). Clearance is governed by several physicochemical properties of the particles, including their molar mass, dimensions, hydrophobicity, and surface charge. The kidneys can excrete particles smaller than 8 nm, whereas the liver and spleen are capable of capturing particles larger than 200 nm (Nel et al., 2009). Degradation of RPNPs into low molar mass components may therefore improve their clearance via renal filtration pathway. Alternatively, degradation may also serve as means to modulate drug release via network decomposition (Cohen et al., 1991).

Imparting degradable properties into RPNPs is a challenge when stimuli responsivity and degradability is sought within single-particle architecture. Degradation is likely to influence a number of characteristics for the particles used in drug delivery, including the network diffusivity and drug release rates and the stability of the particles in the medium.

In the case of nanogels, depending on the type of crosslinker used, two different classes of degradable nanogels can be obtained: (i) nanogels that degrade under physiological conditions over long periods and (ii) nanogels that demonstrate triggered decomposition at rapid rates in response to a stimulus. Whereas slow degradation may enable clearance of the drug delivery particle, triggered degradation enables the release of encapsulated therapeutic agents in a stimuli-specific fashion.

The incorporation of polysaccharides into the nanogel design also promotes biodegradability (Oh et al., 2009). Polysaccharides are promising materials for nanogel synthesis as these polymers are relatively cheap, nontoxic, biodegradable, and abundant in nature. Recently, a new family of thermoresponsive and enzymatically degradable PVCL-based nanogels consisted of highly crosslinked nanogel particles synthesized by using a dextran-based macrocrosslinker was presented (Aguirre et al., 2013). Two types of new thermoresponsive and enzymatically degradable nanogels were produced. Regarding the future use of these new nanogels in drug delivery, after their enzymatic degradation two mechanisms of drug release can be envisaged, smooth and abrupt.

Biocompatible and biodegradable **polymeric micelles** have been successfully employed for the encapsulation of Pt (IV) prodrugs and platinum-based hydrophobic drugs (Callari et al., 2014). The major drawback of physical encapsulation is the limited drug retention in the polymer matrix; this problem is overcome by the crosslinking together with the positive charge of the copolymers forming the micelles, improving the internalization process, giving loading efficiencies of approximately 90%. Mixed micelles temperature- and pH-responsive drug release and enhanced passive targeting (see Chapter 7) have been developed to deliver doxorubicin efficiently to the nuclei of cells in vitro. These carriers also delivered drugs specifically to tumors in vivo prolonging circulation time and reducing size effects. This makes them therapeutically appealing (Chen et al., 2012f; Lo et al., 2009).

Responsive polymeric nanoconjugates are generally synthesized around a polymer with pendant functional groups like –OH, –COOH, or $–NH_2$. Nanoconjugate drugs that are biodegradable are preferred to nonbiodegradable ones that can accumulate in nontargeted tissues and organs. In this case, biodegradability requires metabolic breakdown of the nanoconjugate and final decomposition to water and carbon dioxide. Nanoconjugate delivery systems are considered a highly innovative technology for nanomedicine and they are designed to avoid multiple drug resistance in chemotherapy (Kateb et al., 2011).

In contrast to self-assembled structures like vesicles and micelles, complexation and covalent conjugation are other direct routes to prepare RPNPs. Drug complexes relay on reversible interactions between carrier and drug, whereas drug conjugates utilize covalent interactions. Drug conjugates can be prepared using many chemical pathways, which often depends on the chemical of the drug as well as the carrier. The two major classes of drug conjugates currently under development for theranostic (a combination of therapeutic and diagnostic) use include protein and peptide-associated and polymer-associated drugs. As with vesicular structures, the effectiveness of a drug conjugate is related to its ability to improve therapeutic index relative to free drug, generally by reducing toxicity and/or improving efficacy (Janib et al., 2010).

Another type of PRNPs is **responsive nanocapsules**, which are usually synthesized by colloidal templating, as in the case of the layer-by-layer approach (Callari et al., 2014). van Herk and coworkers (Ali et al., 2011b) demonstrated the effectiveness of a simple reversible addition-fragmentation transfer (RAFT)-based vesicle-templating approach for the synthesis of water-filled pH-responsive crosslinked polymeric nanocapsules. Recently, the synthesis of biocompatible and thermoresponsive nanocapsules based on PVCL through vesicle templating is reported for the first time (Aguirre et al., 2014). Thanks to their hollow morphology, biocompatibility, and capacity of undergoing reversible volume phase transitions in response to temperature, the nanocapsules synthesized can be considered as promising nanocarriers in controlled drug delivery, allowing the pre-encapsulation of hydrophobic drugs before their formation.

Prodrug nanocarriers based on responsive polymers have attracted much attention to perform drug delivery, especially in cancer therapy. There are some drawbacks still remaining with the use of encapsulated anticancer drugs in polymer nanoparticles, among them, "burst release," poor drug loading, and low miscibility of certain drugs with the polymer matrix. To overcome these issues, the prodrug approach has been recently reviewed (Delplace et al., 2014). This approach has some similarities with the previously presented RPNPs consisted of polymeric nanoconjugates. Three main strategies for drug–polymer conjugate-base colloids can be distinguished: (i) conjugating a drug to a presynthesized polymer ("conjugation to"), (ii) conjugating a drug to a monomer prior to polymerization ("conjugation through"), and (iii) growing a polymer chain from a functional drug ("drug initiated"). A number of structures of small organic anticancer drugs and platinum drugs linked to preformed polymers were reported, responsive or not, and the associated coupling strategies. Several structures of functional monomers for the design of polymer prodrug nanocarriers are also reported. Conjugation can be done with amphiphilic copolymers such as poly(ethylene glycol) (PEG)–polyester block copolymers or with temperature-, pH-, or multiresponsive copolymers. Small chemotherapeutic agents linked to responsive macromolecules will lead to RPNPs in the form of micelles, vesicles, or nanoparticles. Compared to the physical encapsulation of drugs into nanocarriers, the prodrug approach is more dependent on the enzymatic biological environment since the activation of the prodrug is a prerequisite for the pharmacological activity to occur. Due to this, the preclinical to clinical translation may be less predictable than with the other nanomedicines.

3.3 Conclusions

This chapter highlights current advances in the design and production of RPNPs for controlled and targeted drug delivery, mainly for the case of micro-/nanogels due to their versatility and for complying with the requirements in biomedical applications. In this field, rapid developments occur almost on a daily basis and they are directed to achieve new nanocarriers. Design and production of new nanocarriers based on more complex nanostructures is not only dependent on a deeper knowledge of how these nanocarriers interact with biological systems but also dependent on the requirements established by the specific applications. In this way, future work on stimuli-sensitive nanoparticles should be directed to produce new families of multifunctional and multitargeted biocompatible and also biodegradable RPNPs to be used for therapeutics and diagnostics and also in theranostics for pharmaceutical and biomedical applications.

Acknowledgments

Jacqueline Forcada thanks the Spanish Plan Nacional de Materiales (MAT2012-36,270-C04-01).

Chapter 4

Preparation of Micelles, Vesicles, and Liposomes

Joan Estelrich
Departament de Fisicoquímica, Facultat de Farmàcia, Universitat de Barcelona, Avda. Joan XXIII, 08028 Barcelona, Catalonia, Spain
joanestelrich@ub.edu

This chapter and the next two chapters will focus on the general preparation methods for micelles, vesicles, liposomes, nanoparticles, nanocapsules, and nanobottles.

In this chapter will cover the preparation of micelles, vesicles, and liposomes because these require only physical methods, compared to most of the preparation methods for nanoparticles and nanocapsules where polymerization reactions are required (except for the layer-by-layer approach, which is treated separately in Chapter 10).

A relatively new method of preparing micelles and vesicles is the polymerization-induced self-assembly (PISA) method, which will be treated in Chapter 5.

Controlled Release Systems: Advances in Nanobottles and Active Nanoparticles
Edited by Alexander van Herk, Jacqueline Forcada, and Giorgia Pastorin

ISBN 978-981-4613-21-7 (Hardcover), 978-981-4613-22-4 (eBook)
www.panstanford.com

4.1 Preparation of Micelles

In principle micelles are formed spontaneously on dissolution of the surfactant above the critical micelle concentration (CMC). For SDS at 60°C the CMC in pure water is around 10^{-2} mol L^{-1}. The CMC is strongly influenced by the presence of salts but also by the presence of hydrophobic molecules that solubilize inside the hydrophobic domain of the micelles. For blockcopolymer micelles the CMC lies at much lower concentrations. Micelles can coexist with other surfactant aggregate morphologies like vesicles and worm-like structures.

There are anionic, cationic, and nonionic micelles. In all case a hydrophobic and a hydrophilic part can be distinguished. Also mixed micelles are possible, for example, catanionic micelles. In general the model of hydrophilic lipophilic balance (HLB) is used to describe the properties of the surfactant (Mitchell, 1981).

4.2 Preparation of Vesicles and Liposomes

Basically there is no difference between a vesicle and a liposome, other than that the liposome is based on lipids and a vesicle is based on any other (synthetic or natural) surfactant.

The basis for the existence of vesicles and liposomes lies in the underlying phase diagram, which can be complex. To force these structures into a certain shape and size several methods are available.

The formation of these structures needs some input of energy, and a number of different approaches have been used to assemble these lipids vesicles, including mechanical methods (thin-film hydration, ultrasonic irradiation, high-pressure homogenization, microemulsification, and extrusion) and methods based on the replacement of organic solvents by aqueous media (injection methods, reverse-phase evaporation, and proliposomes) (Wagner, 2011). Special methods are those based on the use of dense gas in the production of liposomes (supercritical fluid liposome methods). In the literature, one can find other approaches sometimes used to prepare liposomes; however, its use is limited at present. A description of such approaches can be found elsewhere

(New, 1990). In the last 10 years there have been many new developments in liposome formation technology based on microfluidic techniques (Van Swaay, 2013).

4.2.1 Thin-Film Hydration

In this approach (Bangham, 1965) the bilayer-forming elements are mixed in a volatile organic solvent or solvent mixture (such as chloroform, ether, ethanol, methanol, *tert*-butanol, or a combination of these). The solvent is removed, preferably by rotary evaporation under reduced pressure, and all lipid components are dried down onto a solid support (usually the side of the glass container vessel). Then, the dry lipid mixture is hydrated with an excess volume of aqueous medium at temperatures above the main T_m, and, after agitation (the so-called hand-shaken dispersion) multilamellar vesicles (MLVs) form spontaneously. Hydration, more than the other steps, influences the type of liposome formed (number of lamellae, size, and entrapped volume). The hydration process is more efficient when the film of the dry lipid is kept thin. This means that the surface area/lipid volume must be high. In consequence, different-sized round-bottomed flasks must be used for different quantities of lipid. Glass beads and other inert, insoluble particles have been used to increase the surface area available for film deposition. Liposomes produced during hydration are generally heterogeneous in size but can be downsized by extrusion, sonication or other mechanical methods.

4.2.2 Ultrasonic Irradiation

In this method (Huang, 1969), a very high-energy input based on cavitation is applied to the lipid suspension. There are two methods of sonication, using either a probe or a bath ultrasonic disintegrator, depending on the energy needed to achieve the required size. The induced pressure breaks up the larger, MLVs in the sample to form smaller vesicles, usually small unilamellar vesicles (SUVs). The time over which lipid suspensions are sonicated determines the sizes of the vesicles. This method consumes little time, but due to the high-energy input, the obtained SUVs must be used immediately after their preparation since they are potentially unstable.

4.2.3 Homogenization Techniques

Such methods break down the large MLVs either by shear or pressure forces. Examples of these techniques are the *French press system* (Barenholz, 1979), where MLVs are downsized by passage through a small orifice under high pressure (~1400 bar), and *microemulsification*, carried out in a machine that pumps the lipid suspension at 600–700 bar through a 5 μm filter, after which it is forced along defined microchannels, which then direct the two streams of fluid to collide together at right angles at a very high velocity. The fluid collected is recycled through the pump and interaction chamber until vesicles of the required dimensions are obtained (Mayhew, 1984).

One of the most common methods used in the preparation of liposomes is the *extrusion*. Typically, MLVs are forced to pass through a polycarbonate membrane of defined pore size in a specialized vessel at moderate pressure (6–60 bar). After the repeated extrusion results in a relatively homogeneous population of LUVs (Hope, 1985; Mayer, 1986). A small-volume extrusion apparatus have been designed (MacDonald, 1991). Systems with a heating device can either be used with saturated and unsaturated lipids. Liposomes prepared with phosphatidylcholine (PC) and extruded by nominal membrane pores of 100 or 200 nm present a size near the pore size of the membrane used in preparing them, but this is not true with other lipids, especially with charged lipids, and extruded through membranes with sizes different than 100 or 200 nm. For instance, liposomes of soybean-PC presented a size of 180 ± 15 nm (extruded through 200 nm membranes) or 102 ± 5 nm (through 100 nm), but liposomes of PC:PS (50/50 molar ratio) were 134 nm and 106, respectively (Roy, 1997). The considerable advantages of this technique include: it is rapid, it works directly from MLVs, it yields a monodisperse product, it presents a high reproducibility batch to batch, and it is suitable for large-scale production.

Interestingly, Lapinski et al. (2007) have compared liposomes formed by extrusion and sonication and find that, despite apparent differences in average size and size distribution, both methods produce liposomes that exhibit the same molecular-scale environment determined by measuring the rotational motion of a chromophore incorporated into the liposomes, and its translational diffusion in supported bilayers prepared with both kinds of liposomes.

4.2.4 Replacement of Organic Solvents by Aqueous Media

The methods described in this section have in common that organic solvents, either water miscible or immiscible, are replaced by an aqueous solution. This replacement is either performed by injection of the lipid carrying the organic solution into the aqueous phase—the injection methods—or by stepwise addition of the aqueous phase to the organic phase, in particular ethanol—the proliposome–liposome method. In addition, the emulsification methods, namely, the reverse-phase evaporation method and the double-emulsion technique, are based on the replacement of a water-immiscible solvent by an aqueous phase, thus forming liposomes with high encapsulation rates of hydrophilic as well as lipid-phase-soluble substances.

Among the numerous methods available to obtain liposomes, the *ethanol injection method* is one of the most popular to get unilamellar liposomes (Batzri, 1973). It is a fast, simple process and suitable for the production at industrial scale. In this process, a lipid solution of ethanol is rapidly injected through a thin needle into an aqueous solution while stirring and the liposomes are immediately formed. Liposome size, size distribution, and drug encapsulation efficiency are influenced by several parameters: nature of the aqueous solution, injection flow, amount of ethanol injected, and temperature (Kremer, 1977; Pons, 1993; Wagner, 2002; Justo, 2011; Gentine, 2012). The drawbacks of the conventional method the nature of the lipids strongly influencing the vesicle size (20–300 nm), the obtained liposome suspensions being very dilute, the difficulty in removing all ethanol, and the moderate solubility of many lipids in ethanol. Recently, a modification consisting in injecting a hot solution of lipid/ethanol into the hot aqueous phase has been described. The authors affirm that this process overcomes the problems faced when using the conventional method (Gentine, 2013).

In the *proliposome–liposome method* the lipids are dried down on to a finely divided particulate support, generally a polysaccharide. Upon adding water to the dried lipid-coated powder (known as proliposomes), with mixing on a whirlimixer, the lipids swell, while the support rapidly dissolves to give a suspension of MLVs (Payne, 1986). This method is a very useful one for commercial applications where large quantities of proliposomes can be prepared and stored dry before use in aliquots in sealed vials and then resuspended when required to give batches of liposomes reproducible over a long period of time.

The *reverse-phase evaporation method* (Szoka, 1978) is very usual for applications that require high encapsulation efficiencies of water-soluble drugs. In this approach, an aqueous volume is injected into the organic phase, which must be totally immiscible with the aqueous phase, and after sonication a water-in-oil emulsion is formed. Then the organic solvent is removed by rotary evaporation at low vacuum until a gel is formed. Removal of the last traces of solvent transforms the gel into LUVs. As the polydispersity of the resulting vesicles is quite high, liposomes can be downsized by a further extrusion (Szoka, 1980).

4.2.5 Supercritical Fluid Methods

The main advantage of such methods is the low organic solvent consumption (Meure, 2008). The process involves the introduction of a solution of lipids through a solvent pump into a stream of supercritical CO_2 in a high-pressure vessel. This results in the precipitation of the proliposomes in the vessel. The precipitated proliposomes form liposomes on hydration with buffer or water. Liposomes obtained are SUVs with a small fraction of large MLVs (Frederiksen, 1997; Otake, 2001). A modification is the supercritical reverse-phase evaporation (scRPE), in which an aqueous solution is slowly introduced to a cell containing the lipids and the supercritical fluid. When the cell is depressurized, liposomes are formed (Imura, 2003). In the scRPE method, a water-in-CO_2 microemulsion forms as water is introduced into the cell; then when more water is introduced the cell becomes opaque as a macroemulsion forms. A transition occurs from the water-in-CO_2 emulsion to a CO_2-in-water emulsion, with increasing amounts of aqueous phase. Liposomes may then be formed as the CO_2 is evaporated from the aqueous phase upon depressurization. Later, an improved scRPE method has been developed to avoid the use of organic solvent and enhance the stability and the drug loading efficiency (Otake, 2006).

4.2.6 Microfluidic Methods

The term "microfluidic" refers to methods in which fluid-handling procedures are performed in a geometrically constrained volume, typically defined by submillimeter-length scales and low Reynolds numbers. The adoption of microfluidic techniques allows the integration of laboratory procedures into planar chips, reducing reaction volumes and the associated cost of chemical and biological experimentation by several orders of magnitude (Van Swaay, 2013). *Electroformation* is probably the most common microfluidic method for liposome production. It involves spreading lipids dissolved in an organic solvent on the surface of a planar electrode, evaporation of the solvent by vacuum desiccation to form a dry phospholipid film, immersing the coated electrode in an aqueous solution, and finally applying an electric field across the lipid film and surrounding buffer. The lipids interact with the aqueous solution and electric field by "peeling off" the electrode surface in layers and self-assembling into giant but polydisperse, multilamellar vesicles. Later refinements of the method have allowed for giant unilamellar vesicles (GUVs) to be consistently produced (Angelova et al., 1992; Taylor, 2003).

Other microfluidic method is the *flow-focusing method* that consists of a central flow of a phospholipid-containing alcohol solution being intersected on either side by an aqueous solution. As the three flows merge into a main microchannel, the alcohol diffuses into the aqueous solution. As the alcohol dilutes past a critical concentration, the lipids spontaneously self-assemble into liposomes. The laminar flow in the channels enables control over the diffusive mixing and thus the size and size distribution of the liposomes can be manipulated through adjusting the flow conditions. With this method, SUVs can be obtained (Jahn, 2007).

Chapter 5

Synthetic Strategies for Synthesis of Polymer Nanoparticles

Alexander van Herk
Institute for Chemical and Engineering Sciences, 1 Pesek Road, Jurong Island, 627833 Singapore
alexander_herk@ices.a-star.edu.sg

In this chapter an introductory overview of the different polymerization techniques that can lead to polymeric nanoparticles and nanocapsules is presented. We put an emphasis on the preparation of these particles in aqueous systems because for many applications, including controlled drug release, aqueous dispersions of the polymeric nanoparticles are required. Within the narrow definition of the IUPAC nanoparticles are particles with one dimension less than 100 nm. In practice in literature this definition is expanded to particles up to 1000 nm.

This chapter (together with Chapter 3) should facilitate reading the more specialized chapters in Part II where polymerizations are used to produce polymer nanoparticles.

Controlled Release Systems: Advances in Nanobottles and Active Nanoparticles
Edited by Alexander van Herk, Jacqueline Forcada, and Giorgia Pastorin

ISBN 978-981-4613-21-7 (Hardcover), 978-981-4613-22-4 (eBook)
www.panstanford.com

Most of the polymerization techniques are heterophase polymerizations (with the exception of solution polymerization, Section 5.1) and polymer chains can undergo transitions from one phase to another. Precipitation polymerization usually starts off with a solution polymerization but at some point the polymer precipitates (Section 5.2). A dispersion polymerization resembles a precipitation polymerization; the main difference is the presence of a colloidal stabilizer (Section 5.3). Emulsion polymerization, miniemulsion polymerization, and microemulsion polymerization all are producing polymers inside nanostructures (micelles, stabilized monomer droplets, thermodynamically unstable in miniemulsion and thermodynamically stable in microemulsion) (Sections 5.4–5.6, respectively).

For the synthesis of nanocapsules one can resort to using the hydrophobic domain in vesicles instead of micelles for the locus of polymerization or one can use the vesicles as a template to form a polymer layer on the outside of the vesicle, resulting in a nanocapsule (Section 5.7). The latest approach to prepare nanoparticles is the polymer-induced self-assembly (PISA) method (Section 5.8).

We try to stay as close as possible to the definitions of the IUPAC but we also realize that in the literature often the wrong polymerization technique is linked to the particular polymer nanoparticle synthesis at hand.

5.1 Solution Polymerization

At first glance there seems to be a contradiction to produce polymer nanoparticles with solution polymerization. After all if the particles are produced by precipitation of the polymer out of the solution one deals with precipitation polymerization. However, the distinction between large dissolved polymer chains and small polymer nanoparticles becomes unclear in the region of several nanometers (1–20 nm). So in general, when we talk about polymeric nanoparticles prepared by solution polymerization we are in the gray area between dissolved polymer chains and nanoparticles.

An overview of dendritic polymer architectures produced in solution polymerizations is shown in Fig. 5.1 (Gao, 2004a).

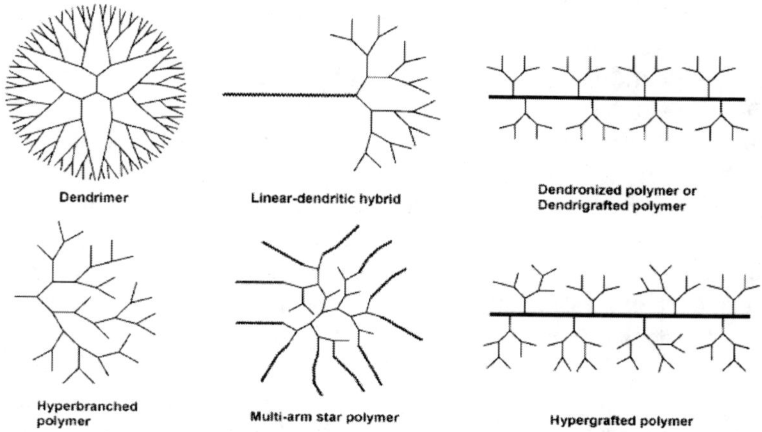

Figure 5.1 Overview of dendritic polymer architectures. Reprinted from Gao (2004a), Copyright 2004, with permission from Elsevier.

An example is the synthesis of hyperbranched polymer nanoparticles via radical polymerization in benzene at 80°C (Sato, 2004). The particles obtained are in the range of 5–25 nm.

Another class of nanoparticles made through solution polymerization are the dendrimers (Esfand, 2001), repetitively branched molecules.

The advantage of solution polymerization is that basically any polymerization mechanism can be applied.

5.2 Precipitation Polymerization

Precipitation polymerization usually proceeds via radical initiation of the monomers/crosslinkers in a homogeneous system followed by propagation, resulting in precipitation of the polymer, leading to polymer particles with a narrow size distribution produced in the absence of any stabilizer or surfactant.

The polymer nanoparticles can be obtained by through varieties; precipitation polymerization, and distillation precipitation polymerization.

In both cases this technique starts as a homogeneous mixture of the monomer(s), initiator(s), and solvent(s). The polymer nanoparticles are formed via polymerization of the monomers/crosslinkers and precipitate out from the homogeneous solution. During the polymerization, the growing polymer chains phase-separate from the continuous phase by enthalpic precipitation (or entropic precipitation in those cases where crosslinking prevents the polymer and solvent from mixing) (Sosnowski, 1996). The group of Stöver reported the precipitation polymerization for the synthesis of monodisperse crosslinked polydivinylbenzene (PDVB) microspheres (Li and Stöver, 2000). Distillation precipitation polymerization has been developed as a powerful and facile technique for the synthesis of monodisperse polymer nano- and microparticles (Bai, 2004). In distillation precipitation polymerization the solvent is partially distilled out during the polymerization. The main difference between the normal precipitation polymerization and distillation precipitation is of course the change in solvent concentration, accompanied by special agitation of the polymerization system. Precipitation polymerization is performed by rolling the reactor on a shaking bed or a rotary evaporator. More recently, "living" radical polymerizations have been utilized for the synthesis of molecularly imprinted polymer microspheres via combination of precipitation polymerization and living radical polymerization, such as atom transfer radical polymerization (ATRP) (Zu, 2009), reversible addition-fragmentation transfer (RAFT) polymerization (Pan, 2011), and iniferter-induced living radical polymerization (ILRP) (Li et al., 2010b). Also nanocapsules can be produced by precipitation polymerization (Li et al., 2013b).

5.3 Dispersion Polymerization

Dispersion polymerization is normally known to produce monodisperse polymer particles on a micrometer scale (1–15 μm). Optimal choice of the reaction conditions is essential to achieve monodispersity of the prepared polymer microspheres. Dispersion polymerization starts as a homogeneous solution polymerization, but as polymer chains grow in size they eventually reach a molecular weight higher than a certain critical value and precipitate from the continuous phase. At this point the difference with precipitation polymerization is set because a stabilizer is present in the system that adsorbs on the surface of the particles. At a certain point most of the stabilizer is adsorbed on the surface of the particles, and at this point, the total number of particles in the system is fixed and the nucleation stage ceases.

In the particle growth stage, polymerization occurs either inside the monomer-swollen particles or in the reaction medium or both, depending on the partitioning of monomer between the two phases. The newly formed polymer chains in the solution will not form additional nuclei but will be captured by existing particles.

The monomer is chosen on the basis of its miscibility in the reaction medium (and the formed polymer should be insoluble in the medium). Styrene (St) and methyl methacrylate (MMA) in alcoholic media being the far most studied. Also supercritical carbon dioxide as well as water has been studied for dispersion polymerizations.

Three types of steric stabilizers are used instead. The first type of stabilizers is homopolymers. During dispersion polymerization a graft forms via chain transfer to polymer. In this process a graft copolymer is produced and this is acting as the stabilizer in the system.

The second type is polymers with two distinct segments, for example, block—or graft—polymers. The stabilizing moieties should be soluble in the continuous phase and the anchor group should have a high affinity to the particle surface.

The third group of stabilizers is macromonomers that can copolymerize with the principal monomer to rapidly generate graft copolymer species, which on their turn act as stabilizers.

Higher concentrations of stabilizers will cover a larger surface area and therefore will lead to smaller particles. However, in general dispersion leads to micron-sized particles.

In the group of Steven Armes aqueous dispersion polymerization with 2-hydroxypropylmethacrylate (HPMA), stabilized with poly(*N*-vinylpyrrolidone) resulted in particles between 100 and 1000 nm and using a RAFT-based reactive steric stabilizer even particles between 20–100 nm could be obtained (Li and Armes, 2010). Also vesicular structures were observed under certain synthesis conditions and later a full phase diagram for a similar system was constructed (Sugihara, 2013) in which besides solid spheres also worm-like structures and vesicles occurred. So indeed it is possible to produce polymeric nanoparticles also with (aqueous) dispersion polymerization.

The self-assembly during polymerization has been coined PISA and will be reported on separately (Section 5.8).

5.4 Emulsion Polymerization

Polymerization whereby monomer(s), initiator(s), the dispersion medium, and possibly colloid stabilizers constitute initially an inhomogeneous system (through the presence of monomer droplets), resulting in particles of colloidal dimensions containing the formed polymer, called a latex. Emulsion polymerization is the most common method of producing polymer nanoparticles, typically in the range of 50 to 400 nm (see Fig. 5.2).

An *emulsion* polymerization comprises water, an initiator (usually water soluble), a more or less water-insoluble monomer, and a colloidal stabilizer, which may be added or may be formed in situ. The main locus of polymerization is within the monomer-swollen latex particles, which are either formed at the start of polymerization or may be added initially (in which case one deals with a *seeded* emulsion polymerization). The term "emulsion polymerization" is a misnomer (arising for historical reasons: the process was originally developed with the aim of polymerizing emulsion droplets, although in fact this does not occur). The clear difference between dispersion polymerizations and precipitation polymerizations is that the polymerization takes place in the micelles and that the monomer is not soluble in the medium. An *inverse* emulsion polymerization is where the continuous phase is organic in combination with an aqueous discrete phase containing a water-soluble monomer (e.g., acrylamide).

An ab initio emulsion polymerization involves the emulsification of one or more monomers in a continuous aqueous phase and stabilization of the droplets by a surfactant. Usually, a water-soluble initiator is used to start the free-radical polymerization. The locus of polymerization is initially in micelles and in the polymer nanoparticles (either formed during the process or added at the start), which are swollen with monomer during the polymerization process and dispersed in the aqueous phase. The final product is a latex comprising a colloidal dispersion of polymer particles in water. If present, micelles are the locus of the commencement of polymerization because they are much more numerous (and thus in total having a much higher surface area) than the monomer droplets and thus much more likely to capture aqueous-phase radicals generated from initiator: *micellar nucleation*. Consistent with this, an increase in surfactant concentration results in an increase in the number of formed particles (leading to smaller particles). If there is no added surfactant, or the surfactant system is below the critical micelle concentration (CMC), a latex can still form, stabilized by entities formed from initiator-derived (charged) radicals that propagate in the aqueous phase to produce surface-active oligomers (often called surfactant-free emulsion polymerization). Particle formation is by the collapse (coil-to-globule transition) of aqueous-phase oligomers to form particles by *homogeneous nucleation*.

One of the advantages of emulsion polymerization is the excellent heat exchange due to the low viscosity of the continuous phase during the whole reaction. Emulsion polymerization is frequently used to produce core–shell particles, which have a layered structure. The preparation of block copolymers in emulsion polymerization through the use of controlled radical polymerization mechanisms is a way to produce structured latex particles. Emulsion polymerization kinetics has important differences from solution and bulk polymerizations. These differences can lead to many advantages: for example, an increase in molar mass can be achieved without reducing the rate of polymerization. Emulsion polymerization is known for its relatively high rates of polymerization and high molar masses as compared to other process strategies. It is also possible to produce nanocapsules through emulsion polymerization, for example, by core–shell structures where the core consists of a polyacid which later on is neutralized and then swells with water to produce a hollow particle.

Besides the morphology also the particle size and particle size distribution are of major importance in the use of these particles in controlled release systems. If we first consider the micellar nucleation mechanism for particle formation we can infer from the theory how we can control particle size. Basically a new particle is formed when a radical, in the form of a surface-active oligomeric radical, is entering a micelle. As new particles are formed, and pre-existing ones grow by propagation, the increase in surface area of the particles will result in progressively more and more surfactant being adsorbed onto the particle surface, and the aqueous-phase surfactant is replenished from that in the micelles. Eventually, the concentration of free surfactant falls below the CMC, and micelles disappear. Once micelles have disappeared, there can be no further particle formation by this micellar nucleation mechanism. Particle formation in a system, which initially contained surfactant above the CMC therefore ceases approximately when the surface area of pre-existing particles is sufficient to reduce the concentration of surfactant below the CMC.

Smith and Ewart (see, for example, van Herk, 2013) set out an elegant mathematical development leading to the particle number (N_P) being given by:

$$N_p = \frac{1}{3}(5a_e)^{3/5}\left(\frac{k_d}{\pi^{1/2}K}\right)^{2/5}[I]^{2/5}[S]^{3/5} \quad (5.1)$$

where $[I]$ and $[S]$ are initiator and surfactant concentrations, respectively, a_e is the saturation value of the area per adsorbed surfactant molecule, k_d is the initiator dissociation rate coefficient, and the volume growth factor of the particles $K = k_p\ M_0\ [M]_p/(N_{Av}d_p)$, where N_{Av} is the Avogadro constant. This work was carried out in the absence of modern mechanistic knowledge, and thus this equation is only rarely quantitatively in accord with experiment. This lack of agreement with experiment includes the dependence on $[I]$ and $[S]$: the predicted exponents of 2/5 and 3/5 may indeed be seen but only over a limited range of conditions. Nevertheless, Eq. 5.1 correctly implies the increase of particle number with both initiator concentration (and radical flux) and surfactant concentration. On the other hand reducing the volume growth factor K also leads to an increase in particle number (and thus a decrease in size). This can be understood by the previous description of the events taking place in the nucleation stage. The adsorption of surfactant by the growing polymer particles is reduced when the volume growth factor is reduced, therefore extending the duration of the nucleation stage and creating more particles. How can we reduce the volume growth factor, which can be done by lowering the concentration of monomer in the growing latex particles ($[M]_p$ in Eq. 5.1)? This is possible by slowly feeding the monomer (starved–fed emulsion polymerization) or by reducing the swelling capability of the latex particles by a high crosslink density.

The surfactants often applied in emulsion polymerization can be anionic (e.g., sodium docecylsulfate), cationic (e.g., cetylammonium bromide), or nonionic (e.g., pluronics, ethylene oxide–propylene oxide block copolymers). The ionic surfactants stabilize the colloidal particles by electrostatic repulsion; the nonionic surfactants stabilize by steric stabilization.

In *homogeneous nucleation* at the start of the polymerization, there are no particles present (contrasting with a micellar system, when they form at the very beginning) and one of the main fates of a newly formed oligomeric radicals is to undergo further propagation until it reaches a critical degree of polymerization where the oligomers become "insoluble," or more precisely undergo a coil-to-globule transition. The resulting collapsed chain is hydrophobic and thus is swollen by monomer. Hence, the radical end is in a monomer-rich environment, rapid propagation ensues and a particle forms. In the absence of added surfactant, charged end groups from the initiator (e.g., $-SO_4^{-\bullet}$) provide colloidal stability. This mechanism is *homogeneous nucleation*. The number of particles is much smaller, and the ultimate size of particles much larger, than in the presence of surfactant.

A full description of all mechanistic aspects of emulsion polymerization is given in the book of Gilbert (1995) and a review paper with additions (Thickett, 2007).

5.5 Miniemulsion Polymerization

Miniemulsion polymerization is the (free-radical) polymerization of submicron monomer droplets that are dispersed in an aqueous environment. A most recent description of miniemulsion polymerization can be found in the thesis of T. Jansen (2014) and the following text is reproduced with permission.

> Miniemulsion polymerization distinguishes from conventional emulsion polymerization by the direct entry of radical species (so-called nucleation), most often originating from the aqueous phase, in submicron monomer droplets which are usually in the order of 50-500 nm (Figure 2). These droplets form a rather colloidally stable, whitish emulsion. In contrast to its conventional sister no continuous stirring is required to prevent creaming. Monomer is present in the monomer

droplets, and no monomer transfer between monomer droplets is necessary. So *in theory*, the droplets are "one-to-one" transformed into particles.

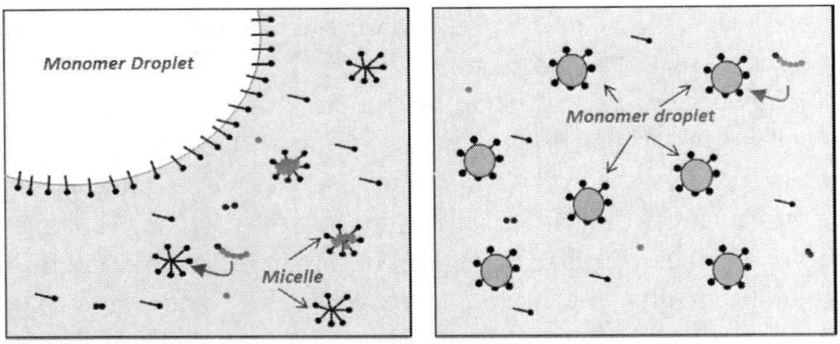

Figure 5.2 Schematic representation of conventional emulsion polymerization (left) and miniemulsion polymerization (right). Displaying radical entry in monomer-swollen surfactant micelles (diameter approximately 5 nm) in coexistence with monomer droplets (diameter approximately 1–10 μm) for conventional emulsion polymerization and radical entry in monomer droplets (diameter approximately 50–500 nm) for miniemulsion polymerization. Entities are not correctly scaled.

In conventional emulsion polymerization, radical entry takes place in monomer-swollen surfactant micelles. Particle growth occurs by simultaneous monomer transfer from relatively large monomer droplets (diameter approximately 1-10 μm) and polymerization. The final product of miniemulsion and conventional emulsion polymerization is a latex, a colloidally stable dispersion of submicron polymer particles in water.

For conventional emulsion polymerization, a relatively large amount of surfactant is necessary. Nucleation is rather difficult to control; small pollutants in the ingredients of the formulation may have a significant effect on the particle size distribution of the final product and the course of the polymerization. For miniemulsion polymerization, *in theory* control of the particle size distribution should be less difficult. In miniemulsion polymerization, less emulsifier is required compared to conventional emulsion polymerization.

Monomer-swollen micelles are much more likely to capture radical species than monomer droplets, due to their abundant presence and large specific surface area. One of the main prerequisites for miniemulsion polymerization is that the surfactant concentration should be below the critical micelle concentration (CMC) – the concentration above which aggregation of surfactant molecules to micelles occurs.

Miniemulsion droplets do not form spontaneously. The increase in interfacial energy accompanied with miniemulsification, results from the interfacial tension and the increase in interfacial area, and has to be supplied to the mixture of organic and aqueous phase by the application of shear. This shear is typically applied by ultrasonication, static mixing, high-pressure homogenisation or by using a rotor-stator mixer. This prefatory emulsification step introduces additional process steps and equipment, as compared with conventional emulsion polymerization, limiting its industrial perspective.

Once miniemulsion droplets have been produced, droplet stability is the foremost threat to miniemulsions. Droplet size and distribution is continuously altered as a result of monomer diffusion- and collision-based processes. Fusion of droplets, as a consequence of collisions by Brownian motion is referred to as coalescence. Coalescence leads to a decrease of the number of droplets and to an increase of the average droplet diameter. The coalescence rate can be reduced by temperature reduction and by surfactant addition. Surface active entities in (mini) emulsion polymerization are therefore not only employed to facilitate nucleation, but also to stabilize monomer droplets and polymer particles against coalescence and coagulation (the aggregation of polymer particles). The process of monomer diffusion between droplets, also known as Ostwald ripening; the difference in surface energy between differently sized particles results in a thermodynamic driving force for monomer transport from small to large droplets, resulting in a broadening of the droplet size distribution and ultimately the disappearance of small droplets. Ostwald ripening can be minimized by the inclusion of a few weight percent of a very hydrophobic component in the submicron monomer droplets. This very hydrophobic component is referred to as the hydrophobe. This hydrophobe is deemed to have a very large time constant of diffusion, due to its low water solubility. Ostwald ripening would consequently result in a hydrophobe concentration difference between large and small particles, resulting in a driving force for mass transfer that counteracts Ostwald ripening.

Miniemulsions adequately stabilized by a combination of hydrophobe and surfactant can be stable for periods in the order of weeks or even months.

The principle of droplet nucleation and the reduced necessity for monomer transfer offers many possibilities for specialty applications. For instance, miniemulsion polymerization gives access to the incorporation of very hydrophobic monomers that have a too low water solubility to be transported from one droplet to another across the aqueous phase. Another example is the encapsulation of polymeric and inorganic material in the droplets, other than as seed latex (although seed latex polymerization resembles a miniemulsion polymerization with an initial monomer content in every droplet). Polymer particles consisting of multiple phases (for instance an organic and inorganic phase or a high molecular weight polymer combined with a low molecular weight resin) are referred to as hybrid particles. Recent examples of hybrid particles synthesized by miniemulsion polymerization comprise the encapsulation of TiO_2, Fe_3O_4 (magnetite), polyurethane resin or oily substances (Weiss, 2010).

5.6 Microemulsion Polymerization

Microemulsion polymerization Emulsion polymerization in which the starting system is a microemulsion and the final synthetic latex comprises polymer colloidal particles dispersed in an aqueous medium.

Microemulsions are thermodynamically stable dispersions and because of their small size they are transparent. The microemulsion contains two immiscible liquids (e.g., oil and water) stabilized by a surfactant and/or cosurfactant (e.g., sodium dodecyl sulfate and *n*-pentanol) (Yan, 2006). The main difference between microemulsion and miniemulsion polymerization lies in the fact that microemulsions are thermodynamically stable and much smaller than miniemulsions.

Microemulsion polymerization can take place in the form of oil-in-water (O/W), water-in-oil (W/O), also known as inverse microemulsion polymerization, or bicontinuous microemulsion polymerization. O/W and W/O microemulsions consist of a number of small globulous oil (or water) droplets over the range of 1 to 10 nm in diameter dispersed in a continuous aqueous (or oily) phase (Chow, 2005). In general, polymers obtained via bicontinuous polymerization have slightly lower molecular weight than those prepared from discrete droplets formed during microemulsion polymerization. On the other hand, besides these three forms of single-phase microemulsion, a variety of other phases known as Winsor systems can coexist in equilibrium in these single microemulsions at low surfactant concentrations (Winsor, 1948). Winsor I is a globular O/W microemulsion in equilibrium with an oil phase, Winsor II is a globular W/O microemulsion in equilibrium with an aqueous phase, and Winsor III is a middle phase with a bicontinuous microstructure that is in equilibrium with both oil and aqueous phases.

Microemulsion polymerization products are usually in the size range between 10 and 100 nm, with a narrow size distribution. The polymer particles obtained via microemulsion polymerization consist of a low number of polymer chains per particle; therefore the resultant polymer commonly has very high molecular weight in the range 10^6–10^7 g mol^{-1}.

For both O/W and W/O microemulsion systems, the amount of monomer is usually restricted to less than 10 wt.% on the basis of the total weight of the microemulsion, while higher amounts of surfactant (7–15 wt.%) are normally required. Microemulsion polymerization lends itself also for controlled/living radical polymerization (Qiu, 2001).

In microemulsion polymerizations of either O/W or W/O a continuous particle nucleation mechanism is accepted. Polymer particles are thereby formed continuously during the overall period of polymerization. Thus the number of particles will increase over time (although slower during the latter stage of the polymerization). The nucleation of the polymer particle that has occurred only accounts for a small fraction ($\sim 10^{-3}$) of the microemulsion droplets initially present in the microemulsion system. The growing latex particles continue to grow in the monomer-containing micelles through the constant supply of monomer from nonnucleated droplets either by diffusion through the continuous phase and/or by collisions between droplets.

Anionic surfactants, like the sulfates (sodium lauryl sulfate), sulfonates (sodium dodecylbenzene sulfonate), fatty acid soaps (sodium or potassium stearate, laurate, palmitate), and the Aerosol series (sodium dialkyl sulphosuccinates) such as AOT (sodium bis(2-ethylhexyl) sulfosuccinate) are the most commonly used surfactants in microemulsion polymerization. A cosurfactant (e.g., a medium-size aliphatic alcohol) and/or electrolyte (e.g., NaCl) is usually present with an anionic surfactant, which, however, is not necessary for double-chain ionic surfactants (e.g., Aerosol OT) and some nonionic surfactants. Some cationic surfactants like dodecyltrimethylammonium bromide (DTAB) and cetyltrimethylammonium bromide (CTAB) are also used. Also phospholipids can be applied in microemulsion polymerizations. Nonionic polymeric surfactants used in microemulsion polymerizations are, for example, Brij, Span-Tween, and Triton X-100.

5.7 Vesicle Polymerization and Vesicle-Templating Polymerization

In vesicle polymerizations the interior of a vesicle structure is polymerized to a nanocapsule. Polymer vesicles, also called polymersomes (Li and Keller, 2009), are vesicle structures usually produced by preformed blockcopolymers (see also Chapter 4).

Vesicle-templating polymerization is using the vesicle as a template and polymerizing on the outside of the vesicle structure.

We will now focus on the different types of templates and the ways of forming a polymer around that template. Although at first sight the morphosynthesis route (Fig. 5.3) seems a logical choice to produce nano- and microcapsules it turns out that this approach often does not lead to the required morphology, because phase separation takes place and the polymer is not forming a uniformous shell but instead is located as a small bilayer attached polymer particle, creating the so-called parachute morphology (Jung, 2000a).

Another approach is to polymerize on the outside of such a vesicle structure (transcriptive synthesis, see Fig. 5.3). This approach in general is more successful and a variety of monomers can be applied. The first approach in the transcriptive polymerization is to adsorb an oppositely charged monomer on the outside of the vesicle, followed by polymerization. Another approach is to adsorb a charged (living) oligomer (opposite in charge to the charge on the outside of the vesicle). The oligomer is produced with RAFT polymerization and after adsorption of this oligomer a new (and different) monomer and initiator can be added to continue polymerization on the surface. With this method several (responsive) polymeric nanocapsules have successfully been synthesized (Ali et al., 2011b).

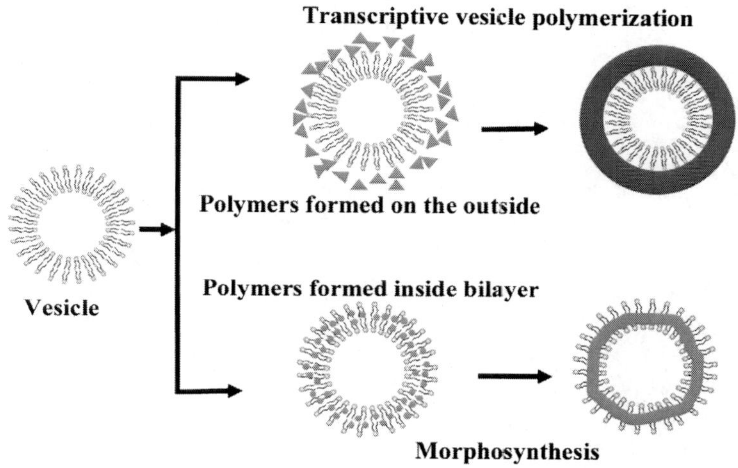

Figure 5.3 Two approaches for vesicle-templating polymerization. Morphosynthesis is polymerizing inside the hydrophobic domain of the vesicle structure. Transcriptive vesicle polymerization is polymerizing on the outside of the vesicle structure.

Biodegradable polymers are usually produced with step-growth polymerizations that are difficult to perform in aqueous systems. A biocompatible polymer, poly(*N*-vinylcaprolactam), has also been used to produce nanocapsules with this approach (Aguirre, 2014).

Another approach is the step-wise adsorption of oppositely charged polyelectrolytes onto a template, the so-called layer-by-layer approach (Decher, 2006; Donath, 1998). The advantage of this approach is that virtually any polyelectrolyte can be used. This method has been applied to vesicles as a template also (Cuomo, 2010).

5.8 Polymerization-Induced Self-Assembly

PISA is the phenomenon where a growing polymer chain changes from soluble to surface active due to the formation of a new block in the polymer chain, creating an amphiphilic structure. These amphiphilic molecules can then self-assemble in micelles, admicelles, worm-like structures, or vesicles. This process usually takes place in water but can also take place in nonpolar solvents.

Although the term "PISA" was coined only a few years ago, the phenomenon itself dates from the initial efforts to conduct controlled radical polymerizations in water. In 2002 an effective ab initio emulsion polymerization under RAFT control was published where initially an acrylic acid block was grown under RAFT control and then this water-soluble RAFT oligomer was used for subsequent polymerization of butyl acrylate (Ferguson, 2002). The amphiphilic RAFT oligomers organized themselves in micelles and the poly-butylacrylate particles were grown to a 60 nm size. Later a full mechanistic description was given including heat traces of the reaction (Sprong, 2006) and a full calorimetric study (Leswin, 2009).

The group of Armes initially focused on aqueous dispersion polymerization of 2-hydroxypropyl methacrylate (HPMA) combined with a poly(glycerol monomethacrylate)-based RAFT agent (Li and Armes, 2010). In similar work using a poly(2-(methacryloyloxy) ethylphosphorylcholine)-based macro-RAFT agent he constructed a full phase diagram which contained besides spheres also worms and vesicles (Sugihara, 2011). Later he also did similar work in nonpolar solvents (Fielding, 2013).

Besides producing block copolymers via CRP separately, Charleux also worked on PISA (Charleux, 2012). Also the group of Pan did similar work, for example, in dispersion polymerization in methanol (Huang and Pan, 2010), and at this point he coined the term "polymerization-induced self-assembling and reorganization" (PISR).

The versatile synthesis of different nanoparticle morphologies makes PISA an interesting method in the area of controlled drug release and a first application is published in 2014 where doxorubicin conjugates were incorporated in the nanostructures, showing significant influence of the nanoparticle carrier morphology on therapeutic efficacy for breast cancer (Karagoz, 2014).

In combination with inorganic seed particles the PISA method can also be used for encapsulation. In this case an admicelle is formed by growing block copolymers on the surface of the inorganic particles. In the group of Claverie first a charged RAFT oligomer was adsorbed on the inorganic surface, followed by polymerization of a hydrophobic block (Daigle, 2008).

In the group of Hawkett *random* RAFT containing copolymers of a charged and uncharged monomer were adsorbed on inorganic particles and subsequently chain-extended (Nguyen, 2008). The same method was also used to prepare nanocapsules by using vesicles as the template (Ali et al., 2011b). Although very related, in these two examples, in the strict sense, one cannot speak about PISA anymore.

Chapter 6

Synthesis of Nanobottles

Alexander van Herk
Institute for Chemical and Engineering Sciences, 1 Pesek Road, Jurong Island, 627833 Singapore
alexander_herk@ices.a-star.edu.sg

In general one can distinguish hydrophobic and hydrophilic active ingredients to be released from nanoparticles. In the case of hydrophobic active ingredients solid nanoparticles can be used (as opposed to nanocapsules). If one applies such a loaded system in aqueous media the driving force for the active ingredient to diffuse out of the particle is limited, especially if the polymer is rigid or even crosslinked. In the case of hydrophilic active ingredients at first instance using a nanocapsule or nanogel seems logical. However, if the concentration of the hydrophilic active ingredient is high inside the nanocapsule there is a strong tendency for water to go in (osmotic pressure). So for the nanocapsule not to be blown up and burst, osmotic pressure matching in the final formulation is needed. In general the diffusion of water through a wall of a nanocapsule is fast and even leakage of bigger active ingredients over a period of months or years from (responsive) nanocapsules is difficult

Controlled Release Systems: Advances in Nanobottles and Active Nanoparticles
Edited by Alexander van Herk, Jacqueline Forcada, and Giorgia Pastorin

ISBN 978-981-4613-21-7 (Hardcover), 978-981-4613-22-4 (eBook)
www.panstanford.com

to prevent. One of the reasons is that responsive polymer walls of nanocapsules are usually not highly crosslinked because else they would lose their responsiveness. First diffusion through polymeric walls will be discussed and then it will become clear that nanobottles as a new concept might be useful.

6.1 Diffusion of Water and Active Ingredients through a Polymeric Wall

To protect water-sensitive compounds often one likes to resort to the nanoencapsulation of these compounds. There are two factors to be taken into account. First of all there is the thermodynamic driving force toward an equilibrium water concentration in the close vicinity of the water-sensitive (hydrophilic) compound. Second there is the rate of water diffusion through the nanomembrane. The two factors combined result in a certain efficacy to protect the water-sensitive compound. For a polymer nanolayer to completely keep water molecules out kinetically is very difficult.

Encapsulation of a hydrophilic water-sensitive compound to protect is from a reaction with water, for example, with a 10 nm thick polystyrene layer, theoretically cannot work.

With a diffusion coefficient (D) of 10^{-15} m^2/s (value for a water molecule) the rate of diffusion through a normal polymer layer of 10 nm (x_1) is about 0.1 sec.

The Fourier number for a flat disk:

$$Fo = \frac{D \times t}{x_1^2} \rightarrow 1 = \frac{10^{-15} \times t}{(10 \times 10^{-9})^2} \rightarrow t = 0.1 \text{ sec} \tag{6.1}$$

With Fo = 1 there is almost complete concentration levelling on both sides of the 10 nm layer within this 0.1 sec.

Even for bigger molecules such a polymer layer cannot keep that inside a nanocapsule for a long time. For a molar mass of 1000 Da the diffusion coefficient is about a factor of 10 lower than for water. So in this case the other side of the wall is reached within 1 sec. More impermeable polymer layers are the ones with high degrees of crosslinking or crystallinity or containing certain inorganic compounds like clay materials. The diffusion coefficients can drop by a factor of 100. But for the same layer thickness of 10 nm this still means that water gets through in 10 sec! However, such layers, at thicknesses of several micrometers can slow down water diffusion to days or months and a highly crosslinked polymer layer can be effective in keeping bigger active ingredients retained in nanocapsules for a considerable amount of time.

In summary, keeping water away kinetically from a hydrophilic active ingredient inside a nanocapsule is almost impossible. But what about the thermodynamics? The rate of reaction of the sensitive molecule with a water molecule depends on the concentration. Strongly lowering the equilibrium concentration of water close to the sensitive compound is a much better approach. In case of a water-sensitive hydrophilic compound, making the molecule itself more hydrophobic or embedding it in a hydrophobic polymer will lead to much better results.

6.2 The Concept of the Nanobottle

So in Section 6.1 it has been made clear that leakage of the smaller active ingredient from nanocapsules is difficult to prevent over long periods of time. The slowest leakage would occur for highly crosslinked, crystalline, or hybrid polymer layers (e.g., polymer with oriented clay inside). It is difficult to combine low leakage of a nanocapsule and fast responsive release (through at most a slightly crosslinked responsive polymeric wall). In general nanocapsule walls are uniformous but it would be interesting to make part of the wall highly impermeable (most likely highly crosslinked) and another small part of the wall responsive (most likely hardly crosslinked). To achieve this, the design of a nanobottle would be interesting (Fig. 6.1).

So the wall of the nanobottle (the container) is highly impermeable but the lid is responsive. In this way leakage of the active ingredient is minimized. Another advantage is that the size of the total structure is changing much less than in the case of a nanocapsule completely constructed out of responsive polymer. The size of the nanoparticles is also important for targeting certain tissue in the human body (Chapter 7 and Part II of this book).

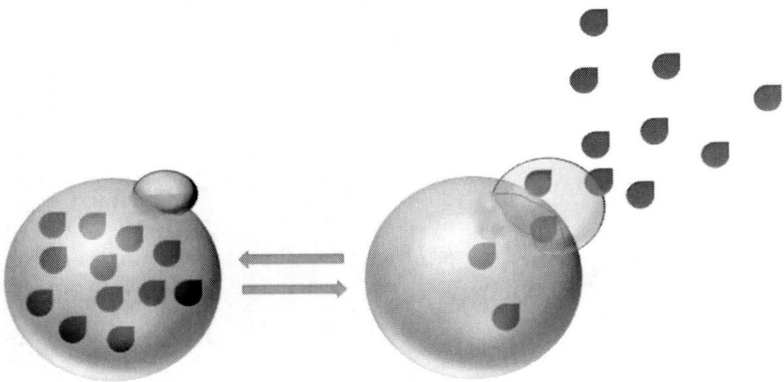

Figure 6.1 The concept of a nanobottle. The active ingredient is retained in the container part (gray), which is highly impermeable, and a small "lid" (bright blue) can expand through an external trigger (light, change in pH, ionic strength, temperature, etc.) to become more permeable and release the active ingredient (in red). The lid can be closed again reversibly, giving the opportunity for pulsatile response.

The following definition for a nanobottle is proposed:

A nanobottle is a responsive nanocapsule where only a part of the wall is responsive (the lid) and the rest of the wall is and remains more impermeable for the active ingredient. The nanobottle can be opened and closed reversibly.

6.3 Nanobottles

Now the question is whether nanobottles have actually been constructed. In a review on progress in multicompartment polymeric capsules from 2013 no structures that follow the definition of a nanobottle were reported (Huang and Voit, 2013). In 2003 silica nanobottles were prepared (Zhang et al., 2003a) but the structures did not contain a lid that can be opened and closes, which in general is the case for inorganic nanobottles mentioned in the literature. In the review on stimuli-responsive polymer vesicles from 2009 (Li and Keller, 2009) the work of Chiu (2008) is mentioned in which multivesicular assemblies were produced via a two-stage double emulsion of poly(AAc-co-DSA). The authors claim that the vesicles are equipped with pH-responsive transmembrane channels (Chiu, 2008). This could be a first structure that can be regarded as a nanobottle according to our definition, although vesicle structures in general are not extremely impermeable. In the elegant work of the group of Reimhult superparamagnetic iron oxide nanoparticles were incorporated in the lipid membrane. By applying alternating magnetic fields the membrane was locally heated and the cargo released repeatedly (Amstad et al., 2011).

In 2013 a new system was published where polymer bottles consisting of a container made out of polystyrene with a single hole in the wall (Hyun, 2013). The lid (cork) consists of a phase change material like 1-tetradecanol. When heating above the melting temperature (38–39°C) the contents of the bottle was released. The size of the bottles was in the range of 1000 nm and the size of the hole could be varied between 50 nm to 600 nm. The typical wall thickness was around 200 nm. Because of these relatively thick walls, leakage of the tracer dye (rhodamine B) was not detected. At this point it seemed that the cork could only be used once. Another interesting feature of the rather large nanobottles was that in theory even nanoparticles could be released from these nanobottles. On the cover of this book such a polymer bottle is shown.

6.4 Conclusion

More advanced nanostructures will be needed for controlled release applications in several application fields. The nanobottles seem to have several advantages over the normal nanocapsules but are not easy to produce. In that respect it is worth mentioning that there is a particular approach that seems promising to produce nanobottles in that is polymerization in the bilayer of vesicles. The early work of Jung has shown that it is relatively simple to produce the so-called parachute morphology in vesicle polymerizations (Jung, 2000b). Recently, although still with one type of polymer for both the wall and the lid, precursors for nanobottles in the form of protrusion structures were observed, which already have promising morphological features related to nanobottles (Fig. 6.2).

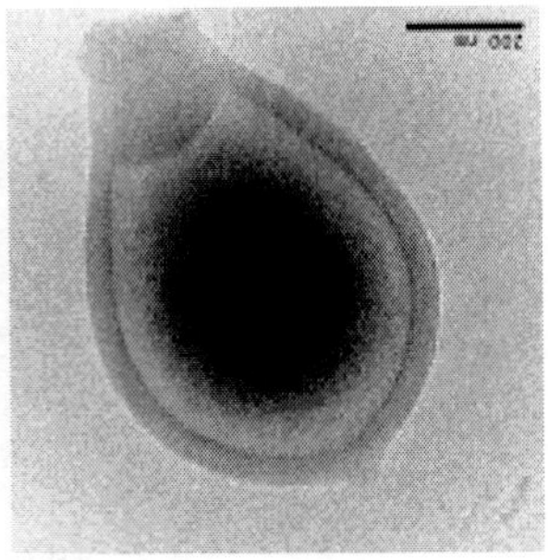

Figure 6.2 Poly(MMA/t-BA/EGDMA) protrusion morphology (Moradi et al., 2015).

Chapter 7

General Criteria for the Selection of Nanoparticles for Certain Applications

Dai Thien Nhan Tram and Giorgia Pastorin
Department of Pharmacy, National University of Singapore, 117543 Singapore
phapg@nus.edu.sg

7.1 Introduction

Nanomedicine, a study of nanoparticles (NPs) and their potential biomedical applications as either therapeutics or diagnostics (or both), is slowly but steadily fine-tuning medical practices in general and cancer treatment in particular. These nanosized objects are more intriguing compared to their macromolecular counterparts due to a host of unique properties, ranging from a tendency to be readily uptaken by cells, the ability to reach deeper and less accessible tissues, avoidance of recognition by the cells of immunity (i.e., stealth effect), and protection from eventual premature elimination from the host (Kotagiri et al., 2014; Singh et al., 2006; Yang et al., 2012b). As a word, "nanoparticles" might be unwittingly overgeneralized. It actually entails multifarious types of particles,

Controlled Release Systems: Advances in Nanobottles and Active Nanoparticles
Edited by Alexander van Herk, Jacqueline Forcada, and Giorgia Pastorin

ISBN 978-981-4613-21-7 (Hardcover), 978-981-4613-22-4 (eBook)
www.panstanford.com

ranging from the organic dendrimers to the inorganic gold NPs, which have proven a valuable asset in scores of medical purposes, one of which is the detection of pathogenic *E. coli* strains (Boken et al., 2013). The shear diversity of NPs might, with no doubt, confound most of those who are new to the field. Therefore, to facilitate a smoother transition into the discipline, this chapter outlines some of the criteria crucial for singling out the NPs most suitable for intended research interest. Even though NPs are varied in terms of structure and properties, it should be noted that they do share similarities in certain aspects, for instance, functionalization. Identity of the biomolecules conjugated onto the external surface of the NPs is just as important and it should definitely not be overlooked. Some of the ligands are capable of endowing NPs with heightened preference for tumor cells. This gives rise to a set of homing strategies known as active targeting, with the other being passive targeting. More details on this will be covered in the following paragraphs.

7.2 Is Passive Targeting indeed "Passive"?

The one most noteworthy characteristic that helps nanomedicine stands firmly at the forefronts of cancer treatment research is undoubtedly tumor selectivity, which is a consequence of passive targeting, active targeting, or even both. Passive targeting as an indicator might be misleadingly "passive." As most would have accurately inferred from the name, "pristine NPs" do not contain any specific domain that can influence accumulation of bioactive species at any particular site in the body. Having said so, it is not completely beyond our control. Functionalizations of NPs with polymers, for example with poly(ethylene glycol) (PEG) (Otsuka et al., 2003), have been shown to optimize passive targeting by lowering plasma clearance and lengthening circulation time of the NPs (Gabizon et al., 1995). Extent of passive targeting does play a crucial role in the choice of functionalized molecules and in some circumstances even the choice of NPs themselves (Table 7.1).

Apart from the all-time favorite PEG, NPs can also be coated with dextran (Laurent et al., 2008), amphipathic polymers (Torchilin et al., 1995), chitosan (Arias et al., 2011), or block copolymers such as poloxamers (Choi et al., 2003) and poloxamine

Table 7.1 Active and passive targeting as strategies to enhance tumor selectivity of NPs

Category	Key points		Authors
Passive targeting	Passive targeting can be optimized by decreasing plasma clearance of the nanoparticle and increasing its stability in circulation		Gabizon et al., 1995
	Polymers for NPs coating that prolong circulation half-life (apart from PEG)	block copolymers such as poloxamers and poloxamine	Choi et al., 2003 Moghimi et al., 2003
		dextran	Laurent et al., 2008
		amphipathic polymers	Torchilin et al., 1995
		chitosan	Arias et al., 2011
	Factors affecting EPR	-Smaller nanoparticles can diffuse through walls of normal vessels ⇒ better distributed to healthy tissues ⇒ less effective passive targeting -NPs larger than 200 nm significantly cleared by the spleen	de Jong et al., 2008 Moghimi et al., 2001
		-NP shape does matter -Nonspherical NPs show prolonged circulation time and greater accumulation in inflammatory tissue	Daum et al., 2012
		Factors that could limit EPR: + smooth muscle cells or pericytes covering tumor blood vessels + dense extracellular matrix + high intercellular fluid pressure + multidrug resistance	Fang et al., 2011
		Vascular endothelial luminal surface is known to carry a negative charge ⇒ basic proteins with positive charges or cationic polymers rapidly bind ⇒ reduced tumor accumulation	Campbell et al., 2002 Nakamura et al., 1998

(Continued)

Table 7.1 (*Continued*)

Category	Key points		Authors
		Reticuloendothelial system in the liver and spleen reportedly showed faster uptake of negatively charged nanoparticles than of neutral particles	Li and Huang, 2008
		Influence of surface charge on passive targeting varies between different types of NPs	Maeda et al., 1983 Keyler et al., 2006
	PEG dilemma	Injection of PEGylated liposomes elicits PEG-specific IgM ⇒ rapid elimination in subsequent introduction of the NPs	Ishida et al., 2006
		Certain PEG-modified NPs experience slower uptake into tumor cells than non-PEGylated molecules	Hatakeyama et al., 2007
		Bonds that could be cleaved by proteases or similar agents in the vicinity of tumor cells help discharge PEG coating after it has fulfilled its purpose, and before it becomes a hindrance	Masuda et al., 2009
Active targeting	Drawback	Targeting ligands could be recognized by the RES ⇒ enhanced clearance from the bloodstream	McNeeley et al., 2007
		No receptors are exclusively expressed on cancer cells ⇒ possible binding and uptake by healthy cells	Loomis et al., 2011
	Heterovalent binding	Synergistic targeting efficacy	Saul et al., 2006

Category	Key points		Authors
		Examples: + EGF + folate (tumor specific) + avb3 integrin + galectin-1 (angiogenesis specific)	Kluza et al., 2010 Saul et al., 2006
		Counterclaim: Laginha et al. found no synergistic binding effect for targeting B cells with (CD19 + CD20)-coated liposome	Laginha et al., 2005
	Subcellular targeting—mitochondria	Examples of mitochondriotropic ligands: + triphenylphosphonium cations (conjugated to liposome) + rhodamine (conjugated to MWCNTs)	Boddapati et al., 2008 Yoong et al., 2014
	Targeting ligands examples	Can be grouped into: affibody + whole antibodies and their fragments + vitaminspeptides + carbohydrates + hormones	Loomis et al., 2011
		Largely applicable to several cancer cell lines (e.g., liposomes conjugated with HER2-specific affibody display enhanced selectivity for HER2+ ovarian, breast, and pancreatic cell lines)	Puri et al., 2008
		Can also target nontumor cells (e.g., hepatic parenchymal cells express higher than normal amount of receptor for galactose)	Wang et al., 2008b
		-Hormones as targeting ligands often target only specific cell lines -Luteinizing hormone releasing hormone (LHRH) conjugated to superparamagnetic iron oxide NP for targeting breast cancer cells ⇒ improve MR imaging	Meng et al., 2009

(Moghimi et al., 2003). Nonetheless, we need to seriously reconsider the popularity of PEG as a coating candidate after the finding that injection of PEGylated liposomes might trigger the production of PEG-specific IgM, giving rise to rapid elimination of PEGylated NPs in subsequent doses (Ishida et al., 2006). Besides, NPs conjugated with PEG have been shown to display slower uptake into tumor cells compared to those without PEG (Hatakeyama et al., 2007). Though posing a setback to the development of a selective and efficacious delivery vehicle for therapeutic cargo in clinical settings, this so-called PEG dilemma has not down rated the popularity of PEG coatings, thanks to a concept commonly seen in triggered release of the cargo (Masuda et al., 2009). The PEG coating can be attached to the NPs via bonds which can be cleaved by biomolecules swarming the extracellular environment of tumor cells. Matrix metalloproteinases (MMPs) have been demonstrated to assists the detachment of the PEG coating from liposomes and facilitate fusion with tumor membrane (Xiong et al., 2012a). In another study, the PEG coating was anchored by disulfide bonds, which are eventually cleaved by endogenous reductants inside tumor cells (McNeeley et al., 2009).

Besides the extent of surface functionalization, several physical properties of the NPs have been determined to have an effect on the enhanced permeability and retention (EPR) effect, the cornerstone of passive targeting. The most obvious factor is surely particle size, since EPR comes about as a direct result of larger pore size on the wall of leaky vasculature found near tumor site (Iyer et al., 2006). Controlling the size of the NPs above a floor value renders them too bulky to diffuse across the wall of normal vasculature. Increasing the size of the NPs neatly limits migration of our nanosized particles from the circulation to healthy tissues but we need to be wary of a ceiling value. Large NPs have trouble infiltrating the tumor, possibly exerting cytotoxicity only at the peripheral (Kuszyk et al., 2001). Then again, the extent to which this compromises antineoplastic and diagnostic effect of NPs, subjected to the way the system is designed, might depend on the size of the tumor. Moreoever, above 200 nm in size, NPs experience a drop in circulation time due to heightened clearance by the spleen (de Jong et al., 2008, Moghimi et al., 2001). In short, a moderate size is optimal for the NPs to be deposited at the site of action, and to diffuse throughout the site while they are at it. Regarding how large or how small can the NPs be called "moderate," it is only a matter of time for a range to be established, as long as sufficient attention is given to it.

The shape of the NPs matters as well. Nonspherical NPs have been demonstrated to exhibit prolonged circulation time and more pronounced accumulation in inflammatory tissues (Daum et al., 2012). With tumor growth and inflammation being closely linked (Rakoff-Nahoum, 2006), it follows that nonspherical NPs (e.g., carbon nanotubes [CNTs]) should display better tumor selectivity via passive targeting than their spherical counterparts (e.g., liposomes).

The last, and arguably the trickiest, factor to be considered when designing NPs is surface charge. Uptake by reticuloendothelial system (RES) in the liver and spleen has been reported to manifest preference for negatively charged liposomes (Li and Huang, 2008). For instance, Price et al., reported the synthesis of a dual pH- and thermoresponsive mixed micellar system, whose hydrophobic segment is composed of poly(*N*-(2-hydroxypropyl) methacrylamide di-lactate)-co-(*N*-(2-hydroxypropyl) methacrylamide-co-histidine). It contains a strong negative surface charge and expectedly proves susceptible to receptor-mediated endocytosis by the liver's nonparenchymal cells (Price et al., 1998). Meanwhile, NPs constituted from cationic polymers readily bind to negative charges found on the luminal surface of vascular endothelia (Campbell et al., 2002; Nakamura et al., 1998). Both phenomena lower accumulation at the tumor site. Then again, one should not narrow down his perspective to just passive targeting while contemplating surface charge. Cationic polymers coating is one of the strategies to improve endosomal escape, which is vital for NPs to retain their chemical integrity before arrival at the targeted intracellular site (Ma, 2014). Yu et al. illustrated that the cationic polymer, poly(2-(dimethylamino) ethyl methacrylate)-block-poly(2-(diisopropylamino)ethyl methacrylate), confers better transfection capability to the micelleplex under investigation (Yu et al., 2011b). Other investigations verified this variability in the way surface charges affect characteristics of NPs in vivo. In a nutshell, there is no refined set of criteria based on which the charge most befitting a particular NP can be selected.

With passive targeting being under the influence of several physical factors, some NPs are bound to be inherently better suited for passive targeting, for instance, polymeric nanoconjugates.

They are, in essence, polymeric backbones equipped with multiple functional groups to which all sorts of chemical moieties can be covalently conjugated. This allows different drug compounds to be concurrently delivered to the tumor site (Pinhassi et al., 2010), hence opening up the myriad possibilities of synergistic anticancer effect among separate classes of antineoplastic agents. In addition, they are not vulnerable to drug resistance, an impediment usually faced by chemotherapeutics (Ryan, 2013), by circumventing multidrug resistance transporters expressed by tumor cells (Lee et al., 2006a; Ljubimova et al., 2008). To revert to the issue of passive targeting, polymeric nanoconjugates are generally smaller than veteran NPs such as liposomes or micelles. Therefore, they are less prone to clearance by the spleen, less immunogenic and as a consequence more long lived in plasma circulation (Dhanikula and Panchagnula, 2005).

Active targeting is not discussed in depth in this chapter since it is already covered elsewhere in the book (e.g., Chapter 14). A rather detailed list of targeting ligands investigated thus far can be found in an article by Loomis et al. (2011). Even though the list is by no means comprehensive, it does a great job in projecting the abundance of targeting ligands out there. Below is a summary table that condenses information on both passive and active targeting.

7.3 Nanoparticles as Delivery Vehicles

In the last few decades, NPs have been extensively investigated as potential transporters for both therapeutic and diagnostic purposes. The type of cargos to be delivered is another criterion vital for NPs selection process (Table 7.2).

7.3.1 The Raid on Chemoresistance: Combination Therapy and siRNA

Genexol-PM™ is a paclitaxel-containing polymeric micelle synthesized from the biodegradable block copolymer, monomethoxy poly(ethylene glycol)-block-poly(D,L-lactide). This formulation has undergone clinical trials on a handful of cancer types such as metastatic breast cancer, solid tumors, and non–small cell lung

Table 7.2 Types of cargo that can be carried by NPs

Header	Key points		Authors
Therapeutic cargo	Paclitaxel	-Genexol-PM is a formulation of paclitaxel encapsulated in a polymeric micelle -Tested in several clinical trials on metastatic breast cancer, solid tumors, and non–small cell lung cancer	Lee et al., 2008b Kim et al., 2007a Lim et al., 2010a
	Cisplatin	CNT Pt(IV) prodrug delivery using folate as targeting ligand	Dhar et al., 2008
	Doxorubicin	Positive results on HeLa, ZR-75-1, MCF-7, and H661 cancer cells	Chen et al., 2012e
	Combination therapy	Widely employed to combat chemoresistance	Agarwal et al., 2003
		-CNT-encapsulated cisplatin prodrug serves as a template to bind doxorubicin -Shown to be highly effective against endometrial adenocarcinoma cells	Chin et al., 2014
		Combination with siRNA suggested to improve the efficacy of chemotherapy	Wang et al., 2009
Imaging agents cargo	Unimodality	Endohedral metallofullerenes are able to solubilize metallic agents ⇒ MRI contrast agent	Kato et al., 2003
		-Trimetallic nitride endohedral metallofullerene NP containing gadodiamide produced as much contrast as the control clinical agent but at much lower concentrations (0.013 mmol compared to 0.050 mmol) -Metallofullerene inhibits the release of toxic metal ions	Fatouros et al., 2006
	Bimodality	Polyamidoamine dendrimer-based nanoprobe for dual modality conjugated with: + Gd(III), MRI contrast agent + Cy5.5, a fluorescent marker	Goldberg et al., 2007

(*Continued*)

Table 7.2 (*Continued*)

Header	Key points		Authors
	Imaging of angiogenesis	Endothelial precursor cells of hematopoietic stem cell origin labeled with dextran-coated superparamagnetic iron oxide NP, monitored by MRI ⇒ noninvasively track tumor vasculature	Arbab et al., 2005
Imaging agents cargo	Uses in diagnostic techniques other than MRI	GNP functionalized with gum arabic ⇒ contrast agent for X-ray imaging	Kattumuri et al., 2007
		-Iodine-containing micelle ⇒ CT imaging -Methoxy-polyethylene glycol and tri-iodobenzoic polylysine micelles ⇒ improved X-ray signal	Trubetskoy et al., 1997
	Hydrophilic cargo favored by nanogel	-Crosslinking prevents dissociation of micelle structures of diblock copolymers	Rapoport et al., 1999
		-Large amount of water at the inner core ⇒ a challenge to encapsulate hydrophobic drugs -Overcome with the use of pyridyl disulfide groups	Liu and An, 2014
Coencap-sulation		Another strategy to allows entrapment of water-insoluble compounds makes use of cholesteryl groups	Morimoto et al., 2012
	Hydrophobic cargo favored by CNT	-Hydrophilic chemicals are not readily incorporated into the inside of CNT ⇒ hydrophobic precursor of a hydrophilic drug helps address the issue	Li et al., 2012d
	Liposome tolerate both	Well suited to codeliver therapeutic and diagnostic agents, which can differ in their physicochemical properties	Torchilin, 2005

cancer (Lee et al., 2008b, Kim et al., 2007a, Lim et al., 2010a). The FDA has hitherto approved some pharmaceutical products which utilize NPs, thus far mainly liposomes, as delivery vehicles (Kateb et al., 2011). The apparent fact that all of the formulations utilize a sole therapeutic agent does not reflect the current research direction. Many researchers have started to shift their interest to combination therapy as a means to subdue chemoresistance (Agarwal et al., 2003). Simultaneous delivery of multiple therapeutic agents helps trigger varied intracellular pathways at once. At any instant, the probability that a particular cell has the genetic makeup requisite for protection of all those drug targets is definitely smaller than when a single therapeutic agent is utilized. This exact line of reasoning suffices to explain why combination therapy has become the new trend. An attempt to model this novel approach can be seen with CNTs functionalized with integrin-targeting cyclic peptide for codelivery of cisplatin and doxorubicin (Chin et al., 2014). Within the CNTs, cisplatin prodrug serves as an anchor to which doxorubicin is tethered. Inside tumor cells, intracellular reduction triggers release of both compounds. The system has been shown to be highly cytotoxic against endometrial adenocarcinoma cells.

Another strategy employed to deal with drug resistance is codelivery of therapeutic agents with short interfering RNA (siRNA). Investigations on mouse lymphoma and lung cancer models have illustrated utility of siRNA. Suppression of error-prone translesion synthesis (TLS) activity in mammalian cells by deactivating either *Rev1* or *Rev3L* genes inhibits drug-induced mutation of the genetic makeup. In effect, relapsed tumors retain vulnerability to said anticancer agents in ensuing treatments (Doles et al., 2010; Xie et al., 2010b). Moreover, siRNA has been shown to synergistically ameliorate cytotoxicity of the companion cargo—therapeutic agent (Wang et al., 2009). Xu et al. developed PLGA–PEG-based NP formulation for codelivery of siRNA and the Pt(IV) prodrug. Even though the xenograft model did not allow for examination of acquired chemoresistance, the observation that this delivery system was capable of silencing target TLS genes for at least three days after the administration is a testimony to its potential as a breakthrough in the escape from the chemoresistance hassle (Xu et al., 2013b).

Thus far, we have provided many cases of drug-encapsulated NPs, but it is not the only manner in which NPs work as a delivery vehicle. Therapeutic agents can be conjugated to the external surface of the NPs as well. For instance, Tripisciano and Borowiak-Palen functionalized cisplatin to the side walls of single-walled carbon nanotubes (SWCNTs) and managed to observe enhanced therapeutic effect compared to cisplatin control (Tripisciano and Borowiak-Palen, 2008). These two delivery methods are not equivalent alternatives. It has been suggested that internalization inside CNTs ensures that the cargo stays impervious to biochemical reactions with endogenous compounds that might undermine chemical integrity of the compounds (Hampel et al., 2008). Nonetheless, encapsulation of therapeutic molecules inside NPs is anything but the ultimate solution. The cargo might not be completely released from its nanocarrier, thereby being a waste of bioactive agent and a potential source of toxicity.

7.3.2 Charismatic Mediators: What It Takes

CPX-351 is a liposomal formulation for codelivery of cytarabine and daunorubicin, recently investigated in clinical trials as a therapy for acute myeloid leukemia (Dicko et al., 2010). Being an anthracycline, daunorubicin is inherently fluorescent. It was confirmed on rats that has the drug not been entrapped inside the liposome, its optical activity would have been lost as the compound gets metabolized by xanthine oxidase in the liver (Dodion et al., 1987). In this way, daunorubicin doubles as an imaging tool for in vivo tracking of the NPs. Unfortunately, there are not that many versatile anticancer agents. An imaging agent commonly needs to tag along the NPs for monitoring of the in vivo distribution of the drug delivery system, when accompanied by therapeutic agents, or purely for diagnostic purposes, when it is administered alone.

The vast majority of anticancer agents are hydrophobic (except some such as first-line therapy cisplatin), whereas imaging agents are mainly hydrophilic. Such contrast in chemical nature is a formidable challenge to the design of NPs. Then again, even in the case of CPX-351, whereby an imaging agent is missing, the rise of combination therapy renders simultaneous entrapment of various compounds, which might very well display polar opposite affinity for water, inevitable. In some other cases, encapsulation is not an option (e.g., polymeric nanoconjugates, dendrimers).

Topping the hierarchy of favorites is none other than liposomes. They are well suited for carrying both hydrophobic and hydrophilic compounds (Torchilin, 2005), requiring no more cargo nor vehicle modifications than the ones normally seen in other NPs intended for monotherapy. In fact, these NPs can trap water-soluble molecules in their inner core, whereas hydrophobic ones remain within the lipid bilayer.

Having said so, we should not give other NPs too little credit. All it takes for them to be up for the task is some elegant tweaks in the design. Nanogel is a type of NPs that especially favor hydrophilic cargos as a consequence of the large water content located in its interior. In the formation of nanogel, crosslinking stabilizes the micellar structure of diblock copolymers and minimizes their dissociation in aqueous environment. In a series of studies (Ryu et al., 2010, Gonzalez-Toro et al., 2012, Matsumoto et al., 2013), a self-crosslinked nanogel was made from copolymer of polyethylene glycol monomethyl ether methacrylate and pyridyl disulfide ethyl methacrylate. Aside from construction of disulfide crosslinks, pyridyl disulfide groups are essential to form hydrophobic regions within the hydrophilic interior of the nanogel (Liu and An, 2014). Cholesteryl groups have proven to be capable of a similar role (Morimoto et al., 2012). Furthermore, nanogels have their own merits to offer. The most outstanding is probably their unusually high drug loading capacity; a nanogel can carry a cargo molecule with a mass eight times its own (Chen et al., 2010b).

Right at the other extremity are CNTs and their highly hydrophobic interior. To encapsulate the hydrophilic cisplatin, many have resorted to the prodrug solution, coupled with enzymatic release (Dhar et al., 2008, Li et al., 2012c). The lipophilic Pt(IV) prodrug can be incorporated into CNTs with ease by nanoextraction. Transported in this dormant form all the way until the vehicle has penetrated into the tumor cells, the compound will then be converted by the endogenous reductants overexpressed inside tumor cells (e.g., glutathione) into the cytotoxic Pt(II) species. In this manner, the activated form is conveniently more hydrophilic and hence promptly released from the CNTs.

7.3.3 Cargo Release on Demand

Triggered release refers to the ability of NPs to release the cargo in response to particular environmental cues (Aaron et al., 2011). It is another hotspot extensively delved into in recent years

(Table 7.3). Having substantial contribution to tumor-specific release of anticancer drugs, triggered release is undeniably one of the key determinants in the selection of NPs.

The stimuli fostering the release of cargo can either come internally from the intracellular environment of the tumor cells or come externally from an artificial source. Among these the following four have attracted much interest: heat, pH, enzyme, and ultrasound.

7.3.3.1 Thermal trigger

Heat for this purpose can be generated by irradiating gold nanoparticles (AuNPs) (Paasonen et al., 2007a, 2007b, Wu et al., 2008) or subjecting iron oxide nanoparticles (FeONPs) to alternating magnetic frequency (Pradhan et al., 2010, Tai et al., 2009). Both of these auxiliary NPs could be coadministered alongside the drug delivery system of interest. In another study, nondestructive high-intensity focused ultrasound was used to induce localized raise in temperature, which accelerated the release of doxorubicin from liposomes compared to nonthermosensitive analogues (Dromi et al., 2007). Despite the additional utility of increasing permeability of vasculature at the area of interest, external stimulation of hyperthermia is discouraged when dealing with tumor situated deep within vulnerable tissues (e.g., brain tissues). In such cases, FeONPs and AuNPs are favored by virtue of their relatively lower collateral damage (Loomis et al., 2011). As a bonus, the heat generated causes cancer cells to ablate and to become more susceptible to damage exerted by chemotherapeutics. Candidates for this category of triggered release include largely the types of NPs constructed from polymers (e.g., nanogel, micelles). Some examples of thermoresponsive polymers include poly(methoxydiethylene glycol methacrylate) (poly- (MeODEGM)) (Heyden et al., 2009), poly(*N*-isopropylacrylamide) (poly(NIPAM)) (Paasonen et al., 2007a, 2007b, Prevot et al., 2006), and poly(*t*-butyl acrylate)-b-poly(*N*-isopropylacrylamide) (Li et al., 2009b). Bhuchar et al. reported the synthesis of a nanogel whose hydrophobic core was composed of MeODEGM and 2-aminoethyl methacrylamide hydrochloride (AEMA) crosslinked with an acid-degradable crosslinker and a hydrophilic shell of poly(2-methacryloyloxyethyl phosphorylcholine) (Bhuchar et al., 2012). AEMA provides cationic character to the nanogel core, which aids the encapsulation of negatively charged proteins (e.g., insulin, bovine serum albumin, and β-galactosidase). Above the lower critical solution temperature, the thermoresponsive MeODEGM core becomes hydrophilic, hence promoting the release of the cargo. Nanogel as a drug carrier has other fortes as well, which were already discussed in Section 7.3.2.

7.3.3.2 pH trigger

It has long been noted that the microenvironment within cancer cells is slightly more acidic than that in normal healthy cells (Song et al., 1993; Webb et al., 1999; Wike-Hookley et al., 1984). This serves as the basis for usage of acid-labile bonds that are readily cleaved under acidic conditions. The PEG stealth coating on nano-carriers can be anchored by acid-labile linker. Once the NPs reach the tumor site, the PEG layer is released under the presence of pH stimulus. This exposes the cellular membrane to membrane-destabilizing complexes previously hiding under the stealth coating and thus does not represent a threat to the cellular plasma membrane of nontumor cells (Kirpotin et al., 1996, Kono et al., 1997). These bonds have also been tapped on in polymer–drug conjugates (e.g., doxorubicin, paclitaxel) (Devalapally et al., 2007; Etrych et al., 2001; Potineni et al., 2003; Prabaharan et al., 2009a). Notable examples of bonds labile to hydrolysis under acidic pH include diorto esters, vinyl esters, disulfide bonds, double esters, and hydrazones (Ishida et al., 2006; Kirpotin et al., 1996). Interestingly, some nanosized systems have been fashioned to be pH sensitive without a need for acid-labile bonds. One such special system is the lipid-coated calcium phosphate NPs (liposome/calcium/phosphate, or LCP), which consists of a CaP inner core and a lipid outer layer (Li et al., 2010a; Li et al., 2012b). At the endosomal acidic pH level, LCP would dissolve to disassemble the NP. The resultant increase in osmotic pressure eventually causes endosomal swelling. Even though the role the pH trigger plays in this situation is not relevant to the ultimate goal of tumor selectivity, given the fact that endosomes can be found in both tumor and healthy cells, it is still worthy of being mentioned here. pH-triggered release within the endosomes actually ensures the escape of the cargo, siRNA in this study, and preserves its chemical integrity, which could otherwise be impaired by lysosomal enzymes encountered along the endocytic pathway.

7.3.3.3 Enzymatic trigger

This strategy exploits the abnormal overabundance of certain enzymes in cancer cells. MMPs are among those most investigated for this objective. In one study, the PEG coating on liposomes was cleaved by MMPs (Terada et al., 2006), the aftermath of which was mentioned in the previous paragraph on pH triggers. Moreover, the enzymes can be used to activate prodrugs (e.g., cisplatin prodrug) (Li et al., 2012c). Other overexpressed enzymes that have been investigated include phospholipase (Nieva et al., 1989) and elastase (Meers., 2001). Interestingly, the enzymatic trigger is applicable not only in the field of chemotherapeutics but also in that of antibiotics. Xiong et al. reported the synthesis of a triple-layered nanogel for delivery of vancomycin (Xiong et al., 2012a). The outer-layer PEG layer fosters dispersity in an aqueous medium and avoidance of opsonization. The inner core is composed of crosslinked poly(phosphoester), which is degradable when exposed to lipase secreted by the bacteria. It also serves as the reservoir for hydrophilic antibacterial drugs. Finally, the intermediate layer made up of hydrophobic lipase-degradable poly(3-caprolactone) is the gatekeeper that prevents premature leakage of the cargo.

7.3.3.4 Ultrasonic trigger

This methodology clearly demonstrates why triggered release is one of the issues imperative to the choice of nanosized carriers. Only a handful of NPs are suitable for controlled release of drugs under the influence of externally applied ultrasound. This small collection of nanosized carriers is epitomized by nanobubbles, which are spherical gas-filled vesicles. They are made up of varying chemical compositions (e.g., denatured albumin, lipid, poly(butyl cyanoacrylate)) (Liu et al., 2006). They can be induced to expand and contract by ultrasound that matches their resonance frequency (Janib et al., 2010). Besides, ultrasound has proven to bring about an increase in permeability of the already-leakier-than-normal vasculature at the tumor site, further raising the potency of passive targeting (Dromi et al., 2007). That being said, an ultrasonic trigger is not without its own shortcomings. In terms of effectiveness, low-frequency ultrasound overweighs the high-frequency counterpart. Unfortunately, it can only be used for superficial tumors near the skin

Table 7.3 Types of stimuli used in triggered release and representative examples of NPs

Header	Point		Author
	Heat can be applied by	hot water bath	Needham et al., 2000
		iron oxide nanoparticles in the presence of alternating magnetic frequency	Tai et al., 2009 Pradhan et al., 2010
		irradiation of gold NP	Paasonen et al., 2007a, 2007b Wu et al., 2008
		high intensity ultrasound	Dromi et al., 2007
	For delivery to sensitive tissues, localized heat application is preferred over external heat		Loomis et al., 2011
Heat trigger	Heat generated from gold and iron oxide NPs shown to: + thermally ablate cancer + sensitize cancer cells to chemotherapeutics		Lowery et al., 2006 Park et al., 2010
	Examples of thermoresponsive polymers	poly(methoxydiethylene glycol methacrylate) (poly(MeODEGM))	Heyden et al., 2009
		poly(*N*-isopropylacrylamide) (poly(NIPAM))	Paasonen et al., 2007a, 2007b Prevot et al., 2006
		poly(*t*-butyl acrylate)-b-poly(*N*-isopropylacrylamide)	Li et al., 2009b

(*Continued*)

Table 7.3 (*Continued*)

Header	Point		Author
	Endosomal fusion	Fusogenic lipids are protonated at acidic pH that is normally found in intracellular endosomes ⇒ change in molecular structure ⇒ destabilize the liposomal membrane with subsequent cargo release	Schroit et al., 1986
pH trigger	Acid-labile bond	Stealth coating on nanocarriers get cleaved under low pH condition ⇒ expose cellular membrane to membrane-destabilizing complexes	Kirpotin et al., 1996 Kono et al., 1997
		These bonds are utilized in polymer–drug conjugate, such as doxorubicin and paclitaxel	Devalapally et al., 2007 Etrych et al., 2001 Potineni et al., 2003 Prabaharan et al., 2009a
		Examples of bonds labile to linkage hydrolysis: diorto esters, vinyl esters, disulfide bonds, double esters, or hydrazones	Ishida et al., 2006 Kirpotin et al., 1996

Header	Point	Author
	-Diblock copolymer comprised a hydrophilic poly(NIPAM-co-acrylic acid) block and a hydrophobic polycaprolactone block -NPs aggregate at pH of 6.9 and body temperature ⇒ triggered release of cargo	Zhang et al., 2007a
	This NP system has more potential than just dual sensitivity to stimuli	Chen et al., 2012g
Dual pH- and temperature-sensitive (micelles)	-Multifunctional mixed micelles utilizing dual responsive diblock copolymer whose hydrophobic segment composed of poly(*N*-(2-hydroxypropyl) methacrylamide dilactate)-co-(*N*-(2-hydroxypropyl) methacrylamide-co-histidine) -Aside from dual sensitivity, NPs size increases with temperature ⇒ restrict extravasation to nontumor vessel ⇒ enhanced passive targeting to tumor site	Na et al., 2006
	Interior structure of NP responds to thermal stimulus by altering microviscosity	Zhang et al., 2007b
	Drawback: strong negative surface charges ⇒ susceptible to uptake by the liver's nonparenchymal cells via receptor-mediated endocytosis	Takakura et al., 1995

(*Continued*)

Table 7.3 (*Continued*)

Header	Point		Author
	Increase the permeability of blood tissue barriers and cell membranes		Price et al., 1998
	High intensity focused ultrasound confers release on low temperature sensitive liposomes as it generates heat		Dromi et al., 2007
Ultrasonic trigger	-Examples of materials constituting nanobubble shell: + denatured albumin + lipid + surfactant layers + poly(butyl cyanoacrylate) -High-intensity ultrasound can rupture the bubbles, releasing encapsulated drugs		Liu et al., 2006
	Drawback	-Low-frequency ultrasound proves a better trigger. However, it is is more difficult to focus than high frequency ultrasound ⇒ only applicable to superficial tumor	Schroeder et al., 2009

Header	Point		Author
	Cleaving of PEG coating	-Matrix metalloproteinases (MMPs) are most exploited enzymes -Cleave PEG coating on liposomes ⇒ fusion with cellular membrane	Terada et al., 2006
Enzymatic trigger	Activation of prodrug	-Once inside tumor cells, glutathione reduce hydrophobic Pt(IV) prodrug of cisplatin ⇒ more hydrophilic Pt(II) species (cyto-toxic species) ⇒ released from hydrophobic interior of MWCNT	Li et al., 2012c
	Other overexpressed enzymes	phospholipase	Nieva et al., 1989
		elastase	Meers., 2001
	A disadvantage of enzyme triggered release is that no enzyme is solely expressed in the targeted region		Loomis et al., 2011

surface due to the difficulty encountered in focusing it (Schroeder et al., 2009). This calls to attention the need for balancing between the desire to access deeper tumors, which generally involves the more serious pathological conditions, and the efficiency of the stimuli in initiating the release of the cargo from the nanobubbles.

7.4 Without the Cargo, NPs Are Still Useful!

Intensively studied for their applicability as delivery vehicles, NPs do have intrinsic utility for both therapeutic and diagnostic purposes. Although inorganic NPs have been less intensively exploited as carriers, they far outweigh their organic counterparts when it comes to cargo-free applicability.

7.4.1 Organic NPs

One of the most recent representatives of organic NPs is undoubtedly CNTs. SWCNTs are endowed with strong optical absorption from ultraviolet to near-infrared (NIR) regions of the electromagnetic spectrum. This unique trait accounts for their utility in photoacoustic imaging (de la Zerda et al., 2008; Xiang et al., 2009) and photothermal therapy (Ghosh et al., 2009; Kang et al., 2009a). Among the range of wavelengths preferably absorbed by SWCNTs, the NIR region is of particular interest to researchers since biological tissues in the body display remarkably low absorbance of the NIR wavelengths. This has much to do with the potential of photothermal therapy for wiping out tumor cells packed with SWCNTs, while selectively sparing healthy cells, all thanks to tumor-selective targeting ligands conjugated to the SWCNTs (Jeyamohan et al., 2013). Research scientists have also examined whether photothermal therapy complements antineoplastic cargo in terms of cytotoxicity. It does indeed, eliminating 95% of the tumor cells at an accelerated rate. This synergistic effect allows us to lower the amount of drugs carried by each CNT without compromising the level of therapeutic efficacy. Notwithstanding rigorous research into tumor selectivity of NPs imparted by targeting ligands, we have not managed to completely screen out collateral damage. As a result, a lower dose of the anticancer drugs required for the same level of cytotoxicity can only be encouraged, if anything. More details on the potentials of CNTs can be found in Chapter 12.

Another type of NPs with exploitable properties is nanobubbles. They are essentially micelles with a spherical cavity for enclosing gases (Gao et al., 2008). The nanobubble shell, which can be made of denatured albumin, lipid, surfactant layers, or poly-(butyl cyanoacrylate) (Schroeder et al., 2009), ruptures in the presence of tumor-directed ultrasound and releases the encapsulated cargo. It is one of the few NPs with which ultrasound works as a stimulus for drug release (Janib et al., 2010). However, this is not the intrinsic use under discussion. It is instead the ability to encapsulate gases that are natural contrast agents for ultrasound imaging (e.g., octofluoropropane, decafluorobutane, and sulfur hexafluoride) (Krupka et al., 2010; Landini et al., 2008). The nanobubbles thus become contrast agents themselves for in vivo visualization of inflammation, angiogenesis, intravascular thrombus, and tumors (Ferrara et al., 2009). Although conventional liposomes and polymeric micelles can be designed to release their cargo under an ultrasonic trigger (Loomis et al., 2011), none other than nanobubbles are endowed with the ability to complement sonography. Exposing the tumor site to ultrasound has yet another effect of increasing permeability of blood vessels (Dromi et al., 2007). Whether this is a boon or a bane in disguise, further investigations are recommended. This caution is derived from the observation that ultrasonically triggered release of drug raises the plasma drug level probably as a result of intravasation back to the vasculature (Schroeder et al., 2009).

7.4.2 Inorganic NPs

Most notable and definitely most relevant to the course of this chapter up until this point is the role of the inorganic NPs in the struggle against cancer. Preferentially binding to proteins bearing numerous cysteine and lysine monomers (Kim and Hyeon, 2014), AuNPs readily attach to recombinant human vascular endothelial growth factor ($VEGF_{165}$) and basic fibroblast growth factor (bFGF) in a male mouse ear model previously injected with adenovirus-producing mouse VEGF (Mukherjee et al., 2005). By distorting the structure and biological functions of these angiogenesis-inducing proteins, AuNPs effectively minimize the blood supply, a source of nourishment vital for tumor growth, and in turn hamper tumor growth, as illustrated in a preclinical mouse model of ovarian cancer (Arvizo et al., 2013). Besides, AuNPs prohibit heparin-binding growth factors from playing their role in cancer metastasis (Cooke et al., 2012; Thiery et al., 2009). In brief, AuNPs have left the tumor stranded. Similarly, silver nanoparticles (AgNPs) also exert their usefulness through inhibition of tumor angiogenesis, having been determined to interfere with VEGF-induced angiogenesis in bovine retinal epithelial cells via blocking of the PI3K/Akt signaling pathway (Gurunathan et al., 2009).

On the other hand, platinum nanoparticles and zinc oxide nanoparticles (PtNPs and ZnONPs, respectively) take a more direct approach. Once inside tumor cells, PtNPs adsorb the glutathione, causing its concentration to drop. Without the suppressor removed from play, production of reactive oxygen species (ROS) goes unchecked until it reaches a level detrimental to the integrity of DNA molecules. Human colon carcinoma cells (HT29) proved vulnerable to PtNPs (Pelka et al., 2009). Cytotoxic effect of PtNPs is boosted when coupled with fast ion radiation or hadron therapy (Porcel et al., 2010). Fast carbon ion irradiation of platinum culminates in the production of free radicals, "partners in crime" of the ROS mentioned above. Meanwhile, ZnONPs exert selective cytotoxicity on cancer cells when tested on human hepatocellular carcinoma HepG2, human lung adenocarcinoma A549, and human bronchial epithelial BEAS-2B (Akhtar et al., 2012). The observation that ZnONPs do not harm primary astrocytes and hepatocyte in rat proffers additional evidence for their tumor-specific activity.

The utility of inorganic NPs as cargo-free particles is not limited to anticancer properties. Some of them have been shown to possess antimicrobial activity as well, though via different mechanisms of action. For instance, AgNPs owe their antibacterial effect to the irreversible inactivation of enzyme phosphomannose isomerase (Wells et al., 1995), whereas ZnONPs attach themselves to the cell walls of both gram-positive and gram-negative bacteria, disrupting membrane structure, increasing its permeability, and eventually culminating in cell damage (Huang et al., 2008b). It has also been suggested that ZnONPs undergo a chemical reaction intracellularly to produce Zn^{2+} ions, which then give ROS production a shove, thus compromising the integrity of bacterial DNA, followed by the bacteria themselves. At the moment, we shed some light on the correlation between NPs' size and their antibacterial activity. To be more specific, their reactivity is inversely proportional to their size (Sotiriou et al., 2010; Zhang et al., 2007c). As mentioned in Section 7.2, 200 nm is a ceiling value that the NP size should not be allowed to exceed. Now, should inorganic NPs be tapped for their antimicrobial properties, there is no longer a fixated ceiling value but rather one to be figured out by estimating the value above which the potency is too low to be of any practical use. The lower value between the two is then regarded as the ceiling.

7.4.2.1 Ceria-NPs

This section rightfully owes its existence to the sheer plentitude of applications cerium(IV) oxide nanoparticles (ceria-NPs) have been shown to have the ability to undergo reversible valence switching between cerium(III) and cerium(IV) oxidation states and high biocompatibility have drawn much attention to ceria-NPs (Imlay et al., 2003; Karakoti et al., 2010). Reports on superoxide dismutase (SOD)- and catalase-mimetic functions of ceria-NPs have earned them the appeal as biological ROS scavengers, and as antioxidants (Korsvik et al., 2007; Pirmohamed et al., 2010).

With this unique antioxidant attribute alone, ceria-NPs have undergone various investigations for possible biomedical applications, many of which yielded positive results. When injected intraperitoneally into mice hosting ovarian cancer, ceria-NPs managed to curb tumor growth and metastasis of SKOV3 ovarian cancer cells (Giri et al., 2013). By scavenging radicals in the wounded region, ceria-NPs protect regenerating tissues, as well as fibroblasts, keratinocytes, and vascular endothelial cells, all of which are employed by the body for wound healing processes (Chigurupati et al., 2012). Ceria-NPs also display protective functions in tissues in which the predominant mechanism of damage revolves around oxidative stress. In transgenic mice expressing monocyte chemoattractant protein-1 in their cardiac tissue, ceria-NPs successfully attenuated complications of ROS-induced myocardial inflammation, which would eventually cause fatal ischemic cardiomyopathy (Grieve et al., 2004; Niu et al., 2007). In addition, ceria-NPs have been reported to exhibit neuroprotective properties via prevention of injury to spinal cord neurons (Das et al., 2007a) and therapeutic efficacy in retinal degeneration (Chen et al., 2006), both of which are mediated by the overabundance of hydrogen peroxide.

Interestingly, they are capable of preserving organ functions following radiotherapy sessions. Many undesirable complications resulted from exposure to radiation could be impeded, ranging from radiation-induced pneumonitis, gastrointestinal epithelial damage, and xerostomia to dermatitis after head and neck radiation (Colon et al., 2009, 2010; Madero et al., 2012). Lastly, there have been implications of coupling ceria-NPs with stem cell therapies in regenerative medicine, whereby the former scavenge free radicals at the implantation site, hence safeguarding progenitor cells against oxidative stress and maintaining a biochemical environment inducive to growth (Pagliari et al., 2012).

7.5 Conclusion

This chapter has more or less served as a general guideline for people in search of a certain type of NPs that suits their academic interest. NPs are increasingly popular as carriers of therapeutic agents and diagnostic ligands. The physicochemical properties of the cargo, especially affinity for water, do play a significant role in the selection of the NPs. Some NPs are well characterized to transport compounds of either hydrophilic or hydrophobic nature, but not both. They would have most likely spiraled down the pecking order if it had not been for innovative modifications of the cargo (e.g., prodrug formulation) or the NPs themselves (e.g., use of pyridyl sulfide groups to construct a lipophilic domain within the intrinsically hydrophilic interior of nanogel), which ensure tolerance for a more varied variety of chemical entities. In some cases, the choice of the NPs can also be made on the basis of their inherent utilities. SWCNTs demonstrate aborption in the NIR region, thus prompting the application of photothermal therapy; AgNPs have antibaterial effects; ceria-NPs are strong antioxidants. And that was just to name a few. Lastly, selection should not only be limited to the identity of the NPs but should also entail their physical characteristics and the sort of ligands attached to their surfaces. Extent of recognition by the RES, rate of elimination by the body, and ultimately the duration with which the NPs remain in the circulation are heavily dependent on such factors.

Chapter 8

Microgels and Nanogels for Drug Delivery

Niels M. B. Smeets and Todd Hoare
Department of Chemical Engineering, McMaster University, 1280 Main Street West, Hamilton, Ontario, Canada L8S 4L7
hoaretr@mcmaster.ca

8.1 Introduction

Microgels or nanogels are colloidal hydrogel particles consisting of a water-swollen, crosslinked polymer network. The distinction between microgels and nanogels is somewhat unclear and is currently a matter of debate. The IUPAC defines a nanogel as a particle of gel, of any shape, with an equivalent diameter between 1 nm and 100 nm, while a microgel is a gel particle with an equivalent diameter in the range of 100 nm–100 μm; however, the prevailing literature definition of microgels demands that the microgels are colloidally dispersible, which practically limits the upper size to the range of 1–5 μm in most cases. For the purposes of this book chapter, we will refer to particles as "microgels" unless the use of the term "nanogels" is clearly consistent with the terminology the

Controlled Release Systems: Advances in Nanobottles and Active Nanoparticles
Edited by Alexander van Herk, Jacqueline Forcada, and Giorgia Pastorin

ISBN 978-981-4613-21-7 (Hardcover), 978-981-4613-22-4 (eBook)
www.panstanford.com

authors of the papers themselves used in that work, recognizing that these definitions may vary between different research groups; for more general statements, we will use the broader term "microgel" to encompass gel particles across the full colloidal size range (10 nm–5 μm) (Pelton, 2012). Regardless of the exact terminology used, the colloidal properties of microgels coupled with their capacity to absorb solvent have resulted in microgels attracting interest for their potential in the design of functional coatings (Hu et al., 2013b), biomedicine (Smeets and Hoare, 2013), wastewater purification (Parasuraman and Serpe, 2011; Snowden et al., 1993), and catalysis (Welsch et al., 2011), among other applications.

Among the cited applications of microgels in the literature, drug delivery is likely the most common. This popularity is not without good reason, as microgels provide many advantages over other nanoparticle-based vehicles. First, many types of microgels can be synthesized in a straightforward manner with excellent control over the particle size, particle size distribution, morphology, and functionality. Second, microgels are highly hydrophilic and present a low surface energy interface to a biological environment, reducing nonspecific interactions with proteins (opsonization) and thus improving their biocompatibility. Third, since the majority of the volume of a microgel typically consists of water (in many cases >95% water), the high internal void fraction of microgels can be used to load high amounts of therapeutics that can be subsequently released by diffusion (i.e., sustained release) or by convection-driven mass transport following microgel deswelling (i.e., burst release). Finally, microgels are highly elastic and can be deformed to pass through pores many times smaller in size. This significantly increases the residence time of microgels in vivo (particularly in circulation), as larger microgels are less likely to be cleared by renal filtration or captured by phagocytosis.

Of particular interest for drug delivery are responsive or "smart" microgels that are able to change their solvation state in response to an external stimulus. Thermoresponsive microgels have captured the most interest in the literature, with microgels based on poly(*N*-isopropylacrylamide (PNIPAAm) (Pelton, 2000), poly(oligoethylene glycol methacrylate) (POEGMA) (Hu et al., 2010a; Lutz, 2011), or poly(vinyl caprolactam) (PVCL) (Ramos et al., 2012) displaying a volume phase transition temperature (VPTT) in aqueous media comparable to the lower critical solution temperature (LCST) of the corresponding linear polymers. These microgels transition from a hydrophilic and highly swollen state to a relatively hydrophobic and deswollen state as the temperature is raised above the VPTT of the microgel. This behavior is of particular interest for drug delivery applications as during the deswelling of the microgel matrix, water and any other molecules present (such as therapeutics for example) are expelled from the microgel interior. Other "smart" environmental triggers have since been engineered into microgels, offering the potential for triggered release of drugs from microgels using changes in ionic strength (Karg and Hellweg, 2009; Bütün et al., 2011; Xiong et al., 2005), pH (Hoare and Pelton, 2004; Imaz and Forcada, 2008a; Imaz et al., 2008; Jiang and Zukoski, 2012; Liu et al., 2011a), molecular binding (Hoare and Pelton, 2008a, 2008b; Ancla et al., 2011; Hoare and Pelton, 2007; Sorrell and Serpe, 2012; Liu et al., 2014), light (Yin et al., 2011; Garcia et al., 2007), magnetic field (Dagallier et al., 2010; Bhattacharya et al., 2007), or enzyme activity (Thornton et al., 2007, 2008; McDonald et al., 2009; Zhang and Wu, 2002), all of which are relevant to the in vivo environment.

However, despite many advantages and the relative maturity of the microgel field from a chemistry perspective, the number of examples of microgel-based drug delivery vehicles with clear clinical relevance remains limited. To the best of our knowledge, no microgel formulations are currently under investigation in clinical trials. Moreover, even in the academic literature, surprisingly few examples of microgel drug delivery vehicles designed and tested with practical in vivo applications in mind exist.

In this chapter, we will provide an overview of the recent developments in the design of microgel-based drug delivery vehicles with an emphasis on the design of responsive microgel delivery vehicles with potential clinical relevance. The various design requirements for the in vivo application of microgels are listed and briefly discussed, followed by recent examples that demonstrate the potential of responsive microgels for in vivo drug delivery applications. It should be emphasized that all the literature discussed will focus on covalently crosslinked microgel/microgel materials, omitting the large body of literature on nanomaterials self-assembled on the basis of hydrophobic interactions, charge interactions, or host–guest interactions that can swell (at least in part) in water but do not contain covalent crosslinks. In many cases, for such noncovalently assembled nanomaterials, the self-assembled core is minimally hydrated such that one of the key requirements of microgels (solvation) is not fully satisfied; in addition, the distinctions between micro/nanogels, micelles, and coacervates become difficult to clearly identify in such cases, with different authors referring to the same material using different terms. We also emphasize that this chapter does not aim to provide a thorough review of the wide body of literature on microgel and microgel synthesis; instead, we aim to highlight research with specific demonstrated promise in drug delivery applications. Methodologies for the synthesis of microgels and nanogels are thoroughly reviewed elsewhere (Pich and Richtering, 2010; Hendrickson et al., 2010).

8.2 Design Aspects

For the rational design of a microgel for the delivery of a specific drug at a specific rate in a specific physiological context, many chemical, physical, and biological parameters must be considered. In Fig. 8.1, we summarize the key design parameters that will be briefly discussed in this section.

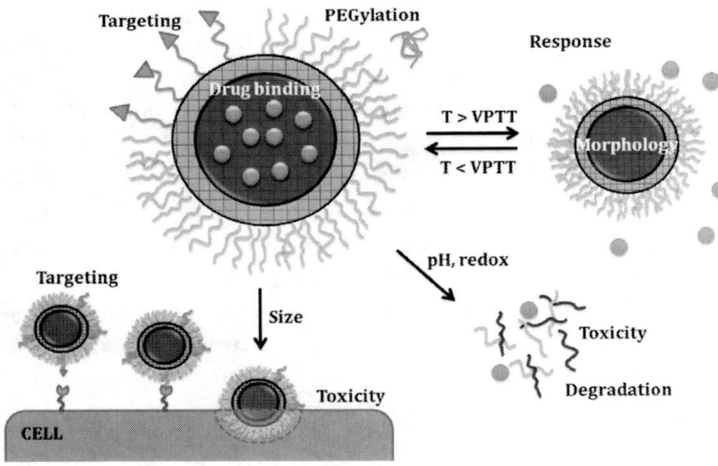

Figure 8.1 Schematic representation of the key design aspects of microgels for drug delivery applications. Reprinted from Smeets and Hoare (2013) with permission from Wiley-VCH.

8.2.1 Microgel Size

The size of a microgel is an important parameter given that it governs many functions in vivo, including circulation time, extravasation, targeting, immunogenicity, internalization, degradation, flow properties, and clearance and uptake mechanisms (Mitragotri and Lahann, 2009). However, the optimum size for a microgel is not a fixed property and is determined by the intended application in vivo. Microgels intended for circulation in the bloodstream are typically prepared within a size range window of roughly 100 to 1000 nm (Duan and Li, 2013), with most microgels providing longer circulation times on the lower end of that size range. Smaller microgels are likely to be cleared from the body by renal filtration (Asgeirsson et al., 2006; Rippe et al., 2007; Salmon et al., 2007), whereas larger microgels are quickly filtered from the bloodstream by the liver and spleen or internalized and digested by phagocytosis (Rejman et al., 2004; May and Machesky, 2001). In addition, many microgel drug delivery systems are required to be internalized by cells or have to pass, for example, the blood–brain barrier in order to achieve clinical efficacy. Such microgels are generally restricted to an approximate size of 100 nm or less (Ogawara et al., 1999). On the other hand, larger microgels of 200–1200 nm can be used for targeting tumor tissue (Hobbs et al., 1998) due to preferential filtration via the enhanced permeation and retention effect (Matsumura and Maeda, 1986). However, as microgels are soft and elastic materials that can deform and have been shown to be capable of passing through pores at least 10 times smaller in size (Hendrickson and Lyon, 2010) it is likely that microgels are subject to less stringent size limits for renal clearance and internalization than conventional (hard) nanoparticles.

8.2.2 Microgel Morphology

For many biomedical applications, it would be highly beneficial if multiple tasks (e.g., drug delivery, sensing, or imaging) can be performed by a single microgel. Compartmentalization of microgels, for example, by creating core–shell-type morphologies, allows for enhanced control over the microgel properties. Physiochemical and (in particular) biological properties of a microgel or any nanoparticle are largely governed by the interface between the nanoparticle and the surrounding environment. Surfaces that would otherwise impart toxicity on the surrounding tissue in vivo or promote opsonization can thus be effectively masked by adding an additional outer layer (Chan et al., 2006). Furthermore, chemical functionalities can be effectively located at the microgel periphery, advantageous for labeling the microgel with specific biomarkers to promote, for example, cell uptake.

8.2.3 Degradability

Degradable linkages in microgels generally serve two purposes, (i) to allow microgels to degrade into sufficiently small fragments to facilitate renal clearance and prevent bioaccumulation and (ii) to offer additional control over the internal pore size of microgels as a function of time, thus regulating drug release rates. Degradability can be introduced into the design of a microgel via the use of degradable crosslinkers containing acetal, ketal, or disulfide bonds (Gaulding et al., 2012a, Oh et al., 2007; Murthy et al., 2002, 2003; Metz and Theato, 2009; Ulbrich et al., 1993) or by incorporating natural-occurring polymers that can be metabolized. Although both approaches can achieve full degradation, the use of cleavable crosslinkers is often preferred as degradation solely through carbohydrate oxidation or protein hydrolysis/enzymatic degradation is typically slower and less controlled in terms of the location of the degradation as well as the potential degradation products.

8.2.4 Drug–Microgel Interactions

Microgels have been used as drug delivery vehicles for low-molecular-weight compounds such as dexamethasone (DM) and bupivacaine (BPV), as well as biomacromolecules such as fluorescently labeled dextran (FITC-DEX) (Snowden, 1992), insulin (Nolan et al., 2006), and bovine serum albumin (BSA) (Huo et al., 2006; Wu et al., 2005). Drug binding is greatly impacted by the local crosslink density and the functional group distribution (e.g., cationic, anionic, or hydrophobic) within the microgel (Eichenbaum et al., 1999a, 1999b). Consequently, the molecular design of microgels and how the chemistry of the microgel interacts with the chemistry of the drug are important parameters governing drug uptake and release. Affinity groups providing a complementary charge or a particular (hydrophobic for example) binding site for drugs may be copolymerized into microgels to increase drug–microgel interactions as desired. An alternative strategy is to covalently attach precursor drugs to the microgel matrix (through a degradable linkage), beneficial for increasing the amount of drug incorporation into microgels and/or precisely controlling the release of the loaded drug (Xiong et al., 2011).

8.2.5 Protein Adsorption

Any biomaterial used in vivo is subject to opsonization, the adsorption of serum proteins that "label" nanoparticles to be visible to the immune system. Microgels have the advantage that as highly hydrophilic, water-swollen particles they are much less prone to opsonization than other classes of nanoparticles. However, for thermoresponsive microgels, if the VPTT is below the physiological temperature, the microgel surface becomes significantly more hydrophobic and opsonization is promoted. Furthermore, microgels generally carry a high charge density resulting from initiator fragments and functional comonomers that can promote opsonization via interactions with oppositely charged protein domains. As a result microgels are often PEGylated to mask the hydrophobic and/or charged surfaces and minimize protein adsorption. Poly(ethylene glycol) (PEG) fragments can be incorporated in the microgel design through the use of ethylene oxide containing monomers (Oh et al., 2007; Pich et al., 2008; Gan and Lyon, 2002) or crosslinkers (Nolan et al., 2005).

8.2.6 Biocompatibility and Toxicity

For the successful implementation of microgels as drug delivery vehicles, it is critical that the microgels are nontoxic and cause only minimal acute inflammation and no significant chronic inflammation upon injection. Generally, low cytotoxicity has been observed in vitro for most microgel systems that have been evaluated using the well-established 3-(4,5-dimethylthiazol-2-yl)-2,5-diphenyltetrazolium bromide (MTT) mitochondrial enzyme assay (Chan et al., 2006; Hoare et al., 2009; Sivakumaran et al., 2011; Hoare et al., 2012a, 2012b, Qiao et al., 2011; Chi et al., 2009). In vivo assays of microgels have been only sparsely reported but have yielded largely positive results. For example, we have shown that injection of high concentrations (600 mg/kg) of poly(*N*-isopropylacrylamide-co-acrylic acid), P(NIPAAm-co-AA), microgels into the peritoneal cavity of mice resulted in no significant impact on animal viability or tissue morphology in the peritoneal wall, liver, or pancreas (Hoare et al., 2012a). Despite these promising preliminary results, an ultimately useful microgel would have to combine low cytotoxicity with complete degradation into products that are nontoxic and (at least for routine administration) clearable following microgel use.

8.2.7 Targeting

The therapeutic efficacy of microgels is governed primarily by the site-specific delivery of their payload. Consequently, microgels are increasingly decorated with targeting ligands that can give increased spatial control and facilitate the preferential delivery of therapeutics at the site(s) of interest and minimize undesirable side effects (Swami et al., 2012). For example, cellular uptake of drug-loaded nanoparticles can be significantly enhanced by conjugating tumor-specific targeting ligands (Danhier et al., 2010), particularly important if the nanoparticle payload consists of biomacromolecules such as DNA or siRNA that require intracellular delivery for bioactivity. A range of ligands has been conjugated to microgels ranging from folic acid (Nayak et al., 2004) and galactose (Choi et al., 2002) to peptide fragments (Blackburn et al., 2009) and aptamers (Navani and Li, 2006). Generally, the incorporation of these specific ligands on the microgel surface resulted in improved cellular uptake for the cell lines overexpressing the corresponding biomarker. For example, uptake of folic acid-labeled microgels was approximately 10 times higher in KB cells (human epithelial cell line typically used as a transfection host) when compared to nonfunctionalized microgels (Nayak et al., 2004).

8.2.8 Stimulus Response

In terms of drug delivery, the majority of microgel delivery systems that use a stimulus response to effect controlled release are based on exploiting natural pH gradients that exist in vivo. Microgels that are pH responsive can, for example, be used to selectively swell or deswell by protonation or deprotonation of acid–base groups to induce drug release in the lower extracellular pH of cancerous tissue (6.5–7) (Danhier et al., 2010) or the acidic lysosome environment (pH ≈ 4.5) (Pillay et al., 2002) after internalization. Furthermore, changes in the microgel environment can also be used as a trigger to selectively degrade microgels though the use of acid- or glutathione-cleavable crosslinkers (Metz and Theato, 2009; Murthy et al., 2002, 2003; Ulbrich et al., 1993). In a drug delivery context, however, the practical utility of many of these triggers is somewhat limited, given that a relatively narrow scope of stimuli that can be used in vivo without significantly affecting cell viability. Consequently, more advanced noninvasive stimuli such as oscillating magnetic fields (OMFs) and near-infrared (NIR) radiation are receiving increasing attention (Gorelikov et al., 2004; Hoare et al., 2009, 2011; Kawano et al., 2009).

In the following section, we review recent progress in the design of stimuli-responsive microgel-based delivery vehicles based on these various stimuli, with a focus on the relevance of the various microgel structures prepared from a potential clinical perspective.

8.3 Stimuli-Responsive Drug Delivery Vehicles

8.3.1 Thermoresponsive Microgels

8.3.1.1 Preparation methods

Typically, thermoresponsive microgels are synthesized via aqueous precipitation polymerization (also referred to as surfactant-free emulsion polymerization) at a temperature that exceeds the LCST of the polymers comprising the microgel. Polymer growth is initiated and proceeds in solution until a critical chain length is reached at which the LCST of the polymer switches below the polymerization temperature, driving phase separation of the polymer chain into a nanoaggregate. These nanoaggregates subsequently aggregate with each other, and the process continues until the surface charge of the resulting particle reaches a critical level such that the particle is electrostatically stabilized (typically achieved by using high concentrations of an ionic initiator, such as a persulfate, to drive polymerization). Given the self-assembly mechanism of particle formation, microgels made via this approach are typically highly monodisperse; indeed, the ease of making monodisperse gel particles using the precipitation approach has led many researchers to use thermoresponsive monomers as the basis for microgels intended to have other types of "smart" functionalities, even if the temperature-responsive aspect of the resulting product microgels is never used in the ultimate application of the materials. In addition, since particles form via sequential aggregation of polymer chains on nuclei formed early in the polymerization reaction, we have demonstrated that compositional drift in polymer chain composition due to differing copolymerization kinetics among the comonomers, and crosslinker(s), is reflected directly in radial compositional distribution in the resulting microgels (Hoare and McLean, 2006). This knowledge allows for simple compartimentalization of functional groups within the microgels if desired directly during synthesis by judicious monomer selection and/or by semibatch addition of monomers desired to be localized at the microgel interface (Sheikholeslami et al., 2012).

8.3.1.2 Poly(N-isopropylacrylamide)

PNIPAAm microgels are by far the most studied microgel platform for drug delivery applications. The first use of PNIPAAm microgels for drug delivery was demonstrated by Snowden and Booty for the release of acetylsalicylic acid (Snowden and Booty, 1993) and FITC-DEX (Snowden, 1992). Since then, significant progress has been made in the design of thermoresponsive microgels to the point that some examples of (predominantly in vitro) drug release have been reported.

Individual responsive microgels have been investigated as drug delivery vehicles for both small- and large-molecule payloads. The loading and release of any drug from PNIPAAm microgels is strongly correlated to the molecular design (i.e., the microgel composition or the microgel morphology) of the microgel. Trongsatitkul and Budhlall (2013) studied the effect of the microgel morphology on the release kinetics of FITC-DEX (70 kDa). These authors prepared a matrix device (i.e., PNIPAAm microgel) and a reservoir device (i.e., PNIPAAm multicore shell microcapsules) and studied thermally induced release at 25°C, 37°C, and 45°C. The microgels and microcapsules were prepared from a simple microarray technique, using a two-phase water-in-oil (W/O) or a three-phase oil-in-water-in-oil emulsion, respectively (Trongsatitkul and Budhlall, 2011). Release from the microcapsules occurs in a step-wise fashion and is generally faster than from microgels. While FITC-DEX release from the microgels occurs through diffusion, release from the multicore microcapsules occurs through a combination of diffusion and convective transport. Diffusion-based drug release from the microcapsule generates an osmotic gradient between the liquid-filled cores that accelerates overall drug release. Consequently, the overall release rate from the multicore microcapsules exceeds that of microgels. At 45°C (above the VPTT) the microcapsules display a very large burst release and all the drug is eluted in 10 minutes as a consequence of a large hydrostatic pressure build-up in the liquid-filled cores as the surrounding matrix de-swells. The authors concluded that the release kinetics of FITC-DEX are dominated by the temperature responsiveness at $T >$ VPTT and by the microgel morphology at $T <$ VPTT (Trongsatitkul and Budhlall, 2013).

We have investigated the binding of water-soluble drugs of different charges and hydrophilicities to carboxylic acid–functionalized PNIPAAm microgels with different functional group distributions (Hoare and Pelton, 2008a). Although drug uptake is largely driven by hydrophobic partitioning, the presence of ionic charges can have a significant impact on the efficacy of carboxylic acid functionalized microgels. Acid–base interactions between a cationic drug and anionic carboxylic acid group cause local deswelling of the microgel due to a reduction in the effective charge density upon binding. Consequently, surface-functionalized microgels displayed much lower drug uptake relative to core-functionalized microgels, as the collapsed layer at the proximity of the surface inhibits the diffusion of additional drug into and from the microgel. These two examples show the importance of the molecular composition and morphology of PNIPAAm microgels for drug uptake and release.

PNIPAAm microgels functionalized with ionic functionalities have demonstrated to be efficient drug delivery vehicles. For example, Bartlett et al. have reported the use of poly(*N*-isopropylacrylamide-co-2-acrylamido-2-methyl-1-propanesulfonic acid), P(NIPAAm-co-AMPS), microgels for the controlled release of a cationic MAPKAP kinase 2 inhibiting peptide KAFAKLAARLYRKALARQLGVAA (Bartlett et al., 2012a, 2012b) The 2-acrylamido-2-methyl-1-propanesulfonic acid (AMPS) monomer was chosen for effective, pH-independent complexation of the cationic peptide. AMPS functionalization was controlled between 0 and 10 mol.%, which increased the average particle size and lowered the average degree of swelling. However, the zeta potential continuously increased with increasing AMPS loading, which proved beneficial for the loading of the cationic peptide. Microgels functionalized with 10 mol.% AMPS have a zeta potential of −22 mV and can be loaded with 60.5 w/w.% (based on the microgel weight) peptide. Release of peptide from unfunctionalized microgels resulted in a significant burst release and complete elution of the peptide after 6 days; conversely, the presence of 10 mol.% AMPS avoided the significant burst release and rather a slow sustained release for at least 21 days. The P(NIPAAm-co-AMPS) microgels are hemocompatible and can effectively shield the peptide from degradation (Bartlett et al., 2012a), which is hypothesized to provide these microgels with sufficient residence time in vivo to carry their cargo to the designated target cells and release the peptide after internalization. In vivo degradability was incorporated into the microgel design by substituting the *N,N*-methylenebisacrylamide (MBAAm) crosslinker for a *N,O*-(dimethacryloyl)hydroxylamine (DMHA) crosslinker (Fig. 8.2) (Bartlett et al., 2012b) The DMHA crosslinker slowly degrades in slightly basic media. MAPKAP release could be controlled to anywhere between one and four days under physiological conditions using these degradable microgels. Cell-specific targeting of these microgels was not considered in this example.

Targeting approaches have been explored by Su et al., who prepared poly(NIPAAm-co-acrylic acid) microgels for the delivery of the anticancer drug doxorubicin (DOX) (Su et al., 2013). The microgels were surface-functionalized with both fluorescent bovine serum albumin encapsulated gold nanoclusters (facilitating imaging of the location of the microgels) as well as the cell-targeting peptide arginine-glycine-aspartic acid (RGD) (allowing cell binding and promoting intracellular delivery of the microgel and its associated payload). The presence of the cell-binding RGD motif significantly promoted cellular uptake of the microgels as well as the cytotoxicity of the delivery vehicle, while the dual thermo/pH responsive nature of the microgel was demonstrated to enhance the controlled release properties of the drug inside the cells. The specific targeting ligand chosen in this study, however, is a general cell-adhesive peptide unlikely to be useful for cancer cell targeting in the in vivo environment. Galactose-targeted PNIPAAm-based microgels have also been reported by Duan et al., who fabricated microgels based on galactosylated chitosan-g-PNIPAAm as carriers of oridonin for tumor targeting (Duan et al., 2011). Enhanced cytotoxicity of drug-loaded microgels to HepG2 (human liver carcinoma cell line) cells was reported in direct correlation with the number of galactose moieties on the microgels, with evidence provided of asialoglycoprotein receptor-mediated endocytosis occurring as a result of the targeting strategy used. Despite the common overexpression of galactose receptors on tumor cells, the ability of this targeting ligand to target cancer cells in vivo was not assessed.

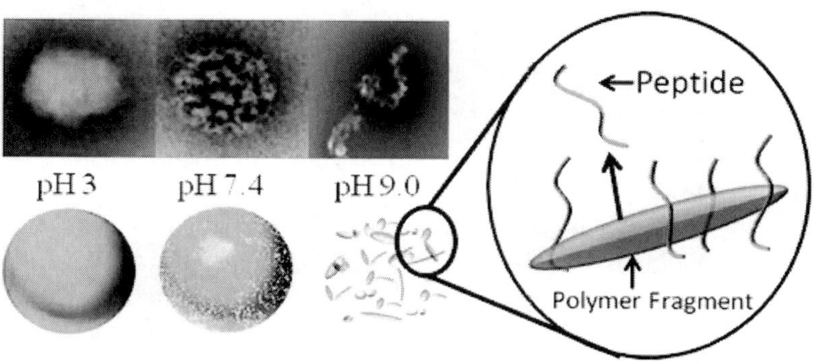

Figure 8.2 pH-controlled delivery of peptides from P(NIPAAm-co-AMPS) hydrogels crosslinked with a degradable crosslinker. The hydrogels are stable at slightly acidic conditions but completely degrade in slightly basic conditions to release the peptide cargo. Reprinted from Bartlett et al. (2012b) with permission from the American Chemical Society.

Dickerson et al. (2010) reported targeted delivery of siRNA using a core–shell microgel design consisting of a poly(*N*-isopropylmethacrylamide) (PNIPMAAm) core and a poly(*N*-isopropylmethacrylamide-co-3-aminopropyl methacrylate), P(NIPMAAm-co-APMA), shell. PNIPMAAm was chosen owing to its higher LCST of 43°C when compared to PNIPAAm, ensuring that the microgel is swollen and hydrophilic at physiological temperature to minimize opsonization. The functional monomer in the shell (APMA) was incorporated to allow for chemoligation of a 12-amino acid YSA peptide (YSAYPDSVPMMS) that allows for specific targeting of EphA2 receptors (ephrin type-A receptor 2) (implicated in mediating developmental events) that are overexpressed by a wide range of tumor cells, allowing for tumor-specific delivery. The core–shell microgels were approximately 100 nm in size, which is sufficient to allow for effective circulation followed by cellular uptake and enclosure in endocytic vesicles. Cell toxicity and viability assays on Hey cells (human ovarian carcinoma cell line with high EphA2 expression) revealed that the core–shell microgel is nontoxic under the studied conditions. The PNIPMAAm core–shell microgels displayed a high efficiency for siRNA loading, with 67% of the siRNA payload remaining bound after 12 hours; this is particularly important as approximately 6 hours of circulation time is required for effective extravasation through the enhanced permeation and retention (EPR) effect (Iyer et al., 2006). Uptake experiments performed on Hey (high EphA2 expression) and BG-1 cells (human ovarian carcinoma cell line with low EphA2 expression) indicated that the YSA protein promoted uptake in both cell lines, with very limited uptake observed in the absence of YSA. Moreover, the use of a fluorescently labeled siRNA (siGLO) revealed that the siRNA was retained in high levels inside the microgel and was successfully delivered inside the target cell, with no cell uptake of the bare siGLO. Observed gene silencing of the EGF receptor resulted in knockdown of that receptor, indicating effective endosomal escape and unpacking of the microgel–siRNA complex in the cytosol. However, no in vivo experiment was performed to confirm this activity.

The use of hydrophobic interactions has also found widespread application for enhancing drug uptake and release with thermoresponsive microgels, with hydrophobic groups introduced into microgels by copolymerizing hydrophobic monomers such as butyl acrylate (BA) (Singka et al., 2010) glycidyl methacrylate (GMA) (Li et al., 2013c), and 2,2,3,4,4,4 hexafluorobutyl methacrylate (HFMA) (An et al., 2012). Poly(*N*-isopropylacrylamide-co-glycidyl methacrylate), P(NIPAAm-co-GMA), microgels were prepared by Li et al. for the controlled release of the anticancer drug 5-fluorouracil (5-FU) (Li et al., 2013c) These microgels were found to be noncytotoxic, as determined from an MTT assay using a MC3T3-E1 (mouse preosteoblat) cell line, and nearly complete release of FU was observed within 200 minutes. Poly(*N*-isopropylacrylamide-co-butyl acrylate), P(NIPAAm-co-BA), microgels were also used for the topical delivery of methotrexate (MTX) (Singka et al., 2010) in which the presence of BA in the microgel enables a MTX loading efficiency of 31.9%. The MTX-loaded microgels (roughly 100 nm at 37°C) are able to penetrate the epidermal layer (demonstrated on porcine epidermal membranes) and release MTX deep in the tissue to modulate the production of prostaglandin E_2. An et al. (2012) prepared a hydrophobically modified microgel by incorporating up to 50 mol.% HFMA. Although copolymerization of the HFMA comonomer had only a minor effect on the VPTT, it did affect the morphology of the microgels, which changed from spherical to golf ball in shape. The presence of the fluorine groups provides for hydrophobic interaction with the tryptophan, tyrosine, and phenylalanine residues of BSA, resulting in significant amounts of BSA adsorbing onto these microgels (An et al., 2012). A more comprehensive study was reported by Blanco et al., who modified P(NIPAAm-co-*p*-nitrophenylacrylate), P(NIPAAm-co-NPA), microgels with folic acid for the controlled release of tamoxifen (TMX) (Blanco et al., 2010). The *p*-nitrophenyl groups were functionalized with 4-aminomethylpyridine to facilitate functionalization with folic acid. The resulting microgels had a folic acid concentration of 1.3 ± 0.8 mmol/g microgel and an average particle size of 600 nm. Cell uptake of these microgels was demonstrated in vitro on human breast carcinoma cell line (MCF7) breast cancer cells and HeLa (human cervical cancer cell line) cells. Microgel uptake was particularly efficient for the HeLa cells, as this cell line is folic acid receptor positive. In vitro studies showed that the unloaded microgels are noncytotoxic and that sustained TMX release can be achieved for 160 hours after an initial burst release. Subcutaneous injection of the microgels does not cause an acute inflammatory response or signs of rejection (i.e., no formation of a collagenous bag after 60 days) in vivo. Furthermore, the bioavailability of TMX is increased when encapsulated in the microgels, resulting in an increased residence time of TMX in the blood plasma.

A different approach for improving drug–microgel interactions was reported by Liu et al., who exploited host–guest interactions (Liu et al., 2009d). Core–shell microgels were prepared by incorporating a beta-cyclodextrin methacrylate (β-CD-MA) in the core of a NIPAAm microgel. β-CD is frequently used in the design of nanoparticles since hydrophobic molecules can be efficiently encapsulated inside the hydrophobic interior. Here, β-CD was used to encapsulate the hydrophobic model drug paeonol (PA). Microgels were loaded with 0, 3, and 6 mmol of the β-CD-MA, which resulted in progressively slower PA release. In the case of 6 mmol β-CD, nearly complete release was achieved after 8 hours at 37°C, although no biocompatibility studies were performed.

In the vast majority of reported cases, the thermal phase transition is used primarily as a means to synthesize monodisperse microgels as opposed to a controlling factor in drug release. However, thermal aggregation of ionically stabilized microgels above their phase transition temperature can be used as a method to immobilize microgels and thus drug at a desired site of action. We have demonstrated the prolonged release of local anesthetics using this strategy, in which microgels with aggregation temperatures above room temperature but below physiological temperature could be delivered as stable, dispersed nanoparticles but subsequently aggregate and thus localize at the sciatic nerve for prolonged anesthetic delivery (Fig. 8.3). Of note, given that the pH of a saline-suspended PNIPAAM-co-acrylic acid microgel is slightly acidic (such that a significant fraction of the –COOH groups are protonated upon injection), the slow titration of the microgel aggregate by physiological fluids increases the degree of ionization of the microgels and, by extension, the critical aggregation temperature over time, ultimately inducing triggered dissociation of the aggregate at tunable rates.

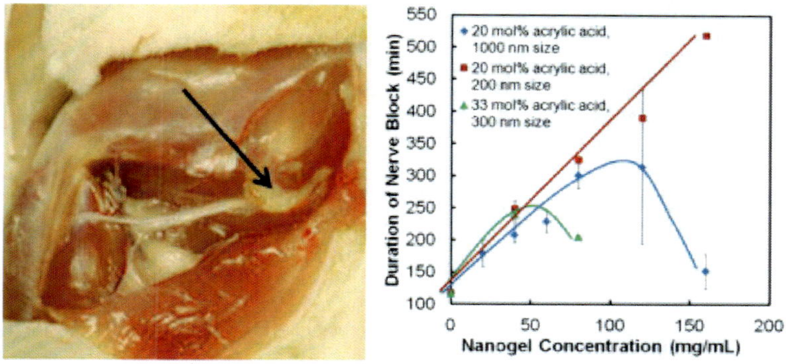

Figure 8.3 Controlled release of bupivacaine using thermoresponsive, in situ aggregating poly(NIPAAm-co-AA) microgels. (a) Picture of a microgel aggregate at the base of the sciatic nerve; (b) duration of an effective nerve block achieved as a function of AA content of the microgel and microgel size. Reprinted from Hoare et al. (2012a, b), Copyright 2012, with permission from Elsevier.

8.3.1.3 Poly(oligoethylene glycol (meth)acrylate)

Poly(oligoethylene glycol acrylate) (POEGA) and the corresponding methacrylate POEGMA microgels represent an alternative to PNIPAAm-based drug delivery vehicles. Their synthesis was pioneered by the group of professor Zibing Hu, who reported the synthesis of P(diethylene glycol methacrylate-co-oligoethylene glycol methacrylate), P(M(EO)$_2$MA-co-OEGMA), microgels in 2007 and POEGMA core–shell microgels in 2009 (Chi, 2009). These microgels are considered analogues to PEG and consequently display all the advantageous biological properties of PEG-based materials in vivo (Hu et al., 2010b).

The use of POEGMA microgels for drug delivery was first reported by Oh et al. (2007). POEGMA microgels were synthesized using atom transfer radical polymerization (ATRP) in an inverse miniemulsion system. The microgels were approximately 225 nm in size, in the correct size range for sufficient circulation in vivo. Degradation of these microgels was facilitated through the use of a disulfide-containing crosslinker. Complete degradation of the microgels in the presence of glutathione (GSH) was demonstrated. Furthermore, as the molecular weight of the primary chains was controlled below 40 kDa through the use of ATRP, the degradation products can be cleared from the body through renal filtration. The microgels can easily be functionalized with biotin for cell targeting and are noncytotoxic as tested against HeLa cells at 0.4 mg/mL. Enhanced release of DOX was achieved by adding GSH (which is present in the body in μM to mM concentration) to effectively degrade the microgel matrix and release the encapsulated DOX. These microgels have also been used for the encapsulation of iron oxide nanoparticles (Dong et al., 2009).

Zhou et al. (2010) have studied POEGMA core–shell microgels for the controlled release of the hydrophobic model drug dipyridamole (DPM). The microgels consist of a $M(EO)_2MA$ core (VPTT ≈ 22°C) and a $M(EO)_2MA:OEGMA_{300}$ shell (1:2, VPTT ≈ 55°C). These microgels were synthesized via the commonly used "seed and feed" precipitation polymerization approach (Chi, 2009; Jones and Lyon, 2000). Under physiological conditions, this microgel will have a collapsed hydrophobic core (facilitating drug loading) and a swollen hydrophilic shell (facilitating favorable biological interactions with the surrounding tissue and reduced opsonization). Interestingly, the loading efficiency of DPM can be controlled by the thickness of the hydrophilic shell; as the thickness is increased from 0 nm to 100 nm, the loading efficiency increases from 20% to 80%. This significant increase originates from the role of the hydrophilic shell restricting the collapse of the core, increasing the accessibility of the core for DPM loading. Release of DPM from these core–shell microgels was investigated under physiological conditions and showed a relatively low burst release of approximately 10% after 1 hour, followed by sustained release of an additional 3% of the loaded drug over the following 30 hours. The unloaded microgels do not impact cell viability at concentrations up to 25 μg/mL; however, DPM-loaded microgels caused a significant decrease in cell viability of B16F10 (mouse skin melanoma cell line) cancer cells. This example illustrates the importance of a core–shell microgel morphology for controlling biological interactions as well as for controlling the uptake of hydrophobic drugs.

Although the core–shell microgels reported by Zhou et al. (2010) proved effective against cancer cells in vitro, the fact that these microgels are not degradable will hamper their in vivo application. Qiao et al. (2011) added in vivo degradability to their POEGA microgels through the use of a disulfide containing crosslinker and copolymerization of an acid-labile orthoester-containing monomer. These microgels can thus swell at low pH, as the orthoester functionality is cleaved to the more hydrophilic ester, or completely degrade in the presence of dithiotreitol (DTT) or GSH. This responsive behavior facilitates different release kinetics under different physiologically relevant conditions. At physiological pH, release of paxlitaxel (PTX) is slow with no significant burst release; at lower pH (e.g., pH = 5, corresponding to the lysosomal pH), release is significantly faster as orthoesters are cleaved and the more swollen microgel network enhances drug diffusion. In the presence of DTT (as a model for GSH), PTX release is also enhanced due to the (partial) degradation of the microgel. These microgels are noncytotoxic, as tested on MCF-7 cells, but display a significant decrease in cell viability once PTX is loaded.

8.3.1.4 Poly(N-vinyl caprolactam)

PVCL is a relatively newer material that is currently being studied for the design of microgel drug delivery vehicles (Ramos et al., 2012). Contrary to PNIPAAm and POEGMA, the LCST of PVCL decreases with both polymer chain length and polymer concentration and can thus be controlled by the molecular weight rather than the polymer composition (Beija et al., 2011). The synthesis of PVCL microgels based on precipitation polymerization has been studied in detail by the group of Professor Jacqueline Forcada (Imaz and Forcada, 2008a, 2009a, 2009b) *N*-vinylcaprolactam (VCL) is susceptible to hydrolysis in aqueous media, and the precipitation polymerization is consequently performed in the presence of a buffer to maintain neutral pH (Imaz et al., 2008). Furthermore, PVCL microgels can be functionalized with other vinyl monomers such as AA (Imaz and Forcada, 2011) or beta-cyclodextrin acrylate (β-CD-A) (Kettel et al., 2011) or a range of inorganic nanoparticles (Pich et al., 2004; Häntzschel et al., 2007; Singh et al., 2012b) to prepare functional microgels.

Since the synthesis of PVCL microgels was only recently investigated in detail, the number of examples of drug delivery remains limited. The first example of PVCL microgels for drug delivery applications was reported by Vihola et al. in 2002, who studied the uptake and release of the beta-blocking agents nadolol and propranolol and the choline-esterase inhibitor tacrine. The hydrophilic nadolol collapsed the microgel and decreased the VPTT, while the hydrophobic drugs propranolol and tacrine swell the microgels have little effect on the VPTT. Fast release of nadolol was observed after six hours due to the hydrophilicity of this drug, while release of propranolol and tacrine was significantly slower. The authors suggested that hydrogen bonding and hydrophobic interactions between the drugs and the PVCL matrix were beneficial for controlling release rates (Vihola et al., 2002). Release of the anticancer drug FU was also reported from PVCL microgels functionalized with acrylamidoglycolic acid (Madhusudana et al., 2013) a monomer facilitating incorporation of pH-sensitive carboxylic acid groups in the PVCL microgels to induce dual temperature–pH responsiveness. Fast release of FU was obtained at 25°C (T < VPTT) and at pH = 7.4 (pH > pK_a), as in both cases the microgel network is highly swollen due to the VPTT of the microgels and the anionically charged carboxylic acid groups. Although both examples demonstrate the potential of controlled drug release from PVCL microgels, no in vitro or in vivo cell work was reported.

Imaz and Forcada reported a comprehensive study on the drug loading, cell viability and biocompatibility of PVCL microgels (Imaz and Forcada, 2010). Three different PVCL microgels were synthesized: (1) PVCL microgel crosslinked with MBAAm, (2) PVCL microgel crosslinked with poly(ethylene glycol dimethacrylate) (PEGDMA), and (3) PVCL microgel crosslinked with MBAAm and further functionalized with a hydrophobic sugar-based monomer. Loading experiments with calcein were performed at pH = 3, at which all the carboxylic acid groups of calcein are protonated. Calcein uptake was highest for the PEGDMA- and MBAAm-functionalized microgels due to H-bonding interactions between the carboxylic acid groups of calcein and the ethylene oxide (EO) and amide groups of PEGDMA and MBAAm, respectively. The PVCL microgels were found to be noncytotoxic to a primary neural cell line after incubation for 24 hours, with only a slight decrease in cell viability observed after incubation for 72 hours.

8.3.2 pH-Responsive Microgels

8.3.2.1 Preparation methods

pH-responsive functionalities have widely been incorporated into a variety of thermoresponsive microgels, as described by a number of the papers cited in the previous section (Hoare and Pelton, 2008a, 2008b; Bartlett et al., 2012a, 2012b; Madhusudana et al., 2013). In this case, the thermoresponsive nature of the base polymer drives the particle formation via the precipitation mechanism, allowing for facile incorporation of acidic or basic functional groups via copolymerization. This method works provided the functional comonomer content is kept relatively low and the functional monomers are polymerized in their protonated state, in which hydrogen bonding of the protic comonomer with NIPAAm residues can in some cases promote the precipitation-driven particle formation mechanism.

Other pH-responsive microgels based on nonthermoresponsive polymers have also been reported, although the fabrication protocols required are typically more complex and/or result in significantly less controlled particle sizes and morphologies than can be achieved with thermoresponsive polymers. Four methods are predominantly used for microgel preparation:

(1) *Shearing of bulk hydrogels:* Given sufficient energy input, bulk hydrogels can be degraded into microgels by shear. Most commonly, probe sonication is used to create microgels from bulk hydrogels (Arunraj et al., 2013). However, this method typically leads to highly polydisperse and often size-irregular particle populations that can complicate the understanding of structure–property correlations in such materials.

(2) *Solution-based methods:* Microgels can be fabricated from (typically dilute) polymer solutions by adding a crosslinking agent in combination with stirring and/or stabilization via the addition of a surfactant. Direct irradiation of a polymer or monomer solution with either gamma rays (Abd El-Rehim et al., 2013) or UV radiation (Xu et al., 2006) (the latter often in the presence of a polymerization initiator) has been demonstrated to induce microgel particle formation through this mechanism. Alternately, chemical crosslinking can be used to fabricate microgel particles. Chang et al. demonstrated such an approach by preparing pectin-based microgels crosslinked with glutaraldehyde, generating pH (and enzymatically) responsive particles (Chang et al., 2007). While this method is often convenient and adaptable to a wide range of potential

chemistries, particle size control is often difficult, leading to broad polydispersities.

(3) *Emulsion-based methods:* Direct emulsion polymerization can be used to prepare microgels in cases in which the pH-responsive monomer in the protonated state is sufficiently hydrophobic. Deka et al. demonstrated such an approach to make microgels based on 4-vinylpyridine crosslinked with divinyl benzene (DVB) for the delivery of short oligonucleotides (Deka et al., 2010), while Tan et al. fabricated P(methacrylic acid-co-ethyl acrylate, P(MAA-co-EA), microgels crosslinked with diallyl phthalate using emulsion polymerization that exhibited pH-responsive release of procaine hydrochloride (Tan et al., 2007, 2010). Alternately, for a more general pathway to microgel synthesis, W/O (inverse) emulsions or microemulsions can be used as templates for microgel synthesis. This method offers significantly better size control than irradiation-based methods but requires the use of organic solvents. Microgels can be formed based on the localized crosslinking of reactive prepolymers (Oh et al., 2013) or via polymerization from monomers in situ (Dadsetan et al., 2013) templated by the inverse emulsion.

(4) *Controlled radical polymerization:* The use of controlled radical techniques to build particles from monomeric solutions has more recently been applied to generate microgels with controlled morphologies (Averick et al., 2012). This method has the advantage that the individual polymer chains comprising the microgels can be well controlled (including limiting their molecular weight below the renal filtration limit for polymer clearance). However, such methods typically involve significantly more complex chemistries that may pose challenges with ultimate clinical approval of the resulting microgels.

8.3.2.2 Cationic (basic) microgels

The majority of pH-responsive microgel-based delivery vehicles presented in the literature that apply the pH-responsive nature of the particle to effect changes in the release kinetics consist of cationic-functionalized microgels designed for intracellular delivery. The presence of a (low) cationic charge on the microgel surface can promote cell uptake by enhancing the interaction between the microgel and the anionic cell membrane; once the microgel is subsequently endocytosed, the lower pH of the endosome (4.5–5) induces additional ionization of the microgel to drive swelling (promoting drug release) as well as potential endosome rupture via the proton sponge effect (Boussif et al., 1995) to drive release of the microgel cargo into the cell cytoplasm.

8.3.2.2.1 *Cationic microgels for gene delivery*

Given the critical importance of the delivery of any genetic material into the cell nucleus for genetic engineering, cationic microgels with the potential for enhanced cellular uptake and enhanced endosomal escape have exceptional potential for gene delivery applications. However, microgels face several challenges in comparison to the more widely applied approach of building coacervates between cationic polymers and anionic nucleic acid–based cargoes. In particular, the large size of DNA/RNA cargoes relative to the generally small and predefined mesh size of the microgel network typically results in significantly lower loading of genetic material into microgels relative to coacervates; alternately, adding the genetic material at the synthesis step introduces issues with DNA/RNA stability in the presence of ionizing irradiation, free radical initiators, and other chemical methods used to drive crosslinking.

Microgels based on poly(dimethylaminoethyl methacrylate) (PDMAEMA) have been most widely applied to gene delivery applications. Nagasaki and coworkers have prepared several morphologies of PDMAEMA-based microgels using EGDMA as the crosslinker and heterobifunctional PEG with one end functionalized with a vinylbenzyl group (enabling incorporation into the microgel) and the other end with a carboxyl group (facilitating functionalization of the microgel for targeting). A microgel–siRNA complex was demonstrated to effectively silence a firefly luciferase gene expressed in human hepatoma (HUH-7) cells, with direct evidence provided of cytoplasmic delivery of the siRNA (Fig. 8.4) (Tamura et al., 2009).

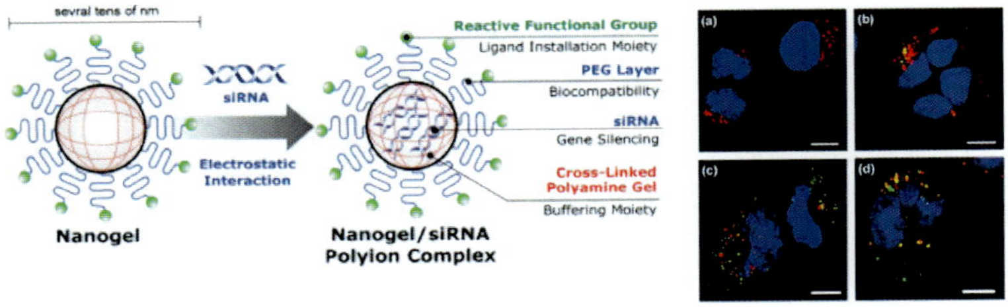

Figure 8.4 PDMAEMA-based microgels for siRNA delivery. Left: Schematic of microgel design. Right: Evidence of intracellular delivery of siRNA to HuH-7 cells after 24 hours of incubation with naked siRNA (a), PEG-b-PDEAEMA/siRNA at N/P = 2 (b), nanogel/siRNA at N/P = 2 (c), and oligofectamine/siRNA (d) containing 1 μM of fluorescein-labeled siRNA (green). The cell nuclei and acidic endosomal compartments were stained with Hoechst 33342 (blue) and LysoTracker Red DND-99 (red), respectively. Reprinted from Tamura et al. (2009) with permission from the American Chemical Society.

The same group demonstrated enhanced efficacy of conventional PEG-b-poly(L-lysine)/plasmid DNA coacervates in conjunction with PDMAEMA-based microgels, with the microgels serving as endosomolytic agents promoting escape of the DNA complex into the cytoplasm (Oishi et al., 2007b). PDMAEMA-based microgels have also been prepared using controlled radical approaches. Averick et al. demonstrated the formation of cationic microgels based on quaternized DMAEMA crosslinked with reducible disulfide bonds produced via atom transfer radical polymerization in inverse emulsion (Averick et al., 2012). The resulting microgels were demonstrated to be effective at delivering both plasmid DNA and short interfering RNA (siRNA) to *Drosophila* Schneider 2 cells with high transfection activity.

Other cationic pH-responsive microgels have been assayed for gene delivery, with varying success even in vitro. Li et al. prepared poly(ethyleneimine) (PEI)-based microgels via crosslinking with PEGDMA via Michael chemistry in an inverse emulsion and demonstrated effective silencing of green fluorescent protein expression in breast carcinoma cells on par with lipofectamine, one of the standard comparison materials in the area (Li et al., 2009c). Khondee et al. prepared poly(vinyl amine)-based microgels via inverse emulsion polymerization of *N*-vinyl formamide (NVF) and controlled hydrolysis of the formamide residues to primary amines, facilitating the generation of microgels with different and well-controlled charge densities (Khondee et al., 2010); lower charge densities facilitated effective low-level but sustained transfection, while limiting the cytotoxicity often observed in conjunction with the use of cationic polymers of any type in vivo (via their potential role in destabilizing the plasma membrane). However, the lack of degradability in many of these preparations limits their use in vivo, both from a clearance perspective as well as in terms of the inherently limited cytoplasmic delivery that can be achieved without triggering degradation of the delivery vehicle.

Raemdonck et al. addressed this issue in part by preparing cationic DEX-based microgels by copolymerizing dextran hydroxyethyl methacrylate, 2-(methacryloyloxy)ethyltrimethylammonium chloride, and dextran–maleic acid (DEX-MA) via inverse emulsion that could degrade via DEX oxidation in vivo (Raemdonck et al., 2009). While the loading capacities for siRNA were reasonably high and the microgels showed minimal cytotoxicity to HUH-7 cells, effective silencing behavior was only achieved when additional stimuli (e.g., photochemical internalization or the incorporation of fusogenic peptides) were applied in concert with the cationic microgel delivery vehicle.

Codelivery of DNA and proteins has also been explored using responsive microgel-based carriers, although more complex "compartimentalized" morphologies are typically required given the differing requirements of each type of cargo. Ahmed et al. demonstrated codelivery of plasmid DNA and various proteins using a microgel containing a hydrophobic core based on $PMEO_2MA$ and a carbohydrate-based shell based on 2-lactobionamidoethyl methacrylamide and 3-gluconamidopropyl methacrylamide prepared using reversible addition-fragmentation transfer (RAFT) polymerization, yielding microgels 30–40 nm in diameter (Ahmed et al., 2012a). The hydrophobic core drove DNA complexation, while the hydrophilic carbohydrate-based shell facilitated protein uptake. The addition of a cationic glycopolymer to the complexes was found to improve cellular uptake and thus gene expression in hepatocytes; furthermore, the activity of β-galactosidase enzyme delivered via the microgel was preserved inside the cell upon internalization.

8.3.2.2.2 *Cationic microgels for intracellular drug delivery*

Given the lower payloads required for impactful delivery, the use of cationic microgels as intracellular drug delivery vehicles for small molecules (particularly for anticancer drugs) offers in our view more promise in the long term. In some cases, thermal phase transitions have been used to drive nanoparticle formation via copolymerization, including combinations of VCL with DMAEMA (Demirel and von Klitzing, 2013), PNIPMAAm with 3-aminopropyl methacrylamide hydrochloride (APMH) (Hu et al., 2010a), and NIPAAm with 1-vinylimidazole (Ogawa et al., 2004). Alternately, similar microgels to those prepared for gene delivery can be used in the context of drug delivery, since many of the key design parameters remain the same. Nagasaki and coworkers have demonstrated the successful use of PDMAEMA-based microgels in this context for the delivery of DOX, with minimal initial burst observed at physiological pH but significant release observed at endosomal pH (Oishi et al., 2007a). The pH-responsive microgels exhibited improved antitumor activity against human breast and liver cancer cell lines relative to nonresponsive PEG-based microgels, indicating the efficacy of the pH responsiveness for driving effective intracellular delivery. Xu et al. have also demonstrated the efficacy of PEI-based microgels for intracellular delivery of cytotoxic nucleoside analogues 5'-triphosphates (Xu et al., 2006). The microgels, prepared by UV irradiation of polyethylene imine solutions, exhibited enhanced transfection efficiency to human lung and liver carcinoma cells and improved localization of nanoparticles in a mouse tumor model.

In other cases, more complex microgels with pH-specific responses have been designed, with promising initial in vitro results. Oh et al. formed microgels based on crosslinking of thiol-functionalized poly(L-aspartic acid)-b-PEG and maleimide-functionalized poly(L-aspartic acid)-g-(3-diethylaminopropyl)-b-PEG inside an oil-in-water emulsion to generate microgels that exhibited triggered swelling at lysosomal pH due to protonation of the diethylaminopropyl residues. The resulting microgels facilitated effective intracellular delivery of DOX and, by extension, enhanced tumor cell cytotoxicity (Oh et al., 2013). Zhou et al. prepared semi-interpenetrating network microgels in which chitosan was interpenetrated with PEG and demonstrated similar efficacy of intracellular delivery of 5-FU (Zhou et al., 2013). Interestingly, a temperature-responsive pH change could be triggered with these materials by which significantly enhanced therapeutic efficiency was achieved via combined chemo–cryo treatments of model B16F10 melanoma cells.

Acid-degradable linkages between microgels and drugs have also been exploited to drive intracellular drug delivery. Xiong et al. demonstrated the intracellular delivery of DOX conjugated to poly(NIPAAm-co-AA) microgels via an acid-labile hydrazone linkage (Xiong et al., 2011). Interestingly, in addition to the enhanced degradation rate of the drug–microgel bond in the endosome, the drug conjugates exhibited a pH-dependent phase transition temperature of ~41°C–43°C at pH < 6.8 (akin to intercellular pH in rapidly metabolizing tumor tissues) but >50°C at pH 7.4; as such, under moderate hyperthermic conditions often used to treat tumours, enhanced cellular uptake (~43% relative to normal physiological temperature) was achieved, facilitating an additional mechanism of site-specific drug delivery based on a pH response (Fig. 8.5). However, it must be noted that no in vivo investigations were performed to confirm the relevance of this observed triggering response under physiological conditions.

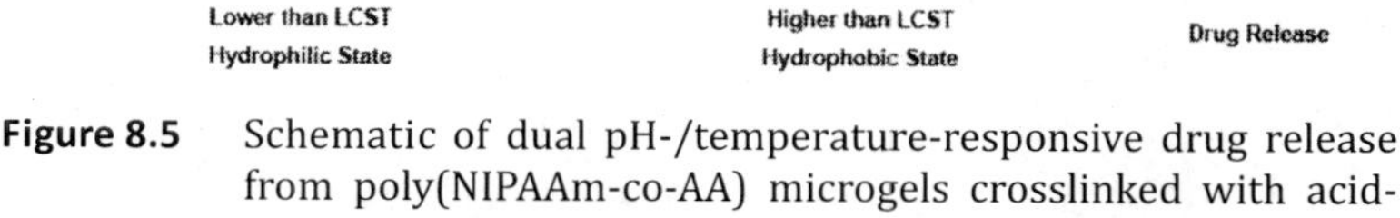

Figure 8.5 Schematic of dual pH-/temperature-responsive drug release from poly(NIPAAm-co-AA) microgels crosslinked with acid-degradable crosslinks. Reprinted from Xiong et al. (2011), Copyright 2011, with permission from Elsevier.

8.3.2.3 Anionic (acidic) microgels

Multiple acid-functionalized microgels have been prepared in order to enhance the uptake and subsequently slow the release of cationic drug payloads via ion exchange, including many papers cited elsewhere in this chapter (Hoare et al., 2012b; Hoare and Pelton, 2008a, 2008b; Bartlett et al., 2012a, 2012b; Madhusudana et al., 2013; Abd El-Rehim et al., 2013; Tan et al., 2007). However, acid-functionalized pH-responsive microgels have also attracted interest from an oral drug delivery perspective, as they can remain collapsed in the acidic stomach environment (protecting a drug cargo and reducing its release rate) but swell in the more basic intestinal environment (promoting drug release). Cirillo et al. designed MAA-functionalized microgels for the controlled release of diclofenac diethyl ammonium salt (DDA) using this strategy, a potent nonsteroidal anti-inflammatory drug that can result in severe side effects such as gastrointestinal disorders when delivered in conventional formulations (Cirillo et al., 2012b). Microgels were synthesized using NIPAAm, MAA, and bovine serum albumin–methacrylate (BSA-MA) in a reverse-phase suspension polymerization, resulting in micrometer-sized microgels that carried both ionic carboxylic acid groups as well as hydrophobic domains originating from the BSA protein. DDA release is consequently a function of both the temperature as well as the pH. Under physiologically relevant conditions, nearly complete release of DDA occurs within 10 hours; furthermore, pulsatile release of DDA was possible by cycling the temperature between 25°C and 40°C, with significantly higher release rates observed at 40°C. Similar behavior was observed when a pH cycle from pH = 1.0 to pH = 7.0 was applied, by which the increase in pH significantly accelerated DDA release. Unfortunately no in vivo cytotoxicity studies were performed to evaluate the encapsulation efficiency and reduced toxicity of the DDA encapsulated microgels. Ashwanikumar et al. (2012) demonstrated the potential of microgels based on MAA and 2-ethylhexyl acrylate (2EHA) (fabricated in inverse microemulsion) to deliver FU for the site-specific treatment of colon cancers. Microgels loaded with FU exhibited significantly higher cytotoxicity to human colon tumor cell line HCT-116 (human colon tumor cell line) than the drug alone, owing to significantly higher cell uptake achieved and the significant pH dependence of the drug release from the microgels. However, to our knowledge, no in vivo trials demonstrating the efficacy of this triggered oral delivery approach with microgel-based delivery vehicles have yet been performed.

8.3.3 Chemoresponsive Microgels

Although the majority of microgel based drug delivery devices are either thermoresponsive, pH responsive, or a combination of both, microgel swelling or de-swelling can also be triggered in response to the presence of specific molecules. A number of chemoresponsive microgels have been developed and tested for their potential as drug delivery vehicles.

8.3.3.1 Glucose-responsive microgels

A physical response from microgels in the presence of glucose can be engineered through the incorporation of phenylboronic acid (PBA) functional groups. PBA can form a covalent bond between the ionic boronate state of PBA and a *cis*-diol group of a polysaccharide. Glucose, for example, can reversibly bind to the tetrahedral ionized boronate, but cannot form a stable bond with the trigonal uncharged boronic acid. As such, glucose binding to PBA-functionalized materials is an equilibrium reaction, which is dependent on the pH (degree of ionization of the PBA moiety) and the glucose concentration. At a given pH value, glucose binding induces the formation of more ionized boronates to maintain equilibrium, effectively reducing the pK_a of the trigonal boronic acid groups; consequently, more glucose can bind to the material, which in turn triggers further ionization until equilibrium is reached at a given glucose concentration.

The glucose responsiveness in microgels has been exploited in a number of different ways. First, the presence of glucose induces a VPTT shift, as the VPTT increases systematically as the glucose concentration is increased. We have demonstrated that the degree of microgel swelling is amplified in the presence of glucose, which can be used for sensing applications (Hoare and Pelton, 2007). Second, PBA-functionalized microgels can be used to achieve controlled insulin release via a feedback mechanism. An increase in the glucose concentration causes swelling and therewith enhanced insulin release owing to an increased microgel pore size; conversely, as the glucose concentration is subsequently lowered by insulin delivery, the glucose-binding equilibrium to PBA groups shifts to reduce the net microgel charge and thus to induce microgel deswelling to reduce insulin release (Hoare and Pelton, 2008b). Third, glucose-induced shrinking of microgels can be achieved if the concentration of glucose binding PBA moieties is sufficiently high so that glucose (which contains two *cis*-diol groups) can form crosslinks within the microgel network (Ancla et al., 2011). In this case, microgels deswell at higher glucose concentrations, actively pushing an insulin cargo out of the network as the water is expelled.

We have exploited the use of PBA groups to develop microgels with engineered responses to the presence of glucose (Hoare and Pelton, 2007, 2008a, 2008b). PNIPAAm microgels were functionalized with aminophenylboronic acid (APBA) using a "grafting to" approach. Binding of glucose to the PBA moieties increases the anionic charge density in the microgel, which drives a swelling response via both Donnan equilibrium and direct charge repulsion effects. Conversely, a feedback controlled de-swelling occurs as the glucose concentration decreases. We demonstrate the enhanced release of insulin from PBA-functionalized microgels when triggered by increases in glucose concentration within the physiologically relevant range, although further tuning is required to achieve adequate on-off insulin release from such materials. Interestingly, if cationic groups are also incorporated into the microgel, the increase in anionic charge density upon glucose exposure can alternately trigger a change in the net charge of the microgel upon glucose exposure, leading to electrostatic-driven release of (net anionic) insulin (Fig. 8.6); a ~50% increase in insulin release rate was observed upon exposure of the microgel to 1 mg/mL glucose in these charge-switching microgels.

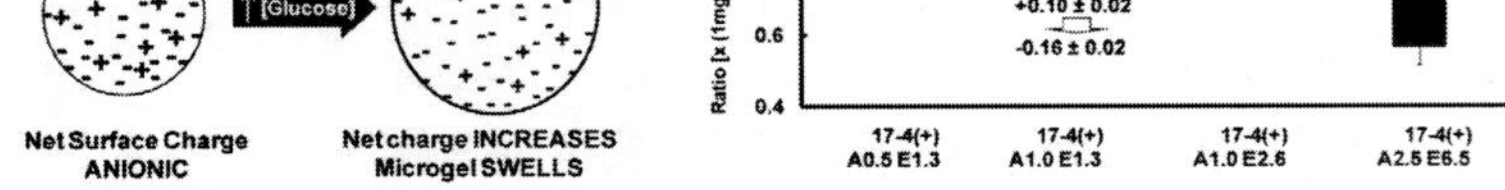

Figure 8.6 Glucose-induced charge-switching mechanism for promoting triggered insulin release from amphoteric PBA-functionalized microgels. Left: Schematic of how net charge densities can drive swelling responses. Right: Glucose-triggered changes in microgel volume and light-scattering intensity as a result of exposure to 1 mg/mL glucose. Reprinted from Hoare and Pelton (2008a) with permission of the American Chemical Society.

Other groups have also prepared PBA-functionalized microgels for this purpose, with the work of Lapeyre et al. particularly noteworthy (Ancla et al., 2011; Lapeyre et al., 2006, 2008; Wang et al., 2011b; Xing et al., 2011b). Sun et al. (2012b) also generated an interesting system based on PBA-driven responses by designing silica nanoparticles coated with a P(NIPAAm-co-APBA)-co-dextran maleic acid shell of roughly 200 nm in size. DEX-MA was incorporated to improve the hydrophilicity, biocompatibility, and glucose sensitivity of PBA under physiological conditions. Insulin release was significantly enhanced in the presence of glucose, resulting in nearly 80% release in 25 hours under physiological conditions. Furthermore, no significant loss in cell viability was observed after two and four days when A549 (human lung carcinoma cell line) cells were exposed to a maximum of 400 μg/mL microgels.

Two other recent examples using other types of glucose-sensing chemistries further illustrate the potential of glucose-responsive microgels for the treatment of diabetes (Sun et al., 2012b; Kawamura et al., 2012; Wu et al., 2012c). Kawamura et al. developed bioconjugated microgels based on DMAEMA and 2-glucosyloxyethyl methacrylate (GEMA) functionalized with concanavalin A (ConA) (Kawamura et al., 2012). The microgels have a core–shell morphology consisting of a hydrophobic PDMAEMA core and a hydrophilic shell containing GEMA–ConA complexes, which act as reversible crosslinks. Upon exposure to glucose, the GEMA–ConA complexes are broken as glucose binds preferentially to ConA. The crosslink density of the hydrophilic shell is thus lowered, which in turn increases the degree of swelling. The degree of swelling (or shrinking) is governed by the glucose concentration, which provides these microgels with the potential to self-regulate the release of insulin (Kawamura et al., 2012). Alternately, Wu et al. (2012c) fabricated poly(acrylamide-co-aminopropylacrylamide), P(AAm-co-AMPAAm), microgels functionalized with a pH sensitive crosslinker that they subsequently loaded with insulin as well as glucose oxidase (GOx) and catalase. Controlled release of insulin was induced by the enzymatic conversion of glucose into gluconic acid, which lowers the local pH, protonates the pendant amine groups in 3-aminopropyl acrylamide (AMPAAm), and thus induces microgel swelling. Furthermore, the acid-labile crosslinks can be cleaved upon local pH decreases to further enhance microgel swelling.

In the absence of glucose, slow release of insulin was observed, resulting in less than 20% release over 30 hours; in comparison, in the presence of 10 or 20 mM glucose, nearly complete insulin release was demonstrated within the same time frame (Wu et al., 2012c). However, unlike the other approaches cited, this material is not completely reversible, as the cleaved crosslinks do not readily re-form as the glucose concentration is again lowered. In addition, both these alternative chemistry systems rely on enzymatic activity, which can be hard to maintain in vivo over extended periods of time or in storage. Indeed, despite the significant amount of work on insulin release from glucose-responsive microgels prepared from multiple potential chemistries, in vivo efficacy has yet to be reported.

8.3.3.2 Redox-responsive microgels

One of the major drawbacks of most microgel synthesis techniques is that incorporating biodegradability in the molecular design is challenging. As mentioned earlier, degradability of microgels is typically achieved by using crosslinkers with cleavable linkages sensitive to, for example, acid (ketals or acetals) or reducing agents (disulfides). Disulfides are interesting from a biological perspective since these covalent bonds play an important role in the stability, folding, and activity of proteins. In the body, the reversible formation and scission of disulfide bonds is controlled by the GSH/glutathione disulfide (GSSG) redox couple. GSH is a tripeptide with a free thiol and is present at μM concentration in the extracellular matrix and blood serum and at mM concentration inside cells (Saito et al., 2003). This natural gradient in GSH concentration between the extracellular environment and the cytoplasm can thus be exploited to degrade microgels that, once internalized, can enable site-specific cargo release (Cheng et al., 2011a).

The use of disulfide-containing crosslinkers for the synthesis of microgels for biomedical applications has been popularized by the group of Matyjaszewski (Oh et al., 2006, 2007; Dong et al., 2009; Dong and Matyjaszewski, 2010). A dithiopropionyl poly(ethyleneglycol) dimethacrylate crosslinker was used to form degradable POEGMA microgels in an inverse miniemulsion polymerization system (Oh et al., 2006, 2007). The resulting microgels are fully degradable in the presence of GSH (Oh et al., 2007) and can be used to release their cargo in a controlled fashion (Oh et al., 2007; Dong et al., 2009).

Qiao et al. (2011) also prepared POEGMA microgels functionalized with cleavable disulfide bonds and acid-labile orthoester groups. These microgels could be fully degraded in the presence of DTT and displayed excellent cell viability when tested on a MCF7 cell line (Qiao et al., 2011). Although complete degradation is a prerequisite in a biological setting, partial degradation of microgels can also have interesting properties. Gaulding et al. (2012b), for example, synthesized PNIPAAm microgels crosslinked with MBAAm and *N,N*-bis(acryloyl)cystamine (BAC). Upon the selective degradation of the disulfides, the microgels swell and expose free reactive thiols that can subsequently be exploited for conjugation of specific (bio) molecules or for inter- and intramicrogel crosslinking to form macroscopic microgel networks (Gaulding et al., 2012b).

DOX release from disulfide crosslinked microgels has been reported by Pan et al. (2012b). Distillation precipitation polymerization of MAA was performed in the presence of BAC to form monodisperse poly(methacrylic acid) (PMAA) microgels ranging from 0.6 μm to 1 μm when swollen in phosphate-buffered saline (PBS) at room temperature. These large microgels could be completely degraded into clearable fragments (M_n = 1200 g/mol) in the presence of 10 mM DTT. These microgels showed excellent drug loading (up to 42.3%) and drug encapsulation (up to 95.7%) due to the strong electrostatic interaction between the carboxylic acid groups of the PMAA polymer and the free amine group of DOX. In the absence of DTT or GSH, DOX is released slowly from the microgel matrix due to electrostatic interactions; the addition of 10 mM DTT or GSH significantly accelerates release from approximately 15% to 30% (DTT) or 80% (GSH) after 24 hours in PBS. Drug release is significantly faster following GSH treatment likely due to interactions between the free amine of GSH and PMAA, which can effectively replace DOX via ion exchange (Pan et al., 2012b). The microgels showed excellent cell viability alone but, once loaded with DOX, proved very efficient in killing glioma tumor cells.

Another interesting example was recently reported by the group of Professor Chun-Yan Hong (Yu et al., 2012d, Wang et al., 2013d). Redox-responsive microgels were prepared from a facile temperature-induced self-assembly and self-crosslinking method (Fig. 8.7) (Yu et al., 2012d). BAC and *N,N'*-((propane-2,2-diylbis(oxy)) bis(ethane-2,1-diyl))acrylamide (an acid labile crosslinker) were polymerized with *N,N'*-dimethyldipropylenepropanetridiamine to form hyperbranched polymers in a water/methanol mixture. The clear solution goes turbid upon addition of water, which indicates the self-assembly of the hyperbranched polymers. Subsequently, the assembled hyperbranched polymers crosslink through intermolecular disulfide exchange reactions (Wang et al., 2013d). The formed microgels are monodisperse in size, are temperature sensitive, and can be completely degraded in the presence of DTT (or acid in cases in which the acid labile BAC crosslinker is used). The microgels contain a multitude of tertiary amine groups that are protonated at physiological pH, making these cationic microgels potentially interesting vehicles for gene delivery. These cationic Redox-responsive microgels have been demonstrated to display good cell viability up to a concentration of 200 μg/mL.

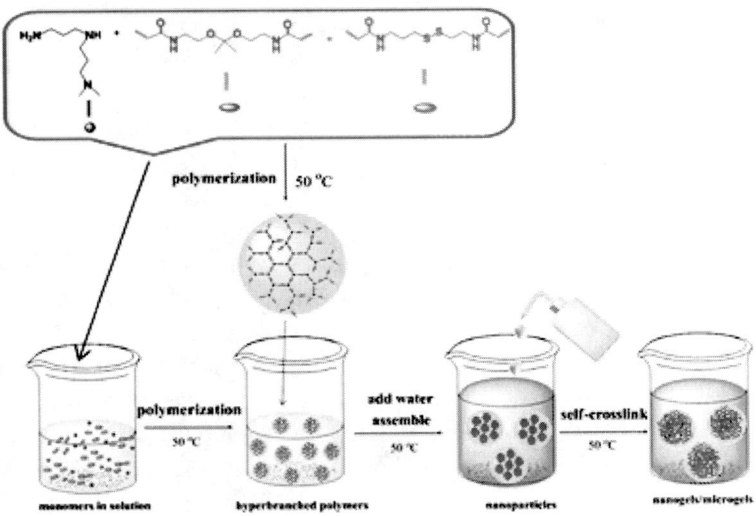

Figure 8.7 Synthesis of Redox-responsive microgels using the temperature-induced self-assembly and self-crosslinking method. Reprinted from Yu et al. (2012c) with permission from the Royal Society of Chemistry.

8.3.3.3 CO_2-responsive microgels

An interesting new category of responsive microgels with the potential to reversibly bind carbon dioxide have recently been reported by Hoshino et al. (2012). Microgels prepared via copolymerization of NIPAAm and *N*-(3-dimethylamino)propyl methacrylamide (NDAPMAAm) exhibited temperature-responsive binding of CO_2, with microgel-bound amines forming ion pairs with absorbed bicarbonate ions below the phase transition temperature (when the microgel is swollen) and almost complete desorption observed above the phase transition temperature (when the microgel is collapsed and water accessibility to the amine groups is more limited). While the microgel has initially been proposed for use in CO_2-harvesting applications, it may also have interesting biomedical applications for triggering selective phase transitions in CO_2-rich (i.e., metabolically active) microenvironments characteristic of infections or malignant tumours, although this concept has yet to be tested.

8.3.3.4 Enzyme-responsive microgels

Enzyme-responsive microgels were first reported by the group of Ulijn (Thornton et al., 2007, 2008; McDonald et al., 2009). POEGMA microgels were prepared from oligo(ethylene glycol) diacrylamide (OEGDAAm) and an amine-terminated oligo(ethylene glycol) acrylate (OEGA)–NH_2 and subsequently modified with diglycine and dialanine enzyme-cleavable linkers (ECLs) separating the aspartic acid and arginine actuating residues using solid-phase peptide synthesis methods (Fig. 8.8) (Thornton et al., 2007). The ECL can be cleaved in the presence of a protease, leaving cationic residues behind. Consequently, the microgels will swell as the protease degrades the microgel and the overall cationic charge of the microgel increases, providing a release mechanism for encapsulated drugs. The PEG microgels produced had an average size of 250 μm and have an uptake molecular weight cutoff of 40 kDa. This could prove to be a potential limitation of enzyme triggered release, as relatively bulky enzymes have to be able to diffuse into the microgel. Elastine (25 kDa) and thermolysin (38 kDa) were found to cleave virtually all the ECLs in the microgel, resulting in a 100 μm size increase. Avidin, a highly cationic globular glycoprotein (63 kDa), was successfully released from the microgels upon the addition of thermolysin. However, BSA (66 kDa) could not be released from the microgels due to charge complexation between the cationic microgel (after enzymatic cleaving) and the anionic protein. In a follow-up study by the same group (Thornton et al., 2008) the initial ECL consisting of a fluorenylmethyloxycarbonyl chloride (Fmoc)-Asp-Ala~Ala-Arg-PEG sequence (which upon cleaving results in a H-Ala-Arg-PEG sequence carrying two cationic charges) was replaced by a Fmoc-Arg-Arg-Ala~Ala-Asp-Asp-PEG; cleaving this peptide results in a residual Ala-Asp-Asp-PEG sequence, which bears a net anionic charge and therefore allows for the release of anionic proteins such as BSA (see Fig. 8.8C,D). Further improvements were made to this microgel design by using branched ECLs, which provide enhanced charge density to overcome electrostatic screening under physiological conditions and therefore induce enhanced microgel swelling (McDonald et al., 2009).

Similar approaches can be taken to induce microgel degradation upon exposure to selected biological fluids for triggered release of a drug payload only upon exposure to that fluid. Clark et al. demonstrated such an approach by preparing hydroxypropyl methacrylamide (HPMAAm)-based microgels crosslinked by *bis*-acrylamide GISSFYSSK peptides, a substrate for the serine protease prostate specific antigen found in semen (Clark et al., 2011). Exposure of the microgels to human seminal plasma induced microgel degradation that triggered the release of entrapped antiviral polymer poly(styrene-4-sulfonate), aimed at degrading human immunodeficiency virus (HIV) on demand. Alternately, Xiong et al. prepared microgels containing a core of polyphosophoester and a shell of PEG or mannose (Xiong et al., 2012b). Phosphatase or phospholipase excreted by bacteria can locally degrade the core and release the entrapped vancomycin drug payload, facilitating local antimicrobial action at the site of a bacterial infection.

8.3.3.5 Antigen-responsive microgels

Swelling responses and thus triggered drug delivery can also be induced by the interaction between antibodies grafted to a microgel and the corresponding antigen for those antibodies. Two approaches can be used: (1) free antibodies are used to crosslink antigen-grafted hydrogels and are released when exposed to free antigen and/or antigens with higher affinities to the antibody to induce swelling (Miyata et al., 2011), or (2) antigens and antibodies can both be grafted to the hydrogel matrix to form affinity pair crosslinks that can be disrupted in the presence of free antigen to

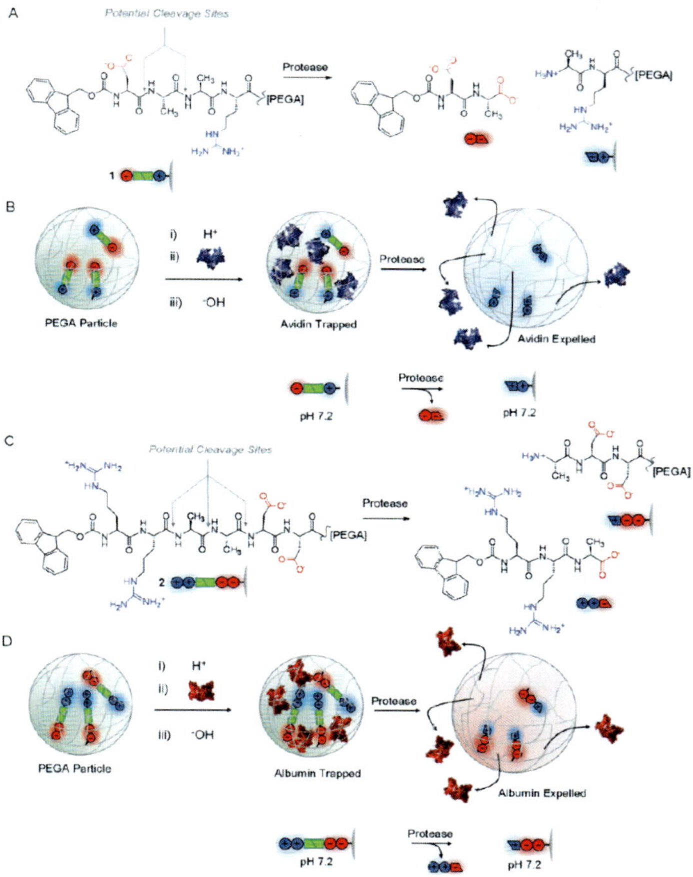

Figure 8.8 (A) The peptide designed for the release of positively charged proteins comprised Fmoc–aspartatic acid–alanine–alanine–arginine, where the amide bond between the two alanine residues is particularly liable to cleavage by our target enzyme. (B) Generation of positive charges by enzymatic cleavage of the bond between alanine residues allows protein molecules to diffuse through the polymer pores for payload release. (C) Peptide designed for the release of negatively charged protein molecules. Two *N*-terminal arginine units are separated from two aspartic acid groups by two alanine residues. A single net negative charge remains on the particle following enzymatic hydrolysis. (D) Exclusion of albumin from the negatively charged swollen particle occurs following hydrolysis of the bond between alanine residues. Reprinted from Thornton et al. (2007) with permission from Wiley.

induce swelling (Miyata et al., 1999). While several examples of such materials have been reported for bulk hydrogels, relatively few examples have been reported for microgels. The work from Andrew Lyon's group on microgel biosensors represents at least a proof-of-concept of this potential on a microgel scale. Biotin (antigen)-conjugated P(NIPAAm-co-AA) microgels were shown to shrink selectively upon the addition of streptavidin or polyclonal antibiotin (antibodies) and subsequently reswell upon addition of free biotin (antigen) (Kim et al., 2005, 2007b). While no direct demonstration of drug delivery has to our knowledge been demonstrated using this or any other antigen-responsive microgel, this type of responsive microgel has considerable potential for site-specific release of drug payloads.

8.3.4 Light-Responsive Microgels

Induction of swelling–deswelling responses via UV/visible light irradiation has been widely reported in the literature to control the swelling of bulk hydrogels (Gong et al., 2008; Tomatsu et al., 2011). When such hydrogels are loaded with drugs, triggered changes in internal pore size can be utilized to induce changes in drug release using light as a stimulus. While the poor penetration of light with UV and visible wavelengths makes such an approach challenging from a clinical perspective in most biomedical applications, ophthalmic applications have the potential to directly utilize light as a stimulus to induce pulsed or externally tunable drug release kinetics.

Several types of light-responsive functional groups can be considered for use in light-responsive microgels. Functional groups including anthracene or other pi-stacking fused aromatic molecules (Wells et al., 2011; Zheng et al., 2002), cinnamate moieties (Gattás-Asfura et al., 2005), coumarin moieties (Chujo et al., 1992), or maleinimido groups (Seiffert et al., 2007) (or derivatives thereof) can form reversible crosslinks, depending on the wavelength at which they are irradiated. Azobenzenes switch from *trans* to *cis* orientations (and vice versa) upon irradiation to drive polarity changes that can also induce changes in swelling, particularly when coupled to thermoresponsive polymers in which the LCST can be altered by the change in azobenzene polarity (Jochum and Theato, 2013). Alternately, photolytically labile functional groups such as nitrobenzyl esters can be used to generate charge with the gel matrix upon irradiation through the cleavage of the ester to expose a carboxylic acid functional group (Zabadal and Klan, 2001), driving a swelling response (albeit one that is not reversible in the same way as the crosslinking methodologies).

Several approaches have been pursued to design photoresponsive microgels, ranging from self-assembly to copolymerization. Kuckling et al. (2006) generated microgels via self-assembly of linear copolymers of NIPAM and 2-dimethylmaleinimido ethylacrylamide followed by light irradiation and demonstrated the formation of covalently crosslinked microgels with sizes controlled by the initial polymer concentration and the density of photocrosslinkable functional groups in the prepolymers (He et al., 2009). Zhao and coworkers reported the formation of microgels by first self-assembling diblock copolymers based on PEG and poly(2-(2-methoxyethoxy) ethyl methacrylate-co-4-methyl-(7-(methacryloyl)oxyethyloxy)coumarin) and subsequently crosslinking the microgels via irradiation of the coumarin groups at a wavelength > 310 nm. Subsequent irradiation at a wavelength < 260 nm led to cleavage of a fraction of these crosslinks and corresponding swelling of the microgel (He et al., 2009), with phototriggered release of DPM and Nile Red demonstrated. The strategy has since been extended to core–shell microgels to localize the photocrosslinking in either the shell or the core (He et al., 2011b). Zhang et al. mixed azobenzene-functionalized PEG with β-CD-A to form a photoresponsive inclusion complex and subsequently crosslinked the pendant acrylate groups via a Michael reaction with PEG-dithiol to form a microgel (Zhang et al., 2012a). UV stimuli could form and break the inclusion complexes that act as a secondary crosslink in the final microgel to drive phototriggerable swelling/deswelling responses. Yin et al. reported the preparation of dual thermo-light-responsive microgels containing fluorescence donor–acceptor ligands suitable for fluorescence resonance energy transfer (FRET) analysis and 2-nitrobenzyl ester groups as photolytic grafts, in this case copolymerized directly into the network during microgel synthesis using the monomer 5-(2'-(dimethylamino)ethoxy)-2-nitrobenzyl acrylate. UV irradiation of the microgels induced the conversion of the nitrobenzyl ester groups to carboxyl groups, inducing an increase in the anionic charge density of the microgel matrix and, by extension, an on-demand swelling response (Yin et al., 2011).

Relatively few studies to date have reported on the potential of photoresponsive microgels for pulsed drug delivery applications. Patnaik et al. demonstrated phototriggered release of a rhodamine and aspirin triggered by the *trans-cis* photoisomerization of azobenzene moieties used as a crosslinker to form DEX-based microgels (Patnaik et al., 2007). Microgels were fabricated by photoirradiation of an azo-DEX solution, with drug release accelerated when the azo moiety was in the *cis* orientation as opposed to the *trans* orientation (Fig. 8.9).

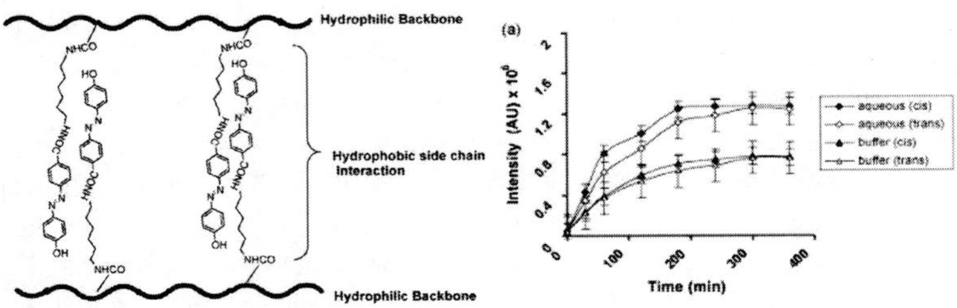

Figure 8.9 Phototriggered release of rhodamine B from microgels prepared with azobenzene linkages. Reprinted from Patnaik et al. (2007), Copyright 2007, with permission from Elsevier.

Lee et al. demonstrated a similar result using coumarin-grafted polyvinyl alcohol nanoparticles prepared in the same way, with an enhanced release rate of carboxyfluorescein upon photodimerization and subsequent collapse of the microgel (Lee and Kim, 2012). However, both these demonstrations have been highly preliminary, with significant tuning of release rate and release payload required prior to clinical application. In addition, the safety of the high intensity low wavelength (~250–350 nm) irradiation required to induce photocleavage of most reported photolabile crosslinks may introduce practical issues with using this approach for drug delivery even in ophthalmic applications, barring the development of photoresponsive groups with activities at higher (visible) wavelengths.

8.3.5 Microgel-Based Nanocomposites

Combining microgels with other polymeric phases and/or inorganic nanoparticles offers additional flexibility in terms of the types of stimuli that can be used to induce triggered drug release and/or the potential rates of release that can be achieved. In particular, incorporation of inorganic nanoparticles allows for remote-controlled, noninvasive, and safe triggering of phase transitions driving smart materials responses in the absence of an inherent stimulus inside the body while incorporating microgels inside solid or gel-based polymeric matrices has been demonstrated to significantly reduce the release rates of drugs relative to the typically fast kinetics observed for small molecules from small and highly swollen microgels.

8.3.5.1 Magnetic microgels

Superparamagnetic iron oxide nanoparticles (SPIONs), typically Fe_2O_3 or Fe_3O_4 iron oxide–based nanomaterials, offer several interesting properties in the context of biomedical applications: (1) the application of a permanent magnetic field can serve to localize particles at a specific location in the body; (2) exposure of the particles to an OMF leads to the generation of heat within the matrix surrounding the SPIONs, due either to hysteresis heating (100–500 kHz oscillating field frequencies) or friction as the nanoparticles move with respect to their matrix (lower frequencies) (Levy et al., 2011); and (3) unlike most inorganic nanoparticles, iron oxide can be reprocessed by the body such that they effectively degrade over time (Levy et al., 2011). Combinations of SPIONs with microgels thus offer potential to localize microgels in vivo and, when a thermoresponsive microgel is used, remotely drive phase transitions for temporal-specific drug delivery induced by a stimulus (an OMF) which is both noninvasive and highly penetrative. Magnetic nanoparticles were incorporated into thermoresponsive PNIPAAm-based microgels by Pichot's group in 1999 (Sauzedde et al., 1999) and in nonresponsive poly(acrylamide) (PAAm) microgels (via inverse emulsion processes) by Menager et al. in 2004. Since these reports, several papers have prepared magnetic microgels with the purpose of using the magnetic properties for guidance of the nanoparticles to a particular target, using a permanent magnet adjacent to or in that target tissue. For example, Wu et al. fabricated poly(oligoethylene glycol methacrylate-co-methacrylic acid), P(OEGMA-co-MAA), microgels by precipitation polymerization seeded on nickel–silver nanoparticles in which the nickel portion of the nanoparticle provides the magnetic property (Wu et al., 2011). Successful uptake of these nanoparticles into B16F10 cells was observed and enhanced release in lysosomal pH was demonstrated. Similarly, Sun et al. reported POEGMA magnetic microgels (prepared photochemically) that exhibited sustained, slow release of DOX (Sun et al., 2009), Dong et al. demonstrated temperature-responsive release of rhodamine B from thermally responsive POEGMA magnetic microgels prepared via atom transfer radical polymerization (Dong et al., 2009), and Deka et al. demonstrated the release of short oligonucleotides mediated by pH changes using 2-vinylbenzene-crosslinked 2-vinylpyridine microgels (Deka et al., 2010). However, while Li et al. confirmed that SPION-loaded microgels undergo OMF-induced phase transitions in 2008 (Li et al., 2008b), there has not yet been a demonstration of magnetically triggered drug release directly from SPION-loaded microgels to our knowledge in the literature. The closest relevant example is the work of Chiang et al., who fabricated hollow microgels derived from the self-assembly of SPIONs with a graft copolymer containing an acrylic acid-co-2-methacryloylethyl acrylate backbone and both PEG (stability) and PNIPAAm (self-association) grafts which can form a vesicle. The vesicle was subsequently stabilized via photoinitiated polymerization of the acryolyl residues in the graft polymer backbone (Chiang et al., 2013). Magnetic guidance using a uniform magnetic field followed by stimulus-triggered release of loaded DOX using an OMF was demonstrated to induce increased pulsatile DOX release relative to that achieved by simple hyperthermia. However, this area largely remains to be explored.

Increasing the phase transition temperature to the temperature range typically used for hyperthermia (43°C–45°C) is essential for the effective use of such dual-magnetic thermoresponsive materials for tumor-specific delivery. To address this, Jaiswal et al. incorporated chitosan into magnetic PNIPAAM microgels by mixing chitosan into the precipitation polymerization recipe to achieve microgels with VPTT $\approx$ 42°C, high specific absorption under alternating magnetic fields, and cytocompatibility with L929 (mouse aneuploid fibrosarcoma) cell lines (Jaiswal et al., 2010). Alternately, Hoare et al. copolymerized NIPAAm with *N*-isopropylmethacrylamide (NIPMAAm) and AAm, achieving phase transition temperatures in the range 39°C–47°C as desired while maintaining relatively sharp and large magnitude phase transitions required for effective on/off or high/low release (Hoare et al., 2011).

8.3.5.2 Near-infrared-responsive microgels

Anisotropic nanoparticles based on gold or carbon are highly effective at absorbing NIR irradiation and thus transducing a NIR signal, with materials such as gold nanorods, gold nanoshells, carbon nanotubes, C_{60} fullerenes, or graphene all reported as effective transducers (Menon et al., 2013). While the bulk of the use of these materials for biomedical applications lies in the imaging (Xing et al., 2012) or theranostics (Menon et al., 2013) areas (i.e., using the heat generation directly for thermotherapy of predominantly cancer tumors), drug delivery materials have been reported which combine the NIR-active nanoparticle with a thermoresponsive hydrogel based on PNIPAAm and/or POEGMA to permit remote-controlled deswelling of the microgel by induction heating of the NIR nanoparticle. Das et al. were the first to demonstrate this phenomenon following the incorporation of gold nanorods inside poly(*N*-isopropylacrylamide-co-maleic acid), P(NIPAAm-co-MA), microgels, with the MA residues used to localize the nanorods inside the microgel (Das et al., 2007b). An eightfold decrease in size was achieved upon irradiation at 809 nm that was fully reversible. Kawano et al. extended this idea to significantly smaller microgel particles by using a single gold nanorod as a nucleus for PNIPAAm microgel growth (Kawano et al., 2009). However, neither paper showed effective therapeutic drug delivery.

Shang et al. demonstrated tunable drug release from NIR-responsive microgels by preparing gold nanorod-loaded poly(*N*-isopropylacrylamide-co-oligoethylene glycol methacrylate), P(NIPAAm-co-OEGMA), microgels for NIR-mediated delivery of zinc phthalocyanines, which can generate singlet oxygen in situ to kill off HeLa cells (Shang et al., 2013). Wang et al. demonstrated similar efficacy with chitosan-modified reduced graphene incorporated into a PNIPAAm microgel for the delivery of DOX, with rapid and repetitive drug release possible upon irradiation with 808 nm NIR light (Wang et al., 2013a). The improved tumor cell killing observed upon NIR irradiation was attributed both to the pulsatile drug release achieved as well as the local increase in temperature achieved upon NIR irradiation. However, while thermotherapy using such nanoparticles has been investigated in vivo, no such demonstrations have yet been conducted using these dual drug release/thermotherapy composite microgels.

8.3.5.3 Microgel pore-filled membranes

Incorporation of microgels into the pores of otherwise impermeable polymeric membrane materials offers an alternative methodology to use the responsive functionality of microgels for tuning drug release. In this case, the microgels are used as gates that can be reversibly opened and closed by the stimulus. Relative to the use of microgels alone, such a construction typically allows for significantly slower release in the "on" state and much lower baseline release rates in the "off" state, as the limiting factor for drug release is now diffusion of drug through the membrane rather than diffusion of drug through a highly swollen hydrogel network. Zhang and Wu were the first to use this approach for drug delivery, entrapping a methacrylic acid (MAA)-functionalized PNIPAAm microgels inside an ethylcellulose matrix for glucose-responsive insulin release (Zhang and Wu, 2002). Hoare et al. extended this approach to more general on-demand drug release applications by incorporating both SPIONs and copolymer microgels based on NIPAAm (LCST ≈ 32°C), NIPMAAm (LCST ≈ 43°C), and AAm (no LCST) with VPTT values in the range of 37°C to 46°C inside an ethylcellulose matrix (Fig. 8.10) (Hoare et al., 2009; Hoare et al., 2011).

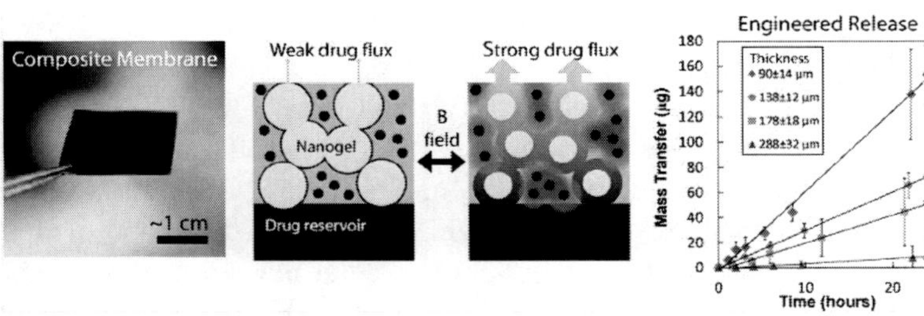

Figure 8.10 Composition and function of microgel-filled magnetic composite membranes. (a) Photograph of the microgel and SPION-loaded ethyl cellulose membrane, (b) schematic illustration of drug diffusion at temperatures below and above the VPTT of the microgels, and (c) mass transfer rates as a function of the "on" triggering time and membrane thickness. Reprinted from Hoare et al. (2011) with permission from the American Chemical Society.

The SPIONs permit remote heating that can drive opening of the membrane pores via induction of a microgel phase transition; removal of the magnetic field subsequently induces pore closure as the microgels cool and reswell. Drug release is reduced to slow or zero through the interconnected microgel channel in the swollen, "off" state, with repeated remote controlled drug release demonstrated for both sodium fluorescein and FITC-DEX (40 kDa molecular weight). Up to 20-fold enhancements in "on" state relative to "off" state drug release were achieved, significantly better than any reported contrast in release with microgels alone. Furthermore, subcutaneous implantation of the membrane in Sprague–Dawley rats indicated that the membranes induced only minimal and disorganized capsule formation and mild inflammatory responses, suggesting its potential use in an implantable triggerable drug delivery device.

Yu et al. reported an analogous approach which would not require the additional step of fabricating the membrane into a drug delivery device, forming a porous ethylcellulose shell via microfluidics and subsequently using those pores to template the formation of PNIPAAm microgels in situ (Yu et al., 2012c, 2012d). Triggerable release of vitamin B12 was demonstrated upon heating, although no transducer to facilitate remote-controlled heating was included in this design.

8.3.5.4 Hydrogel–microgel composites

To address the dual challenges of slowing down drug release rates from free microgels as well as ensuring the facile delivery of any resulting composite construct via injection, the incorporation of microgels inside in situ gelling hydrogels offers an alternative to the solid membrane-based approaches, while maintaining significantly higher fabrication simplicity than the pore-filled hollow shell approach previously described. In particular, the properties of the bulk hydrogel phase (e.g., charge, crosslink density, phase transition behavior) can be independently manipulated to effect changes in drug release kinetics relative to those possible with responsive microgels alone. Sivakumaran et al. have demonstrated such a phenomenon by physically entrapping poly(NIPAAm-co-AA) microgels inside an injectable hydrogel based on the in situ gelation of aldehyde-functionalized DEX and hydrazide-functionalized carboxymethyl cellulose following co-injection through a double-barrel syringe (Sivakumaran et al., 2011). BPV, a cationic anesthetic, was included in the formulations and was observed to selectively partition into the highly anionic microgel phase as opposed to the weakly anionic bulk hydrogel phase. Relative to the hydrogel or microgel alone, in which BPV release is almost quantitative within two to three days and exhibits an expected large burst at early times, BPV release from these composite hydrogel structures could be sustained for up to 40 days with minimal burst, significantly longer than releases typically reported from hydrogel-based drug delivery vehicles for analogous drugs (Fig. 8.11). Microgels with higher charge densities (i.e., higher affinities for the cationic drug) promoted lower burst releases and longer overall release times. In addition, microgels with higher crosslink densities and/or microgels crosslinked directly to the bulk hydrogel facilitated significantly slower drug release than microgels with lower internal crosslink densities or microgels that were simply physically entrapped inside the bulk hydrogel network, an observation attributed to the reduced convective driving force for drug release from microgels as increased crosslinking inhibits deswelling upon heating (Sivakumaran et al., 2013). The composites also elicited only moderate acute and minimal chronic inflammatory responses, suggesting their potential use as injectable local drug delivery vehicles with prolonged release kinetics.

8.4 Future Outlook

The promise of using responsive microgels and nanogels for drug delivery applications is clear; the potential to target delivery to specific biological microenvironments as well as facilitate remote-controlled drug delivery with pulsatile release kinetics using only materials (i.e., without requiring implanted electronics or sensitive biological ligands) is in particular applications in which responsive microgels offer clear and specific advantages from a clinical perspective. However, as has been consistently noted throughout this chapter, there continues to exist a significant gap between the claims of utility of various microgels for drug delivery applications and the actual in vivo demonstration of such utility in the literature. This is particularly problematic, given that the major smart responses discussed in this book chapter rely on phenomena

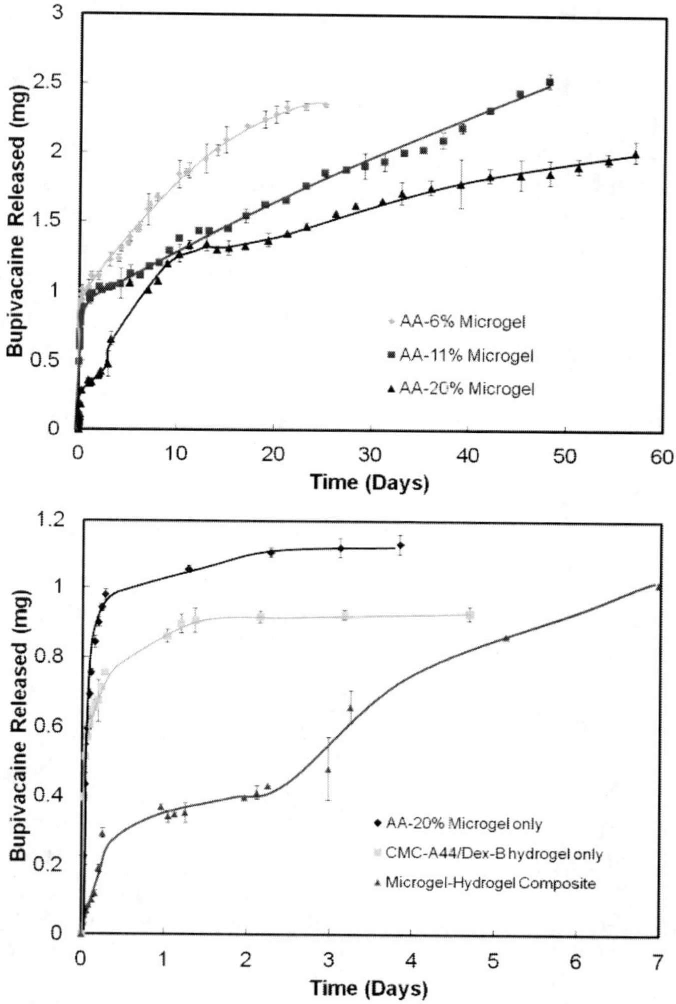

Figure 8.11 Drug delivery from acid-functionalized thermoresponsive microgel–hydrogel composites. (a) BPV release from nanocomposite hydrogels consisting of a 50:50 (wt.%) mixture of AA-NIPAAm or NIPAAm microgels with various acrylic acid loadings (6 mol.%, 11 mol.%, and 20 mol.%) embedded inside an injectable carboxymethylcellulose–DEX hydrazone crosslinked hydrogel. (b) Comparison of drug release from the hydrogel, microgel, and a 50:50 mixture of the hydrogel and microgel tested at the same overall polymer concentration. Reprinted from Sivakumaran et al. (2011) with permission from the American Chemical Society.

such as hydrophobic–hydrophilic balances (thermal responses), pH ionization equilibria (pH responses), or specific binding responses (chemical responses) that may be significantly influenced by the chemical complexity of the in vivo environment (e.g., protein/lipid adsorption, specific ion interactions, or competitive binding of chemoresponsive receptors with chemically similar ligands with lower affinities but present at higher concentrations). As such, testing in cell-free serum (at minimum) is advisable to ensure that the phase transitions observed in idealized conditions can be replicated in the physiological milieu. However, true in vivo testing is required in order for any of these smart microgel-based approaches to ultimately be translatable to the clinic. Challenges also persist with effectively engineering microgels to exhibit large magnitude phase transitions within the narrow range of stimuli either naturally observed or safely inducible in the physiological environment. For example, if local pH gradients are used to facilitate site-specific delivery, it is essential to design microgels with the largest possible pH transition in the range $6.5 < pH < 7.4$, typical of tumoral or infected tissue pH; analogously, thermal transitions in the temperature range 37°C < temperature < 43°C (above which healthy cell hyperthermia can occur) must be optimized in order to maximize the difference in on-off drug release due to thermal triggering. Finally, developing new crosslinking chemistries or strategies to replace the nondegradable crosslinks used in many of the examples cited with linkages demonstrating controllable and tunable degradation rates would ensure the safe clearance of microgels at appropriate rates. Successfully solving these challenges, coupled with well-designed in vivo testing of promising delivery vehicles, is anticipated to lead to microgel and nanogel-based delivery vehicles that uniquely address existing clinical challenges.

Chapter 9

Magnetic Particles for Cancer Diagnosis and Drug Delivery

Naveed Ahmed,[a,b] Ahmad Bitar,[a] Michele Karoline Lima,[a] Nasir Mahmood Ahmad,[c] Hatem Fessi,[a] and Abdelhamid Elaissari[a]

[a]*University Lyon-1, UMR-5007, LAGEP-CPE, 43bd 11 Novembre 1918, Villeurbanne, France*

[b]*Department of Pharmacy, Quaid-i-Azam University, Islamabad 45320, Pakistan*

[c]*Polymers and Surface Engineering Laboratory, Department of Materials Engineering, School of Chemical and Materials Engineering (SCME), National University of Sciences and Technology (NUST), NUST H-12 Campus, Islamabad 44000, Pakistan*

elaissari@lagep.univ-lyon1.fr

Various types of hybrid magnetic nanoparticles (MNPs) under the dispersion aspect have been applied successfully for the development of novel diagnostic techniques and therapeutic approaches for many diseases, including cancers. In this direction, the newer diagnostics and therapeutic agents based on MNPs are continuously developed. A unique feature of these MNPs is that these can be encapsulated in other materials such as functional polymers as well as a silica matrix. Furthermore, intense efforts are being made to

Controlled Release Systems: Advances in Nanobottles and Active Nanoparticles
Edited by Alexander van Herk, Jacqueline Forcada, and Giorgia Pastorin

ISBN 978-981-4613-21-7 (Hardcover), 978-981-4613-22-4 (eBook)
www.panstanford.com

employ the MNPs to deliver these to the targeted organ or effective site and test their performance. Thus hybrid MNPs and their colloidal forms with desirable functional and reactive characteristics are very promising because of their several intrinsic properties such as their high specific surface area, shape, chemical composition, size and surface functionality control, surface charge nature and density, degradation, internal morphology, and responsiveness to stimuli of pH, light, and temperature. Due to the dual role of the MNPs in diagnostics and therapy, now a new subfield known as theranostics has emerged and generated high interest in the treatment of diseases and is primarily applied for cancer. One principal advantage is evident that due to dual functions of MNPs: these can enable us to save critical time for treatment of fatal diseases like cancer where the lapse between diagnosis and therapy is very important. In addition, the controlled release of therapeutic agents for cancer treatment to the tumor site is also a challenging field at present. Because of novel development in the MNPs, these particles can be tailored, as required, to serve as therapeutics drug delivery vehicles. In view of the above-mentioned importance of MNPs and their role in diagnostics, therapy, and drug delivery, the current chapter provides an overview on some of the latest developments about MNPs such as their role as biomedical materials, encapsulation and release from silica matrix, in vitro and in vivo diagnostics of cancer, pH- and temperature-responsive drug release, magnetic resonance imaging (MRI) and tumor diagnostics, and theranostics. In addition, a short description of nanomaterials toxicities is also given at the end.

9.1 Introduction

Biomedical nanotechnology has given a variety of opportunities to fight against many diseases. Drug and gene delivery, protein and peptide delivery, and recent advancements of theranostics are the important subfields along with other applications. Diseases, which are in dire need of complete solutions to save humanity, include diabetes mellitus, cancer, and neurodegenerative and cardiovascular diseases. Every year millions of people lose their struggle for life against these well-known illnesses. Cancer is important and has been well investigated for a cure but still there is no valuable success.

Along with other branches like nanonephrology, advancements in proteomics, and genomics, nanotechnology has also emerged as one of the most fruitful applications in oncology and can be understood as a definite medical boon for diagnosis, treatment, and prevention of cancer disease. In the last decade, the application of nanotechnologies for drug delivery in cancer has been extensively explored with hope to improve the efficacy and to reduce side effects of chemotherapy (Douzeich-Eyrolles et al., 2007). It might be applied in cancer through molecular tumor imaging, early detection (like high-throughput nanosensor devices for detecting the biomarkers of cancer), molecular diagnosis, targeted therapy, and cancer bioinformatics. Considering the preference is intratumoral administration, the various methods of direct introduction of anticancer drugs should kept in mind, including injection of drugs directly into the tumor, tumor necrosis therapy, injection into the arterial blood supply of cancerous tissues, local injection into tumor for radiopotentiation, localized delivery of anticancer drugs by electroporation (electrochemotherapy), and local delivery by anticancer drug implants. Further specification may include heat-activated targeted drug delivery, tissue-selective drug delivery, use of vascular-targeting agents, use of a carrier, or selective permeation of the anticancer agent into the tumor. The choice of the method depends upon the stage and location of the tumor, drug specifications, and the importance of treatment. In the present chapter, we have tried to shed some light on the cancer nomenclature, diagnosis, treatment, classification of the nanomaterials used at present for the treatment, their use in controlled release for both diagnosis and treatment purposes, and, at the end, toxicities of nanomaterials in short.

From the perspective of diagnosis and therapy of cancer, there is a great focus on the nanomaterials at the moment. One main characteristic, which distinguishes nanomaterials from their bulk counterparts, is the scale of various materials. Figure 9.1 presents an elegant view to compare the size range of various materials. Furthermore, various nanomaterials have been developed and include photonic particles (such as quantum dots [QDs]) as optical probes to monitor DNAs and the interactions of proteins; metallic nanoparticles as plasmonic nanoprobes; magnetic nanoparticles (MNPs) as unique probes to detect proteins, bacteria, cells, and viruses; nanotubes, nanowires, and nanodevices as ultrasensitive probes to detect cancer biomarkers; and multifunctional nanomaterials as multiplexed probes, as shown in Fig. 9.2. All these nanomaterials are vital for the future of bionanotechnology. Nonetheless, among these, diagnostics and therapy of cancer by employing MNPs are an exciting area and attracting great attention at the moment.

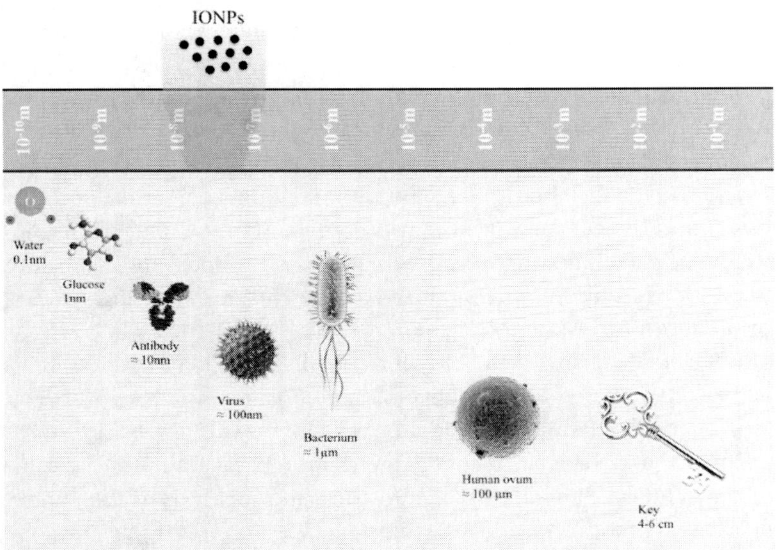

Figure 9.1 Scale of various nanoparticles and biological entities. The iron oxide nanoparticles (IONPs) discussed in this can range from around 5 to 10 nm in diameter.

Magnetic particles with sizes ranging from nanometers to a few microns have been used in various applications for years because of their unique properties such as superparamagnetism, high-field irreversibility, and high saturation field. Nanoscale magnetic materials have their own advantages that provide many exciting opportunities in biomedical applications.

First, the size and the magnetic properties (ferromagnetic, paramagnetic, superparamagnetic, etc.) can be controlled via the synthesis process, and then the size can be varied from 10 nm to a few micrometers.

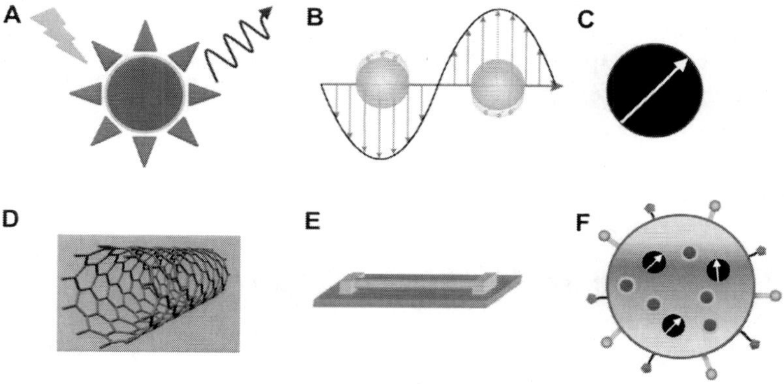

Figure 9.2 Scheme of representative nanomaterials and nanodevices as probes for in vitro diagnostics: (A) QDs, (B) plasmonic nanoparticles, (C) magnetic nanoparticles, (D) carbon nanotubes, (E) nanowires and nanodevices, and (F) multifunctional nanomaterials containing QDs, magnetic nanoparticles, and targeting moieties. Adapted from Chi et al., Copyright 2012, with permission from Elsevier.

Second, due to the magnetic properties of these particles, the motion, the local concentration, and immobilization, or basically the manipulation, of these particles can be performed using an external magnetic force, which provides tremendous advantages for many applications involving biological detection, diagnosis, and drug delivery.

Third, MNPs have the ability of magnetic resonance contrast enhancement because the signal of magnetic moment of a proton around MNPs can be captured by resonant absorption (Purushotham and Ramanujan, 2010). Due to their unique and promising properties, the exploiting of magnetic nanomaterials as nanoprobes to detect bacteria, virus, proteins, and other active biomolecules has attracted high research interests (Lee et al., 2008a; Landmark et al., 2008; Chujo, 2007; Ramakrishna et al., 2001).

MNP-based probes involve the use of nanotechnology to meet the demands of diagnostics and therapy for increased sensitivity, rapid detection, and treatment efficacy. In the present chapter, selected examples are discussed to summarize the recent progress made in this particular field and provide the assessments and problems encountered. Herein, we focus on employing MNPs as unique probes to detect cancer biomarkers and multifunctional nanomaterials as multiplexed probes. The goal is to provide the readers with a rapid state of the art on particles in drug delivery, magnetic particles in both cancer diagnosis and treatment, and also a rapid aspect on theranostic-based nanoparticles with special attention dedicated to cancer.

9.2 Nanoparticles in Drug Delivery

Examples of medical devices utilizing nanotechnology, which are already on the market, are surgical tools with enhanced properties. A few examples include the nanosized contrast agents (CAs) for molecular imaging, bone replacement materials constructed from nanostructured materials, pacemakers and hearing aids of reduced size and increased power, lab-on-a-chip devices for in vitro diagnostics (IVD), wound dressings containing nanocrystalline silver particles, microcantilevers, and microneedle-based systems for minimally invasive drug administration. Over the next 10 to 20 years nanotechnology may fundamentally transform science, technology, and society, offering a significant opportunity to enhance human health in novel ways, especially enabling early disease detection and diagnosis, as well as precise and effective therapy tailored to the patient.

Numerous investigations are being reported on nanotechnology and different materials are being used in different research works to develop suitable procedures to detect, cure, or mitigate the disease. These researches are dedicated to the applications of this new field on different types of disease such as diabetes, nephrology, and most importantly different types of cancer, along with applications in surgery, detection, and diagnostic imaging. Nanotechnology is a suitable tool for delivery of genes, drugs, and proteins. It is generally observed that nanoparticles are promising tools for the advancement of drug delivery and medical imaging and as diagnostic sensors. Every research group is working on a specific nanomaterial according to its final requirement that can be improvement in detection, targeting a drug to specific cells, cell repair, etc. Surveying the literature, a long list of colloidal materials, including metals, nonmetals, metal oxides, semiconductors, and polymer-based particles with some hybrid materials (i.e., combinations of organic and inorganic materials), were found being used in/as nanomaterials (Cao, 2008).

In general colloidal particles for the biomedical applications include different forms of the nanomaterials designed for a specific purpose. Final applications can play a vital role in designing new shapes and protective layers for the nanomaterials. In Table 9.1, there are different forms/tools of nanotechnology used for cancer treatment.

Table 9.1 General classification of nanomaterials with applications

S.No.	Name	Class	Applications	References
1	Nanocapsules		Drug delivery applications	Imaz and Forcada, 2008b, Zhu et al., 2007
2	Nanospheres		Drug delivery applications	Ahmed et al., 2012c
3	Quantum dots		Diagnostics, tumor targeting, and live cells study	Chang et al., 2011 Gan et al., 2011 Wu et al., 2012b
4	Liposomes	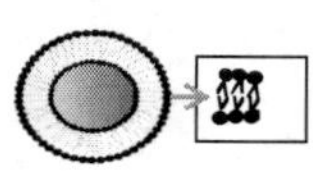	Drug delivery applications	DeVita et al., 2000 Stocchi and Nelson, 1998
5	Polymeric micelles	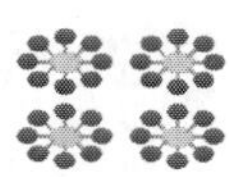	Drug delivery applications	Xu et al., 2005
7	Dendrimers		Oral, intratumoral, and transdermal drug administration	Dastidar and Sharma, 1989

9	Carbon nanotubes	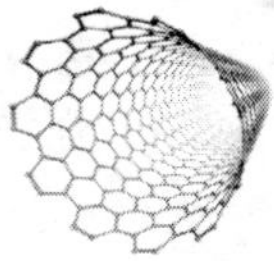	Drug delivery, diagnostic thermal, gene, and photodynamic therapy	Xie et al., 2010a
10	Niosomes	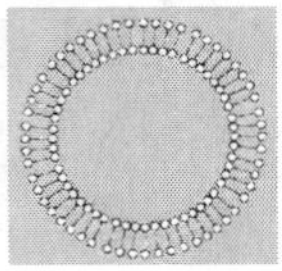	Drug delivery applications	Hamoudeh et al., 2008
11	Fullerenes	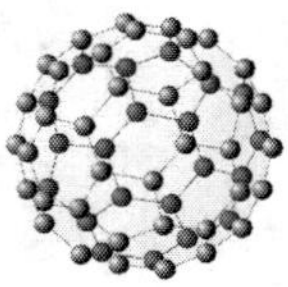	Drug delivery applications	Richard et al., 2008
12	Janus particles		Biosensing and drug delivery	Ahmed et al., 2012c

The different forms of nanomaterials used include nanoparticles, nanotubes, nanopores, dendrimers, etc., but nanoparticles have been used as a research subject in many cases. Further nanoparticles that are being used are of many types because of their different applications. Various types of nanomaterials include carbon nanotubes, nanocrystals, QDs, gold, silver, titanium oxide, zinc oxide, manganese oxide, silicon dioxide, iron oxide, cadmium sulfate, etc. Recent studies also show that nanoparticles like CdS can be prepared by microorganisms such as fungi and yeast (Arbab et al., 2004) to give rise to a new direction to nanoparticles research. Another interesting aspect is that nanostructures can be conjugated to biological molecules, including hormones and antibodies, which enables their targeting to tissues expressing their cognate receptors (Devalapally et al., 2007; Van Vlerken and Amiji, 2006; Farokhzad et al., 2006). For example, fluorescent QDs conjugated to various peptides specifically target either the vasculature of normal tissues or alternatively of cancer cells (Akerman et al., 2002). The rapidly increasing activity of nanotechnology in oncology may be because of factors (Fortina et al.,

2007) like (i) the discovery of the new forms of matter, buckminister fullerenes (buckyballs), and nanotubes made of carbon (Li et al., 2005; Kroto et al., 1985); (ii) the increasing ability of material scientists to produce reagents on a small but controlled size scale; (iii) the availability of instruments like the atomic force microscope and the scanning tunneling microscope to complement traditional instrumentation for viewing and characterizing nanoparticles (Woolley et al., 2000); (iv) the discovery of quantum effects, such as size-dependent fluorescence emission in small particles; (v) nanoparticulate delivery systems providing better penetration of therapeutic and diagnostic substances within the body at a reduced risk in comparison to conventional cancer therapies (Praetorius and Mandal, 2007); and (vi) the stimulation provided by new funding initiatives from worldwide government agencies.

9.2.1 Magnetic Particles

Particles having magnetic characteristics are applied in the biomedical field and important examples of these particles include manganese, iron oxide, gadolinium etc. Synthesis of iron oxide nanoparticles (IONPs) with a specific size and shape is considered as a challenge. However, due to extensive research efforts and emergence of different synthetic techniques have enabled to synthesize IONPs with good control over sizes, functionalities, biocompatibility, structures, and shapes. In general, the synthesis of IONPs can be divided into physical, chemical, and biological methods. Gas-phase deposition and electron beam lithography that mainly constitute physical methods are unable to control the size of the particles (Gupta and Gupta, 2005)). Biological methods can also be considered as alternate ways for the preparation of IONPs but most widely used methods are chemical methods. Among these, the coprecipitation method is the most widely employed. Generally, in this method ferric and ferrous ions are mixed in a molar ratio of 1:2 in a basic medium at room or elevated temperature. The factors affecting the size and shape of the resulting nanoparticles are the type of salts used, reaction temperature, and the pH and ionic strength of the media. In one of the studies, the effect of the $[Fe^{3+}]$/ $[Fe^{2+}]$ molar ratio and iron salts' concentration on the properties of superparamagnetic IONPs in the water/ethanol/toluene system has been reported (Jiang et al., 2011). However, other factors like stirring speed and time have also been considered by researchers for the production of low-sized nanoparticles. Higher size distribution of the IONPs produced by coprecipitation can be minimized using decantation or filtration mechanisms. In one of the recent works, IONPs were prepared by a modified coprecipitation method using ammonium hydroxide as a basic medium. The successful preparation of IONPs and further dispersion in aqueous and organic media were also obtained.

Magnetic particles, as composites or hybrids, are widely used to design drug carriers, which are composed of a magnetic core, an inorganic shell, and a stimuli-responsive polymer shell. Generally, iron oxides are used to prepare magnetic particles and the stimuli-responsive shell is made of a polymer (Zhang and Misra, 2007; Zhang et al., 2007a) or silica. Using silica to encapsulate the magnetic core (Fig. 9.3) has several advantages such as ease to elaborate, low cost, more biocompatibility, colloidal stability, core protection ability, and a very reactive surface. The silica surface can be easily modified and functionalized by several types of functions such as amine (–NH_2), carboxyl (–COOH), phenyl (–Ph), and phosphonate (–PO_3^-) (Chang et al., 2010a).

Figure 9.3 Illustration of a composite magnetic core and a silica shell.

9.2.2 Magnetic Silica Particle Preparation and Drug Delivery

Mesoporous silica particles were proposed to carry and deliver targeted drugs (Vallet-Regi et al., 2000). Therefore, a variety of synthesis methods were applied to investigate and control the load and release of drugs from these particles. In silica magnetic particles, the pores of silica, size of 1–10 nm, are used for controlled drug release and iron is used for targeting and directing the particles under the control of an external magnetic field.

First, synthesis was performed via the preparation of mesoporous silica particles, historically called mobile crystalline materials (MCM-41), and then the iron is produced, into the pores, from iron nitrate (Arruebo et al., 2006). MCMs were prepared via hydrothermal synthesis of tetraethyl orthosilicate (TEOS) in KOH and using Cetyl trimethyl ammonium bromide (CTAB) as surfactant. Metallic iron was loaded through three steps: (i) impregnation of MCMs in Fe^{3+} solution to accumulate iron oxide in silica pores, (ii) oxidation, thermic under air condition, of iron oxide to a-Fe_2O_3, and (iii) Fe_2O_3 reduced under H_2 to metallic iron. This cycle was repeated four times to obtain a magnetic moment as high as 3.40 emu/g at 2T. Hu et al. (2007a) prepared magnetic-sensitive silica nanospheres using the same approach. They supposed that the iron particles in silica nanospheres could be controlled by a high-frequency magnetic field (HFMF), which accelerates the movements of MNPs, and then they will create pores in the silica shell to release the drugs.

Furthermore, the sol–gel process is applied to encapsulate iron oxide particles by a mesoporous silica shell (Zhou et al., 2007). $CaCO_3/Fe_3O_4$ composite nanoparticles were prepared and used as a template for a silica shell using a stabilizing surfactant, which can control the pore size by adjusting its amount in the system. Then, these particles were calcined at 550°C to remove the template surfactant and treated by diluted acetic acid to remove the $CaCO_3$ core. A third preparation method based on aerosol-assisted synthesis was reported (Ruiz-Hernández et al., 2007).

9.2.2.1 Controlled release from silica magnetic particles

Efficient drug carriers must respect several norms and conditions, as their uses will be in in vivo, such as size, biocompatibility, toxicity, and targeting. First of all, the mechanism of delivery depends on the kind of the liaison between the drug and the carrier, which should have two different states of connection with the drugs. A strong liaison state provides a high drug load, and a weak liaison state ensures an efficient release at the target level. The nature of this connection generally depends on the external conditions such as medium pH, temperature, and salinity. All drug carriers have favorite conditions for load and favorite conditions for release, and for in vivo uses the preferable conditions for release should be akin to the biological systems. Furthermore, particles size is an important factor that attracts more attention during particle synthesis. Particles having a size more than 100 nm can't easily circulate through the blood vessels and their uptake and elimination by the reticuloendothelial system (RES) is more favorable (Drummond et al., 1999). In addition, drug release is preferable at the target site; therefore an additional function is required on the particle surface to correctly direct the particle to its target.

In general, a polymeric shell could be added on the silica magnetic particles, which are designed for drug delivery and controlled release. These polymers play many roles such as to increase particles stability (Lee et al., 2010) and biocompatibility (Huang et al., 2009) or particle detection (Chen et al., 2010a). Moreover, stimuli-responsive polymers, temperature, and pH could have an essential role in controlled release:

(i) Thermal drug release using magnetic silica particles for release: Temperature-responsive materials have received great attention in biological applications, especially temperature-responsive polymers such as poly(*N*-isopropylacrylamide) poly(NIPAM) (Pelton and Chibante, 1986) and poly(*N*-vinylcaprolactam) poly(NVCL) (Imaz and Forcada, 2008b, 2010). On the basis of the combination of magnetic silica particles and temperature-responsive polymers, many works reported the temperature-controlled release. Zhu et al. (2007) synthesized ordered mesoporous silica (SBA-15) with MNPs and poly(NIPAM) as the temperature-responsive polymer (Fig. 9.4).

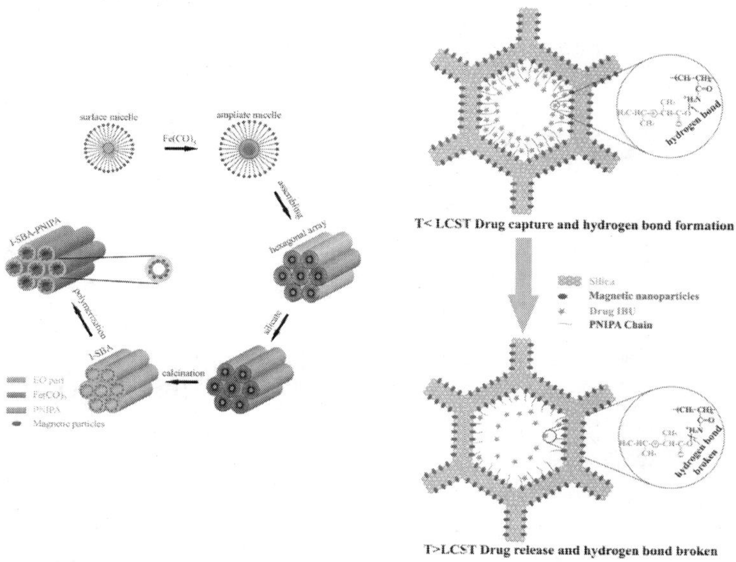

Figure 9.4 Schematic representation of the SBA-15 synthesis process and the drug capture and release responding to temperature. Adapted from Zhu et al. Copyright © 2007 WILEYVCH Verlag GmbH & Co. KGaA, Weinheim.

SBA-15-based systems were obtained with a high pore diameter (7.1 nm) and pore volume of 0.41 cm^3g^{-1}. Poly(NIPAM) polymerized inside the pores acted as a drug holder, which is responsive to the environmental temperature. Ibuprofen (IBU, anti-inflammatory drug) was used as the drug model to study the load and release efficiency. In another work, *N*-hydroxymethyl acrylamide (NHMA) was polymerized with NIPAM onto the surface of silica magnetic particles to control the release temperature (Liu et al., 2009a). It is well known that the volume phase transition temperature (VPTT) of a copolymer changes by changing the copolymerization ratio; thus the release temperature can be easily adjusted.

(ii) Drug release using pH-responsive polymers: Like temperature-responsive polymers, pH-responsive polymers were used to encapsulate silica magnetic particles prepared for drug delivery. Chang et al. (2011) prepared silica magnetic particles coated with a copolymer of poly(NIPAM-co-methacrylic acid), or poly(NIPA-co-MMA). This shell of poly(NIPA-co-MMA) is thermos- and pH sensitive, and it has a swollen state in pH 7.4 or temperature lower than the VPTT and collapsed state in pH 5 or temperature higher than the VPTT. This behavior, which depends on temperature and pH, enables use of these particles as drug carriers, whereas drugs could be loaded above pH 7.4 or temperature below the VPTT and released below pH 5 or temperature above the VPTT. Doxorubicin hydrochloride (DOX), a commercial anticancer drug, was used as a model drug to study the profile of load and release. Gan et al. (2011) synthesized pH-responsive silica magnetic particles for drug delivery, but they used the MNPs as caps for the silica pores (Fig. 9.5), silica particles were functionalized with (3-aminopropyl) trimethoxysilane (APTMS), and Fe_3O_4 nanoparticles were functionalized with boronic acid, which interacts with *cis*-diol molecules forming a reversible boronate esters linker, at natural pH. This mechanism of reversible interaction between silica pores and Fe_3O_4 ensure the drug residence inside the silica pores so very high efficacy of drug delivery.

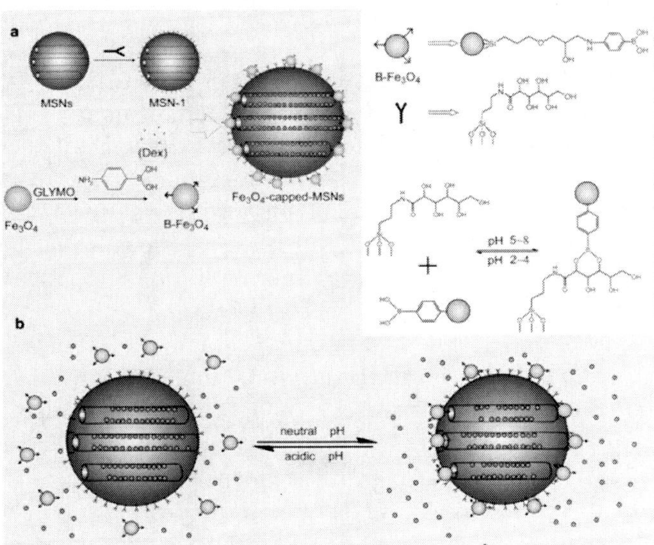

Figure 9.5 Drug release from pH-responsive mesoporous silica particles capped by magnetic nanoparticles. Reprinted with permission from Gan et al. Copyright 2011, Elsevier.

Another pH-responsive polymer, polyacrylic acid (PAA), was used to design pH-responsive silica magnetic particles for controlled release (Wu et al., 2012b). DOX was used to study the load and controlled release as a function of pH. In addition, the particles showed insignificant toxicity to HeLa and L02 cells in vitro.

9.3 Cancer Diagnosis

Before starting the treatment of any disease a good diagnosis is necessary, and in the case of cancer, the importance increases because of the complexity of the disease, having similarity of symptoms with other diseases, different types of cancer, and sometimes different protocols for the diagnosis. Cancer can be declared on the basis of different facts, the most important of which are physical examination and the results of screening tests. Generally X-ray examinations for other purpose such as an injury, showing some abnormalities, which can result in cancer, could be considered as a basis of diagnosis. For confirmation purpose, further diagnostic tests are required. After the diagnosis, the next step is staging, that is, finding how advanced the cancer is, or in other words understanding how big the tumor is or if it has spread to other neighboring tissues or not. A good and timely diagnosis can be helpful for successful cancer therapy and improves the survival rates (DeVita et al., 2000; Stocchi and Nelson, 1998). A similar term to diagnosis used in the medical field is "prognosis," which means the prediction of outcome of a disease. In cancer, it's the disease duration, chances of complications, possible outcomes, recovery aspects, recovery period, death rates, survival rates, and other possible results. At present prognosis of cancer has improved because of the modern early detection techniques and due to public awareness.

The basis of cancer diagnosis is the difference in the morphological changes of a normal tissue from that of a cancerous tissue, for example, in structure: cancer cells develop a different DNA or gene structure or acquire abnormal numbers of chromosomes as compared to normal cells. Another such a useful characteristic of cancerous cells (although not seen is all cancer tissues) is the Warburg effect (Xu et al., 2005; Dastidar and Sharma, 1989) an observation is that most cancer cells predominantly produce energy by a high rate of glycolysis followed by lactic acid fermentation in cytosol, rather than by a comparatively low rate of glycolysis followed by oxidation of pyruvate in mitochondria like most normal cells (Kim and Dang, 2006). The diagnosis of cancer can be done by various assays and tests. For the simplification of understanding all these assays and tests are divided into two broad categories, in vivo and in vitro diagnosis, and both will be reviewed here shortly.

9.3.1 In vivo Tumor Diagnosis with Special Focus on Magnetic Resonance Imaging

Tumors are of so many characteristics and can arise and develop in different ways in various body parts. This implies that a single method cannot be sufficient for accurate diagnosis of tumors. Hence, for confirmation of the presence of cancer in a person, a time-consuming series of diagnostics tests is required (Golman et al., 2006). Various modern techniques can be used for localization and diagnosis of tumors on the basis of characteristics of the malignant tissue. The different morphology of tumor tissues as compared to normal tissues provides a basis for their detection (Xie et al., 2010a). The location of a tumor in the body might be a deciding factor in the choice of techniques, but until now mostly imaging techniques used for in vivo detection have a problem of limited information about tumors, such as problems related to the background ratio in conventional tumor targeting (Weissleder et al., 1999). One of the major focuses in recent years is to apply imaging techniques for the purpose of diagnosis. General imaging studies include dynamic or static imaging and in vivo as well as in vitro functional tests. Dynamic imaging provides data about biological turnover of radioisotopes in different body compartments and organs (Hamoudeh et al., 2008). Imaging might be optical imaging, MRI, radionuclide-based imaging, computed tomography (CT), and ultrasound. In case of radionuclide imaging, signals might be obtained at low background signals and require less signal amplification. MNPs are being used as CAs for the RES. Two generally divided techniques of radionuclide imaging are positron emission tomography (PET) and single-photon emission computed tomography (SPECT). The MRI technique on the other hand is based on the principle that its signal is the precession of water hydrogen nuclei in an applied magnetic field. After application of radiofrequency pulses, the relaxation process through which the nuclei return to the original aligned state can be exploited to produce an image (Richard et al., 2008). With the progress in research, numerous materials are devised to be used in imaging techniques, especially as CAs. Due to the unique characteristics, MNPs are frequently employed in MRI to acquire important imaging analyses of several types of cancers for the diagnosis and early detection of the diseases. MRI was first developed by Paul Lauterbur in 1973 (Lauterbur, 1973), and in 1985, it got Food and Drug Administration (FDA) approval for clinical use. By this technique, information on healthy or diseased internal organs, cells, and tissues of the human body is easily obtained with the highest resolution and the fastest collection times. By employing magnetic compounds known as CAs, the contrast of images in MRI can be enhanced due to the variable distribution of the nuclei's spin density of hydrogen in the body. Gadolinium, manganese, and IONPs are used mostly as CAs, although metallic MNPs were not considered earlier to be applied in the biological field because of their chemical instability. Nonetheless, recently MNPs are being used after coatings with different materials to solve the problem of instability, for example, IONPs could be coated with oleic acid (Morales et al., 2005; Faridi-Majidi et al., 2006), amine, or dextran (Burtea et al., 2005; Pirko et al., 2004). Since MRI enhances the image contrast between the normal and diseased tissues of the body (Hong et al., 2008), therefore researchers have found that MRI based on MNPs is vastly superior to other noninvasive methods to identify lymph node metastasis from solid tumors and histologically positive lymph nodes outside the field of resection (Högemann et al., 2000). The MRI technique could be used in the detection of apoptosis in an early stage and has the advantages as compared to the radionuclide technique and magnetic resonance therapy (MRS) because of the detection of apoptotic regions with relatively high spatial resolution. Two of the best examples of IONPs include cell targeting and cell tracking. The first cellular imaging studies were performed with no functionalized IONPs for labeling leukocytes, lymphocytes, etc. (Moore et al., 2000; Schulze et al., 1995). There are other examples of usage of these MNP materials that include hyperpolarized carbon imaging for in vivo diagnosis, protease-activated imaging, etc.

Continuing the discussion on CAs, when these CAs reach in the body region, they act to reduce the relaxation times of the nuclei present in that region as compared to their natural value (Merbach and Toth, 2001). Furthermore, depending on the type, CAs are classified as positive (or T1 relaxing) and negative (or T2 relaxing) (Casula et al., 2010). The positive CAs are characterized by relatively bright spots, and the negative yield dark zones where they in the regions where these are delivered. The positive CAs are paramagnetic compounds such as gadolinium-based compounds, whereas the negative ones are in general compounds of magnetic cores synthesized from superparamagnetic ferrites. There is further differentiation of negative CAs that is based on superparamagnetic iron oxide (SPIO) particles and ultrasmall superparamagnetic iron oxide (USPIO) particles with respective hydrodynamic radius in the range of >15–20 nm and <15 nm (Fig. 9.6). Gadolinium-based MRI CAs are still extensively employed but there are drawback like that these suffer from an inherent high threshold of detection. On the other hand SPIO particles of ferrites compounds can be detected even at low concentrations of iron in the range of micromolar to exhibit sufficient sensitivity for T2-based weighted imaging. The SPIO CAs were constituted from an iron oxide–containing colloidal dispersion. Upon injection, these CAs reduce the T2 signals of absorbing tissues to provide contrast imaging. It is important to mention that the SPIO CAs have been now used successfully employed in certain cases, including enhancement of tumor image of liver (Lauterbur, 1973; Nakamura et al., 2000). Biocompatibility, economy, and easy preparation are characteristics that encourage iron oxide use in vivo but their most important property is superparamagnetism, which is a form of magnetism, and a superparamagnetic material has small magnetic points in its structure but with undefined directions so the material is not magnetic. But on the application of an external magnetic field, these small points change their direction accordingly and then the material can become magnetic. This is possible in materials made of small parts, maybe ranging in size below 30 nm. For researchers, the superparamagnetic property increases interest in use of iron oxide particles. Superparamagnetic iron oxide nanoparticles (SPIONs) are highly sensitive to magnetic resonance detection, are nontoxic, and also have been approved for clinical use as a blood agent for MRI and hence can be used as a CA to differentiate the diseased tissue from healthy ones.

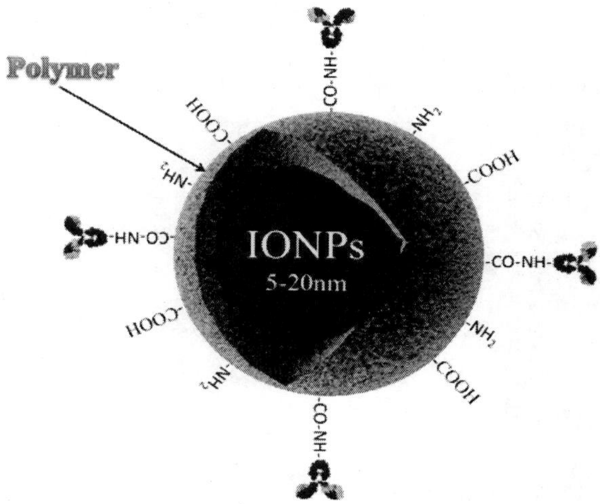

Figure 9.6 Polymer-coated IONPs specifically antitumor-functionalized for MRI diagnostics. The specificity is induced by the chemically grafted antibody-antitumor.

Although SPIO- and USPIO-based ferrites have been now approved for use in MRI as a CA, to label and detect the cells or organs of interest to acquire enhanced imaging, nanomaterials of the type of SPIO have now considered as one of the most promising CAs. Many products are either approved or in clinical trials to obtain imaging of macrophage activity for tumor staging of the liver and lymph nodes. Extensive research efforts are being undertaken to monitor migration of cells and their trafficking by using the ability to label nonphagocytic cells in culture using derivatized particles of SPIO. Such efforts are also vital to understand transplantation or transfusion in living organisms as SPIO-based CAs provide increased imaging of cellular biodistributions. Until today, most of the studies are simply demonstrated as the "proof-of-principle" type. However, to convert this into a practical solution to diagnose, a much deeper insight of imaging characteristics determined by the dynamics of in vivo cell biology is required.

Boutry et al. (2006) successfully performed the targeting of endothelial inflammatory adhesion molecule E-selection by MRI with a superparamagnetic CA in the context of in vitro and in vivo models of the inflammation. Folic acid is also used for this purpose and is grafted on magnetic particles for folate receptor targeting

(Stella et al., 2000). Cell tracking in vivo by MRI can also be done by IONPs (Arbab et al., 2004; Bulte et al., 2001). Cell tracking with a resolution reaching cell size can be obtained by MRI when cells can be loaded sufficiently with magnetic particles (Modo et al., 2005). Also to enhance the cellular uptake of magnetic IONPs, particles can be vectorized with various peptides, fragments of proteins (Zhao et al., 2001) or can be coated with dendrimers (Strable et al., 2001). Burtea et al. mentioned the cellular magnetic-linked immunosorbent assay (C-MALISA), a modified form of the enzyme-linked immunosorbent assay (ELISA), which has been developed for in vitro clinical diagnosis and can be considered as another application in MRI (Burtea et al., 2005).

9.3.2 In vitro Diagnostics

The developments of nanobiotechnology have grown spectacularly to employ the MNPs for IVD of various types of cancers (Douzeich-Eyrolles et al., 2007; DeVita et al., 2000; Stocchi and Nelson, 1998). IVD uses the MNPs to detect and quantify the complex biological specimens. Main advantages of the MNPs for IVD include biocompatibility, safety, high sensitivity, quantitative, and unique features of point-of-care, early and accurate detection (Xu et al., 2005; Dastidar and Sharma, 1989; Kim and Dang, 2006; Golman et al., 2006; Xie et al., 2010a; Weissleder et al., 1999; Hamoudeh et al., 2008; Richard et al., 2008). The MNPs have been developed as probes for practical conditions of IVD to efficiently detect the analytes of interest such as cancer cells from minute amount of clinical samples of various types such as tissues, blood, serum, and urine. Due to the novel functions of MNPs, specifically their large surface area enables attachment of high density of target-specific molecules for ultrasensitive in vitro detection. Moreover, the intrinsic magnetic characteristics of MNPs can be combined with other functionalities of optical, and electrical to offer additional opportunities to detect cancer cells. MNPs can also be used in hand held IVD magnetic resonance (DMR) systems to analyze unprocessed biological samples with great potential of real time ultrasensitivity, and quantification of small numbers of cancer diseases in future (Morales et al., 2005).

Similarly, IVD based on giant magnetoresistive (GMR) probes such as spin valve (SV) probes has also been developed to monitor the very small amount of MNPs (Faradi-Majidi et al., 2006; Burtea et al., 2005). GMR probes technology has demonstrated the diagnostics of multiplex cancer markers at sub picomolar concentration level (Pirko et al., 2004). Because of several advantages of MNP-based GMR protein assay technology is versatile for vitro diagnostics in terms of its viability from bench top to bedside (Lauterbur, 1973).

Since the MNPs take advantage of the shift in the magnetic properties to act as magnetic relaxation switches (MRS) for ultra-fast and ultrasensitive detection of cancer disease. Therefore, MRS process involves the dephasing the spins of surrounding water protons and decreasing spin–spin relaxation time (T2 relaxation time and dT2) (Hong et al., 2008). Because of this property, the MNPs are capable of detecting a variety of biological targets such as circulating tumor cells (CTCs) as well as viruses and proteins (Högemann et al., 2000; Nunn et al., 1996). Antibody-conjugated MNPs of adenovirus (ADV) and antiadenovirus 5 (ADV-5) and their incubation resulted in the formation of nano-assemblies in solution (Fig. 9.7A). This approach is useful to perform the PCR and ELISA assays for in vitro analysis of tumor cells and can find applications in point-of-care IVD in complex biological media (Moore et al., 2000). On the basis of MNPs, a miniaturized diagnostic magnetic resonance (DMR) (Schulze et al., 1995) system has been used for multiplexed, quantitative, and rapid bioanalysis for the tumor cell (Sipe et al., 1999). One advantages of the DMR system is its capability of detection limit approach even to a single cell and thus indicates promising applications to analyze tumor cells. In vitro cytotoxicity studies of the multifunctional vesicles toward prostate tumor cells and antiproliferative effects have been reported (Boutry et al., 2006). As shown in Fig. 9.7B,C, incubation of PC3 cells with $scAb_{PSCA}$-Dtxl/SPIO nanoparticles resulted in relatively darker regions in the T2-weighted intensity as compared to polymeric vesicles that do not have $scAb_{PSCA}$. This suggests that the scAbP-SCA-Dtxl/SPIO nanoparticles can be effective as MRI CAs for prostate-targeted diagnostics as well as therapeutics effects.

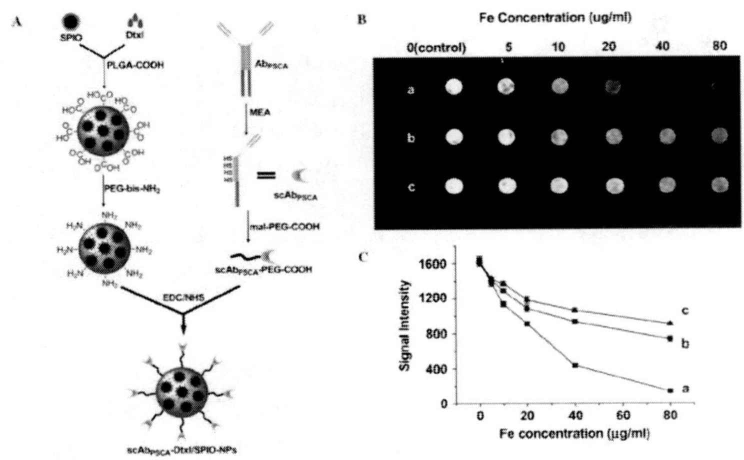

Figure 9.7 (A) Schematic representation of the formulation of $scAb_{PSCA}$-Dtxl/SPIO nanoparticles, (B) T2-weighted imaging, and (C) the MRI signal intensity of PC3 cells (1×10^6) after 2 h incubation with (a) $scAb_{PSCA}$-Dtxl/SPIO nanoparticles and (b) PEG-PLGA-Dtxl/SPIO nanoparticles using a 1.5 T MRI machine. Reproduced with permission from Ling et al. (2011). Copyright © 2011, Elsevier.

MNPs have also been employed to carry out IVD and in vitro development of novel biomarkers for various types of cancers. In a novel work, CSNIDARAC peptide-conjugated liposomal DOX (Lipo-DOX) was synthesized with a binding affinity for lung tumors identified by in vitro screening of a phage display peptide library (Chang et al., 2010a). In another work, human breast cancer was expressed using the HER2/neu protein (Stella et al., 2000). Examination of the expression levels of the HER2/neu protein using commercially available streptavidin-conjugated SPIO of MNPs has also been demonstrated (Arbab et al., 2004). In vitro T2-weighted MRI contrast analysis was used to determine the levels of HER2/neu expression on the membrane of breast cells. This noninvasive imaging method found to be highly sensitive with a detection limit in the range of 5×10^4 receptors per cell and in comparison to the general expression levels of around 10^5 to 10^6 per cell of the HER2/neu receptor.

Serum albumins were linked to oleic acid–coated manganese-doped magnetic-engineered iron oxide (MnMEIO) nanoparticles and this type of platform was also used to develop all-in-one target cell-specific MNPs for imaging and delivery of short interfering RNA (siRNA) (Bulte et al., 2001). To target $\alpha_v\beta_3$-integrin-positive cancer cells and to inhibit specific protein expression, SH-PEG-RGD and SH-siRNA-Cy5, respectively, were conjugated to MnMEIO via an SPDP crosslinker. In vitro evaluation of MRI and gene suppression demonstrated that these multimodal nanoparticle systems selectively silenced genes, confirming their theranostic behavior. Several in vitro studies have reported maximization of a nanoparticle system targeting efficacy due to the critical role of the targeting ligand density on the nanoparticle surfaces. This is an important consideration since the higher concentrations of the targeting ligands on the nanoparticles often increase cellular uptake (Modo et al., 2005; Zhao et al., 2001). In one of the related works, 0%, 5%, or 16% RGD ligand–immobilized DOX-loaded polymeric micelles were developed to investigate their in vitro targeting and therapeutic properties (Strable et al., 2001). The 16% RGD ligand–bearing nanoparticles displayed the highest cellular uptake and cytotoxicity in $\alpha_v\beta_3$-integrin-overexpressing SLK cells.

It has become increasingly clear that the tumor-targeting properties of nanoparticles optimized in vitro can be helpful to predict in vivo performance. In one of the works, identification was made for the maximally targeted and maximally stealth surface engineering conditions for in vitro and in vivo performance using PSMA-targeted aptamer-conjugated Dtxl-loaded self-assembled nanoparticles (Espinosa et al., 2003). Increasing the ligand density to 5% significantly increased the nanoparticle uptake by the target cells (LNCaP), whereas further increase in aptamer density modestly increased the nanoparticle uptake. These results indicated that the optimum ligand density for PSMA-specific endocytosis in vitro was 10–80 nmol aptamer per μmol nanoparticle. LNCaP xenograft mouse models injected with the targeted nanoparticles showed that increasing the aptamer density from 0% to 5% significantly increased nanoparticle retention in tumors, but the retention decreased for aptamer densities beyond 10%. The authors suggested that higher aptamer densities may have reduced the nanoparticle stealth properties, resulting in rapid clearance by the liver.

The above mentioned IVD examples describe the important role of various MNPs. However, another desirable IVD area, which likely will lead to novel applications in biomedicine related to cancer, requires the development of multifunctional MNPs to enable detection of the biological samples using multiplex detection modalities (Trevor et al., 2009; Brunton et al., 2010). Intrinsic diagnostic capabilities in the context of magnetism-related applications can be integrated to include polymers, metals oxides, enzymes, optical, pH, thermal, or other desirable functionalities to tailor them with the required functions for the cancer-specific in vitro delivery of ultrasensitive molecular diagnostics and therapeutics (Harvey, 2011). This strategy of multivalent targeting can significantly enhance the binding affinity of MNPs toward a target cell (Jain, 1987b). This is possible due to the large surface area of MNPs that allows the incorporation of diverse and large numbers of therapeutic and diagnostic ligands on a single particle surface area. Surface modifications of MNPs with targeting moieties can be introduced through conjugation, such as antibodies, fluorescence dyes, peptides, molecules, genes, or even drugs, to deliver multimodal functionalities (Brigger et al., 2002; Jain et al., 2005).

9.3.3 Treatment of Cancer and Classification of Cancer Therapeutic Agents

Various methods have been discovered and many more are in the process of discovery for the treatment of cancer. Major treatment techniques are based on chemotherapy, hormonal therapy, radiotherapy, and surgery (Espinosa et al., 2003) (Fig. 9.8). All these techniques provide physicians various options so they can choose a suitable method according to the stage and type of disease and more importantly patient condition. There is a wide range of drugs being used for cancer treatment and with every passing day numerous new drugs are being added to the list. A need of proper classification was felt, and the proper classification of all anticancer agents is a big task because every drug differs from another in its design, reaction mechanism, structural classification, and target tissues. An effort has been made to classify the most commonly used drugs for better understanding and this classification is based on different literature resources (Trevor et al., 2009; Brunton et al., 2010; Harvey, 2011) (Table 9.2).

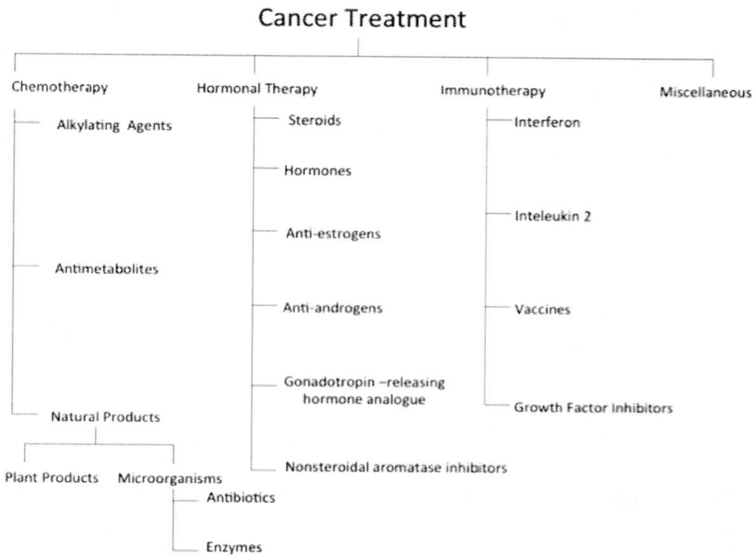

Figure 9.8 A chart for the classification of cancer therapeutic agents.

Considering the recent therapies being used for cancer, they might be divided broadly in two parts, conventional therapies and nanoparticle-based therapy.

9.3.3.1 Conventional therapies

Various modalities to treat cancerous tissues include radiation therapy, photodynamic therapy, surgery, chemotherapy, and various traditional therapies used in different regions of the world. Chemotherapy was considered as the most reliable among all. But the anticancer drugs can fail to kill cancer cells for various reasons. One of them can be the transport of drugs to the cancerous cells. The transport of an anticancer drug is normally governed by physiological and physicochemical properties of the target cell and of the drug itself (Jain et al., 2005). These properties include pressure, charge, size, configuration, electrochemical properties, hydrophilicity, etc. Drugs are usually given systemically and are therefore subject to variations in absorption, metabolism, and delivery to target tissues that can be specific to individual patients. Thus for the therapeutic agents' delivery to the tumor cells, the problems that should be addressed are:

Table 9.2 Representation of anticancer treatment options with examples

Class	Subclass	Sub-subclasses with examples		
Chemotherapy	Alkylating agents	**Nitrogen mustards**, e.g., mechlorethamine, melphalan, cyclophosphamide, ifosfamide, chlorambucil		
		Nitrosoureas, e.g., carmustine, lomustine		
		Alkylsulfonates, e.g., busulphan		
		Ethyleneimines, e.g., thiotepa		
		Triazenes, e.g., dacarbazine, temozolomide		
		Methyl hydrazines, e.g., procubaine		
		Platinum voordination complexes, e.g., cisplatin, carboplatin, oxaliplatin		
	Antimetabolites	**Folate antagonists**, e.g., methotrexate, trimetrexate		
		Purine antagonists, e.g., 6-mercaptopurine (6-mp),6-thioguanine (6-TG)		
		Pyrimidine antagonists, e.g., 5-fluorouracil (5-FU), cytarabibe, capecitabine, gemicitabine, arabinosyleystosine, decitabine		
	Natural products	Plants	**Vinca alkaloids**, e.g., vincristine, vinblatine	
			Taxanes, e.g., paclitaxel, docetaxel	
			Epipodophyllotoxins, e.g., etoposide, teniposide	
			Camptothecins, e.g., topotecan, irinotecan	
		Microorganisms	Antibiotics	**Anthracyclines**, e.g., daunorubicin, doxorubicin hydrochloride, doxorubicin hydrochloride pegylated, liposomal epirubicin, hydrochloride, idarubicin hydrochloride, valrubicin

Class	Subclass	Sub-subclasses with examples		
				Actinomycins, e.g., dactinomycin
				Various cytotoxic antibiotics, e.g., bleomycin sulfate, mitomycin, mitotane, mitoxantrone hydrochloride
			Enzymes, e.g., L-asparaginase	
Hormonal therapy	**Steroids,** e.g., prednisolone, dexamethasone			
	Hormones	**Estrogens**, e.g., diethylstilbestrol sodium diphosphate		
		Progestogens, e.g., medroxyprogesterone acetate, megestrol acetate		
	Antiestrogens, e.g., tamoxifen citrate			
	Antiandrogens, e.g., bicalutamide, cyproterone acetate, flutamide, nilutamide			
	Gonadotropin-releasing hormone analogs, e.g., Buserelin acetate, Goserelin acetate, leuprolide acetate			
	Nonsteroidal aromatase inhibitors, e.g., anastrozole, exemestane, letrozole, aminoglutethimide			
Immunotherapy	**Interferons**, e.g., type 1, 2, and 3 **Interleukin 2**, e.g., aldesleukin			
	Vaccines, e.g., melanoma theraccine			
	Growth factor inhibitors, e.g., cituximab, gefitinib, erlotinib, bevacizumab, rituximab, imatinib mesylate, dasatinib			
Miscellaneous anticancer agents	**Retinoic acid derivatives**, e.g., all–*trans* retinoic acid			
	Others, e.g., porfimer sodium, amifostine, arsenic trioxide, hydroxyurea			

(i) Drug resistance at the tumor level (noncellular-based mechanisms, for example, a decreased uptake of water-soluble drugs such as folate antagonists, nucleoside analogues, and cisplatin, which require transporters to enter cells)
(ii) Drug resistance at the cellular level (cellular-based mechanisms, for example, various changes in cells that affect the cytotoxic capacity of drugs to kill cells, including alterations in cell cycle, increased repair of DNA damage, reduced apoptosis, and altered metabolism of drugs)
(iii) Pharmacokinetic properties of the anticancer agent in the body (Brigger et al., 2002)

In the last decade the research was inclined toward drug targeting, which is a very good idea, and numerous studies are under process in oncology. Targeting of the cancerous tissues within body is a unique approach in cancer treatment and it was also desirable because mostly anticancer agents demonstrate nonspecific toxicities, significantly limiting their therapeutic efficiencies. There are some factors that can affect the process by modifying the physical and physicochemical properties of the drug and can hinder for the good targeting. The important factors that can be considered for good targeting include size, shape, density, and charge to affect the transport of the particles through the body, such as to cross the biological membranes.

With extensive research in oncology, the transplantation phenomenon got fame and is still actively used. Bone marrow transplant and blood transplantation are novel applications of this technique for the treatment of cancerous cells and blood cancer.

Gene therapy is another therapeutic technique for cancer where genes are introduced in the cells to fight against a specific disease, and with the division of the cells, the genes move to other cells and the disease can be mitigated.

Hyperthermia includes the heating of cancerous tissues to a temperature above the normal body temperature, so cell death is caused by heating. It can be local, regional, or whole-body heating up to 45°C, thus killing a sufficient number of cells. However, it can be used as a combined therapy with radiation therapy (Johannsen et al., 2010).

Microparticles were introduced and anticancer agents/drugs were converted to microparticle formulations for better efficacy. The concept of microparticles arose due to the fact that smaller-sized particles allow a higher volume-to-area ratio and thus permit higher absorption of the drugs in a specific tissue. In the last decade the microparticle concept is also applied for hyperthermia, radiation therapy, etc.

9.3.3.2 Nanoparticle-based therapies

The struggle for obtaining higher adsorption of pharmacological agents in the body tissues achieved nanosized agents. The nanoparticle evaluation opened new horizons for all the fields, including drug delivery, and brought about the term "nanomedicines." The nanoparticles are used as diagnostic tools or carriers for drug or gene delivery. MNPs with magnetic characteristics were also used for vectorization of the drugs; however, the use of MNPs for drug delivery vehicles must address issues such as drug-loading capacity, desired release profile, aqueous dispersion stability, biocompatibility with cells and tissues, and retention of magnetic properties after modification with polymers or chemical reaction (Jain et al., 2005) (Fig. 9.9).

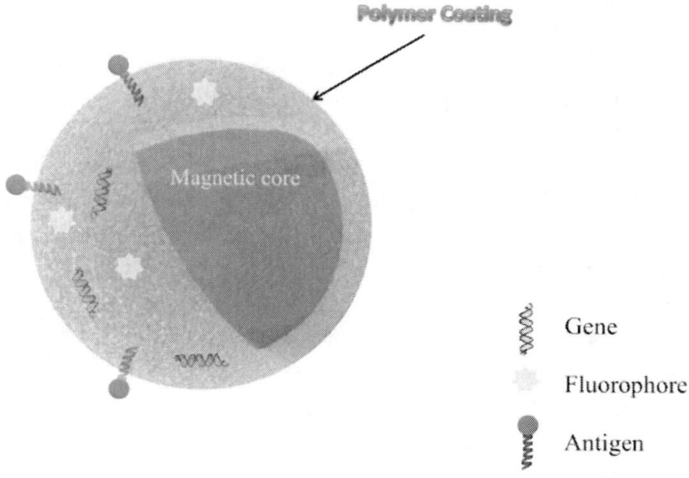

Figure 9.9 Schematic illustration of multifunctional polymer-coated magnetic particles and surface-functionalized particles for diagnostic and therapy applications.

9.4 Theranostics Application of Nanoparticles

"Theranostics" is a new term coined for drugs or methods that are used for diagnosis and treatment at the same time (Lee et al., 2008a; Warner, 2004). Diagnostics and therapy combination leads to theranostics. Theranostic nanomedicine is an important area of nanotechnology and has important implementations for the development of the next generation of drugs. This term can be defined as "an integrated nanotherapeutic system that can diagnose, deliver targeted therapy and monitor the response to therapy." This is a useful idea in designing nanotechnology-based imaging CAs and imaging-guided therapeutics (Sumer and Gao, 2008). It is assumed that this combined technique will result in the acceleration of drug development and improvement of disease management, reducing risks and cost. The diseases that are the main cause of morbidity and mortality at present in the world are in need of such type of new trials for quick diagnosis and treatment, because some of these diseases require rapid diagnosis and long treatment duration, so theranostics can be proved as a revolutionary technique in this specific aspect. One of these diseases is cancer and by chance the earlier research on theranostics is mostly inclined toward oncology. The important consideration in the creation of a theranostic is the requirement of a wide knowledge and deep understanding of mechanisms of both detection and therapy. This knowledge requirement includes an understanding of molecular mechanisms, diagnostic strategies, therapeutic efficiency, toxic and side effects of the materials, and nanoparticle preparation techniques for the dual purpose of diagnosis and therapy. Hence, a suitable theranostic is expected for all the diseases, especially for cancer, in the future, although a lot of work still remains to be done. A general schematic of a nanoparticle to present the concept of theranostics for cancer disease is given in Fig. 9.10.

The aim of theranostics is to prepare a nanoparticulate system that can serve simultaneously the purpose of diagnosis and therapy (Fig. 9.10). The factors influencing all processes starting from preparation of nanoparticles until the elimination of the metabolites of active molecules and other used materials are important to be considered. A study of these factors requires a keen interest, full concentration, and deep investigation of every step before employing it. In a broader sense, these factors can be the compatibility of the chemicals used, preparation conditions, formulations/modifications according to the route of administration selected, biocompatibility and biodegradability of the materials, toxicity of the materials and their metabolic products, pharmacokinetics and pharmacodynamic parameter evaluation, and, at the end, the benefit/disadvantage ratio of all processes.

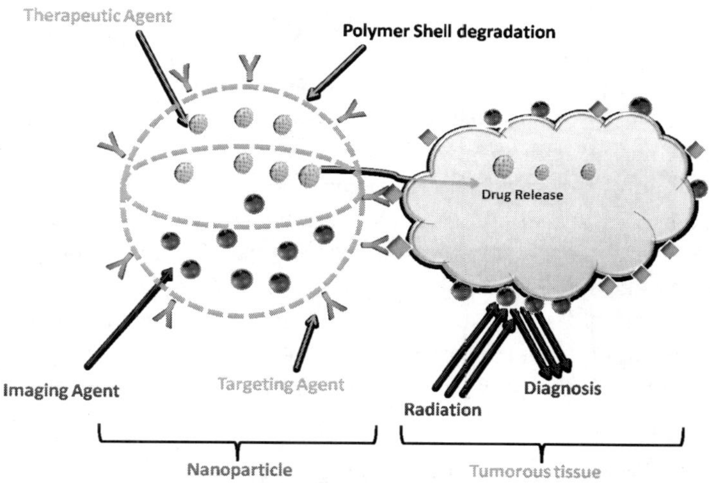

Figure 9.10 Schematic of a nanoparticle to present the concept of theranostics for the diagnosis and therapy of cancer disease. Reprinted with permission from Ahmed et al. Copyright © 2012 Elsevier Ltd. All rights reserved.

The nanotechnology involvement made it easy to realize the theranostic concept. Minute-sized particles containing two different agents have been elaborated. Therapeutic and diagnostic agents have been prepared in nanosized particles and used for theranostic applications. The magnetic particles helped in the realization of this concept, having magnetic properties that can be exploited as diagnostic tools. The choice of the diagnostic tool depends upon the target site. Normally the magnetic particles are involved in imaging of the diseased tissues in the body. Nevertheless, they are involved in other biomedical applications like treatment via hyperthermia (killing of diseased cells using heat), drug delivery, IVD techniques (like bioassays), etc. Typical examples include works that include manganese oxide nanoparticles carrying both drugs and CAs (Liong et al., 2008) and silica nanoparticles with magnetic and fluorescent tags (Insin et al., 2008). In the last decade, a lot of attention focused on magnetoplasmonic nanoparticle preparation by combining both the magnetic component and metal nanoparticles or shells (Lim et al., 2009; Moriggi et al., 2009; Levin et al., 2009; van Berkel et al., 2009; Zhao and Gao, 2009; Kamei et al., 2009; Zeleňáková et al., 2008). As mentioned above, MRI is the mostly employed diagnostic technique using magnetic materials as CAs for the visualization/differentiation of diseased cells/tissues from healthy cells/tissues. The researchers are using different CAs and especially particles having magnetic properties are being well researched, for example, gadolinium, iron oxide, gold, silver, and other metals are in the process to find a suitable one with less toxicological effects but iron oxide particles are of great interest because of superparamagnetic activity and good research, as mentioned earlier. Cancer being the main focus of theranostics, nanoparticle agents incorporating the possibility of cancer diagnostics and therapeutics (Li et al., 2004; Morel et al., 2008; Souza et al., 2008) had been developed.

9.5 Toxicities of Nanomaterials

Perhaps the most critical parameter to determine the application of nanomaterials, including nanoparticles, is the understanding of the fate of these particles in the body. In general pharmacokinetic and pharmacodynamic studies are vital but special focus should be on the possible toxicity effects of the nanomaterials themselves. Attention should be paid on the material being added for diagnostic purposes, as a therapeutic agent is normally well studied for pharmacological effects.

In nanotechnology application, the toxicology profile of the nanomaterials being used is also expanded to the environment after their removal from the body or before administration. Thus a broader division of nanomaterial toxicity includes environmental, genetic, and cyto- and pharmacokinetic toxicity. A broader aspect of different types of toxicological effects can be found elsewhere

(Buzea et al., 2008). Here, an attempt is made to discuss in general the salient features of nanoparticles and their toxicity (Fig. 9.11).

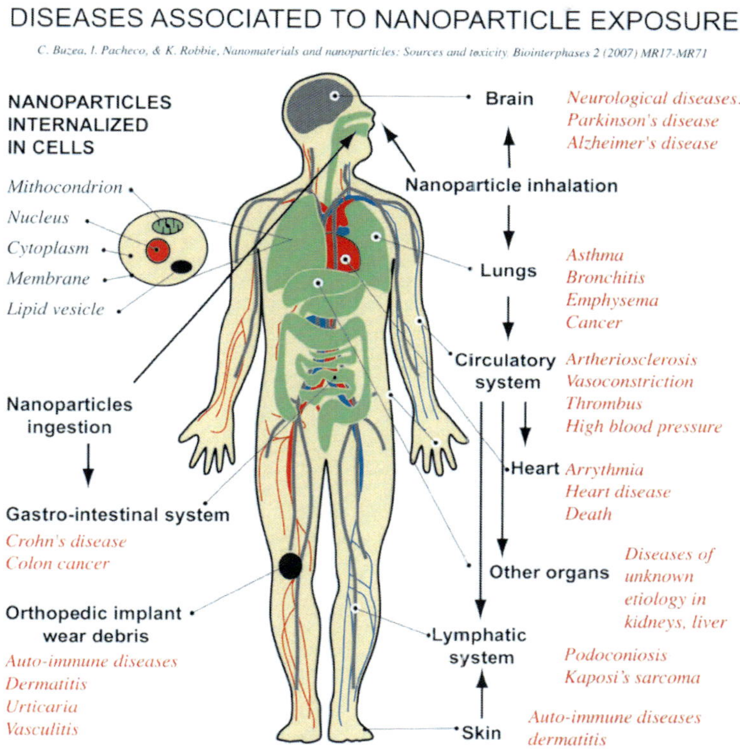

Figure 9.11 Different forms of nanoparticle toxicity. Reprinted with permission from Buzea et al. Copyright © 2007, American Vacuum Society.

The main source of the pharmacokinetic toxicity is the modifications brought to the nanoparticles for enhancement of their bioavailability or delivery to the target. It could be possible by using the minute size of the nanoparticles that will circulate for a long time in the main circulation. However, stealth mechanisms are mainly responsible. The charges on the particles are also an important factor to be considered and could cause the long circulation in the blood. One such study was done by Levchenko et al. (2002) on liposomes administered to rats, mentioning the effects and reasons of long circulation.

9.6 Conclusion and Future Perspectives

The application of nanotechnology in the biomedical field accelerated the research in its various subfields, especially in diagnostic mechanisms, drug and gene delivery, etc. Nanomaterials of various origins have been applied in these subfields. Cancer being a fatal disease received nanotechnology application at the earliest stage. As diagnosis is important for any disease, nanomaterials are applied in the diagnosis of cancer, both in vivo and in vitro. For treatment, therapy consisting of nanoparticles has been discussed along with various nanotechnology tools that are being used in oncology. Progress in research developed a new term of "theranostics" relating to diagnosis and therapy at the same time. Various materials are now applied in different ways after the arrival of this specific subfield of biomedical applications. Another such term for dual functionalities has been coined but not related to cancer. It is "cosmetotherapy," where nanoparticles can be used for therapy and cosmetics simultaneously. The particles contain the moieties for therapeutic effects and could be used or the cosmetics. Focusing on the theranostic aspects of nanotechnology applications in cancer, the diagnostic part in vivo is mostly dedicated to the imaging techniques such as MRI in combination with other techniques. The nanomaterials are involved in the enhancement of contrast in these imaging mechanisms. But with the progress and development in this field, it would be hoped that theranostics would bring about an important breakthrough in the diagnosis and treatment of cancer. The diagnosis and therapy will move toward monitoring of response and hence head toward personalized medication.

Chapter 10

Micelles, Liposomes, Bubble Liposomes, the Layer-by-Layer Approach, and Nanocapsules

Joan Estelrich[a] **and Alexander van Herk**[b]

[a]*Departament de Fisicoquímica, Facultat de Farmàcia, Universitat de Barcelona, Avda.Joan XXIII, 08028Barcelona, Catalonia,Spain*

[b]*Institute for Chemical and Engineering Sciences, 1 Pesek Road, Jurong Island, 627833 Singapore*

joanestelrich@ub.edu; alexander_herk@ices.a-star.edu.sg

10.1 Micelles

A micelle is a particle of colloidal dimensions (typically with a diameter around 5 nm) that exists in equilibrium with the molecules or ions in solution from which it is formed. The segregation of the hydrophobic parts of the surfactant molecules from the aqueous phase is the driving force for the formation of micelles. To retain/solubilize actives inside micelles in aqueous systems the active itself should be of considerable hydrophobicity.

Controlled Release Systems: Advances in Nanobottles and Active Nanoparticles
Edited by Alexander van Herk, Jacqueline Forcada, and Giorgia Pastorin

ISBN 978-981-4613-21-7 (Hardcover), 978-981-4613-22-4 (eBook)
www.panstanford.com

10.1.1 Small-Molecule Micelles

Micelles are dynamic structures where the average residence time of a surfactant molecule in the actual micellar structure is in the order of microseconds. Diffusion of a hydrophobic active in the interior of a micelle to the exterior water phase and vice versa also takes place in microseconds (Grätzel and Thomas, 1973). This means that micelles made out of low-molecular-weight surfactant molecules like sodium dodecyl sulfate (SDS) or cetyl trimethylammonium bromide (CTAB) cannot be regarded as static containers for hydrophobic actives. The active can obviously be solubilized with micelles but, for example, cannot be protected against hydrolysis or interaction with detrimental molecules in the bulk phase.

10.1.2 Block-Copolymer Micelles

With the advent of controlled radical polymerization many studies appeared on the formation and use of block copolymer–based micelles, including pH- and temperature-responsive micelles (Han et al., 2013) and multicompartment block copolymer micelles (Moughton et al., 2012). Compared to micelles formed by small surfactant molecules, polymeric micelles are generally more stable and have a slower rate of exchange of the molecules, allowing retention of actives for a longer period of time (Kataoka et al., 2012).

In the application of drug delivery the choice of the hydrophilic block often is for poly(ethylene glycol) (PEG) for its known effects in avoiding opsonization, followed by engulfing the micelles and transporting them to the liver or spleen for degradation and excretion. PEGylation of nanostructures interferes with opsonin adsorption and extends the blood circulation times.

The ability to control the period of time over which active release takes place is rather limited with micelles. Usually 100% release of actives from block-copolymer micelles is achieved within hours (Hong et al., 2005). An approach to delay the release is by crosslinking the micelles (Hong et al., 2005). Although formally these structures can no longer be called micelles (no free-molecular-block copolymers in the aqueous phase anymore), these nanoparticles can show release rates delayed to days instead of hours. For release of actives over longer periods of time micelles are less suitable.

10.1.3 Preparation of Micelles

In principle micelles are formed spontaneously on dissolution of the surfactant above the critical micelle concentration (CMC).For SDS at 60°C the CMC in pure water is around 10^{-2} mol L^{-1}. The CMC is strongly influenced by the presence of salts but also by the presence of hydrophobic molecules that solubilize inside the hydrophobic domain of the micelles (see, e.g., Janssen et al., 1993). For block-copolymer micelles the CMC lies at much lower concentrations.

However, two effects need to be mentioned.

- Grätzel and Thomas (1973) observed a slow equilibration of simple micelles made out of SDS. An increase in order within the micelles with time, on time scales of hours. This is similar to the much more pronounced "hydration" times of vesicles and liposomes (see the next section).
- In many cases block copolymers (including di-, tri, and gradient block copolymers) can show several morphologies simultaneously, like micelles, nanowires and vesicles. So visualization of the morphologies by (cryo)-TEM is essential to assure that indeed only micelles are present (see, for example, Huang and Pan, 2010).

10.2 Liposomes

Liposomes, first described by Bangham et al. in 1965, are a kind of colloidal system formed by the self-assembly of amphiphilic lipids, which self-associate to enclose an aqueous compartment surrounded by one, or multiple concentric lipid bilayers (Bangham et al., 1965). The assembly into closed bilayered structures is mainly driven by the hydrophobic effect which organizes amphiphilic molecules so as to minimize entropically unfavorable interactions between hydrophobic acylchains and surrounding aqueous medium (Tanford, 1980). This effect is further settled by various intermolecular forces as electrostatic interactions, hydrogen bonding, as well as van der Waals and dispersion forces (Lasic, 1993). Liposome properties vary substantially with lipid composition, size, surface charge, and method of preparation. Once formed, liposomes are typically 20 nm to 10 μm in diameter with the bilayer being about 4 to 5 nm thick (Jesorka and Orwar, 2008). Their charge and surface properties can be tuned simply by varying the composition of phospholipids or adding new components to the lipid mixture during the preparation (Sawant and Torchilin, 2010). Because of the presence of hydrophobic and hydrophilic domains, both water and lipid soluble drugs can be successfully entrapped into the liposomes. The lipophilic drugs are entrapped within the membrane whereas water soluble drugs can be encapsulated in the aqueous core (Torchilin, 2005).

Some of the major drawbacks of liposomes used as carriers for therapeutically active compounds are the rapid degradation due to the macrophage–phagocyte system (MPS). As with all foreign colloidal particles, liposomes are quickly recognized as "nonself" and taken up by the cells of the MPS, chiefly macrophages in the liver and spleen. This fact produces the inability to achieve sustained drug delivery over a prolonged period of time (Torchilin, 2005).

10.2.1 Classification of Liposomes

According to their properties, the most usual liposomes can be classified into the following types:

(i) Conventional liposomes
(ii) Stimuli-responsive liposomes
(iii) Cationic liposomes
(iv) Targeted liposomes
(v) Long-circulating liposomes

Conventional liposomes are liposomes with nonspecific reactivity. *Stimuli-responsive liposomes* are reactive liposomes that can change their structure under the action of an external stimulus. In this way, a number of different stimuli have been exploited to trigger the release of the encapsulated cargo from liposomes, including temperature (Ponce et al., 2006; Han et al., 2006), pH (Simões et al., 2004), light (Troutman et al., 2009), redox potential (Ong et al., 2008), near-infrared (Volodkin et al., 2009), and ultrasound (US) (Schroeder et al., 2009), each with its specific advantage, depending on the application (Sawantand Torchilin, 2010). *Cationic liposomes* are liposomes that disintegrate and restructure during complexation with nucleic acids (Lasic, 1997; Smith, 1997). *Targeted liposomes* are liposomes that have been functionalized with different types of targeting moieties such antibodies, peptide, oligopeptide, and receptors (Martin et al., 1990; Storm, 1999). *Long-circulating liposomes*, also known as sterically stabilized liposomes or stealth liposomes, are liposomes that incorporate cholesterol (CHOL) and, mainly, biocompatible, hydrophilic polymers with a flexible main chain, such as PEG (Lasic, 1997; Needham et al., 1992).

Other types of multicompartmentalized assemblies based on liposomes have been reported: double liposomes (Ebato et al., 2003), multivesicular vesicles (Talsma et al., 1987) or verosomes, where small unilamellar liposomes are entrapped within a large liposome (Kisak et al., 2004), and capsosomes, which unite two fundamentally different systems—polymer capsules and liposomes (Chandrawati and Caruso, 2012). When iron oxide nanoparticles are encapsulated in liposomes, the general term "magnetoliposomes" is used for the resulting structures (Margolis et al., 1983; Nappini et al., 2010). This kind of liposomes can be sensitive to an external magnetic field (Fattahi et al., 2011; García-Jimeno et al., 2012; Amstad et al., 2011).

Liposomes are versatile systems that can combine several multifunctional properties: they can encapsulate drugs or magnetic particles, can bear a fluorescent probe in the outer monolayer, can be protected by chains of polyethylene glycol, and can be targeted to specific receptors by the presence of a ligand at the terminal end of the chain of polyethylene glycol (Fig. 10.1).

As vesicle size is a critical parameter in determining many of the properties of liposomes (e.g., their circulation half-life), and both size and number of bilayers influences the extent of drug encapsulation within liposomes, a classification based on size and number of bilayers is also used. In this case, liposomes are classified into four classes, and the different types of liposomes are referred to by a three-letter acronym:

(i) Multilamellar vesicles (MLVs), with an onion-like layered membrane with several bilayers
(ii) Large unilamellar vesicles (LUVs), with a unique bilayer and size between 100 and 1000 nm
(iii) Small unilamellar vesicles (SUVs), with a bilayer and a diameter less than 100 nm
(iv) Giant unilamellar vesicles (GUVs), larger than 1000 nm

Figure 10.1 Multifunctional liposome showing a unique bilayer.

10.2.2 Materials Used in the Preparation of Liposomes

As indicated previously, the main component of liposomes is amphiphilic lipids with the ability to self-assemble in solution into more or less complex aggregates, provided their concentration exceeds a certain CMC, which depends upon their chemical structure and the ions present (Hunter, 1987; Hiemenz and Rajagopalan, 1997). A typical CMC value for bilayer-forming lipids ranges from 10^{-10} to 10^{-6} mol L^{-1}, whereas the CMC of ordinary amphiphiles is 4 to 5 orders of magnitude bigger, meaning that the water solubility of these bilayer-forming lipids is extremely low (Israelachvili, 1991). In an aqueous environment the hydrophobic segments of the amphiphiles will tend to attract one another, whereas the head groups will be attracted to water molecules and may also be charged, both factors leading to repulsion. A combination of these attractive and repulsive forces leads to an energy minimum at an area per molecule, A_o, called the optimal surface area per molecule.

The nature of the structures that form when amphiphiles assemble together is determined to a large extent by the packing parameter P (Israelachvili et al., 1980):

$$P = \frac{v}{A_o \times l_c}$$

where v is the volume of the hydrophobic part of the amphiphile, assumed to be fluid and incompressible, A is the optimal headgroup area, and l_c is the critical chain length (a semi-empirical parameter somewhat shorter than the length of the extended chains).

For a given molecule in a given solvent both the minimum interaction energy and the packing parameter will be fixed, but there is not usually one unique shape that satisfies these requirements. Several structures are possible so the determining factor will be the entropy change of the system on aggregation, and this will normally mean that the smallest possible structure consistent with the energy and structural parameters of the amphiphile will be favored (Israelachvili et al., 1980). The basic bilayer phase consists of a lamellar liquid crystal made up of bilayers packed in large stacks and separated by a solvent, usually water. In an ideal situation, the bilayers would extend indefinitely in the lateral direction. In reality, finite samples show defects, and bilayers tend to close up on themselves in onion like structures, the liposomes (Evans and Wennerström, 1994).

The amphiphiles that can form curved bilayers are molecules with large headgroups and large tail volumes giving a truncated, almost cylindrical, cone shape with a packing parameter from 0.5 to 1. Examples of these amphiphiles are the phospholipids and dialkylammonium surfactants (e.g., the didodecyldimethylammonium bromide). The most common phospholipid is phosphatidylcholine (PC), which is a molecule in which a glycerol bridge links a pair of hydrophobic acyl hydrocarbon chains with a phosphate group which carries a hydrophilic head group, the choline. Other headgroups present in naturally occurring phospholipids are ethanolamine (in phosphatidylethanolamine, PE), glycerol (in phosphatidylglycerol, PG), and serine (in phosphatidylserine, PS). The fatty acids in naturally occurring phospholipids usually have an even number of carbon atoms with 16 and 18 carbon atoms dominating and may be saturated or unsaturated (Cullis et al., 1996). Unsaturation is mostly in the *cis* conformation and, where present, is generally found in the acyl chain attached to the central carbon of glycerol. Even when these two chains are identical, the tilted orientation of the glycerol group to which they are attached means that one chain extends further from the phosphate group than the other, and this effect may be larger with the many phospholipids where the two chains are different. PC can be derived from both natural and synthetic sources. They are readily extracted from egg yolk and soya bean. PC from natural sources is, in fact, a mixture of phosphatidylcholines, each with chains of different lengths and varying degrees of unsaturation. PC from plant sources has a high level of polyunsaturation in the fatty acyl chains, while that from mammalian sources contains a higher proportion of fully saturated chains (Betageri et al., 1993).

Phospholipids bilayer undergo a melting-like phase transition at a temperature-designed transition temperature, T_m. Below T_m, saturated lipids in the bilayer exist in the gel or "solid state" with linear all-trans chains. Above T_m, the bilayer exists in the more disordered liquid crystalline or "liquid state" with several gauche C–C bonds in each carbon chain. The transition is accompanied by decrease in membrane thickness along with enhanced membrane permeability. Values of T_m increase with chain length and decrease with chain unsaturation (Menger et al., 2005).

CHOL is usually present in the composition of many liposomes. It does not, by itself, form bilayer structures, but it can be incorporated into phospholipid membranes in high concentrations (up to 1:1, molar ratio). CHOL, as an amphipatic molecule, inserts into the membrane with its hydroxyl group oriented toward the aqueous surface and with the aliphatic chain aligned parallel to the acyl chains in the center of the bilayer. The effect of CHOL appears to be a reduction in the fluidity of the acyl chains in the liquid-crystal state and an increase in their fluidity in the gel state. CHOL produces a condensation effect; upon the addition of CHOL, the area per lipid decreases more than one would expect for ideal mixing. This effect is maximal close to the T_m (de Meyer and Smit, 2009). Mixtures of phospholipids and CHOL are also able to form liquid-ordered phases (Marsch, 2010).

The predominant component of liposomes is usually a neutral phospholipid such as PC. CHOL, up to 40–50 mol.% is often included to provide greater stability in biological fluids. A charged species may be added (normally 5%–20%) to prevent aggregation. Common natural acidic lipids include PS, PG, phosphatidylinositol (PI), phosphatidic acid (PA), and cardiolipin. Small amounts of antioxidants such as α-tocopherol are also included when polyunsaturated lipids are used.

For an overview of liposome preparation methods see Chapter 4 in Part I of this book.

10.2.3 Characterization of Liposomes

To assess the liposome quality and to obtain quantitative measurements that allow the comparison among different batches of liposomes five main parameters (four if the lipids used are neutral) may be monitored: size and size distribution, surface charge, lamellarity, quantitative lipid analysis, and encapsulation efficiency (Edwards and Baeumner, 2006).

Particle mean size and size distribution are the most important parameters to consider when describing any liposome dispersion (Jesorkaand Orwar, 2008). These parameters are usually determined by dynamic light scattering, but transmission electron microscopy at cryogenic temperature (cryo-TEM), freeze-fracture TEM, static light scattering, field-flow fractionation, and gel chromatography can be also used (New, 1990).

When some of the lipid constituents of the bilayer carry charge (due to the adsorption of ions or dissociation of their ionizable groups), the surface charge can be determined by Doppler electrophoresis via measuring the electrophoretic mobility of liposomes subject to an external electric field. The electrophoretic mobility is related to the ζ-potential, and as ζ-potential is related to the colloidal stability, the determination of the ζ-potential has a double aim: as a characterization tool and to predict stability (Hunter, 1988).

The lamellarity of liposomes is determined by freeze-fracture electron microscopy, nuclear magnetic resonance (^{31}P-NMR), and small angle X-ray scattering (SAXS)(Bouwstra et al., 1993). In the resonance technique, the signals are recorded before and after the addition of a nonpermeable broadening agent such as Mn^{2+}. Manganese ions quench the ^{31}P-NMR signal from phospholipids on the outer leaflet of the outhermost bilayer. The degree of lamellarity is determined from the signal ratio before and after Mn^{2+} addition. Thus, a 50% reduction in NMR signal means that the liposome preparation is unilamellar, whereas a 25% reduction in the intensity of the original signal means there are two bilayers in the liposome (Fenske, 1993).

Several wet-chemistry techniques are commonly used for the determination of lipid components present in the bilayer. Phospholipids can be indirectly determined by the Bartlett method (Bartlett, 1959) or by the Stewart method (Stewart-Marshall, 1980). In the Bartlett method, phospholipid phosphorus in the sample is first acid-hydrolyzed to inorganic phosphate. This is converted to phosphomolybdic acid by the addition of ammonium molybdate, and the phosphomolybdic acid is quantitatively reduced to a blue-colored compound by amino-naphtyl-sulphonic acid. The intensity of the blue color is measured spectrophotometrically at 830 nm. In the Stewart method, the ability of phospholipids to form a complex with ammonium ferrothiocyanate in an organic solution is utilized. The complex is read at 485 nm. CHOL gives a purple complex with iron upon reaction with a combined reagent containing ferric perchlorate, ethyl acetate, and sulfuric acid. Absorbance at 610 nm is directly proportional to the quantity of CHOL present, in the range 0–80 μg (Wybenga et al., 1970). An alternative method is the enzymatic CHOL oxidase assay (Allain et al., 1974). The method relies on hydrolysis of CHOL esters with CHOL ester hydrolase, followed by oxidation of the CHOL by CHOL oxidase and subsequent production of hydrogen peroxide. This product also oxidatively couples 4-aminoantipyrine to phenol in the presence of peroxidase to yield a blue-colored quinomeimine dye, which shows strong absorption at 500 nm.

The encapsulation efficiency can be defined as the molar amount of encapsulated material per mole of lipid. Methods for determining the amount of material encapsulated within liposomes depend on the nature of the encapsulant and include spectrophotometry, fluorescence spectrophotometry, enzyme-based methods, and electrochemical techniques.

Limiting factors for some applications are physical, chemical, and biological liposome stability. Characteristic factors of physical instability are fusion (coalescence) or aggregation of the vesicles. Both processes imply an increase of the size and polydispersity. Liposomes that lack a net electrical charge tend to be less stable toward aggregation than charged liposomes. Thus, aggregation can be prevented or can slow down by incorporation of a charge-carrying lipid into liposomal formulation. Chemical instability arises primarily from oxidation and hydrolysis of lipid molecules. Oxidation of phospholipids mainly takes place in unsaturated fatty acyl chain–carrying phospholipids. But saturated fatty acids can also be oxidized at high temperatures (Grit and Crommelin, 1993). Peroxidation of phospholipids in liposomes can be minimized by the use of high-quality raw materials. Storage at low temperatures and protection from light and oxygen will exclude the chance of oxidation. To further enhance protection against oxidation, antioxidants such as α-tocopherol can be added. The four ester bounds present in a phospholipid molecule may be subject to hydrolysis in water. The carboxy esters are hydrolyzed faster than the phosphate esters. The hydrolysis of carboxy esters produces acyl-lysophospholipids. Further hydrolysis produces a glycerophospho compound.

When liposomes had to be administrated to live organisms, a preceding study of toxicity is necessary. Main components of liposomes (phospholipids and CHOL) are considered safe. However, the modification of the chemical structure and inclusion in liposomal formulations of additional compounds may enhance the toxic potential of the resulting liposomes. Thus, it is necessary to perform a toxicity screening of liposomes (Parnham and Wetzig, 1993), particularly following parenteral administration. This includes test for hemolysis, thrombosis, and cytotoxicity. Moreover, it is essential to rule out possible contamination with bacterial endotoxin (pyrogenicity). Liposomes cannot usually be sterilized by exposure to high temperatures and are also sensitive to various types of irradiation, as well as chemical sterilizing agents, so the only method available for sterilizing after manufacturing is the filtration through 0.2 μm filters. Clearly, this procedure is only applicable to liposomes sufficiently small in size to pass easily through the pores of such filters. For larger liposomes, every stage must be carried out under aseptic conditions.

10.3 Bubble Liposomes

Bubble liposomes (BLs) are gas-encapsulated liposomes that are developed for both therapeutic and diagnostic application (Suzuki et al., 2007a). The size of such liposomes is of several hundred nanometers in diameter.

US imaging is a widely used diagnostic technique that offers high spatial resolution, allows for real-time imaging, and combines the advantages of noninvasiveness with the lack of ionizing radiation, easy access to the public, and low cost. US contrast agents are gas-filled, echogenic microbubbles that remain exclusively in the vascular compartment. Microbubbles enhance the specificity and sensitivity of imaging for various types of diseases, especially with tumors. Therefore, liposomes containing a US imaging gas can be used as contrast agents. To this end, AG73 peptide–modified BLs have been developed as neovascular-targeting BLs to enhance the contrast image (Negishi et al., 2013). The AG73 peptide is a ligand for syndecans, which are expressed in neovascular vessels. On the other hand, to enhance the antitumor effect of doxorubicin-encapsulated AG73 peptide–modified liposomes, these were combined with BLs and US. This combination does not enhance cellular uptake of doxorubicin, but it did promote drug release in the cytoplasm. Thus, this strategy may be useful for cancer therapy as a dual-function drug delivery system with targeted and controlled release (Hamano et al., 2013).

Among physical nonviral gene transfer methods, it has been shown that therapeutic US enables genes to be delivered into cells and tissues (Fechheimer et al., 1987). In the process of cavitation, the ultrasonic field interacts with gas in the liquid, leading to their growth and implosion. This collapse initiates shear stress, shock waves, and microjets that affect the cells in the nearby vicinity. These effects lead to the permeability of the plasma membrane and reduce the thickness of the unstirred layer of the cell surface. Thus, a combination of microbubbles and US has been proposed as a less invasive and tissue-specific method for gene delivery (Suzuki et al., 2007a, 2007b). This combination produces transient changes in the permeability of the cell membrane and allows for the site-specific intracellular delivery of molecules. However, microbubbles have problems with size, stability, and targeting function. BLs solve these problems and can be used as novel gene delivery tools when they coupled with US exposure. Plasmid DNA (p-DNA)-loaded BLs have been prepared to enhance the US imaging effect and the transfection efficiency (Endo-Takahashi et al., 2013). In other study (Negishi et al., 2012), sterically stabilized BLs loaded with p-DNA have been injected into the tail vein of hind limb ischemia model mice, and transdermal US exposure was applied to ischemic hind limbs. Results have indicated that this combination induced various angiogenic factors and improved the blood flow.

Phospholipids used in the preparation of BLs are phospholipids in the gel phase (e.g., dipalmitoyl phosphatidylcholine (DPPC)), since it seems that bubbles of BLs with the shell in the gel phase have a high level of static stability. If BLs are used as gene delivery agents, they contain cationic lipids. The method used in the preparation is reverse-phase evaporation. Once the liposome suspension is formed and introduced in sterilized vials, they are filled with perfluoropropane gas, capped, and then pressurized with more gas. The vials are placed in a bath-type sonicator (42 kHz) for five minutes. During this process, portions of the gas were trapped inside the liposomes, resulting in the suspension becoming clouding.

10.4 Layer-by-Layer Approach

The layer-by-layer approach to produce nano- and microparticles and capsules for controlled release is a rapidly growing area. The layer-by-layer approach was only invented in 1998 (Donath et al., 1998) but a review in 2011 on the use of this approach, only in drug delivery, already contained 250 references (Delcea et al., 2011).

The layer-by-layer assembly method comprises the sequential coating of cores or templates with polyelectrolytes of opposite charges.

The core or template can be an inorganic particle, an organic particle (Delceaet al., 2011) or even surfactant assemblies like vesicles (Cuomo et al., 2010). To obtain nano- or microcapsules in the first two types of template, the core needs to be removed by using strong acids or organic solvents respectively. In the latter case the void in the capsule is already introduced through the vesicle template itself. The adsorption of the polyelectrolytes (PEs) is based on electrostatic interactions, initially with the template and later with the previous PE layer. Thickness of the capsule wall can be controlled by the number of layers. After each PE layer, removal of excess PEs is important to avoid aggregation and formation of nontemplated PE complexes.

Typical templates are polystyrene and silica, as well as $CaCO_3$. Using porous templates like $CaCO_3$ already opens up the possibility to preload the template with the active to be encapsulated

The elegancy of the method lies in the fact that any PE can be used, and because the PEs are preformed, compatibility of the template and/or actives to be encapsulated with the polymerization conditions for the formation of the PEs are not required. If a polymer layer is formed around a template through polymerization a lot of restrictions can apply on the choice of the polymer (see the next section).

10.5 Nanocapsules

Controlled release of molecules from nano- and microcapsules finds many applications in bulk and specialty products ranging from release of medicine, vitamins, actives for hair treatment, fragrances, and biocides to crosslinkers and molecules involved in self-healing processes of coating layers. The preparation of these capsules often relies on relatively complicated and expensive chemistry. The synthesis of dendrimers, hyperbranched polymers, assemblies of block copolymers, etc., often requires multiple steps and a variety of (expensive) chemicals. On the other hand the preparation of solid nanoparticles by emulsion polymerization in simple micelles basically represents an example of nanotechnology on the kiloton scale. The synthesis of submicron latex particles for applications in coatings, adhesives, carpet backing, etc., is a large-scale industry where the cost price of one kilo of wet latex is often close to a couple of dollars only. There are a few capsule products on the market with relatively low prices (e.g., Ropague™ and Spindrift™). These particles are used because of their air voids, making them suitable as synergistic pigments together with, for example, titanium dioxide. However, the method of synthesis of these particles does not easily allow the incorporation of substances in the interior of the particles. Whereas normal emulsion polymerization comprises the polymerization of hydrophobic monomers in the interior of a micelle, leading to solid polymeric nanoparticles, vesicle polymerization takes place in more or less spherically shaped bilayers, leading in principle to nanocapsules (Fig.10.2).

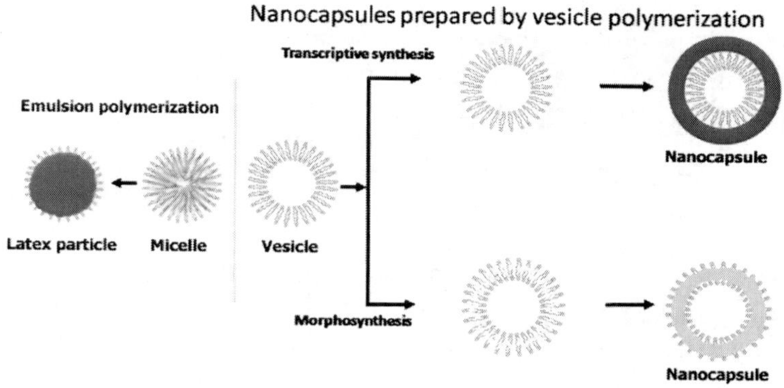

Figure 10.2 The two approaches (transcriptive synthesis and morphosynthesis) to utilize vesicle polymerization to prepare capsules from vesicles, compared to latex particles prepared from micelles in an emulsion polymerization.

The vesicles are obtained by dissolving specific synthetic amphiphiles, like dioctadecyl ammonium bromide (DODAB), or naturally occurring phospholipids, like phosphatidylcholine, in water. Above the critical bilayer concentration (analogue to the CMC for micelles) bilayer structures are formed. By extrusion through membranes the more or less spherical unilamellar vesicles are formed that can than subsequently be used to produce capsules (Fig.10.3).

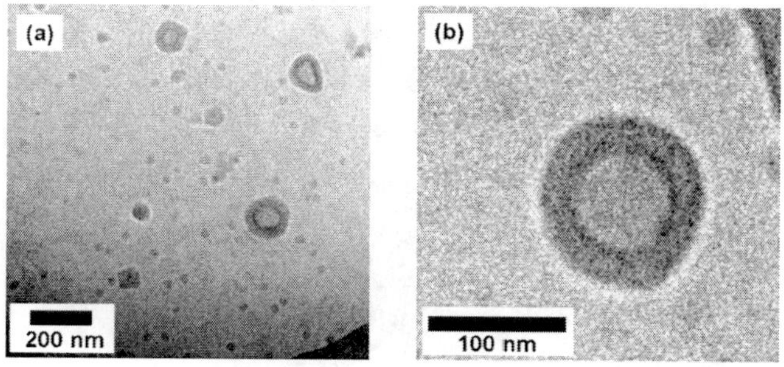

Figure 10.3 (a) Cryo-TEM micrograph of nanocapsules obtained by transcriptive synthesis, encapsulating DODAB vesicles (obtained by extrusion through 100 nm pore size filters) using a RAFT copolymer BA_6-co-AA_9 and a feed composition ratio of MMA:BA=10:1 for encapsulation. (b) High-magnification image of a single nanocapsule. Reprinted from Ali et al. (2011b) with permission from Royal Society of Chemistry.

Two approaches are possible, *transcriptive synthesis*, where the outer shape of the charged vesicle structure is used as a template for polymerization, for example, by adsorbing oppositely charged reversible addition-fragmentation transfer (RAFT) oligomers on the outside followed by further polymerization with monomers like methylmethacrylate (MMA), and *morphosynthesis*, where the hydrophobic domain inside the bilayer is used as the locus of polymerization (Jung et al., 2000a, 2000b). Whereas the first approach has recently been successfully utilized (Fig. 10.3), the latter approach often leads to phase separation of the polymer formed inside the bilayer, creating the so-called parachute morphology (Fig. 10.4). One still has to utilize more complicated chemistry to create nanocapsules with a uniformous polymer wall through morphosynthesis (Junget al., 2000a).

When working with a positively charged vesicle structure (like DODAB) we would use negatively charged RAFT oligomers (acrylic acid containing), but it is also possible to work with negatively charged or neutral vesicle structures.

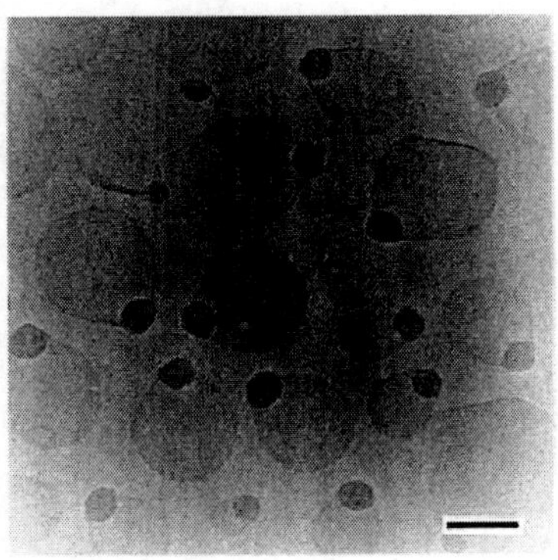

Figure 10.4 DODAB–polystyrene parachutes (scale bar represents 100 nm), outcome of a typical morphosynthesis experiment. Reprinted with permission from Ali et al. Copyright © 2005 Royal Society of Chemistry.

10.5.1 Size of the Capsules and Composition and Thickness of the Wall

The size of the capsules basically is controlled by the initial size of the vesicle and can range from 50 nm up to several micrometers.

The thickness and composition of the wall determines diffusion rates and responsive release rates.

The vesicle-templating approach can also be used to produce inorganic capsules by polycondensation of silanes (Hentze et al., 2003).

Loading of nanocapsules can be done before or after the actual formation of the capsules. In case of the use of a polymerization to create the nanocapsule (e.g., vesicle-templating polymerization) the active to be loaded should withstand the polymerization conditions (including, for example, radicals). In many cases therefore postloading is applied.

Chapter 11

Inorganic Nanoparticles in New Upcoming Biomedical Applications

Manuel António Martins, Sónia Oliveira Pereira, and Tito Trindade
Department of Chemistry-CICECO, University of Aveiro, 3810-193 Aveiro, Portugal
tito@ua.pt

The last decade has seen an unprecedented progress on new nanotechnological tools applied to medicine. In this context, inorganic nanoparticles have been used, either as single phases or as components of more complex structures, for a number of applications. This chapter is an overview of recent advances on bioapplications of single-phased inorganic nanocrystals. The systems have been selected for their current relevance and by the potential impact in forthcoming nanomedicines that include faster clinical diagnosis methods and target-selected therapies. These applications will be briefly mentioned in order to understand the potential offered by the unique properties of inorganic nanocrystals and relevant aspects concerning biofunctionalization procedures.

Controlled Release Systems: Advances in Nanobottles and Active Nanoparticles
Edited by Alexander van Herk, Jacqueline Forcada, and Giorgia Pastorin

ISBN 978-981-4613-21-7 (Hardcover), 978-981-4613-22-4 (eBook)
www.panstanford.com

11.1 Introduction

The progress made in the last two decades in nanoscience and nanotechnologies has been driven by several factors, among which is undoubtedly the synthesis of new nanomaterials. In this context, research on inorganic nanoparticles (NPs) occupies a prominent place for the contribution they have had in understanding phenomena intrinsically dependent on particle size and surface effects. For example, the recognition that magnetic and optical properties of some materials depend markedly on particle size effects is one of the distinctive aspects of nanoscience applied to inorganic materials (Trindade and Thomas, 2013). The applications of such knowledge in developing new technologies and devices, involving materials that can be manufactured from the molecular scale to nanoscopic crystalline fragments, are currently perceived as nanotechnologies.

Biomedical applications have been early recognized as a field in which there was great potential to use inorganic NPs (Sapsford et al., 2013; Trindade and da Silva, 2011). Although this is a field of application where the introduction of new materials and processes should be necessarily complemented with health impact studies, there is no doubt that in many cases innovative forms of diagnosis and therapy using inorganic NPs are underway. These applications take advantages of the unique properties observed in inorganic NPs, resulting from their dimensions, and that in much contexts are comparable to the biological systems of interest. Interactions of biological systems and NPs, which are materials of a high specific surface area, are necessarily mediated by interface phenomena. Thereby, processes of surface chemical modification and chemical functionalization of NPs are of great relevance.

In this chapter, we intend to review well-established knowledge on the chemical synthesis and properties of inorganic NPs, together with recent developments that take into account potential biomedical applications. Inorganic NPs form a very diverse class of materials. We focus here on colloidal inorganic NPs whose importance for research on bioapplications is unquestionable. Thereby, in this chapter, the focus will be on fundamental aspects related to the chemistry of metal NPs, quantum dots (QDs) and metal oxides, and bioapplications thereof. The subsequent chapter is devoted to processes of biofunctionalization of surfaces that highlights their relevance in applying inorganic NPs in biological systems.

11.2 Gold and Silver Nanoparticles

11.2.1 Gold Nanoparticles

Gold nanoparticles (Au NPs) have been widely studied, namely, due to their optical properties associated to *surface plasmon resonance* (SPR) behavior. The SPR frequency corresponds to the frequency of the collective oscillation of free electrons in the surface of a metal as consequence of the interaction with electromagnetic radiation of equal frequency that result in strong absorption and scattering of light. The SPR band of Au NPs is observed in the visible spectrum and is sensitive to particle size and shape. The location and profile of the SPR band also depends on the dielectric constant of the surrounding medium and on interparticle interactions that might arise, for example, in colloids or when deposited on solid substrates (Liz-Marzán, 2004; Dreaden et al., 2012; Saha et al., 2012).

In fact, the optical properties of Au NPs can be tuned by controlling the particles' size and shape. Hence, while spherical NPs in water show a single SPR band (typically around 520 nm) in the absorbance spectrum, anisotropic Au nanostructures such as nanorods exhibit two absorption bands due to a transversal optical mode and a longitudinal optical mode. The wavelength of the latter can vary from the visible to the near-infrared region, depending on the aspect ratio of the nanorod. Typically, the preparation of Au NPs is based on the reduction of a Au(III) salt in the presence of a stabilizing agent. The synthesis can be performed either in water (Turkevich et al., 1951) or in an organic medium (Brust et al., 1994). In case bioapplications are envisaged, the latter strategy requires subsequent ligand exchange procedures in order to disperse the NPs in water (Gittins and Caruso, 2001; Gandubert and Lennox, 2005). A number of publications have reported chemical methods that allow control over the nanospheres' size (Jana et al., 2001a; Hussain et al., 2005; Njoki et al., 2007; Bastús et al., 2011; Ziegler and Eychmüller, 2011) and the synthesis of other morphologies such as nanorods (Jana et al., 2001b; Nikoobakht and El-Sayed, 2003; Gao et al., 2003; Gole and Murphy, 2004; Perez-Juste et al., 2005; Murphy et al., 2011) nanostars, nanobelts, nanoplates, cubes and dodecahedra (Draeden et al., 2012; Xiao and Qi, 2011).

Due to their optical properties, Au NPs have been widely investigated in bioapplications such as sensors (SPR and surface-enhanced Raman spectroscopy [SERS] platforms) as well as biolabels in imaging, detection and targeting procedures (Draeden et al., 2012; Jans and Huo, 2012; Doane and Burda, 2012; Kumar et al., 2013; Dykman and Khlebtsov, 2012). Some of these applications will be briefly discussed below. These applications require particles' surface functionalization with specific biomolecules (Giljohann et al., 2010). For example, it is well known that Au NPs have high affinity to sulfur donor ligands (e.g., thiols) and many of the surface functionalization methods for Au NPs involve this type of Lewis bases (Sato et al., 2003; Zanoli et al., 2012; Gittins and Caruso, 2002; Daniel et al., 2004). The surface modification of Au NPs with polyelectrolytes mediated by electrostatic interactions can also be employed and this topic has been reviewed very recently (Pereira et al., 2013). An interesting possibility in this strategy is the use of natural polyelectrolytes such as DNA and RNA (Zhao et al., 2012; Elbakry et al., 2009; Lee et al., 2011c), proteins (Ma and Ding, 2009; Caruso et al., 2000) and polysaccharides (Tan and Zhang, 2005), besides the use of diverse synthetic polyelectrolytes (Elbakry et al., 2009; Lee et al., 2011c; Scheider et al., 2009). The functionalization of the NP surfaces as mentioned above might involve the growth of silica shells over the Au NPs (Liz-Marzán, et al., 1996a, 1996b). The silica coat not only confers robustness to the Au NPs but also allows surface derivatization using well-known chemical protocols.

11.2.1.1 Au NPs as sensors

The SPR band is a reliable optical feature of Au NPs with well-defined morphological characteristics useful in the development of nanodevices such as biosensors. For example, binding of molecules to the Au surface can itself lead to a variation of the plasmon frequency, which is detected by scattered light phase contrast microscopy (Raschke et al., 2003). The frequency observed for the SPR dependents on interparticle distance (Jain et al., 2006). This plasmon coupling effect has been used for colorimetric detection of analytes and is a well-known example of an optical sensor based on Au NPs (Mirkin et al., 1996; Elghanian et al., 1997; Moller et al., 2007). This type of sensor was developed for the detection of DNA fragments. For instance, dispersed Au NPs functionalized with oligonucleotides maintain the characteristic ruby-red color of a sol of Au NPs. The addition to the Au sol of the complementary strand of DNA leads to hybridization, forming aggregates of Au NPs that causes a variation in the frequency of the SPR and the colloidal solution turns to blue-violet color (Sato et al., 2003; Mirkin et al., 1996; Storhoff et al., 2000). This concept has been largely improved in the sense that quantitative detection of DNA sequences is possible at very low concentration and also for other analytes (Nam et al., 2005; Hussain et al., 2013). Au NPs can be linked by DNA sequences in a configuration that ensures that interparticle distance prevents plasmonic coupling. The choice of DNA sequences which change its specific conformation after binding to metals (Liu and Lu, 2007) or proteins (Zhao et al., 2008; Wang et al., 2008c) of analytical interest causes a decrease in the interparticle distance and hence a change in color of the colloid from red to violet-blue.

Light emission of many fluorophores is quenched when in the vicinity of metal surfaces, such as Au substrates (Dulkeith et al., 2002 and 2005; Lakowicz, 2006; Gueroui and Libchaber, 2004; Yun et al., 2005; Jennings et al., 2006; Sapsford et al., 2006). This effect can be used for the development of other types of optical sensors, namely, based on competitive displacement (Demers et al., 2000). In this case, quantitative detection of a substance requires metal NPs previously modified with organic ligands conjugated with fluorophores. Fluorescence emission is suppressed in this situation due to the proximity of the metallic NPs. The addition of the functionalized Au NPs to the analytical solution might cause competition of the analyte of interest to the surface of the Au NPs, hence displacing the fluorophore to bulk solution, thereby originating fluorescence emission that can be detected. A variation of this modality involves the use of Au NPs as quenchers of highly fluorescent QDs (Oh et al., 2005). A distinct detection scheme makes use of molecules that change its conformation upon binding to the analyte and act as a spacer between the fluorophore and the Au NPs. In the absence of the analyte, fluorescence is observed because the spacer molecule adopts an extended conformation that keeps away the fluorophore from the metal surface. However, when the analyte binds to the spacer, it changes the spacer's conformation such that fluorophore and NP approach each other, thus causing fluorescence quenching. The same principle can be used in the inverse mode when analyte binding causes distension of the spacer molecules (Dubertet et al., 2001; Wang et al., 2004; Ray et al., 2007).

11.2.1.2 Au NPs as biolabels

Au NPs are very attractive contrast agents allowing a wide variety of different techniques for the visualization of biological entities. Indeed, Au NPs have been widely used as optical labels by providing contrast to regions of the specimen that have been enriched with the metal. The most common detection techniques are based on the interaction between light and Au NPs (Huang et al., 2007). Au NPs greater than 20 nm allow direct visualization of specimen regions with an optical microscope in phase contrast mode or differential interference contrast (DIC) (Tkachenko et al., 2004). In dark-field microscopy only light scattered by NPs can be detected by optical microscopy. Nanoengineered Au NPs can be used for optical labeling because the color of the scattered light depends on the size and shape of the NPs (Sönnichsen et al., 2002).

Light absorption in Au NPs causes local heating because excited electrons relax by transferring energy to the lattice in the form of heat, also increasing temperature in the surroundings. This local phenomenon can also be used as a detection method. Photothermal recording is possible by using DIC microscopy, which allows the detection of density fluctuations (i.e., local variations of refractive index) in the liquid surrounding the particles (Boyer et al., 2002; Berciaud et al., 2004). Another possibility is by photoacoustic imaging that takes advantage of the expansion of the liquid due to local heating. A pulse of local heat due to light absorption leads to expansion of the liquid from the surrounding environment, and thus the creation of a sound wave that can be detected by a microphone (Mallidi et al., 2007; Kim et al., 2007d). The above methods involving photoexcitation (phase contrast/interference contrast microscopy, dark-field microscopy, photoacoustic and photothermal imaging) provide sufficient sensitivity to allow the detection of particles.

Besides interaction with visible light, Au NPs can also be detected by electromagnetic radiation in other wavelength regions, such as X-rays, and using electron beams of a microscope. For example, Au NPs provide high contrast in transmission electron microscopy (TEM) imaging due to high atomic weight (Roth, 1996). Au NPs also disperse X-rays efficiently and thereby provide contrast X-ray images

(Hainfeld et al., 2006). In comparison to methods of fluorescence emission labeling, for these cases the Au NPs have the advantage of not suffering from degradation during analysis.

11.2.1.3 Au NPs in drug delivery

Au NPs have also been investigated for drug/gene delivery either as a vector or for controlled drug/gene delivery via photothermal strategies. Hence, Au NPs incorporated in thermosensitive polymer capsules, that also contain the drug, can act as nanoheaters (Angelatos et al., 2005; Skirtach et al., 2005, 2006; Salgueiro et al., 2013). The drug release can be triggered via disruption of the polymeric capsule due to local heating caused by light absorption using a laser. Noteworthy, other nanostructures have been exploited as nanoheaters aiming remotely controlled drug release, such as carbon nanotubes (Estrada et al., 2013).

11.2.2 Silver Nanoparticles

Silver nanoparticles (Ag NPs) present a characteristic SPR band in the visible spectrum, which for nanospheres dispersed in water is typically observed at about 420 nm. Likewise, the SPR band is sensitive to size, shape, surrounding environment, and interparticle distance (e.g., aggregation) (Cobley et al., 2009; Wiley et al., 2007b; Sun and Xia, 2003). Chemical preparation of silver hydrosols is normally based on the reduction of a silver(I) salt using reducing agents such as citrate, borohydride, or glucose in water or using ethylene glycol in an organic medium. The latter is a versatile method to obtain different shapes of Ag NPs, as well as a convenient strategy to produce nanofillers compatible with synthetic polymer composites (Cobley et al., 2009; Wiley et al., 2007b; Pillai and Kamat, 2003; Yu et al., 2008; Pinheiro et al., 2013). The synthesis occurs in the presence of a stabilizing agent to prevent aggregation as well as to help control particle growth.

Ag NPs have been explored for diagnostics and therapeutic applications. Methods in diagnostics include the use of Ag NPs for imaging (McFarland and Van Duyne, 2003) and have also been widely used in SERS for the detection of vestigial amounts of adsorbates, namely, by achieving single-molecule detection levels (Vo-Dinh, 1998; Moskovits, 2005; Alvarez-Puebla and Liz-Marzán, 2010). The SERS effect depends on chemical and electromagnetic mechanisms whose contributions are still a matter of debate. However, NP size and shape have influence on the observed signal enhancement of molecular adsorbates, namely, because the SPR frequency can be shifted by using these particle characteristics in relation to the frequency of the excitation line (Cobley et al., 2009). Moreover, it has been observed that aggregated NPs can form nanojunctions with strong local electromagnetic field enhancement, leading to Raman signals more intense for molecules adsorbed on those "hot spots" (Alvarez-Puebla and Liz-Marzán, 2010; Xu et al., 1999; Michaels et al., 2000; Kneipp et al., 2006). The development of highly sensitive SERS substrates for biodetection can involve the coating of Ag NPs with polymers to detect single molecules, for example, using chitosan for detection of DNA bases such as adenine (Monica et al., 2012) or using poly(ethylene glycol) (PEG) for detection of the anticancer drug mitoxantrone (Shkilnyy et al., 2009). Furthermore, a surface modification step in order to confer functional groups aiming molecular recognition can be performed. For example, Ag NPs modified with sugar residues (e.g., lactose) have affinity to the lectin concanavalin A (ConA), allowing detection of the interaction between the protein and the sugar residue (Craig et al., 2013).

11.2.2.1 Ag NPs for antimicrobial applications

The use of silver as an antimicrobial material is known since long ago. For example, in the Roman civilization, the immersion of silver coins in barrels containing water would be a common practice in order to conserve drinking water (Alexander, 2009). Silver as an antimicrobial agent fell into disuse with the introduction of antibiotics produced in the laboratory. However, its use in specific contexts has been placed back in the scientific agenda because some strains of microorganisms can develop resistance to the administration of commonly used antibiotics. On the other hand, the current uses of nanodispersed silver in various products have also contributed to an increasing interest in this research topic (Wong and Liu, 2010). The exact mechanisms that explain the antimicrobial properties of Ag NPs and potential environment/health impacts are still a matter of controversy. Cationic release due to metal oxidation of the exposed surfaces play a major role and therefore Ag NPs have gained huge importance due to their high surface area in comparison with bulk silver (Chernousova and Epple, 2013; Eckhardt et al., 2013). Thereby, relative low amounts of

Ag per volume of final material can be used, prompting also interest in developing new nanocomposites for antimicrobial coatings and surfaces (Pinto et al., 2009).

11.3 Quantum Dots

11.3.1 Cadmium and Zinc Chalcogenides

Nanosized semiconductor materials (QDs), such as cadmium and zinc chalcogenides (CdE and ZnE, E = S, Se, Te), show quantum size confinement effects for sizes close or below the Bohr radius of the exciton (electron–hole pair) of the bulk material. An extensively investigated quantum size effect has been the widening of the band gap of nanosized semiconductor particles for decreasing particle sizes. This dependence of the optical properties (e.g., absorption and luminescence) with particle size has been observed for a range of colloidal NPs and various synthesis methods are now well established that allows fine-tuning of particle size distributions (Smith and Nie, 2010; Medintz et al., 2005). An important consequence of quantum size effects observed in colloidal NPs is the ability to adjust their light emission (wavelength maximum and quantum yield) by controlling particle size and surface chemistry, besides the chemical composition (Fig. 11.1). Taking into account the variety of semiconductor nanocrystals available, it is now possible to select materials within a wide spectral range for emission (Fig. 11.2) (Smith and Nie, 2010; Medintz et al., 2005).

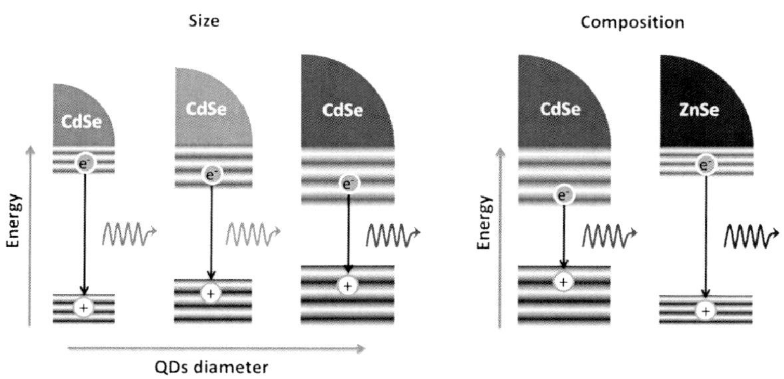

Figure 11.1 Strategies to tune the band-gap energy of QDs. Reprinted with permission from Smith and Nie (2010). Copyright 2010 American Chemical Society.

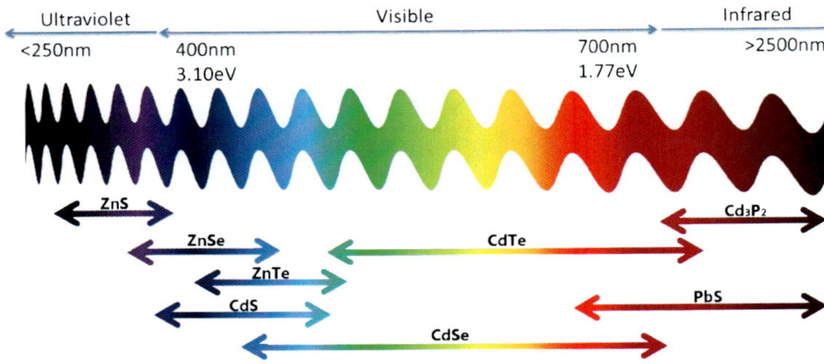

Figure 11.2 Optical emission wavelength range using a variety of semiconductors with distinct particle size distributions. Reprinted by permission from Macmillan Publishers Ltd: Nature Materials (Medintz et al., 2005), copyright 2005.

Photoluminescence (PL) due to electron–hole recombination in QDs (excitonic emission) is size dependent and depends strongly on the surface characteristics. Because surface defects can act as traps for excitonic emission, QDs cores have been passivated with organic coatings and inorganic shells of a wider-band-gap semiconductor. Hence, CdSe cores have been passivated with ZnS or CdS shells. In addition to the higher PL yields, these shells protect the core against fast oxidation (Medintz et al., 2005). Common colloidal QDs have a core/shell structure with the surfaces coated with organic coatings that render these particles stable in organic solvents and compatible with polymer matrices. Although polymer nanocomposites will not be reviewed in this chapter, it should be noted that they are themselves of great interest in bioapplications, namely, due to their PL properties. In particular, polymer beads containing QDs can retain size-tuned PL of the starting QDs and offer a platform for chemical functionalization, both as solid substrates and as emulsions (Fig. 11.3). An interesting bioapplication of these PL beads is the fabrication of bar codes for bioimaging, as will be discussed below.

Chemical methods to prepare QDs of various semiconductors have been extensively reviewed (Medintz et al., 2005; Trindade et al., 2001; Murray et al., 2000; Rogach et al., 2007). High-quality colloidal QDs are commonly prepared by solution-phase techniques using surfactants as solvents. These methods are based in a large extent on pioneering work of Murray et al.

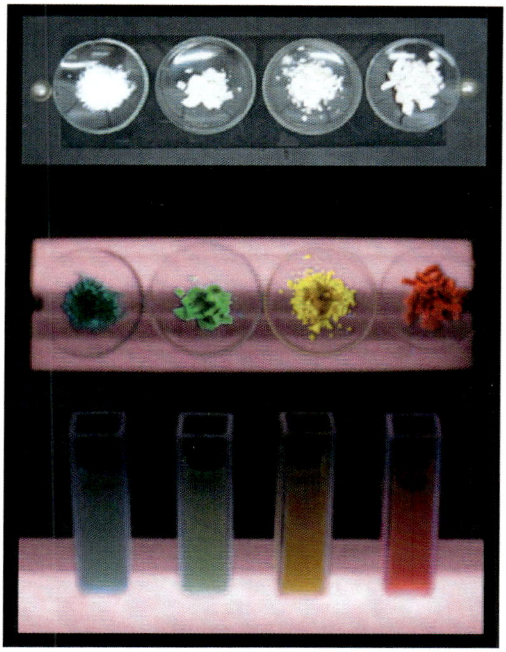

Figure 11.3 Samples of nanocomposites made of poly(*t*-butylacrylate) and CdSe/ZnS quantum dots of increasing particle size (left to right): solid samples under solar exposure conditions (top) and UV light irradiation (middle); aqueous emulsions under UV light exposure (bottom).

(1993) that described the synthesis of CdSe NPs by the reaction of organometallic precursors in a surfactant solvent at high temperature—trioctylphosphine oxide (TOPO)—also known as the "TOPO method." This first use of the TOPO method has the disadvantage of involving hazardous compounds (e.g., $(CH_3)_2Cd$) at high temperature. An alternative to this approach that overcomes this limitation is the use of single-molecule precursors that contain in the same compound the metal and the chalcogenide (S or Se) (Trindade and O'Brien, 1996; Trindade et al., 1997). This approach also offers the possibility to extend this type of synthesis to other semiconductor nanocrystals by judicious choice of the precursor compound (Pradhan and Efrima, 2003). Peng and Peng (2001) have adapted the TOPO method to the preparation of CdE QDs using common commercial sources in the presence of ligands with high coordinating ability. For example, using hexylphosphonic acid (HPA) or tetradecylphosphonic acid (TDPA), $Cd(CH_3)_2$ was immediately converted into the complexes Cd-HPA and Cd-TDPA, respectively, which can be used as the solute precursors. Fortunately, these Cd(II) complexes can be obtained using CdO or $Cd(CH_3CO_2)_2$ as the cadmium sources. Using the above methodologies and related adaptations, a wide range of QDs have been synthesized with different sizes by varying the experimental conditions, such as reagents, temperature, and reaction time (Medintz et al., 2005; Peng and Peng, 2001; Cozzoli et al., 2005; Malik et al., 1998). It should be noted that by adapting the TOPO method, it is also possible to synthesize semiconductor shells (ZnS or CdS) that passivate the surface of a semiconductor core (Dabbousi et al., 1997; Hines and Guyot-Sionnest, 1996; Peng et al., 1997; Neves et al., 2008).

Surface modification of hydrophobic QDs is normally required for their use in biomedical applications. Hence, the organic surfactant molecules can be replaced or chemically modified with functional molecules that allow the QDs to be dispersed in aqueous media and subsequently functionalized with biomolecules. Several strategies have been developed either via encapsulation or via ligand exchange at the particles' surfaces. Encapsulation involves the use of amphiphilic polymers or surfactants (e.g., cetyltrimethylammonium bromide, CTAB), as well as silica encapsulation by hydrolysis of tetraethyl orthosilicate (TEOS) onto the surfaces of the QDs (Fig. 11.4). Ligand exchange methods usually involve the use of thiol-ligands such as mercaptoacetic acid or cysteine (Fayi and Warren, 2011; Jin and Hildebrandt, 2012; Bailey et al., 2004; Ghasemi et al., 2009; Mattoussi et al., 2012; Gezeke-Moritz et al., 2013).

Methods for the aqueous synthesis of metal chalcogenide QDs have been reported at least since the eighties (Rogach et al., 2007; Gaponik et al., 2002). In fact, these reports have contributed to establish the basis for subsequent improvements on QD science, either in terms of theoretical background or in terms of experimental improvements. Although those aqueous synthesis methods fail in fine control of particle size, some recent methods have been developed with interesting results. For example, thiol-capped CdTe QDs have been synthesized in water, leading to high-quality optical properties without the need of an inorganic coating (Gaponik et al., 2002). Furthermore, the use of thiols with functional groups such as thioglycolic acid (TGA) and mercaptopropionic acid (MPA), which also contain free carboxylic groups, allows further biofunctionalization of the surfaces (Rogach et al., 2007).

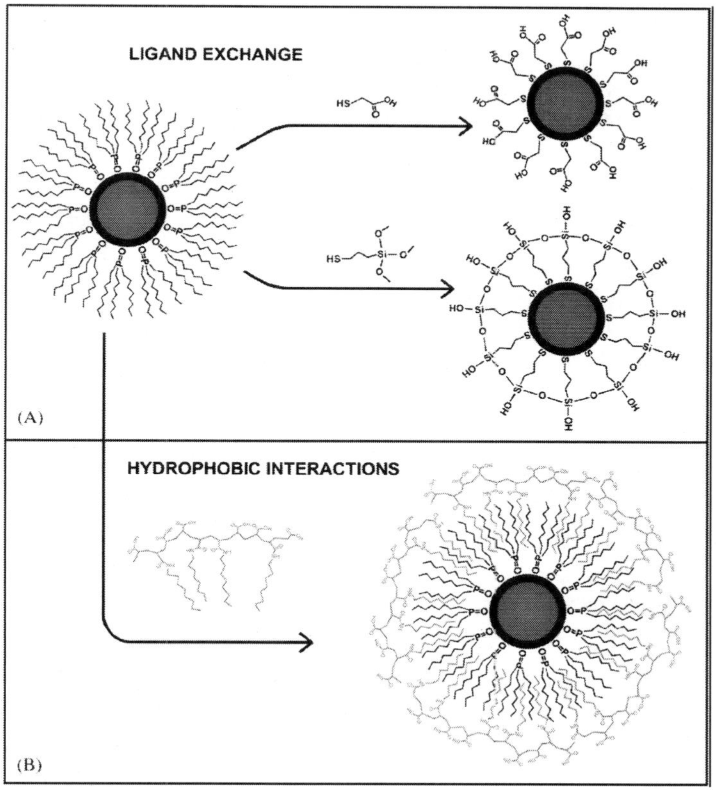

Figure 11.4 Scheme of two general strategies used to make organically capped QDs in water-compatible particles. Reprinted from Bailey et al. (2004), Copyright 2004, with permission from Elsevier.

QDs have been widely investigated for bioapplications due to their remarkable optical properties that are of great interest for bioimaging, detection, and targeting (Medintz et al., 2005; Jin and Hildebrandt, 2012; Bailey et al., 2004; Ghasemi et al., 2009; Mattousi et al., 2012; Gezke-Moritz et al., 2013; Peirobon et al., 2012; Valizadeh et al., 2012). Illustrative bioapplications of these nanomaterials are summarized below.

11.3.1.1 QDs as biolabels

QDs have been established as optical tags in a variety of analytical protocols of biological components, such as immunocytochemistry, cell motility assays, detection of DNA hybridization, and cell markers (Chan and Nie, 1998; Bruchez et al., 1998; Alivisatos et al., 2005). Biological systems tagged with QDs can be then monitored by means of techniques such as confocal microscopy, total internal reflection fluorescence microscopy (TIRFM), and fluorescence microscopy (Michalet et al., 2005). Although fluorescent organic dyes have been used for these purposes when associated to these techniques, it should be noted that QDs offer some additional advantages (Resch-Genger et al., 2008). Hence, QDs have wider excitation profiles, that is, optical excitation can occur at a wavelength region below the absorption maximum wavelength of the dots associated to the first electronic transition of the exciton. In practice, such broad absorption profile allows the appropriate choice of the excitation wavelength to enable good separation between excitation and emission profiles. The emission bands of QDs are narrow and can be size-tuned. The possibility of using simultaneously QDs with different emission "colors" perfectly distinguishable under a single excitation wavelength enables high-sensitive multiplex labeling. Thus, different cellular organelles labeled with distinct-sized fluorescent QDs result in multicolored images in which the various components of the cell can be differentiated (Medintz et al., 2005). Finally QDs, and as compared to organic dyes, have superior resistance to photobleaching under prolonged times of light irradiation, which ensures the acquisition of images with good contrast (Wu et al., 2003). Figure 11.5 depicts rat fibroblast cells (3T3) whose nuclei and microtubules were labeled alternately with organic dye AlexaFluor 488 (green) and QDs (with emission at 630 nm, red) conjugated with streptavidin. The organic dye degrades losing color emission, while the QDs have no significant loss of luminescence after continuous exposure (Wu et al., 2003).

The photostability of QDs is especially praised in 3D visualization using optical sections imaging. In this case, photobleaching of dyes is a problem during the acquisition time of successive sections, which compromises the correct reconstruction of the 3D structure. The advantage of using QDs in 3D high-resolution confocal microscopy has been demonstrated in a study of the distribution of protein in the cell membrane of erythrocytes involved in anion transport protein (TRAY3) (Tokumasu and Dvorak, 2003; Tokumasu et al. 2005) and *p*-glycoprotein, typically overexpressed in the membrane of breast adenocarcinoma cells (Sukhanova et al., 2004). The great photostability of QDs associated with the ability for size-tuning the wavelength emission has enabled the development of methods for multicolor labeling in biological assays. The multicolor code can be obtained by encapsulation of distinct-sized QDs in polymeric microspheres and controlling the relative amount of dots of various sizes in the capsule. This has some similarities to the barcode system (Wang et al., 2007; Han et al., 2001) and the number of possible combinations is huge. For example, it is possible to obtain a library of million codes by using 10 intensity levels (using different QDs concentrations) and 6 different colors (Fig. 11.6) (Han et al., 2001).

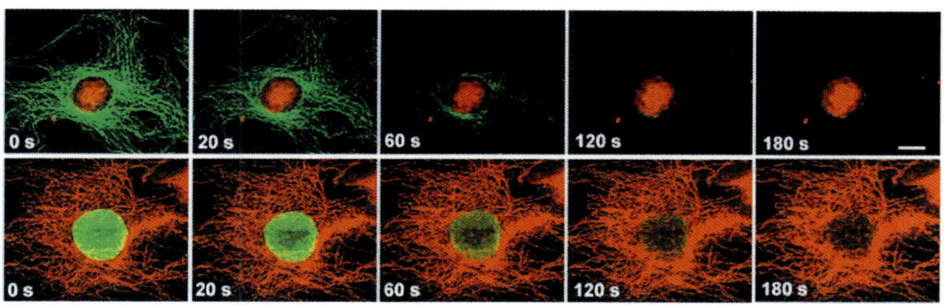

Figure 11.5 Comparison of the stability of QDs with an organic dye subjected to optical excitation. In the upper row, in a 3T3 cell, nuclear antigens were labeled with QDs (with emission at 630 nm, red) conjugated to streptavidin and microtubules were labeled with an organic dye, AlexaFluor 488 (green). On the bottom row microtubules were labeled with streptavidin-QDs (630 nm, red) and nuclear antigens with green dye Alexa 488. Reprinted by permission from Macmillan Publishers Ltd: Nature Materials (Medintz et al., 2005), copyright 2005.

Time-resolved fluorescence spectroscopy is of great relevance in biological labeling. Organic dyes exhibit fluorescence lifetimes typically of the order of a few nanoseconds, which are generally too short for efficient temporal discrimination of labeled objects (Resch-Genger et al., 2008). QDs have longer lifetimes (typically five to hundreds of nanoseconds) for the emitting state, which allows better temporal discrimination of the labeled objects from the autofluorescence of the background, thereby increasing sensitivity (Dahan et al., 2001; Grecco et al., 2004). However, due to their optical behavior dependent on size, surface, wavelength, and time, originates bi- or multiexponential decay kinetics (Schlegel et al., 2002; Zhang et al., 2006), which limit the identification of species by time resolved fluorescence measurements. The use of QDs for lifetime multiplexing acquisition has been demonstrated (Grabolle et al., 2009) and used in fluorescence lifetime imaging microscopy (FLIM) in DNA hybridization monitoring (Giraud et al., 2009).

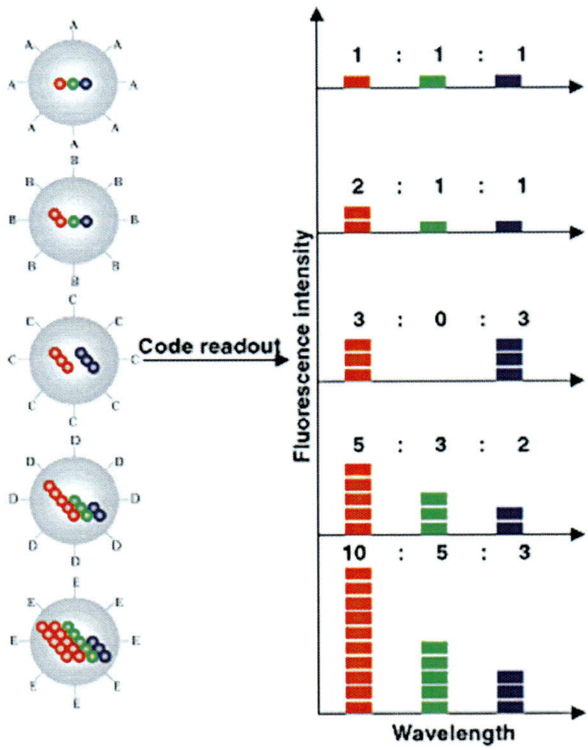

Figure 11.6 Schematic representation of the library construction of the QDs codes using three colors and different concentrations of QDs. Reprinted by permission from Macmillan Publishers Ltd: Nature Materials (Han et al., 2001), copyright 2001.

11.3.1.2 QDs in PDT

QDs have been investigated in photodynamic therapy (PDT) to improve the quantum yield of organic photosensitizers or acting themselves as photosensitizers (Geszke-Moritz and Moritz, 2013; Bakalova et al., 2004). For example, CdTe QDs and methylene blue (MB), a model drug, when light-irradiated promote charge transfer and energy (via fluorescence resonance energy transfer [FRET]) from the QDs to MB molecules, leading to singlet oxygen production that causes the death of malignant cells. Noteworthy, the efficiency of MB in these experimental conditions is much lower in the absence of CdTe QDs (Rakovich et al., 2010). A similar experiment was performed but using the QDs as photosensitizers, which once functionalized with folic acid allow tumor cells to be targeted for PDT (Morosini et al., 2011).

11.3.2 Carbon-Based Nanomaterials

Since their discovery, the use of heavy metals containing QDs for biological applications has been the subject of controversy, especially when in vivo applications are envisaged. This is an issue still under debate, thus requiring deeper insights into the science involved, but that nevertheless has prompted research for alternative optical biolabels based on inorganic NPs. As a consequence considerable efforts have been devoted to the development of heavy-metal-free PL nanoprobes (Hardman, 2006; Lin et al., 2008; Geys et al., 2008). Among these materials, "zero dimensional" carbon-based nanomaterials such as carbon NPs or carbon dots (C-dots) (Cao et al., 2007; Wilson et al., 1993) are of particular interest. These materials share some advantageous characteristics of semiconductor QDs, such as high photostability, tunable emission, and large two-photon excitation cross sections. (Resch-Genger et al., 2008; Cao et al., 2007; Wilson et al., 1993) Unlike semiconductor QDs, they exhibit nonblinking fluorescence and are water compatible (Cao et al., 2007; Wilson et al., 1993).

Although not fully understood, the mechanism of PL emission of C-dots seems to arise from surface traps that turn emissive upon stabilization as a consequence of surface passivation processes. This type of behavior shares similarities with that proposed for the luminescence emission of Si nanocrystals (Wilson et al., 1993; Sun et al., 2006a). C-dots can be produced by a wide variety of methods, which can be divided into top-down or bottom-up methodologies (Baker, 2010a). In the top-down approaches the C-dots are formed by using larger carbon structures under arc discharges (Xu et al., 2004c), laser ablation (Cao et al., 2007), and electrochemical oxidation (Lu et al., 2009b). In the case of bottom-up approaches, the C-dots are built from molecular precursors by combustion/thermal decomposition (Bourlinos et al., 2008), template-supported synthesis (Bourlinos et al., 2008), and microwave methods (Zhu et al., 2009).

Fluorescent nanodiamonds (FNDs) comprise an important class of carbon-based nanomaterials with potential application as fluorescent biolabels (Mochalin et al., 2012; Say et al., 2011). Nanosized diamonds have been obtained by a variety of methods known for a long time, such as chemical vapor deposition, and in detonation synthesis (Baker and Baker, 2010b). FNDs show bright fluorescence in the 550–800 nm range caused by point defects in the particles, some of which occur naturally, while others can be artificially incorporated during synthesis or using high-energy ion beam irradiation and subsequent thermal annealing (Gruber et al., 1997; Chang et al., 2008; Barnard, 2009). The surface of FNDs can be readily modified with various functional groups for covalent conjugation with bioactive molecules (Baker and Baker, 2010b; Kruger et al., 2006; Huang and Chang, 2004). Among these methods, surface treatment with strong acids is a common method to render carboxylic surface groups aiming bioconjugation procedures (Baker and Baker, 2010b). In addition to the carboxylic groups, FNDs possess hydroxyl groups that can be reacted with halides, amines, cyanides, azides, and thiols, thus offering diverse chemical platforms for subsequent functionalization (Barnard, 2009).

11.4 Metal Oxides

11.4.1 Zinc Oxide

Zinc oxide is a very important technological semiconductor that crystallizes with the *wurtzite* structure and has a direct band gap of 3.3 eV. For nanocrystalline ZnO, quantum size behavior occurs for particles having dimensions comparable or below the Bohr radius of the bulk exciton, which in this case is about 2.4 nm. Although polycrystalline ZnO powder does not pose great toxicity concerns, for example, it has been widely used in cosmetics and pharmaceutics, there have been concerns related to its toxicity when composed of very small nanocrystallites. Also drawbacks related with lower quantum yield emission and poor water stability have not been yet completely solved.

Popular methods to prepare ZnO QDs are based on sol–gel chemistry (Spanhel, 2006; Zhang et al., 2009b). ZnO QDs prepared by these methods are normally unstable in aqueous environments because water molecules can displace surface ligands, leading to larger aggregates and luminescence loss. To obtain water-stable ZnO QDs, surface modification methods are employed that also allows for subsequent surface functionalization. Several surface modification strategies have been reported that include the use of organosilanes (Jana et al., 2007), oleic acid (Fu et al., 2007), inorganic shells such as SiO_2 (Muhammad et al., 2011a; Vaishnavi et al., 2008; Zhang et al., 2012b) or ZnS (Li et al., 2009a), and polymer coatings (Xiong, 2010). In the latter case, either homopolymers (Guo et al., 2000; Xiong et al., 2007) or amphiphilic copolymers (Ziong et al., 2008) have been employed. In this type of polymers, the hydrophobic part protects the core (ZnO QD) from water and the hydrophilic part allows the dispersion of the particles in water as well as their functionalization (Xiong et al., 2007).

Concerning biological applications, nanosized ZnO QDs have been investigated for their antimicrobial properties (Martins et al., 2013b) and, similarly to Cd-containing QDs, which exhibit size-tuned PL bands of interest for biolabeling, luminescence behavior of ZnO QDs can be tuned via surface modification. For example, Zhang et al. (2009b) have reported the synthesis of ZnO@SiO_2 NPs with blue, green, and yellow emission, and Xiong et al. (2008) have reported the preparation of ZnO@polymer with green and yellow emission, in which the polymer shell comprises an hydrophobic internal layer (polymethacrylate) and an external hydrophilic layer (PEG–methyl ether methacrylate). While ZnO@oleic acid QDs exhibit blue emission (Fu et al., 2007). Muhammad et al. have reported the use of ZnO QDs for bioimaging and simultaneous pH responsive drug carriers. These QDs are sensitive to pH and in acidic conditions they dissolve. These features can be used for pH responsive drug carriers since in some intracellular compartments the pH is slightly acid (Muhammad et al., 2011a, 2011b).

11.4.2 Iron Oxides

Ferrimagnetic iron oxides such as magnetite (Fe_3O_4) and maghemite (γ-Fe_2O_3) might exhibit superparamagnetism for sizes usually below 20 nm. Superparamagnetic iron oxide nanoparticles (SPIONs) are already in use in a number of bioapplications, such as magnetic separation of biomolecules and cells, magnetic resonance imaging (MRI), drug delivery, and hyperthermia therapy (Akbarzadeh et al., 2012c; Krishnan, 2010; Jeong et al., 2007; Lee and Hyeon, 2012; Pan et al., 2012a). Control over the size and surface of the particles is crucial for these purposes (Fig. 11.7).

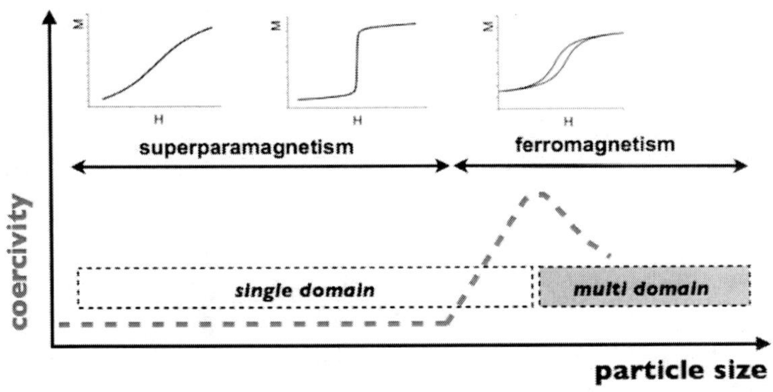

Figure 11.7 Variation of coercivity with nanoparticle size and typical magnetization curves indicating different magnetic behavior of the particles. Adapted from Lee and Hyeon (2012) with permission of The Royal Society of Chemistry.

Common methods to prepare magnetic iron oxide NPs, such as γ-Fe_2O_3 and Fe_3O_4 NPs, are the co-precipitation of aqueous ferrous/ferric ions and the thermal decomposition of iron coordination compounds in organic medium (Krishnan, 2010;

Jeong et al., 2007; Massart, 1980; Kang et al., 1996; Rockenberger et al., 1999; Hyeon et al., 2001; Jana et al., 2004). Laurent et al. (2008) have reviewed the chemical synthesis and functionalization of magnetic iron oxide colloids for bioapplications. Post-synthesis surface functionalization is required to stabilize the NPs against aggregation, as well as to achieve hydrophilic surfaces for the case of the thermal decomposition method, which yields hydrophobic NPs (Tassa et al., 2011; Ling and Hyeon, 2013). Lewis bases coordinate easily at acidic sites located at the particles' surfaces. In general, iron oxide NPs can be modified with anionic stabilizers, inorganic shells or polymer stabilizers (Laurent et al., 2008; Gupta and Gupta, 2005). Monomeric anionic stabilizers are usually phosphates, sulfates or carboxylates (e.g., citric acid) (Liu et al., 2009c). Silica and gold shells are common inorganic materials used in the surface modification of iron oxide NPs (Yang et al., 2009; Lim and Majetich, 2013). Two main approaches can be identified for performing a polymer coating: in situ coating, NPs are prepared in the presence of the polymer; *post-synthesis:* the polymer is grafted to the surfaces of previously prepared NPs (Boyer et al., 2010).

11.4.2.1 Iron oxide NPs in bioseparation

The separation of biological entities (DNA, proteins, and cells) from their native medium is sometimes critical in bioanalysis and clinical diagnosis. This may be due to difficulties that arise from the reduced concentration of the analyte and/or surroundings' interference. Magnetic separation techniques have some advantages compared to traditional methods of separation (Feczkó et al., 2008). Magnetic separation is simple and few steps are required over the entire process using a nonexpensive technology. SPIONs are especially suited for these applications because the magnetization can be switched "on/off" by application/removal of an external magnetic field, without causing irreversible agglomeration and thus allowing sample redispersion.

For magnetic separation of biological entities, NPs are functionalized with receptors of the analyte to be extracted. For example, NPs functionalized with specific antibodies can be used to separate tumor cells (Liberti et al., 2001; Zigeuner et al., 2003; Kularatne et al., 2002), erythrocytes (Tibbe et al., 1999), Golgi vesicles (Mura et al., 2002), bacteria (Morisada et al., 2002), and proteins (Lin et al., 2006). An interesting challenge in magnetic separation technology is the implementation of devices that also allow the detection of the target analyte. Thus, miniaturization of processes for handling magnetic separation, transport, and detection of magnetic NPs in microfluidic systems or lab-on-a-chip systems has aroused great interest. These systems limit sample consumption and the magnetic NPs provide a chemically active substrate with high surface area, which may allow integration of all the functions in a miniaturized system, from sample treatment to biomonitoring (Gijs, 2004).

11.4.2.2 Iron oxide NPs in bioimaging

Magnetic NPs can be used as markers for the detection of biological entities (Burtea et al., 2011; Mahmoudi et al., 2010). In this respect, magnetic NPs have some advantages; their magnetic properties are stable over time and rarely affected by chemical etching or photodegradation when in physiological conditions. Because biological samples are weakly diamagnetic, the magnetic field is not attenuated by aqueous solutions, biological fluids, or tissues, thus eliminating background signals that could interfere with the magnetic nanoprobe (Kriz et al., 1996; Larsson et al., 1999).

A number of detection devices, highly sensitive to the magnetic field, have been developed in recent years, such as sensors based on giant magnetoresistance (GMR) (Baibich et al., 1988). Spin-valve-type magnetic sensors (Dieny et al., 1991; Freitas et al., 2000) allow the detection of extremely weak magnetic fields generated by a single magnetic particle. Although these methods are still limited due to size constraints, there has been interest in this area (Tondra et al., 2000; Li and Wang, 2003). These devices can detect magnetic NPs coated with dextran and functionalized with streptavidin with an average diameter of 250 nm (Nanomag®-D, Micromod) for detecting biotinylated DNA fragments (Germano et al., 2009).

SPIONs also show interest as image contrast agents in MRI. Their sensitivity is still lower than the conventional methods, but enhancement of the contrast can be obtained by doping these NPs with transition metal ions. Experimental tests either in vivo as well as in vitro have already been performed using iron oxides functionalized with antibodies as contrast agents in MRI, allowing specific recognition of cells, namely, tumor cells (Krishnan, 2010; Lee and Hyeon, 2012; Weinmann et al., 2003; Corot et al., 2006).

11.4.2.3 Iron oxide NPs in drug delivery

Drug delivery systems have been widely studied and are very promising because a drug can be targeted to the diseased tissue/cell (e.g., cancer cells). Thus, healthy cells are protected against side effects and the drug dosage required can be reduced (Akbarzadeh et al., 2012c; Laurent et al., 2008; Mahmoudi et al., 2010). For example, Fe_3O_4 magnetic NPs modified with poly(D,L-lactic-co-glycolic acid) and PEG and an anticancer drug (doxorubicin) showed high potential in the release of doxorubicin in lung cancer cells. The results of in vitro drug release showed antigrowth effects on the cells showing potential as chemopreventive and chemotherapeutic for in vivo drug release (Akbarzadeh et al., 2012a, 2012d).

11.4.2.4 Iron oxide NPs in hyperthermia

SPIONs have proven high efficiency in hyperthermia therapy due to their superparamagnetic properties, which can also be combined with MRI (Hayashi et al., 2013). Heat is generated when an alternating magnetic field is applied on SPIONs (magnetocaloric effect). In this way, local increase of temperature can be induced in tumor cells, leading to their damage or death (Krishnan, 2010; Laurent et al., 2008; Silva et al., 2011).

11.5 Biofunctionalization Strategies of Inorganic Nanoparticles

This section describes important methods of biofunctionalization of inorganic NPs aiming their conjugation with biological entities. This is an important step for most of the bioapplications of nanoengineered NPs mentioned above and that can be employed on already surface-modified particles. The biofunctionalization of NPs can be carried out by a number of methods (see Fig. 11.8 for some examples). These methods have been classified according to the type of main interactions occurring at the biointerfaces and will be briefly described in the following sections.

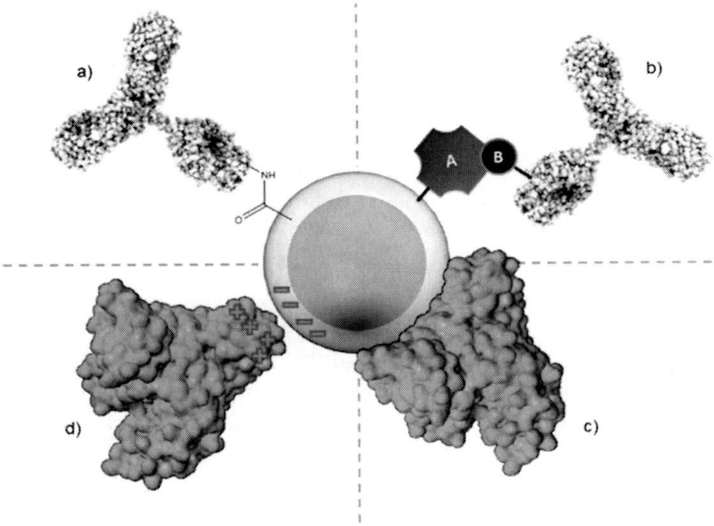

Figure 11.8 Schematic representation of the different bioconjugation strategies: (a) covalent chemical bond, (b) bonding using organic intermediates (A: avidin, B: biotin), (c) hydrophobic interaction (adsorption), and (d) electrostatic interaction. Models for antibody IgG and albumin were obtained from (http://www.rcsb.org) and displayed as surfaces models.

11.5.1 Conjugation via Covalent Binding

11.5.1.1 Coupling using carbodiimides

A common method involves the formation of amide linkages by reacting carboxyl groups and amino groups, usually in the presence of an activating agent. Thus carboxyl groups can be activated by reaction with a carbodiimide (e.g., 1-ethyl-3-(3-dimethylaminopropyl)carbodiimide hydrochloride, EDC), forming a reactive *o*-acylisourea intermediate (Fig. 11.9). The *o*-acylisourea intermediate may undergo nucleophilic attack by an aminic group forming an amide bond. This method has been used for conjugation of terminal carboxylic acid groups on the surface of NPs and free amino groups of proteins (Chan and Nie, 1998) and has been applied to couple NPs to enzymes, with retention of up to 50%–80% of enzymatic activity (Koneracka et al., 1999).

An advantage of this procedure is that there is no need of a long-chain coupling agent, thus not causing significant changes on the hydrodynamic radius of the NPs (Gao et al., 2004b). On the other hand, the high-reactive *o*-acylisourea can readily hydrolyze, regenerating the carboxylic acid groups (Fig. 11.9). To limit this reversible reaction, activated carboxyl groups can be reacted with *N*-hydroxysulfosuccinimide (NHS) or sulfo-NHS to form a succinimide ester (Fig. 11.9) (Sperling and Parak, 2010). The active ester group has a half-life greater than that of the intermediate *o*-acylisourea and also allows for reaction with primary amines (Sperling and Parak, 2010). NHS can be used without addition of the carbodiimide activating agent, thus allowing the combination of two amine groups due to the presence of two ester groups in NHS (Lin et al., 2006). The combination of the hydroxyl groups with activated carboxyl groups is still possible via a reaction of dehydration, resulting in an ester bond as opposed to amide bond formation (Gupta et al., 2003; Ho et al., 2004; Lin et al., 2005).

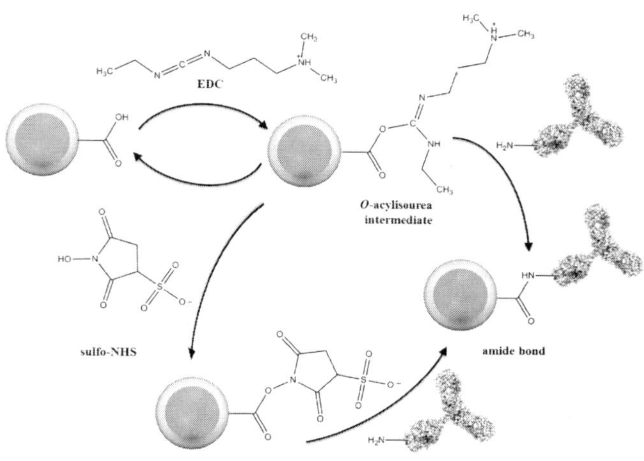

Figure 11.9 Schematic representation of a bioconjugation route involving carbodiimides. Adapted from Sperling and Parak (2010) by permission of the Royal Society.

11.5.1.2 Coupling using maleimides

Maleimides can be used for the conjugation with thiol groups forming a thioether bond using, for example, sulfosuccinimidyl-

4-(N-maleimidomethyl)cyclohexane-1-carboxylate (sulfo-SMCC). This is a bifunctional agent because the sulfo-NHS group reacts with the primary amine on the surface of NPs and the maleimide function at the other end is reactive in the presence of thiol groups (Fig. 11.10) (Brinkley, 1992). Maleimide coupling has been used for the conjugation of NPs with biological systems, such as DNA (Dubertret et al., 2002), drugs (Herceptin) (Huh et al., 2005), and proteins (Bonasio et al., 2007; Gindy et al., 2008). Lee et al. have used maleimide for coupling antibodies to multifunctional nanostructures of iron oxide and fluorescent-doped silicas (Lee et al., 2006c).

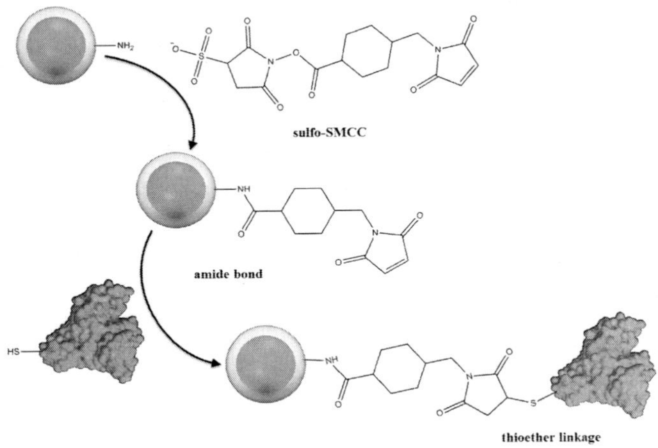

Figure 11.10 NPs with primary amine groups on the surface can react with active ester groups forming amide bonds. Adapted from Sperling and Parak (2010) by permission of the Royal Society.

11.5.1.3 Coupling by "click chemistry"

Azide–alkyne cycloaddition reactions catalyzed by Cu(I) ("click chemistry") have gained importance as a method of biofunctionalization of NPs. These involve the reaction of an azide with an alkyne, giving rise to a triazole ring, resulting in the formation of strong covalent bonds between NPs and biological entities (Fig. 11.11). This process has proven to be quite versatile and suitable for conjugation of a variety of biological species.

This versatility is still reflected in the possibility of expressing the group alkyne or azide group in the biological entity using the complementary functional group on the surface of NPs. Combining this fact with the variety of functional groups available for the formation of NP–ligand binding, this procedure has great potential for the coupling of biological material and the introduction of multiple functionalities (White et al., 2006; Fleming et al., 2006). This procedure has the disadvantage that the methods of preparation of the alkyne or azide group are often time consuming and in low yield (Sun et al., 2006b). Thus, while the final stage of cycloaddition proceeds rapidly and with high productivity, the overall process of functionalization may be long and inefficient.

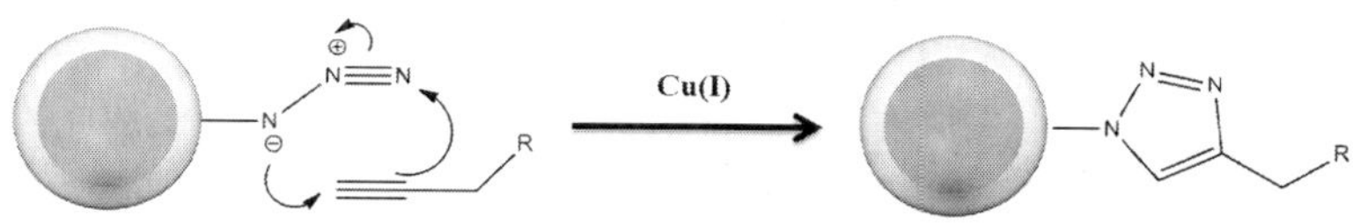

Figure 11.11 The "click" reaction occurs by azide–alkyne cycloaddition catalyzed by Cu(I), resulting in the formation of a cyclic 1,2,3-triazole bond. Adapted from Sperling and Parak (2010) by permission of the Royal Society.

Although the need of Cu(I) ions as a catalyst prevents the use of this type of reaction in the presence of living cells, due to the toxicity of the cations, the cycloaddition of azides to cyclo-octyne that does not require the use of Cu(I) catalyst has been reported in click chemistry methods (Chang et al., 2010b).

11.5.1.4 Coupling via disulfide

The chemical coupling of NPs to biological entities can also be achieved through the formation of disulfide bridges (Hilliard et al., 2002; Lee et al., 2009b). The binding of NPs using disulfide can be advantageous because of the reversibility of the reaction. Tian et al. (2007) have demonstrated the formation of disulfide bonds between silica NPs and Fe_3O_4 NPs by an oxidation process mediated by the enzyme glutathione (formation of oxidative coupling) and subsequent bond cleavage using dithiothreitol (DTT) (reductive cleavage). This approach can be adapted for the controlled release of drugs due to the presence of natural cleavage agents such as glutathione reductase (GSH) (El-Sayed et al., 2005) in appreciable concentrations in vivo.

11.5.2 Specific Interaction between Biological Intermediates

11.5.2.1 Coupling mediated by avidin–biotin binding

The use of molecular recognition processes found in nature has inspired the development of new methods for bioconjugation. Perhaps the best-known example is the avidin–biotin system (Wilchek and Bayer, 1988, 1989). The conjugate system comprises the molecule biotin (vitamin H) and a receiver, the protein avidin. This globular protein contains four identical subunits with specific sites, which recognize and bind biotin (Fig. 11.8b). The association constant of the conjugate is high at room temperature, which accounts for its stability in a wide pH range and in the presence of some chaotropic agents. In addition to avidin, analogous proteins can be used, such as streptavidin and neutravidin. Biotin is a relatively small molecule that has a free carboxylic acid group. This group can be used to form a covalent bond with functional groups on the surface of NPs, or bind directly to the surface of NP, after appropriate surface modification (Aslan et al., 2004).

11.5.2.2 Coupling using histidine residues

The introduction of a terminal peptide fragment consisting on six histidine residues (*His-6-tag*) to a protein allows coordination to a metal cation whose coordination sphere may accommodate other ligands, such as nitrilotriacetic acid, that can be attached to the surface of NPs (Fig. 11.12).

This type of coupling has been used in magnetic separation of proteins extracted directly from the cell lysate by using magnetic NPs (Xu et al., 2004a, 2004b) and in the binding of specific enzymes to the surface of Au NPs, with nearly complete retention of the enzymatic activity (Abad et al., 2005). This interaction is able to detect proteins with a terminal histidine residue at concentrations as low as 0.5 pM, avoiding the problems associated with nonspecific binding of undesirable proteins (Xu et al., 2004b).

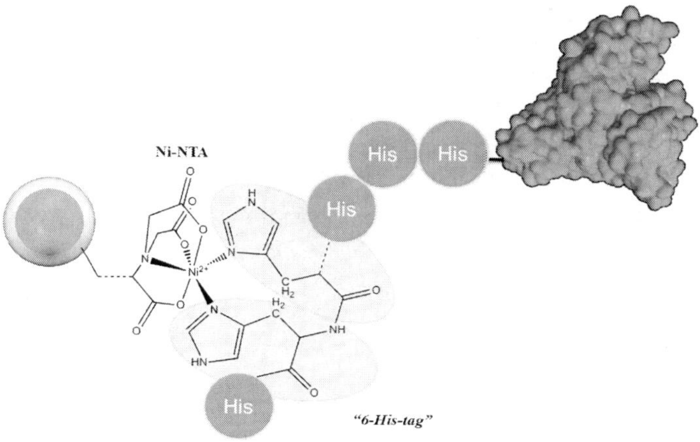

Figure 11.12 Scheme of bioconjugation via histidine-tag. The bond to the surface of NPs is mediated via a coordination complex with a metal cation (usually Ni(II) with a group of nitrile acetate Ni-NTA). Adapted from Sperling and Parak (2010) by permission of the Royal Society.

11.5.3 Nonspecific Binding of Proteins

Proteins can interact at the NPs' surfaces via van der Waals forces, hydrogen bonding, gold–thiol linkages (through cysteine residues), or hydrophobic interactions. In this case, the protein can be irreversibly immobilized at the NPs' surfaces via one type of binding or by a combination of such chemical interactions (Gole et al., 2001). Partial or complete denaturation of the protein can occur in this process (Gao et al., 2002). The denaturation of the protein causes a conformational change that might reveal the interior of the protein, inducing a greater hydrophobic interaction and/or causing an increase of the contact area between the protein and NPs. The electrostatic contributions can be minimized by adjusting the pH of the medium to a value close to the isoelectric point of the protein.

11.5.4 Electrostatic Interaction

Surface-charged colloidal NPs can be coupled to biological and polymeric species of opposite charge (Potapova et al., 2003) or to oppositely charged NPs (Galow et al., 2000). Examples include the coupling of DNA (Ha et al., 2006; Li et al., 2008a) or negatively charged liposomes with positively charged NPs (Shinkai, 2002; Nagatani et al., 2000). Careful adjustment of the pH allows coupling of a variety of proteins to NPs having opposite surface charge. For example, immunoglobulin G (IgG) is positive at pH 6 and may be immobilized at the surface of Au NPs stabilized with citrate ions, at the same pH (Ho et al., 2004).

11.6 Conclusions

Inorganic NPs have appeared in the last decades as new platforms for a number of biomedical applications that include their use in new nanomedicines. These applications take advantage of a number of unique phenomena that are observed in nanosized materials and that scale with particle dimensions. Furthermore, there is a vast interfacial area that mediates the interaction between the biological system and these NPs. Although the importance of biointerfaces has long been recognized, our understanding of fundamental mechanisms at this level is very incomplete. Among the vast number of inorganic NPs with potential for bioapplications, most current developments have focused on those that from a chemical composition point of view are simple, for example, single-phased inorganic cores. While substantial knowledge exists already for these nanomaterials in the biological context, considerably less is known for more complex nanostructures. These include multiphase and doped NPs, and nanocomposites that offer additional possibilities such as multifunctionality. The emergence of new nanomedicines based on these more complex nanostructures is not only dependent on deeper knowledge of how inorganic surfaces interact with biological systems but also dependent on upcoming requirements established by the specific contexts of applications.

Acknowledgments

S. Pereira and M. Martins thank FCT (Fundação para a Ciência e a Tecnologia) for the grants SFRH/BD/80156/2011 and SFRH/BD/29475/2006, respectively. The authors thank funding by

FEDER through COMPETE-Programa Operacional Factores de Competitividade and national funding from FCT, in the frame of the project FCOMP-01-0124-FEDER-037271 (Ref. Pest-C/CTM/LA0011/2013).

Chapter 12

Carbon Nanotubes as Nanocarriers for Controlled Drug Delivery

Cécilia Ménard-Moyon[a] and Tatiana Da Ros[b]
[a]*CNRS, Institut de Biologie Moléculaire et Cellulaire, Laboratoire d'Immunopathologie et Chimie Thérapeutique, UPR 3572, 67000 Strasbourg, France*
[b]*Dipartimento di Scienze Chimiche e Farmaceutiche—Università degli Studi di Trieste, 34127 Trieste, Italy*
c.menard@ibmc-cnrs.unistra.fr; daros@units.it

12.1 Introduction

Since their discovery in 1991, carbon nanotubes (CNTs) have attracted a lot of interest due to their extraordinary properties, including high electrical and thermal conductivities, high aspect ratio and surface area, mechanical strength, and ultralight weight (Iijima, 1991). The combination of these characteristics has made them an exciting material with the potential for various applications. A CNT can be represented as a graphene sheet rolled up to form a hollow cylinder constituted of sp^2-hybridized carbon atoms (Fig. 12.1). Depending on the number of layers, they can be classified

Controlled Release Systems: Advances in Nanobottles and Active Nanoparticles
Edited by Alexander van Herk, Jacqueline Forcada, and Giorgia Pastorin

ISBN 978-981-4613-21-7 (Hardcover), 978-981-4613-22-4 (eBook)
www.panstanford.com

as single-walled carbon nanotubes (SWCNTs) or multiwalled carbon nanotubes (MWCNTs). They have already been incorporated in many commercial products, such as composite bicycle frames, antifouling paints, printed electronics, and electrostatic discharge shielding (De Volder et al., 2013). Potential applications can also be found in the biomedical field (Kostarelos et al., 2009); they can indeed be components of medical devices and biosensors such as detectors of chemical and biological molecules and electrodes for signal transmission (Lu et al., 2009a; Saito et al., 2009; Yang et al., 2010, 2011b). They can be used as electromechanical actuators for artificial muscles (Fennimore et al., 2003). CNTs can also find applications in the field of biomaterials, for instance, as scaffolds for tissue engineering (Heister et al., 2013). Due to their cell-penetrating properties, they have been investigated as a drug delivery system (Ménard-Moyon et al., 2010). In addition, CNTs possess interesting physicochemical properties that can be exploited for imaging, including fluorescence, ultrasonography (Delogu et al., 2012), Raman, or photoacoustic imaging (Hong et al., 2009a; Liu et al., 2011c), and for therapy through heating using radiofrequency or near-infrared (NIR) radiation to induce hyperthermia (Iancu and Mocan, 2011).

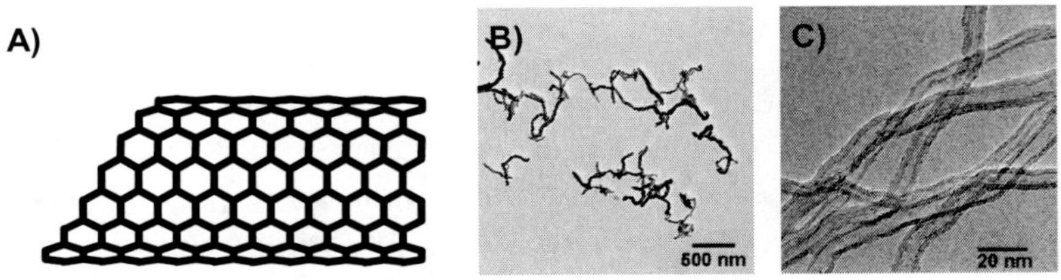

Figure 12.1 (A) Schematic representation of a SWCNT. (B) Transmission electron microscopy (TEM) and (C) high-resolution transmission electron microscopy (HRTEM) images of MWCNTs.

Drug delivery systems have been investigated to increase the efficacy of therapeutic molecules without adversely affecting the surrounding tissues, resulting in the minimization of undesirable side effects. Current efforts focus on cancer therapy to overcome poor distribution and penetration of anticancer drugs into tumor tissues, nonspecificity, and rapid elimination. In addition, the limited solubility and stability of some drugs render their administration a challenge. The drug release kinetics is determinant for maintaining high drug concentration at the desired site but not in the bloodstream to avoid systemic toxicity. Therefore, the major challenges lie on high drug-loading capacity, high drug retention during the delivery process, and the ability to release the drug on command.

In this context, CNTs present many advantages as a drug delivery system, in particular due to their high surface area and empty cavity available for loading of drugs. Their structural properties, tubular shape, and mechanical strength associated with flexibility allow them to easily cross biological barriers and to be internalized in cells with minimal cytotoxicity (Kostarelos et al., 2007). When CNTs are functionalized with suitable molecules to make them dispersible in physiological solutions, to disaggregate and shorten them, their possible toxicity is alleviated (Dumortier et al., 2006; Ali-Boucetta et al., 2013). Moreover, their electronic and optical properties bring additional functionalities to CNTs as carriers in comparison to other drug delivery systems. These properties can be exploited for diagnostic, imaging, and therapeutical purposes, including photothermal therapy (Iancu and Mocan, 2011). Functionalized CNTs can be eliminated via renal and/or fecal excretion, depending on the type of functionalization (Singh et al., 2006; Liu et al., 2008b).

Therapeutic molecules can be conjugated to CNTs on their sidewall via covalent bonding (Singh et al., 2009) or noncovalent interactions (Karousis et al., 2010). As covalent grafting of drugs onto CNTs may change their pharmacological activity, strategies have been developed where cleavable bonds are inserted to allow their release from the nanotube sidewall triggered by enzymes (Chaudhuri et al., 2010; Kam et al., 2005a; Liu et al., 2008a; Samorí et al., 2010). Supramolecular interactions between adsorbed molecules and the nanotube surface can be modulated by pH or external stimuli such as NIR irradiation and heating, leading to drug release. Alternatively, the inner hollow cavity of CNTs offers space for encapsulation of molecules (so-called endohedral functionalization), provided that their ends are open to make the empty core accessible (Khlobystov et al., 2005). In this case, the loaded drugs are protected from deactivation before reaching the target sites. External boosts, including NIR irradiation or a magnetic field, can induce unloading of entrapped molecules. A few approaches have been designed to seal the ends of CNTs with fullerenes or gold nanoparticles (NPs) and control the removal of the caps.

In this chapter, we aim to give an initial overview on endohedral functionalization of CNTs for controlled release of drugs. Different strategies have been explored to trigger the release of therapeutic molecules loaded inside CNTs. Another part is dedicated to the release of drugs adsorbed on the nanotube surface by different stimuli such as pH, NIR light, and heat. Finally, the last part focuses on the sidewall functionalization of CNTs with drugs via enzyme-cleavable linkers.

12.2 Endohedral Functionalization of Carbon Nanotubes for Controlled Release of Drugs

Encapsulation of molecules inside the empty cavity of CNTs presents many advantages such as protection from deactivating agents or conditions that could impact their stability. In addition, there is no need of covalent bonding as it can be the case for sidewall functionalization; therefore the structural integrity of the drug compounds is not affected. It has been demonstrated theoretically and experimentally that the inner core of CNTs has a higher binding energy for adsorption of molecules compared to the external sidewall (Kondratyuk and Yates, 2007; Simonyan et al., 2001). Although a series of organic molecules (Koshino et al., 2007; Liu et al., 2007b; Takenobu et al., 2003; Yanagi et al., 2006), inorganic crystals (Hong et al., 2010; Meyer et al., 2000), and NPs (Castillejos et al., 2009) has been inserted inside the nanotube cavity, investigations on the encapsulation of drugs are still in an early stage. MWCNTs seem to be more appropriate for filling with bioactive compounds as they possess a larger internal diameter in comparison with SWCNTs, although loading of molecules inside SWCNTs has already been performed (Kang et al., 2009c; Monthioux, 2002; Morgan et al., 2002; Tripisciano et al., 2009; Yao et al., 2011). Furthermore, the external surface of MWCNTs can be functionalized without inducing defects that could lead to the leakage of the embedded molecules as it could be the case for SWCNTs. Nevertheless, the loading and release are rather difficult to control. It is crucial that the cargo molecules are released at the desired location. The molecules have to possess low surface tension to facilitate their loading inside CNTs by capillary or van der Waals forces. Due to high-energy barrier at the open ends of nanotubes, the encapsulation process is often irreversible (Hilder and Hill, 2009). However, slow diffusion of incorporated drugs has been observed in some cases and exploited for sustained release in cells. In a few studies, smart strategies have been developed to block the nanotube ends and/or trigger the drug release.

12.2.1 Sustained Release of Encapsulated Drugs

The anticancer drug *cis*-dichlorodiammineplatinum(II) (cisplatin, CDDP) has been loaded in the inner cavity of MWCNTs via capillary forces (Tripisciano et al., 2010). CDDP clusters could be visualized inside CNTs by HRTEM, which was confirmed by energy-dispersive X-ray spectroscopy (EDX) revealing characteristic signals of the drug. Raman and infrared (IR) spectroscopy excluded the presence of CDDP on the external surface of CNTs. The release of the drug started to be effective between 12 and 48 hours, resulting in ~95% recovering of CDDP. Wilson and coworkers have used ultrashort (US) SWCNTs for the encapsulation of CDDP (Guven et al., 2012). US-SWCNTs have a diameter of ~1.4 nm and a length in the range of 20–80 nm and they are produced by fluorination of SWCNTs followed by pyrolysis. They have been already filled with clusters of Gd^{3+} as a magnetic resonance imaging contrast agent (Sitharaman et al., 2005), I_2 as an X-ray contrast agent (Ashcroft et al., 2007), and $^{211}AtCl$ as an α-radiotherapy agent (Hartman et al., 2007). CDDP@US-SWCNTs released the drug in phosphate-buffered saline (PBS) solution at 37°C at a much slower rate when they were wrapped with a Pluronic-F108 surfactant. The wrapping of the CNTs was expected to block the ends and the sidewall defects where the drug could be released from the nanotubes. However, this strategy has some limitations, as the release of CDDP was faster in vitro in comparison with PBS, probably due to instability of the Pluronic coating in vitro. The CDDP@US-SWCNTs displayed a higher efficacy against cancer cells after 24 hours compared to free CDDP. Further improvements could be envisaged, for instance, the wrapping with cancer-specific enzyme-activable peptide sequences instead of a surfactant.

Carboplatin (CP) (diammineplatinum(II)-cyclobutane-1,1-dicarboxylate) has been loaded inside MWCNTs and carbon nanofibers (CNFs) (Arlt et al., 2010). CP is a second-generation platinum agent with higher solubility in water and fewer adverse side effects compared to cisplatin. The loading was estimated to be 20 and 13 wt.% for CP@CNTs and CP@CNFs, respectively. The release of CP from the CNFs was very low (10% after 4 days with no more release within 3 more days), while CP was constantly released from the CNTs (up to 68% after 14 days without reaching a plateau). Due to sustained release, CP@CNTs exhibited a higher anticancer activity in tumor cell lines than CP@CNFs and free CP.

A third-generation platinum analog of the 1,2-diaminocyclohexane derivatives, oxaliplatin, has been loaded in the inner core of MWCNTs (~10 nm diameter) using nanoextraction (Wu et al., 2013). Oxaliplatin has a broader spectrum of antineoplastic activity than cisplatin. Two types of nanotubes were studied, oxidized and poly(ethylene glycol) (PEG)ylated MWCNTs. The loading of oxaliplatin was estimated to be 52% and 44% for oxidized and PEGylated, respectively. The drug release from oxidized MWCNTs was very fast, within 3 hours. After 6 hours, 50% of the compound was recovered and a plateau was reached after 24 hours. Finally, 89% of oxaliplatin was recovered after 5 days. PEGylation of MWCNTs delayed the drug release, as only 34% of oxaliplatin was recovered within 6 hours, followed by a mild ramp between 18 hours and 120 hours, leading to ~80% release after 5 days. It was hypothesized that the long PEG chains (PEG 600) grafted at the defect sites of the nanotubes, mainly at the tips (Bonifazi et al., 2006), could partially block the open ends and thus retard the drug release. Oxaliplatin entrapped inside oxidized MWCNTs and PEGylated MWCNTs showed a slightly decreased cytotoxic effect after 12 and 24 hours' incubation compared to the free drug, most probably due to retarded release. However, cell viability was drastically affected after 48 and 96 hours due to the cell-penetrating properties of CNTs and sustained release of oxaliplatin. Cytotoxicity was even higher for oxaliplatin@PEG-CNTs.

Carvedilol, used for the treatment of hypertension, has also been encapsulated inside MWCNTs (pristine or oxidized) using three different procedures: fusion, incipient wetness impregnation, and the solvent method (Li et al., 2011c). Each process led to different drug loading, distribution, dissolution, and physical state (molecular or crystalline form). The release of carvedilol was affected mainly by the distribution and physical state of the drug, and it was delayed when loaded inside oxidized CNTs.

All these studies show that it is possible to fill CNTs with different types of drugs. However, the release of the embedded molecules is difficult to control. In most cases, a sustained release was observed, leading to an improved therapeutic effect. When the open ends of the nanotubes were blocked by a polymer (Pluronic surfactant or PEG 600), the release was even slower.

12.2.2 Nanobottle Systems

Encapsulation of drugs in the inner space of CNTs has other limitations, as the open ends leave the encapsulated drugs exposed to the environment, which could result in possible degradation. In addition, the release of the drug can occur before reaching the desired target. To overcome these issues, a few studies have focused on blocking the tips of CNTs to seal them, followed by controlled removal of the caps to liberate the drug.

In this context, the group of Pastorin has introduced the concept of "carbon nanobottles" that relies on the sealing of SWCNTs after encapsulation of guest molecules (Ren and Pastorin, 2008). As a model compound, hexamethylmelamine (HMM), an antitumor agent, was inserted inside CNTs with an average diameter of ~2 nm, namely, a mixture of SWCNTs and double-walled carbon nanotubes (DWCNTs), while a C_{60} fullerene was employed as a capping agent of the nanotube ends. It is known that C_{60} has a strong affinity to enter SWCNTs (Smith et al., 1998). C_{60} fullerenes have a diameter of 0.7 nm, while the size of HMM is 0.8 nm. To open the tips of the nanotubes, heat treatment was performed at 550°C in air. This method was preferred to oxidation in acidic conditions as the latter also generates oxygenated species (e.g., COOH). These functions can block the ends and thus prevent the loading of the nanotube inner core (Kuznetsova et al., 2000). After the heat treatment, the opened CNTs were dispersed in a saturated aqueous solution of HMM. The poor affinity of HMM for water favored the encapsulation of the drug. The HMM@CNTs were then dispersed in a saturated ethanol solution of C_{60}. Extensive washing of the sample using ethanol allowed to remove HMM adsorbed on the nanotube surface, but not C_{60} due to its poor solubility in ethanol. The fullerenes were observed by TEM at the ends of the nanotubes, forming caps able to close the tips. No C_{60} was found in the middle part of the CNTs. As control, CNTs filled with only C_{60} showed a uniform fullerene distribution all along the whole length of the nanotubes. Therefore, the absence of C_{60} in the middle part of HMM@CNTs confirmed successful loading of the drug. Another control experiment was performed by filling the CNTs with C_{60} before encapsulation of HMM. In this case, the loading of CNTs with HMM was low. The HMM@CNTs were characterized by Raman and X-ray photoelectron spectroscopy (XPS). A blue shift of the radial breathing mode peaks in the Raman spectrum was observed, as it has already been described in other studies (Yanagi et al., 2007). Nitrogen was detected by XPS, which confirmed the presence of HMM as it is rich in nitrogen atoms. Dichloromethane was used to remove the fullerene caps as C_{60} and HMM are soluble in this solvent. The release of HMM was monitored by ultraviolet-visible (UV-Vis) spectroscopy. Even if the conditions used to unload HMM are not compatible with a physiological environment, this approach, based on removable caps, is very promising.

Green and coworkers have used dimethylamino-functionalized C_{60} fullerenes as removable "corks" of SWCNTs in aqueous media (Luksirikul et al., 2010). As proof-of-concept they filled open-ended SWCNTs with copper acetate or uranyl acetate as copper can be detected by UV-Vis spectroscopy in the washings, while uranium can be easily visualized inside CNTs by HRTEM. The release of the encapsulated material was triggered by lowering the pH of the aqueous solution (in 6 M HCl). Even if such a low pH is not compatible with biological conditions, the selective release of the encapsulated acetate derivatives was achieved, which demonstrated proof-of-principle of this concept. Further studies are required to adjust the pH of release to biological one. However, it will be possible to exploit acidic conditions also in vivo as it is known that the pH of primary tumors and regions of inflammation and infection is lower in comparison with physiological pH.

Recently, Pastorin and coworkers have reported another approach by sealing MWCNTs with gold NPs, which are compatible with a biological environment. In this study, CDDP was encapsulated inside MWCNTs that were subsequently capped at their ends with gold NPs (Li et al., 2012d) (Fig. 12.2). To tune their size and surface properties, the NPs were functionalized with 1-octadecanethiol. The amount of CDDP corresponded to 44% of the total weight of CDDP@Au-NPs/CNTs. TEM-EDX confirmed the presence of CDDP all along the inner cavity of the nanotubes. Unlike CDDP, the gold NPs (4 nm diameter) were found mainly at the tips. It was estimated that at least 30% of the MWCNTs were capped with the NPs on the basis of the statistical analysis of TEM images. The internal diameter of the MWCNTs was ~10 nm; therefore the NPs narrowed the open ends, which were partially blocked, resulting in a reduced release rate. Indeed, while ~90% of CDDP was released from uncapped MWCNTs after one hour, ~40% of the drug still remained inside CNTs capped with the gold NPs after eight hours. Therefore, the presence of the caps significantly decreased the drug release rate especially in the first hour. The release of CDDP was comparable at pH 5.5, mimicking the tumor environment and at physiological pH (7.4). The hydrophobic nature of the gold NPs, due to the presence of C_{18}-alkyl chains, prevented their removal from the nanotube ends in water, as confirmed by TEM after 72-hour drug release experiments. Cancer cell viability assay showed that IC_{50} (half-maximum inhibitory concentration) of CDDP inside uncapped MWCNTs was similar to that of free CDDP, whereas IC_{50} of CDDP@Au-NPs/CNTs was significantly lower. This study demonstrated that it is possible to use a lower amount of CDDP to reach similar cytotoxic activity, which should induce fewer side effects. Thus, the carbon nanobottle concept allows us to improve the pharmacological effects of drugs. But, one limitation of this strategy is the incomplete release of the encapsulated drug. To overcome this issue, an alternative strategy would be to remove the caps under specific conditions of pH or temperature, instead of permanently blocking the ends of CNTs.

Still with the objective of controlling the release of drugs from the nanotube internal cavity, Pastorin and coworkers encapsulated an inert and highly hydrophobic platinum(IV) complex as prodrug inside MWCNTs through hydrophobic–hydrophobic interactions (Li et al., 2012c). The release of the Pt(IV) complex was induced by chemical reduction, which converted the prodrug to its hydrophilic and cytotoxic form. Thus, a drastic reversal in hydrophobicity of drug molecules is a promising and efficient strategy to control their release from CNTs.

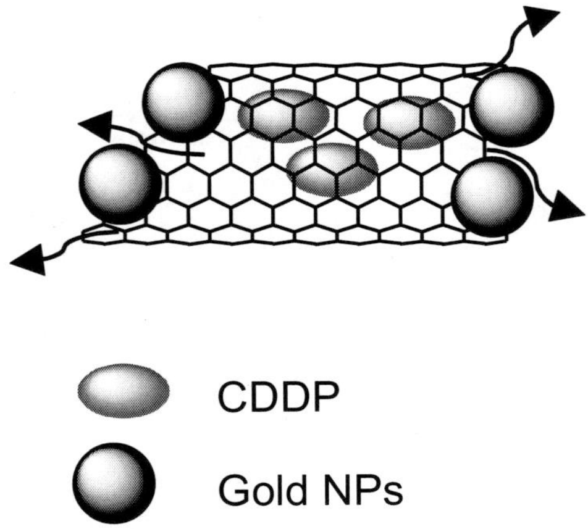

Figure 12.2 Schematic representation of CDDP@MWCNTs capped with gold NPs. Arrows show possible ways for the release of encapsulated CDDP.

12.2.3 Magnetically Induced Release of Encapsulated Drugs

Inductive heating generated upon application of a magnetic field, which can reach deep parts of the body, can trigger the release of drugs. Indeed, one study has reported the encapsulation of a combination of chemotherapeutic agents, paclitaxel (Taxol), and C_6-ceramide, followed by magnetically induced unloading, which led to an efficient killing of multidrug resistant cancer cells (Wu et al., 2012a). It is known that C_6-ceramide can significantly increase the efficacy of many chemotherapeutics; however, its use is limited because it is hydrophobic and susceptible to metabolic pathways. In this study, vertically aligned MWCNTs with ~40 nm inner diameter were grown on an anodized aluminum oxide template and were then subjected to a chemical or mechanical treatment to open both ends while embedded in the template. Gelatin, a temperature-sensitive hydrogel, was mixed with the drugs in order to keep the mixture inside the CNTs and prevent water from entering the nanotube cavity. The hydration would displace the mixture of gel and drugs due to viscosity and surface tension (Fig. 12.3). The exposure to an alternating current (AC) magnetic field produced inductive heating, thereby causing a gel–sol phase transformation of the hydrogel, which dissolved in the medium and released the payload. It should be noted that the MWCNTs used in this work possess high electrical resistivity, which is beneficial to generate resistive heat via magnetic field induction. Preliminary experiments were performed to demonstrate the feasibility of filling the nanotube array with an aqueous solution of gelatin and quantum dots (QDs) by applying vacuum suction at the bottom. TEM images clearly showed QDs encapsulated inside the nanotubes. The release of the gel/QD mixture was induced only when subjected to an AC magnetic field, as indicated by the green luminescence observed in the solution withdrawn from the MWCNTs. For the loading of paclitaxel and C_6-ceramide, droplets of a solution containing the drugs and the gelatin were deposited on the top of the nanotube array membrane while applying vacuum suction at the back side of the template. The individual loaded MWCNTs were released from the alumina template by a basic treatment and treated in acidic conditions to reduce their length down to 200–1000 nm. The nanotubes were then sonicated with amine-terminated phospholipid–poly(ethylene glycol) (PL-PEG). Incubation of cancer cells with the paclitaxel/C_6-ceramide/gelatin@CNTs led to ~70% cell death by apoptosis while applying an AC magnetic field pulse. To achieve a comparable effect, the concentration of exogenous Taxol and C_6-ceramide (i.e., with no CNTs as carrier) had to be 100 times higher. Furthermore, at the same low concentration of drugs loaded inside the nanotubes, both drugs did not induce cell death when added exogenously, probably due to poor internalization into the cells. Control experiments performed with nonloaded CNTs under an AC magnetic field showed high cell viability, therefore indicating that the nanotubes and the magnetic field had no detectable adverse effect on the cells. In addition, when no magnetic field was applied, the paclitaxel/C_6-ceramide/gelatin@CNTs were not cytotoxic as the drugs remained inside the nanotubes. The advantage of this study relies on the low electromagnetic field (25 kHz) required to release the drugs by induction heating, which is safe for the cells. This is in contrast to the magnetic NPs that require high magnetic fields (up to megahertz frequency) due their superparamagnetic properties and low saturation magnetization in order to release their payloads (Hu et al., 2007b; Rovers et al., 2012).

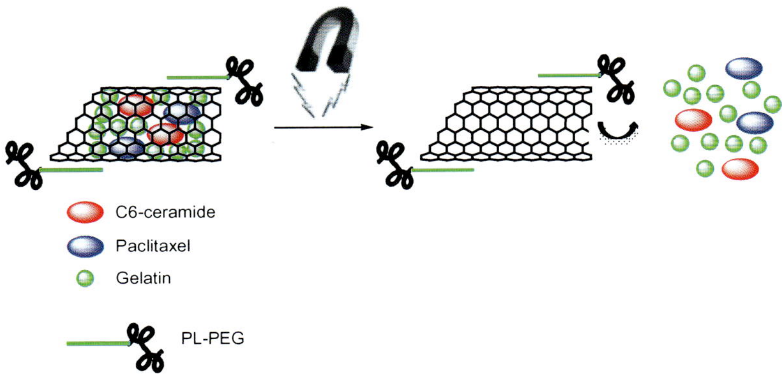

Figure 12.3 Schematic representation of encapsulation of a mixture of gelatin, C_6-ceramide, and paclitaxel inside MWCNTs, followed by coating with PL-PEG. The release of the embedded molecules was triggered by exposure to an AC magnetic field.

12.2.4 NIR-Induced Release of Encapsulated Drugs

Unloading of encapsulated molecules from the nanotube cavity can also be driven by the use of NIR irradiation, as supported by a theoretical study (Chaban et al., 2010). In the context of therapeutic applications, CNTs are very efficient at converting NIR energy into heat in solution. The optical properties of CNTs have been exploited for thermal ablation of tumors (Kam et al., 2005b; Chakravarty et al., 2008). In a study reported by Honek and coworkers, indole has been loaded inside DWCNTs and released upon NIR-induced heating, which probably favored molecule diffusion (Su et al., 2011). Indole can disrupt the membranes of cells, resulting in cell growth inhibition. The indole@CNTs were prepared by oxidation of DWCNTs in acid conditions, followed by mixing with indole at 70°C in ethanol (protocol A) or under high pressure (1500 psi) using supercritical CO_2 at 70°C (protocol B). To remove nonencapsulated indole molecules, the mixtures were washed with ice-cold ethanol. Indole@CNTs prepared by protocols A and B contained about 3.5 and 7.6 mg indole per milligram of DWCNTs, respectively. The encapsulation of this molecule was confirmed using different characterization techniques. Comparison of Fourier transform infrared attenuated total reflectance (FT-IR ATR) spectroscopy of indole, indole@CNTs before and after washing highlighted the disappearance of indole peaks after thorough washing. In addition, the fluorescence of the heteroaromatic bicycle adsorbed on the nanotube surface was completely quenched by addition of sodium iodide, while it was not the case for indole@ CNTs, where the molecules were protected from interaction with NaI. Indole encapsulation was also evidenced by ^{1}H nuclear magnetic resonance (NMR) spectroscopy via hydrogen–deuterium exchange experiments. DWCNTs protected the N–H proton of encapsulated indole from H–D exchange with a deuterated solvent. One-dimensional ^{15}N heteronuclear single quantum coherence (HSQC) NMR spectroscopy was performed on ^{15}N-indole@CNTs to enhance the NMR signal of the N–H proton. The chemical shift of the N–H proton in the encapsulated ^{15}N-indole was downfield-shifted by 0.27 ppm, which could be explained by molecular interactions between indole and the nanotubes. To impart targeting capability, oxidized DWCNTs were conjugated through amidation to the TNY peptide, which is constituted of 15 amino acids and targets EphB4-expressing cells. The EphB4 receptor is a common tumor suppressor in prostate, breast, and colorectal cancer. The TNY/CNTs were labeled with a fluorescent probe, Lucifer yellow (LY), via the reaction of LY iodoacetamide with the thiol group of a cysteine inserted at the N-terminus of the TNY peptide. Indole was encapsulated in LY-labeled TNY/CNTs following protocol B. Irradiation of EphB4-expressing cells incubated with indole@TNY/ CNTs using pulsed NIR light caused significant cell death. Pulsed laser light was preferable to continuous irradiation to avoid local overheating that could result in cell damage. Control experiments were performed to show the efficacy of indole, the importance of NIR irradiation, and the targeting effect. Indeed, when L-ascorbic acid was loaded inside TNY/CNTs instead of indole, the nanotubes bound to EphB4-expressing cells but these cells remained healthy under NIR irradiation. The same effect was observed for TNY/CNTs without indole. Finally, indole loaded inside DWCNTs conjugated or not to a control peptide that does not bind to EphB4 did not induce inhibition of cell growth. This work demonstrated that indole@CNTs were efficient in killing cancer cells under NIR irradiation when combined to a targeting peptide.

Heating-induced unloading of molecules from CNTs has been reported using fluorescein@MWCNTs (Chen et al., 2011a). In this study, MWCNTs with a mean diameter of 15 nm were first heated in air at 400°C to open the nanotubes and then oxidized in acidic conditions. Oxidation of CNTs led to the generation of oxygen-containing species, mainly carboxylic acids. In this study, the COOH functions were derivatized with S-(2-aminoethylthio)-2-thiopyridine to form thiopyridine-functionalized CNTs (CNT-S-S-pyridine). After filling the nanotubes with fluorescein, thiol-silica nanospheres having a mean diameter of 15 nm were attached to the ends of the nanotubes, thus sealing their tips. The release of encapsulated fluorescein was triggered by heating at temperatures up to 80°C. Alternatively, exposure to a disulfide reducing agent, 1,4-dithiothreitol, also cleaved the S–S bond between the silica nanospheres and the functionalized MWCNTs, resulting in the release of the dye.

12.2.5 Electrically Controlled Release of Encapsulated Drugs

Electrical stimulation is an alternative approach for triggering drug release from the nanotube cavity. Electrically controlled release has found many applications (Svirskis et al., 2010), in particular for implantable devices. Indeed, to avoid undesirable tissue reaction due to implantable neural electrodes, one strategy relies on the delivery of anti-inflammatory drugs or neurotrophic factors in the area of the implant.

One study has reported the electrically controlled release of drugs from nanotubes (Luo et al., 2011). In this work, MWCNTs with dexamethasone (Dex), an anti-inflammatory drug, loaded in their inner core were covered by a film of a conducting polymer, polypyrrole (PPy), through electropolymerization. This coating resulted in the sealing of the nanotube ends, thus preventing unwanted release of the drug molecules. Conducting polymers, such as Ppy, have been widely studied as a drug delivery system, as the release of drugs can be achieved under electrical stimulation (Abidian et al., 2006; Wadhwa et al., 2006). However, one main limitation is the low drug-loading capacity, which can be overcome by the use of CNTs. In the study of Luo et al., oxidized MWCNTs were dispersed in an aqueous solution of Dex 21-phosphate disodium salt under sonication. The presence of Dex inside CNTs was confirmed by electron energy loss spectroscopy. Phosphorous was detected at the core area of the drug-loaded MWCNTs but not at the edges. After electropolymerization of the pyrrole monomers, the effective sealing of the nanotubes was confirmed by diffusion tests in PBS, showing no significant release of Dex. The unloading of the drug was electrically induced. The CNT/PPy films exhibited a more linear and sustainable release profile compared to PPy films and a higher drug-loading capacity. Two types of MWCNTs were tested and the nanotubes with the higher inner diameter were more efficient in loading and releasing Dex. The mechanism involves a de-doping process during negative bias, resulting in an electrochemical reduction of PPy, which is positively charged, to a neutral state. This reduction led to the release of the negatively charged Dex from the CNT/PPy film, causing a reduction of lipopolysaccharide-induced microglia activation. This biological effect indicated that Dex retained its anti-inflammatory activity after the loading and electrically induced release experiments.

12.3 Stimulus-Induced Release of Drugs Adsorbed on Carbon Nanotubes

As alternative to encapsulation, the external surface of CNTs can be exploited for adsorption of therapeutic molecules. The noncovalent interactions can be modulated using different factors including pH, NIR light, and temperature.

12.3.1 pH-Induced Release of Drugs

The release of drugs from the nanotube surface can be triggered by pH. As already mentioned, the microenvironments of intracellular lysosomes, endosomes, or cancerous tissues have lower pH (5.5) than physiological pH (7.4). Drugs such as doxorubicin (Dox) (Lodhi et al., 2013) and daunorubicin have strong affinity for the sidewall of CNTs through π–π and hydrophobic interactions. At slightly acidic pH the amine function of these drugs is protonated, thereby their solubility is enhanced due to higher hydrophilicity, resulting in lower interactions with CNTs and desorption of the molecules from the nanotube surface.

The group of Dai has adsorbed Dox on the surface of PEGylated SWCNTs (Liu et al., 2007a). Two types of SWCNTs were used in this study: SWCNTs grown by the laser ablation technique with a mean diameter of ~1.9 nm and SWCNTs synthesized by the HiPco method having a diameter of ~1.3 nm. The release rate of Dox was found to be slower for the larger diameter SWCNTs due to stronger π–π stacking interactions of the drug onto larger nanotubes. The loading of Dox on PEGylated SWCNTs was in the range of 50–60 wt.%, which is drastically higher compared to 8–10 wt.% for conventional liposomes (Storm et al., 1998). There was no detectable release of Dox at pH 7.4 over several hours, while ~40% of the drug was recovered at pH 5.5 after one day. In vitro studies showed that Dox adsorbed on PEGylated SWCNTs exhibited a similar cytotoxicity compared to the free drug, whereas the toxic effect was increased when RGD peptide was grafted on PEG for selective tumor targeting. Dox loaded on PEG/SWCNTs afforded enhanced therapeutic efficacy in vivo with lower toxicity in comparison with the free drug and Doxil, which is a PEGylated liposomal formulation of Dox (Liu et al., 2009e).

McFadden and coworkers have adsorbed Dox on SWCNTs functionalized with bovine serum albumin derivatized with fluorescein and a monoclonal antibody recognizing the tumor marker carcinoembryonic antigen (Heister et al., 2009). Intracellular tracking of Dox and CNTs was possible by confocal microscopy due to the fluorescent hydroxy-substituted anthraquinone chromophore of Dox and fluorescein, respectively. The complex was efficiently internalized into cancer cells with subsequent intracellular release of Dox due to lower pH inside endosomes. The drug was then translocated to the nucleus to exert its cytotoxic action, while the fluorescently labeled SWCNTs remained in the cytoplasm. Colocalization of Dox and SWCNTs was found in the nuclear envelope and on the cell membrane. These observations have been confirmed by another research group using SWCNTs wrapped with fluorescein-labeled chitosan and loaded with Dox via π–π interactions (Kang et al., 2012). After incubation with living cells Dox detached from the nanotube surface inside the lysosomes, escaped into the cytoplasm, and eventually reached the nucleus. The SWCNTs were trapped inside lysosomes and remained in the cytoplasm.

In another study the group of McFadden has investigated the release of Dox or mitoxantrone, another anticancer drug, from the surface of SWCNTs covalently functionalized with PEG (Heister et al., 2012). After 3 days 44% of Dox and 55% of mitoxantrone were released at pH 5.5, while only 7% of Dox and 8% of mitoxantrone were released at pH 7.4 after 72 hours. The drug release was also studied in cellular growth medium. In these conditions, a significant amount of both drugs was released: 23% of Dox and 21% of mitoxantrone after three days, probably because of desorption of the drugs from the nanotube surface induced by serum proteins. This highlights the possible premature extracellular release of the drugs for long blood circulation times (one to three days). Cell viability assays indicated that the drugs loaded onto CNTs were less cytotoxic than the free ones. However, when folic acid (FA), a targeting agent for many cancer cells, was grafted on the PEG chains of functionalized SWCNTs, the construct efficacy was significantly enhanced.

To control the loading and pH-induced release of Dox, two polysaccharides, alginate (ALG) and chitosan (CHI), have been adsorbed on the surface of SWCNTs (Zhang et al., 2009a). Dox was loaded on the polysaccharide/CNTs and FA was covalently grafted to CHI. The system was stable in physiological conditions (pH 7.4), while at lower pH (5.5) Dox was released from the nanotube surface over three days, inducing cell death with a much higher efficiency compared to free Dox. The loading efficiency and release rate of Dox could be controlled by modulating the surface potential of the functionalized SWCNTs. Indeed, for ALG/CNTs the loading efficiency was higher compared to CHI/CNTs, while the drug release was slower (Ji et al., 2012). The CHI/ALG/CNT complex was therefore a good compromise to achieve both high loading and release rate. ALG/CNTs could be used if slow release is necessary, whereas CHI/CNTs would be useful for faster release. Varying the ratio of ALG and CHI polysaccharides indeed allowed to tune the release rate of Dox.

A smart CNT conjugate has been designed for in vitro simultaneous cancer-targeted optical imaging and magnetically guided drug delivery (Chen et al., 2012d). MWCNTs filled with Fe_3O_4 NPs have been conjugated to SiO_2-coated CdTe QDs and transferrin. Dox was then adsorbed onto the nanotube surface.

The release of the drug from the nanotube surface was highly pH dependent: at pH 7.4 Dox release was very limited, while 50% of the drug was released at pH 5.5 after 24 hours. Upon incubation with cancer cells, the complex was internalized into the endosomes, where the low pH environment induced release of Dox. The drug subsequently migrated into the nucleus, resulting in cell death.

More recently, Dox has been loaded on MWCNTs functionalized with the generation 5 poly(amidoamine) (PAMAM) dendrimer modified with fluorescein isothiocyanate (FITC) and FA (Wen et al., 2013). In this study, the drug release rate was also pH responsive, with fast release of Dox in an acidic environment and slow release at physiological pH. The complex displayed an efficient therapeutic efficacy, similar to that of free Dox, but with the capacity to target cancer cells overexpressing the FA receptor, leading to growth inhibition.

The same strategy based on pH-induced release of drugs has been applied to SWCNTs noncovalently functionalized with the sgc8c aptamer and daunorubicin (Taghdisi et al., 2011). The aptamer targets a leukemia biomarker, while daunorubicin is an anthracycline antitumor antibiotic widely used in the treatment of leukemia. After three days, 12% of daunorubicin was released from the nanotube surface at pH 7.4, whereas ~60% of the drug was released at pH 5.5. Similar to Dox, at slightly acidic pH the hydrophilicity of daunorubicin was enhanced due to protonation of the amino group, resulting in an increased solubility of the drug and lower interactions with the CNTs. The conjugate was selectively internalized into lymphoblastic leukemia T cells and allowed to exert its cytotoxic effect.

As alternative to π-stacking, Dox has been covalently conjugated to CNTs via a pH-responsive hydrazone bond, which is more sensitive to the tumor microenvironment than other covalent linkers (Gu et al., 2011). In this study, oxidized SWCNTs were first functionalized with a PEG diamine derivative through amidation (Fig. 12.4). The amino groups were then derivatized with hydrazinobenzoic acid. Finally, Dox was covalently attached on the hydrazine-modified CNTs by reaction with its C_{13} ketone moiety through the formation of a hydrazone bond.

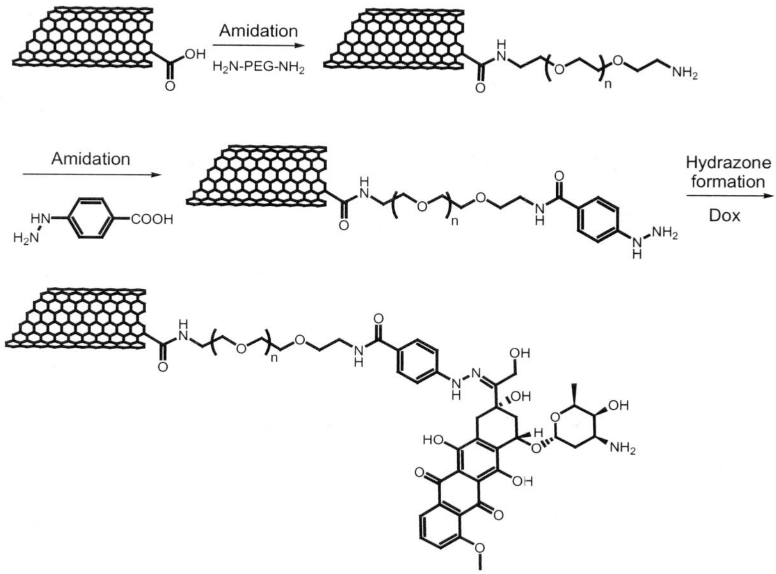

Figure 12.4 Functionalization of CNTs with Dox via a hydrazone linker.

In parallel, another Dox/CNT conjugate was prepared for comparison by adsorption of the drug on the sidewall of PEGylated SWCNTs via supramolecular interactions. Both constructs were stable in physiological conditions. At pH 5.5 the Dox release was higher for the covalent conjugate compared to the noncovalent complex: 51% versus 37% after 12 hours and 73% versus 50% after 60 hours. This result indicated that hydrazone bonds may be more sensitive to slightly acidic pH than π–π interactions. The covalent construct displayed higher cytotoxicity against cancer cells as compared to the noncovalent complex.

In a completely different approach, polyethyleneimine (PEI)-coated DWCNTs loaded with the antimalarial drug chloroquine have been used for gene delivery (Sanz et al., 2011). By mixing open-ended oxidized DWCNTs with an aqueous solution of chloroquine diphosphate salt, the drug molecule was found inside the DWCNTs, but also adsorbed onto the nanotube surface and between bundles. Its release was triggered in the lysosomes as it became protonated due to the slight acidic pH, which resulted in the rupture of the vesicle and the release of DNA/PEI/DWCNTs into the cytoplasm, thus avoiding possible degradation of DNA inside the lysosomes. The presence of chloroquine was found to increase the ability of PEI-coated DWCNTs to deliver the genetic material into the cells.

Overall, these studies show that the release of drugs from the nanotube external surface can be induced at slightly acidic pH found in the tumor microenvironment, endosomes, and lysosomes. This specificity results in better therapeutic efficacy with concomitant reduction of the systemic toxicity of the drugs.

12.3.2 NIR-Induced Release

CNTs can absorb NIR light and dissipate the energy as heat. This process can be used to locally increase the temperature of cells that have internalized CNTs. This effect, called hyperthermia, leads to cellular death.

An appropriate modulation of the irradiation can be exploited to trigger desorption of molecules from the CNT surface via an endothermic process. This methodology is also particularly useful to deliver drugs encapsulated into CNTs (see Section 12.2.4) or that stack on the carbon sidewalls by low energy π–π interactions. To the best of our knowledge there are no theoretical studies on the NIR-induced release of drugs from the external surface of CNTs, but only on encapsulated ones (Chaban et al., 2010). However, it can be envisaged that the same stimulus could easily detach molecules adsorbed on the carbon surface by means of supramolecular interactions.

One of the first studies in this field has been performed by Kam et al., who investigated the effect of NIR irradiation on CNT-treated cells (Kam et al., 2005b). Although the work was focused on the use of irradiation to trigger hyperthermia and consequently cell death, some experiments on the release of DNA wrapped around SWCNTs were also reported. Using Cy3-labeled single-stranded DNA (ssDNA), it was possible to monitor the fluorescence of Cy3 into HeLa cells. The exposure of these cells to Cy3-ssDNA/SWCNT complex resulted in an internalization of the construct, revealed by confocal microscopy. This analysis permitted also to see that the complex accumulated into the cytoplasm with a poor localization in the nucleus. When the same experiment was repeated under irradiation with NIR light for a short period, without inducing cytotoxic responses, it was possible to observe localization of Cy3-ssDNA in the nucleus. This demonstrated that the cargo was efficiently released into the cells and was able to cross the nuclear membrane. A further confirmation of the release of the genetic material from the nanotube surface was obtained in a cell-free system. Upon NIR irradiation of a solution containing Cy3-ssDNA/SWCNTs, the dye fluorescence apparently increased because of the detachment of Cy3 from the CNTs, while initially the close contact between both entities quenched the luminescence of the dye.

SWCNTs have been functionalized with an antibody (Ab) against the P-glycoprotein (P-gp) and used to overcome multidrug resistance in human leukemia cells (Li et al., 2010c). P-gp is involved in the extrusion process related to drug resistance: when P-gp is overexpressed, the cells are more efficient in eliminating drugs from cytoplasm. The treatment with Ab/SWCNT derivatives led to a specific recognition of leukemia cell line, with the opportunity to selectively deliver drugs into these cells. Dox was loaded onto the CNT surface and its release was studied in vitro using a cell-free system. The drug liberation was triggered by irradiation using a NIR laser source. Once the light was absorbed by the nanotubes, a drug release endothermic process was induced. As previously mentioned, it is important to control the process in order to avoid an excessive increase of the temperature, which would damage or kill the cells if applied on tissues. Therefore, cycles of 5-minute irradiation every 35 minutes were used. The difference in the drug percentage release was significant. Indeed, 60% of Dox was found in solution after 24 hours, whereas only 20% of the drug was recovered when NIR irradiation was not applied. In the experiment on cells, an irradiation of 5 minutes every 8 hours was performed. The cell survival was 40% after 24 hours for the Dox/Ab/SWCNT construct, while in the presence of the free drug 60% of cells survived after 24 hours, indicating a higher treatment efficacy in presence of SWCNTs.

More recently the same authors have proposed a similar approach using iron filled MWCNTs (Fe@MWCNT) (Li et al., 2011a). CNTs were covalently functionalized with FA as the targeting agent and/or FITC. Dox was subsequently adsorbed onto the nanotube surface by supramolecular interactions. HeLa cells were incubated for 48 hours and irradiated by NIR light for 5 minutes every 8 hours to induce the release of the drug (Fig. 12.5). The cytotoxicity studies were performed in the presence or the absence of a magnetic field, applied to enhance the cellular uptake of the magnetic species.

The best results were obtained with Fe@Dox/FA/MWCNTs in the presence of the magnetic field (about 15% of cell survival).

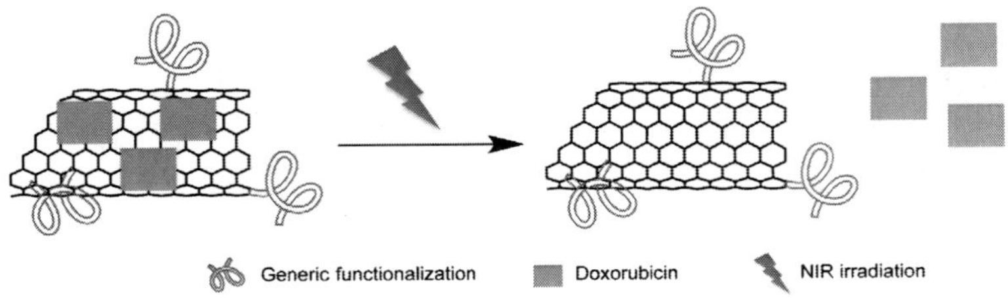

Figure 12.5 Schematic representation of NIR-induced drug release.

The capability of Fe@MWCNTs to adsorb Dox has also been exploited to design an analytical tool to determine the drug content in different tissues after administration of Doxil in rats (Shen et al., 2011). The magnetic Fe@Dox/MWCNTs were used to extract Dox and were easily separated from the cellular preparation by applying a magnetic field. The recovered material was then irradiated by NIR light to release Dox, which was eventually dosed by high-performance liquid chromatography.

SWCNTs have been embedded into hydrogels of *N*-isopropylacrylamide (NIPAM) or *N*-isopropylacrylamide/poly(*N,N*-dimethylacrylamide) (PDMAAM), as anchoring points for Dox molecules, interacting—also in this case—with the CNT surface by π–π bonds (Fujigaya et al., 2011). These systems were used as a drug reservoir and their ability to release the antitumor molecule was studied by changing pH and by photoirradiation. The NIR irradiation at 1064 nm of NIPAM/SWCNT reservoirs induced a gel shrinking due to a phase transition of the polymer. This effect inhibited the release of the drug. The NIPAM/PDMAAM/SWCNT material showed a different behavior as after 30 minutes of irradiation almost 90% of Dox was released. This combination would permit us to create an interesting drug reservoir platform.

12.3.3 Thermosensitive Release

The use of different polymers such as PEI and polyvinyl alcohol (PVA) has allowed the preparation of CNT-based derivatives where PEI was covalently linked to the carboxylic functions of oxidized MWCNTs (Mashat et al., 2012).

Both polymers interacted by hydrogen bonds leading to the so-called zipper effect, with alternate deposition of layers of H-bond donors and acceptors. These interactions are thermally sensitive as an increase of temperature generally induces the dissociation of the polymers. The so built-up complex can be used to embed a cargo that can be released after thermal dissociation of the polymers (i.e., the unzipping effect). In this study, Dox was loaded onto PEI/PVA/CNTs. The stability of the system was tested at various temperatures (25°C, 37°C, 40°C, 70°C). The release of the drug was obtained when heating at 40°C or at higher temperatures, while at physiological temperature (or lower) the system exhibited a low release or complete stability (Fig. 12.6).

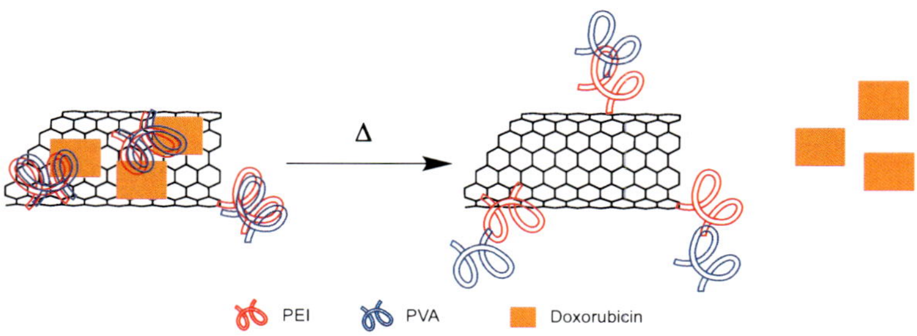

Figure 12.6 Schematic representation of thermosensitive drug release.

Lung fibroblast (LF), breast adenocarcinoma (BA), and HeLa cells were incubated for 48 hours with this construct, positive (Dox) or negative (PEI/PVA/CNTs) controls at 35°C, 37°C, and 40°C. Even though the results varied depending on the cell lines, Dox-loaded PEI/PVA/CNTs were cytotoxic when heated at 40°C, while poor or null effects were reported at 37°C and 35°C. LF cells were the most affected, while the BA cells seemed be the most resistant.

12.4 Functionalization of Carbon Nanotubes with Drugs via Enzyme-Cleavable Linkers

Not only encapsulation or noncovalent interactions have been used to design CNT-based drug delivery systems, but in fact drugs can also be conjugated by covalent bonds directly on the sidewall of CNTs or on molecules wrapping the nanotubes. In both cases the release of the active molecules must take place after the vector has reached the target and must be efficient. The use of enzyme cleavable linkers could be very useful to achieve controlled and full release.

12.4.1 Disulfide Bonds

The cleavage of disulfide bonds in the cytosol and endoplasmic reticulum is performed by enzymes belonging to the thioredoxin family. Moreover, also the endocytic pathway is involved in this process, considering the reducing environment of the endosomes and the lysosomes. Thus, disulfide bonds can be successfully used to reversibly link drugs to vectors. However, the CNT literature reports only few examples of this strategy. The first study was published in 2005, describing the linking of oligonucleotides (Fig. 12.7) (Kam et al., 2005a).

A more recent example has been reported by Delogu et al. (2009). CNTs were wrapped using 1,2-distearoyl-sn-glycero-3-phosphoethanolamine-*N*-amino(poly(ethylene glycol))$_{2000}$ (DSPE-PEG 2000) and few synthetic steps were necessary to obtain the pyridyldithio-functionalized PEGylated CNTs. The latter was allowed to react in presence of 5′-mercapto-oligonucleotide, a sequence inhibiting the protein tyrosine phosphatase N22 (PTPN22) expression. This protein is an important target for autoimmune diseases and negatively regulates growth and activation of T lymphocytes. The release of the oligonucleotide was proved by agarose gel electrophoresis after incubation of the construct in presence of dithiothreitol. The oligonucleotide-S-S-CNT derivative was used to deliver the antisense sequence into T cells and the CNT conjugate was effective in silencing PTPN22 expression at messenger RNA (mRNA) and protein levels.

Analogously, small interfering RNA (siRNA) inhibiting murine double minute clone 2 (MDM2) has been linked to SWCNTs following the previously reported synthetic pathway (Chen et al., 2012b). MDM2 protein binds to the p53 protein, inhibiting its activity in the apoptosis pathway. MDM2 gene overexpression has been correlated to many cancers. Breast cancer B-Cap-37 cells were efficiently transfected using this construct showing significant apoptosis effect after 48 hours.

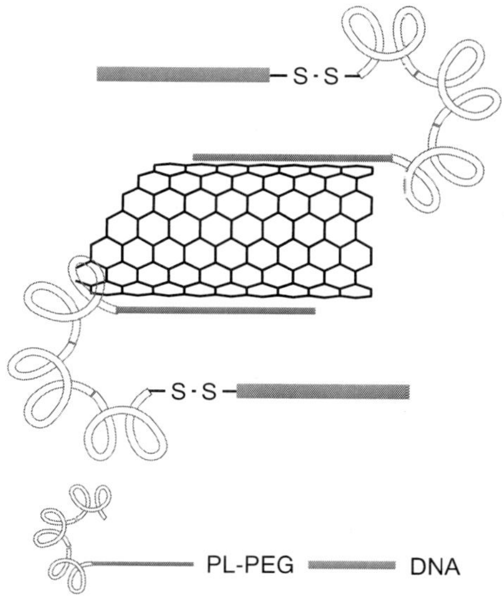

Figure 12.7 Schematic representation of the oligonucleotide-S-S-CNT platform.

A combination of disulfide and ester bond cleavage has been proposed by Chen et al. (2008). The authors prepared SWCNT functionalized with a targeting agent (biotin) and a taxoid derivative linked as ester of 2-mercaptophenylacetic acid, forming a disulfide bond with the CNTs. The reducing activity of glutathione (GSH) and/or thioredoxin induced the cleavage of the disulfide bond, resulting in the release of taxoid (Fig. 12.8).

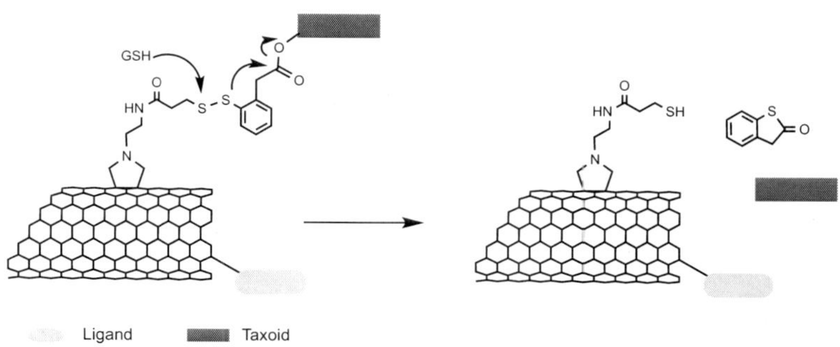

Figure 12.8 Structure of a taxoid/SWCNT derivative and the products released after cleavage of the disulfide bond.

These reactions take place only in cells, where the concentration of glutathione is higher than in the plasma; therefore the construct is stable in the bloodstream. The taxoid/SWCNTs IC_{50} was about 0.36 μg/mL in leukemia cells, able to uptake the CNT derivatives by endocytosis mediated by biotin receptors. The IC_{50} in murine leukemia and human fibroblast cell lines, not overexpressing biotin receptors, was >50 μg/mL.

12.4.2 Peptide Linkers

At the best of our knowledge, there is only one report on the use of specific peptide linkers to conjugate drugs on CNTs (Samorí et al., 2010). The tetrapeptide glycine-phenylalanine-leucine-glycine (GFLG) is known to be cleaved by lysosomal cathepsin B, overexpressed in many human cancers. This peptide sequence has been already utilized as linker to release polymer-bound Dox (Wedge et al., 1991). GFLG was used to link methotrexate (MTX) to MWCNTs, previously functionalized by introducing pyrrolidine rings, bearing a hydrophilic chain on the nitrogen (Fig. 12.9). The MTX-GFLG/MWCNT conjugate caused a cytotoxic effect on human breast carcinoma cells, inducing 90% cell death, whereas other constructs where MTX was bound to CNTs through a noncleavable linker or 6-hydroxyhexanoic ester, an esterase-sensitive linker, were almost ineffective.

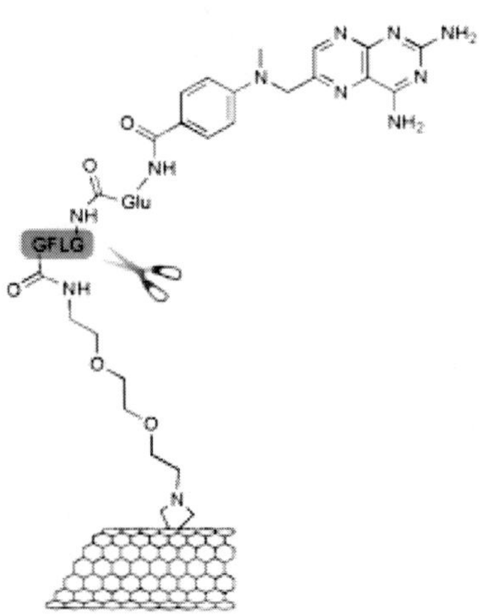

Figure 12.9 Structure of an MTX-GFLG/MWCNT derivative.

12.4.3 Ester Bonds

The use of oxidized CNTs, presenting many carboxylic groups on their surface, is a convenient synthetic way to prepare covalently functionalized nanotubes for drug delivery, using a direct attachment of the molecules to the carboxylic acids or by insertion of a spacer. The most used chemical procedure consists in the activation of the carboxylic moieties as acyl chlorides, followed by reaction with nucleophilic derivatives such as amines or alcohols, leading to the production of amides or esters, respectively. The rather low stability in cellular environment of the latter can be exploited to obtain a proper release of the attached units into cells. The cleavage can be induced by esterases or by the lower pH in lysosomes. Many CNT-based derivatives have been produced following this synthetic pathway; however, only in a few cases the release of the drug was specifically studied.

The release of dipyridamole, an antithrombotic compound, linked to SWCNTs via an ester bond, and of betahistine, an antivertigo drug, conjugated by amidation, has been investigated. In both cases the total release of the drugs was observed, but almost 100% of dipyridamole was liberated from the CNTs in three hours, whereas for betahistine it took more than eight hours in the same conditions (Khazaei et al., 2010).

One of the first successful strategies for conjugation of drugs to CNTs by means of ester bonds has been reported in 2008 (Liu et al., 2008b). Paclitaxel (PTX) was linked to DSPE-PEG (PEG 5000) adsorbed on the nanotube surface through a bifunctional linker (succinic acid) via the formation of an ester bond. The murine 4T1 breast cancer model was generated in BALB/c mice and the animals were treated with PTX-DSPE-PEG/SWCNTs, obtaining a significant reduction of the tumor volume with respect to controls such as PTX-DSPE-PEG.

An analogous procedure has been reported for cyclosporine A (Hadidi et al., 2013). In vitro studies on the drug release were performed at pH 5.5 to mimic the lysosomial environment and a slow release was reported. In fact only about 20% of cyclosporine was freed in 48 hours and 40% after 72 hours.

DSPE-PEG has also been used to covalently link pirarubicin through an ester bond. Pirarubicin is an analog of Dox, known for being less cardiotoxic and for presenting activity versus some

Dox-resistant cells (Chen et al., 2012b). Cytotoxicity experiments demonstrated that the pirarubicin-DSPE-PEG/SWCNTs were more active on human bladder cancer cells than the free drug. Interestingly, a completely different behavior was observed in human myofibroblast cells. In this case the CNT vector bearing the drug was much less toxic than pirarubicin alone. But, the in vivo tests evidenced a therapeutic activity—trigger of apoptosis in *N*-methyl-*N*-nitrosourea induced bladder carcinoma—without side effects.

A different functionalization approach has been exploited to link PTX to CNTs by using a hyperbranched poly(citric acid) (PCA) derivative obtained by polymerization of citric acid in the presence of oxidized MWCNTs. The so-obtained PCA/MWCNTs were esterified with PTX (Sobhani et al., 2011). Also in this case, a lower pH induced a more efficient release of the drug: 12% at pH 6.8 and 22% at pH 5 with respect to 10% at pH 7.4 in 24 hours. These results were a good starting point to proceed with in vivo experiments, considering that in cells also the esterases are actively involved in releasing PTX. Cytotoxicity assays on ovary cancer cells and lung cancer cells proved that the conjugate was more effective than free paclitaxel, with better results on the lung cancer cell line.

The presence of a hydroxyl function on the 10-hydroxycamptothecin (HCPT) renders this antitumor compound suitable for conjugation by means of ester bonds (Wu et al., 2009). The hydrolysis in fetal bovine serum was much faster than at pH 5, leading to 80% release in 2.5 hours versus 10% in acetate buffer, confirming the important activity of esterases in releasing the drug. In vitro cytotoxicity experiments on human gastric carcinoma cells showed that the activity of HCPT/MWCNTs was enhanced with respect to the free drug, probably thanks to a better penetration of the nanotube derivatives into the cells. In addition, in vivo experiments on an animal model of hepatic tumor confirmed the activity of HCPT/MWCNTs, which induced a significant reduction of the cancer growth in comparison to free HCPT at the same concentration.

12.4.4 Carbamate Bonds

Carbamate bonds, characterized by an intermediate stability between ester and amide bonds, can be cleaved by chemical or enzymatic hydrolysis. This characteristic allows us to obtain a good biodistribution of the drug-bearing vector and the release of the active molecule takes place intracellularly, in the lysosomes. Up to now there is only one CNT-based system where a drug has been linked to the vector by means of a carbamate bond (Chaudhuri et al., 2010). In the proposed system Dox was connected to a pyrene unit used to supramolecularly interact with the CNTs via π-stacking (Fig. 12.10). The drug release was studied in melanoma cell lysate or in PBS. Dox was cleaved from the pyrene unit in the cellular environment, while the system was pretty stable at pH 7.4. In vitro experiments indicated that the Dox-pyrene/CNT derivatives were internalized by cells and were localized into lysosomes. Dox was released from the vector and could exert its activity, as demonstrated by cytotoxicity studies. In vivo therapeutic effects were proved using mice bearing melanoma as model. The Dox-pyrene/CNTs reduced the tumor growth as observed for the free drug. Furthermore, contrary to free Dox, the systemic toxicity of the conjugate was significantly decreased.

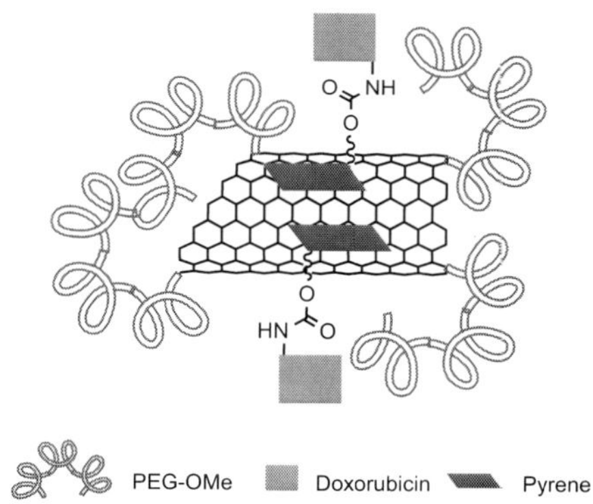

Figure 12.10 Structure of a Dox-pyrene/CNT derivative.

12.5 Conclusion

The use of CNTs, both single- and multiwalled, can be really of interest for the delivery of drugs. In fact the capability of CNTs to penetrate into cells is an important driving force for delivering the therapeutic molecules in selected cells, especially if conjugated to targeting agents. However, for many drugs the activity can be exerted only if they are released from the delivery systems and this can be achieved by many different methods, which have been reviewed in this chapter.

Depending on the methodology used to connect the active biomolecules to nanotubes, various means can be used, such as the opening of the tips, which can be temporally blocked, to release materials encapsulated into the CNT cavity. Alternatively, the change of pH can be exploited to protonate compounds and detach them from the carbon surface. The differences between intra- and extracellular environments (e.g., pH or presence of specific enzymes) are essential to develop engineered platforms with a stability that changes, depending on the biological compartment where they accumulate.

The possibility to induce these processes using external stimuli, such as NIR irradiation or electrical impulses, permits a precise control of the drug release. The presence of the vector and the triggering agent (either exogenous or endogenous) is necessary to induce a more selective and efficient therapeutic effect, avoiding undesired side effects in cells and tissues that are not specifically targeted.

Chapter 13

Dendrimers

Karnaker R. Tupally, Ganesh R. Kokil, Sachin S. Thakur, Prachi Singh, and Harendra S. Parekh
Pharmacy Australia Centre of Excellence, University of Queensland, 20 Cornwall Street, Brisbane, Queensland 4102, Australia
h.parekh@uq.edu.au

Nature has inspired humankind for centuries and on many fronts, with translation of its beauty, functionality, and innate processes having been carefully studied and transformed into products and applications that shape the fields of science and engineering well into this century, and beyond. Structures of immense strength and functionality, in particular, such as the honeycomb and the spider's web are classic examples of nature's contribution to humankind. Here, chemically derived dendrimers are no exception, being highly branched macromolecules with a plethora of applications that are elaborated on in this chapter.

Controlled Release Systems: Advances in Nanobottles and Active Nanoparticles
Edited by Alexander van Herk, Jacqueline Forcada, and Giorgia Pastorin

ISBN 978-981-4613-21-7 (Hardcover), 978-981-4613-22-4 (eBook)
www.panstanford.com

13.1 Introduction

13.1.1 The Origin of Dendrimers

The term "dendrimer" originates from the Greek word *dendron*, meaning "tree," and as such their architecture closely resembles the branches (and roots) of a flourishing tree. Dendrimeric architecture can be seen across the plant/animal kingdom, for example, geckos possess a network of many millions of microscopic foot hairs arranged in a highly ordered and branched (dendrimeric) fashion, which serve as remarkable adhesive anchors.

From a chemistry perspective, the first such library of synthetic dendrimers, or "cascade molecules," as they were then termed by their creator Fritz Vögtle, was reported in the late 1970s (Buhleier et al., 1978). Since then, dendrimers have come into their own as a unique class of monodisperse polymers, differing from conventional polymers, given their precise structure, size, and surface-active functionality. Today over 100 varieties of dendrimers have been prepared, with some of the most commonly utilized examples depicted in Fig. 13.1 below (Mintzer and Grinstaff, 2011; Tomalia, 2005).

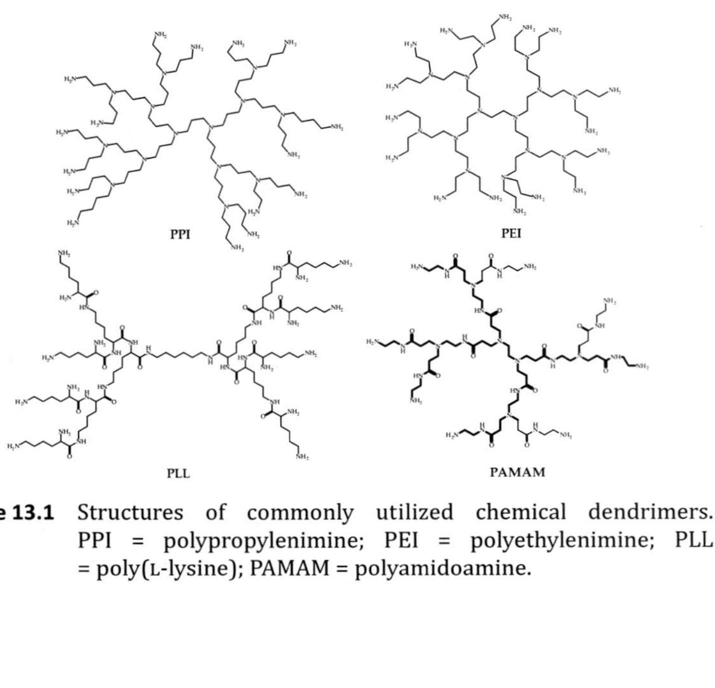

Figure 13.1 Structures of commonly utilized chemical dendrimers. PPI = polypropylenimine; PEI = polyethylenimine; PLL = poly(L-lysine); PAMAM = polyamidoamine.

13.1.2 Dendrimer Architecture

The architecture of these principally symmetrical macromolecules can be described as originating from a bifunctional core (i.e., G_0; Fig. 13.2), to which consecutive layers of a monomer are chemically tethered in a controlled fashion, ensuring that a successive generation yields twice the number of surface-active functional groups as its predecessor.

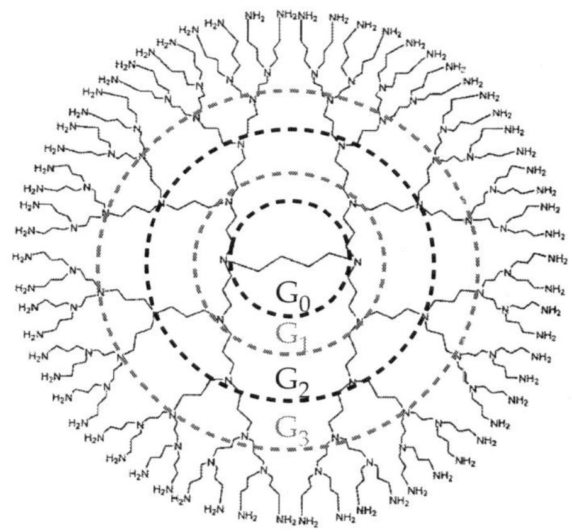

Figure 13.2 A PPI dendrimer with the first three generations (G_0–G_3) highlighted. Reprinted from Jain et al. (2010), Copyright 2010, with permission from Elsevier.

The structure of a dendrimer may be divided into its core, interior cavities and surface-active groups (Mintzer and Grinstaff, 2011). The core determines the ultimate shape of a dendrimer, be this spherical, ellipsoidal, or cylindrical. The interior cavities on the other hand are able to accommodate species through encapsulation, most often through hydrophobic interaction. Finally, the surface-active groups play a pivotal role in their ability to directly interact with molecules through complexation (i.e., ion pairing), while also being available for further polymerization or functionalization, for example, for targeting purposes.

13.1.3 Dendrimer Synthesis

Two well-established strategies are principally employed in the chemical synthesis of dendrimers, these being convergent or divergent in nature. In the divergent approach, a (bi)functional core is reacted with partially protected monomers. This is followed by a deprotection step that liberates a precise number of surface-active groups that are further reacted with additional monomer units, and the process is continued until dendrimers of desired generation are prepared (Fig. 13.3a) (Parekh et al., 2007). Structural flaws to dendrimers prepared via the divergent approach can result from each deprotection step, which, if incomplete, results in the creation of fractured rather than precisely symmetrical systems.

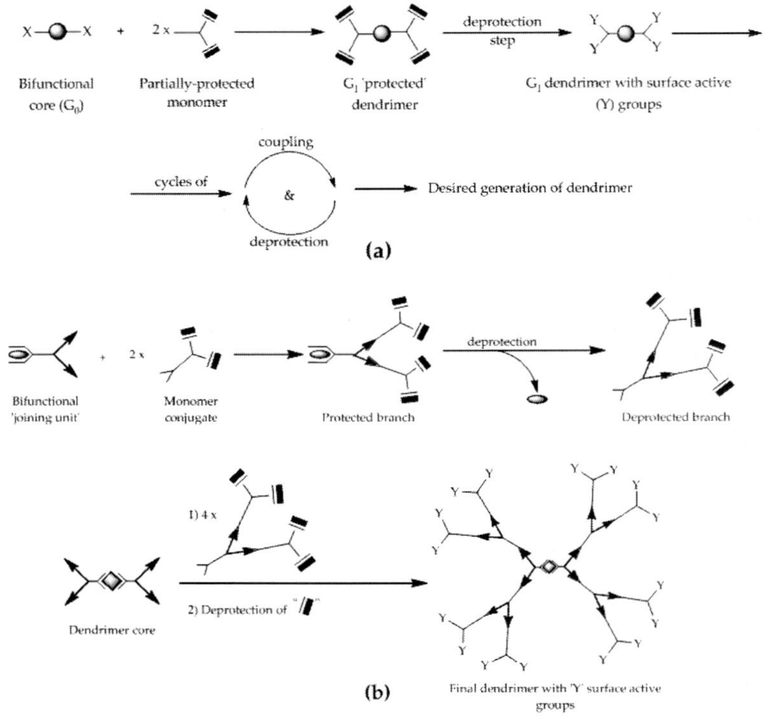

Figure 13.3 Schematics representing (a) divergent and (b) convergent dendrimer synthesis. Reprinted from Parekh et al. (2007), with kind permission of Dr. N. Udupa and Dr. P. D. Gupta.

On the other hand, the convergent approach first requires preparation of well-defined branched fragments, which are then collectively tethered to a central (bi)functional core to deliver the final dendrimer (Fig. 13.3b) (Hawker and Frechet, 1990;

Parekh et al., 2007). An obvious drawback to this approach is inaccessibility to the core by highly branched deprotected fragments, which limits the generation of dendrimer that can be produced by a convergent approach.

13.2 Dendrimers: Fundamentally Interacting Architecture

Prior to the development of a biocompatible dendrimeric system, it is crucial to understand and appreciate the propensity of these molecules to form associations with various components of biological systems (Al-Dosari and Gao, 2009; Tian and Ma, 2013). Interaction and affinity toward cell membranes, for example, is considered vital and it is most notably the surface functionality of a dendrimer that dictates this fate (de Jong and Borm, 2008; Fröhlich, 2012; Mecke et al., 2005). A considerable amount of effort has been applied toward unravelling and interpreting the physicochemical properties of dendrimers so as to better predict their behavior upon encountering biomolecules, including cellular lipid membranes, proteins, and genes (Gonzalez-Nilo et al., 2011; Tian and Ma, 2013).

13.2.1 Dendrimer–Lipid Membrane Interactions

Cationic dendrimers, upon coming into contact with a cell, will first encounter its membrane, which comprises $\approx 10^9$ lipid molecules rich in negatively charged carbohydrates and phospholipids (Alberts et al., 2002). The interaction between dendrimer and cell membrane (Fig. 13.4A) is the first step toward the carrier being internalised and studies indicate that large cationic molecules, such as dendrimers, disrupt lipid membranes (Fig. 13.4B), enhancing delivery of biomolecules into endosomes (Fig. 13.4C) (Karoonuthaisiri et al., 2003; Mecke et al., 2005; Zhang and Smith, 2000). The phenomenon of dendrimer-induced membrane pore formation has been widely studied using a range of techniques including atomic force microscopy, solid-state nuclear magnetic resonance (NMR), and dynamic light scattering (DLS) (Erickson et al., 2008; Hong et al., 2009b; Kelly et al., 2008a; Mecke et al., 2004). On the other hand, in silico studies using atomistic, coarse-grained, and mesoscale simulations have also served to improve our understanding of the forces that drive dendrimer-mediated internalization, enabling the design and synthesis of systems with optimal delivery characteristics alongside reduced cytotoxic potential (Kelly et al., 2008b; Kelly et al., 2009; Lee, 2012; Lee and Larson, 2011).

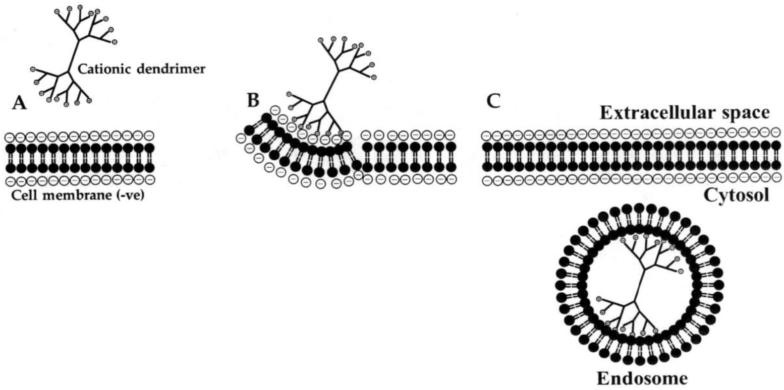

Figure 13.4 Dendrimer encountering a cell's (lipid) membrane.

13.2.2 Dendrimer–Protein Interactions

Dendrimers have the potential to significantly alter the 3D conformation of proteins they are capable of binding with, and in the case of human serum albumin they have demonstrated an ability to shrink α-helices, while also increasing random coil formation (Froehlich et al., 2009). Not surprisingly, such conformational changes of proteins can lead to diminished or total loss of activity through aggregation/precipitation, mediated through electrostatic and/or hydrophobic interaction with dendrimers (Gabellieri et al., 2006; Giehm et al., 2008; Heegaard et al., 2007; Shcharbin et al., 2006). Conversely, such interactions may also be utilized for therapeutic benefit. For example, G_1 gallic acid–triethylene glycol dendrimers have been proposed as inhibitors of dimerization, between the capsid proteins present on HIV particles (Domenech et al., 2010); this purportedly circumvents virulence through interaction of the hydrophobic regions of the capsid with large hydrophobic moieties (benzoates) that decorate the dendrimer surface. Furthermore, dendrimer–protein interactions are immensely dependent on dendrimer generation and surface groups, as well as the pH, ionic strength, and composition of the surrounding media/environment (Giehm et al., 2008; Shcharbin et al., 2005).

13.2.3 Dendrimer–Gene Interactions

The vast majority of dendrimer-based applications appearing in the literature in recent years demonstrate them to be promising carriers of genetic material in vitro, and albeit with more limited success, claims of in vivo delivery have also been reported (Biswas and Torchilin, 2013; Dutta et al., 2010; Juliano, 2006; Patil et al., 2008; Ravina et al., 2010; Taratula et al., 2009; Yu et al., 2012b). Limitations of a dendrimer-mediated gene delivery stem from their inherent cytotoxicity, driven by their cationic charge density, a factor also crucial to effective complexation (Fig. 13.5), while poor transfection efficiency is chiefly a result of premature degradation in endolysosomes (Jain et al., 2010; Lechardeur and Lukacs, 2002).

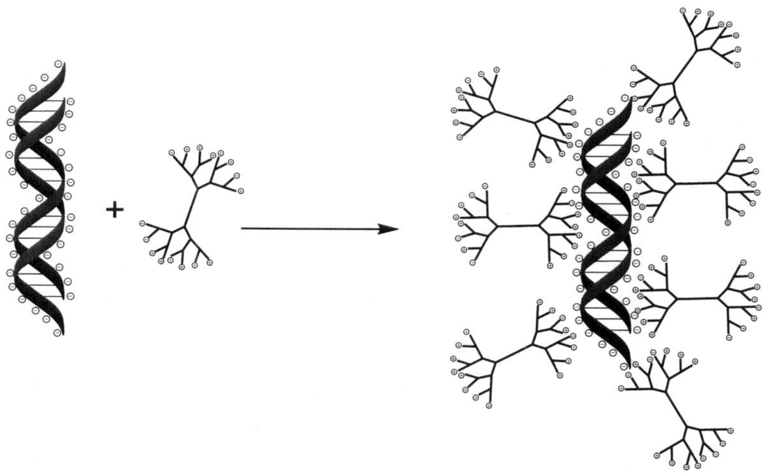

Figure 13.5 Anionic gene–cationic dendrimer interactions based on charge density–driven complexation.

Degradation within the intracellular endolysosomal compartment is a major bottleneck to gene delivery where chemical vectors, such as dendrimers, are employed (Lechardeur and Lukacs, 2002). Previously reported in vitro studies demonstrate that not only are carriers possessing a range of amine chemistries (e.g., PPI, PEI; see Fig. 13.1) able to more successfully condense genetic material; they can also buffer the acidifying environment of maturing endosomes, through gradual protonation of internal (3°) amines upon lowering pH (Boussif et al., 1995; Mintzer and Simanek, 2009). Thus, dendrimers possess fundamental prerequisites of carrier systems, such that they complex/condense genetic material, rendering them more resistant to nucleases and amenable to membrane translocation, while with the appropriate chemistry they can also be engineered toward timely escape from endosomes. From these findings, it is evident that condensing genes using dendrimers is a critical step that not only shields the gene against degradation but also facilitates transfection. Thus, a better understanding of the molecular mechanisms underlying gene condensation/protection and timely liberation of a gene from its carrier are crucial in realizing the potential of dendrimers as in vivo gene delivery agents (Ouyang et al., 2010a, 2010b).

To date, various visual (e.g., transmission electron microscopy [TEM], scanning electron microscopy [SEM]) as well as spectroscopic (e.g., DLS, NMR) techniques have been trialed to qualify the gene condensation process, and typically interpolyelectrolyte complexes bearing a net positive charge are needed for complete compaction (and transfection), this often being quantified through calculation of charge (N:P) ratios, with N denoting the cationic charge on the dendrimer-nitrogens and P the anionic charge of gene phosphate groups in the final mixture (Jensen et al., 2011; Nandy et al., 2012; Pavan et al., 2010; Shah et al., 2011; Shan et al., 2012). Trialing different generations of dendrimer as well as altering the pH of the surrounding media is useful in regulating and optimizing the charge ratio (Karatasos et al., 2012; Ouyang et al., 2010a, 2010b; Ouyang et al., 2011). Here, it is only at neutral-acidic pHs that amine-terminated dendrimers are ionized due to the protonation of basic amine moieties, and the resultant strong electrostatic interaction with genes that results is responsible for overcoming the bending elastic energy of that genetic material, allowing it to collapse onto the dendrimer (Nandy and Maiti, 2010). Collectively, these findings have also inspired the creation of numerous simulation-based approaches, including atomistic molecular dynamics, the bead-spring model, and analytical model studies aimed at examining the conformational dynamics of gene carrier condensates as well as quantifying their affinity for one another; after all, separation of gene and carrier mustn't be overlooked as it is an essential prerequisite to transfection following cellular internalization (Ouyang et al., 2010a, 2010b; Pavan et al., 2011; Tian and Ma, 2013).

The dissociation of genetic material from their vector once in the target cell is an important consideration impacting the overall efficiency of gene delivery systems. The release mechanisms at play are still poorly understood and probing modalities that enable controllable release using practical and theoretical/computational models will be required to establish better dendrimer design protocols that fulfill the many demands for improved cytosol-/nuclear-targeted delivery and timely release of therapeutic agents.

13.3 Applications of Dendrimers: A Bird's-Eye View

13.3.1 Delivery Vectors for Therapeutics: Drugs and Genes

The unique features of dendrimers such as modifiable surface functionalities along with a well-defined size and shape makes these polymers exceptionally attractive candidates as carrier systems (Patri et al., 2002). Various strategies exploiting the ability of dendrimers to encapsulate, complex with, and conjugate to molecules have been utilized to enhance the therapeutic efficacy of a variety of bioactives.

13.3.1.1 Encapsulation and complexation

The spatial architecture of a dendrimer allows it to form associations with a broad range of molecules. While void spaces within the interior of the dendrimer permit encapsulation of small molecules (e.g., drugs), the branched surface allows for formation of association complexes with larger molecules (e.g., genes). The driving forces for these interactions include hydrogen bonding, hydrophobic interactions, and also electrostatic intermolecular interactions between charged groups of oppositely charged entities (D'Emanuele and Attwood, 2005).

Stemming from such interactions dendrimers have facilitated the solubilization (of poor-soluble drugs), offered increased stability via protection from degradation, enhanced circulation time, imparted sustained release properties, and allowed for targeted delivery (Mintzer and Grinstaff, 2011). In fact, the ability of dendrimers to assist with therapeutic targeting has led to them drawing great interest in fields such as chemotherapy

(Ali et al., 2011a; Stanczyk et al., 2012). This section will next focus on the encapsulation of therapeutics, while targeting strategies will be discussed later.

The rationale for utilizing dendrimers in the delivery of chemotherapeutic agents can be primarily attributed to the leaky vasculature of tumor tissue; this being accompanied by ineffective lymphatic drainage leads to comparatively higher uptake of macromolecules, which plays in the favor of dendrimer-mediated anticancer therapies (Fox et al., 2009; Mintzer and Grinstaff, 2011). Camptothecin is an anticancer agent whose efficacy is limited by its poor aqueous solubility and a debilitating adverse effect profile; this led to development of 10-hydroxy camptothecin, which has subsequently been investigated when encapsulated in a biocompatible polyester dendrimer composed of glycerol and succinic acid (Morgan et al., 2006). The dendrimer–drug complex demonstrated a highly acceptable IC_{50} (nmol/L range) when applied to a range of human cancer cell lines. The high efficacy attributed to an increase in cellular uptake of the complexes when compared to native drug. Similarly, dendrimers based on poly(etherhydroxylamine) or 1,3,5-triazine have also been trialed and found to significantly enhance the solubility of camptothecin and 10-hydroxy-camptothecin, respectively (Tomali et al., 2006; Zhang et al., 2003b). Various anticancer drugs have been encapsulated into dendrimers with varying degrees of success (in vitro) and these include cisplatin, doxorubicin, docetaxel, paclitaxel, 5-fluorouracil, and methotrexate (MTX) (Benito et al., 2004; Bhadra et al., 2003; Dhanikula et al., 2008; Malik et al., 1999; Ooya et al., 2004; Papagiannaros et al., 2005).

Dendrimers have also been extensively investigated for their ability to augment delivery of anti-inflammatory agents (e.g., nonsteroidal anti-inflammatory drugs [NSAIDs]). While this class of agent has long been used to treat symptoms of acute/chronic pain, efficacy and compliance are commonly limited by poor water solubility and gastrointestinal side effects. Studying the impact of dendrimer generation on piroxicam solubility led to the firm conclusion that G4-PAMAM significantly increased drug solubility as well as half-life of the drug, this in comparison with the G_3-PAMAM dendrimer as well as the native drug (Prajapati et al., 2009).

Dendrimer complexation with genes has been applied as a means of delivering small interfering RNA (siRNA), antisense oligonucleotides (ODNs), and plasmid DNA (pDNA) for a variety of therapeutic applications (Biswas and Torchilin, 2013; Parekh,

2007; Ravina et al., 2010; Šebestík et al., 2012). However, as alluded to earlier (Section 13.2.3), a high charge density is typically required for the efficient loading of these molecules, which in turn increases their cytotoxic potential. As a result, studies have investigated various modifications to dendrimeric architecture in order to circumvent toxicity. To this end siRNA has been loaded into internally quarternized and surface-acetylated PAMAM dendrimers, these dendrimers forming condensed spherical particles that retained the siRNA within their internal cavities (Patil et al., 2008). Such modified dendrimers demonstrated far reduced cytotoxicity compared to conventional PAMAM, while also being able to enhance fluorophore-labeled siRNA uptake in vitro (Patil et al., 2008). Another successful approach toward limiting dendrimer toxicity is through the engineering of dendrosomes (dendrimers encapsulated within liposomes). Such systems retained the silencing efficiency of dendrimer/siRNA complexes, while all but eliminating their cellular toxicity (Dutta et al., 2010). Further modifications to enhance plasma stability and intracellular bioavailability of dendrimer/gene complexes have been achieved through encapsulation of the complexes into PEGylated-luteinizing hormone releasing hormone ligand containing dimethyl-3-3′-dithiobispropionimidate-HCl-caged nanoparticles (Taratula et al., 2009).

13.3.1.2 Chemical conjugation

While physical interactions between dendrimers and drugs serve as a useful approach toward enhancing drug delivery, the strategy is not without its significant flaws. Namely, physical association only allows for relatively low drug loading, while there is very little control over the kinetics of drug release from dendrimer (Cheng et al., 2007; Svenson, 2009). On the other hand chemical conjugation permits multiple drug copies to be covalently tethered to the dendrimer surface; this allows for a more reliable and higher loading capacity, which in turn results in greater payloads being delivered to the target (Cheng et al., 2007; Svenson, 2009).

Similar to the aforementioned drug encapsulation strategies, chemotherapeutic agents have been the most extensively investigated by way of chemical conjugation to dendrimers. Doxorubicin is one of the most widely used anticancer agents having been encapsulated in liposomes (Doxil®) to enhance its solubility and reduce dose-related toxicity. Despite its commercial success, concerns regarding the long term stability as well as efficacy of Doxil® still remain. In an attempt to counter this, doxorubicin has been investigated as a prodrug, through chemical conjugation to dendrimers. With doxorubicin bound to an asymmetric, biodegradable polyester-based dendrimer (Fig. 13.6), drug toxicity was reduced, tumor uptake improved, and complete tumor regression (comparable to Doxil®) was observed with a 100% survival rate (over 60 days) being recorded in a mouse model of colon cancer (Lee et al., 2006b). In a similar fashion to doxorubicin a range of anticancer drug–dendrimer conjugates such as paclitaxel, epirubicin, and MTX have been also investigated, with similar findings reported (Gurdag et al., 2006; Majoros et al., 2006; Pasut et al., 2005).

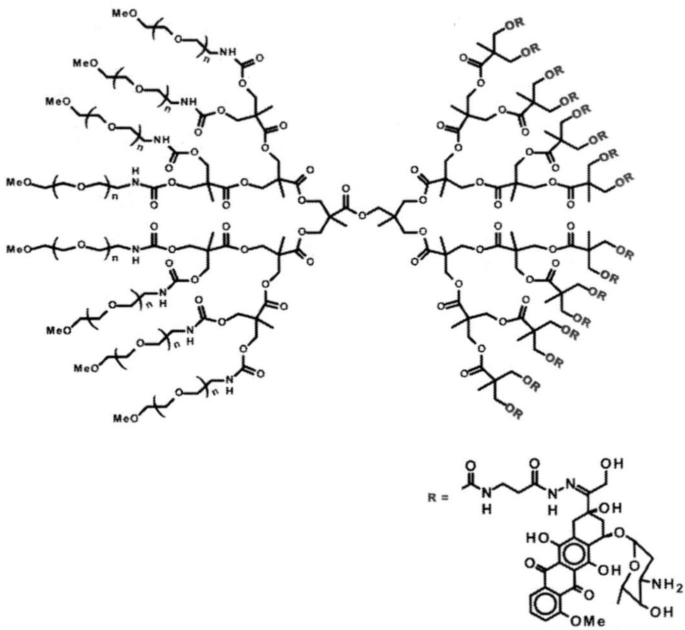

Figure 13.6 Chemical structure of an asymmetric polyester–doxorubicin dendrimer conjugate. Reprinted with permission from Lee et al. (2006b), Copyright 2006, National Academy of Sciences, U.S.A.

Therapeutic success, at least in vitro and through chemical conjugation has not been limited to chemotherapeutic agents; anti-inflammatory agents including ibuprofen and naproxen have also drawn attention, although here efforts have focused on studying the specific linkages through which drugs are tethered to dendrimers

(Kurtoglu et al., 2010; Najlah et al., 2007; Najlah et al., 2006). In the case of the PAMAM-(ibuprofen)$_n$ system where amide as well as ester linkages were employed, the drug was unable to release from PAMAM when incubated with human serum. In contrast, conjugation through a PEGylated linkage led to the conjugate being more susceptible to human serum-based esterases, releasing ~2/3 of the total drug present on the dendrimer within 12 hours (Kurtoglu et al., 2010). In the case of naproxen, while all dendrimer–drug linkages trialed enhanced drug solubility, the amide linkage was found to be more resistant than the ester to enzymatic degradation in human serum. To elaborate subtleties further, two varieties of ester linkages resulted with varied enzymatic cleavage profiles, with the lactide ester derivative releasing approximately 25% of naproxen within 24 hours, whereas a diethylene glycol ester resulted in 100% release within just 6 hours (Najlah et al., 2006).

13.3.2 Target Specificity

In 1906, Paul Ehrlich first postulated the concept of targeted drug delivery through his well-known "magic bullet" theory. The concept proposed delivering therapeutic molecules to the tissues in the body where they are most required, while limiting off-target delivery (Bertrand and Leroux, 2012; Muller and Keck, 2004; Strebhardt and Ullrich, 2008). Targeted delivery remains the holy grail over a century later, especially so for complex disease states such as cancer, where chemotherapy is a primary treatment modality (Vasir and Labhasetwar, 2005). Dendrimeric systems have demonstrated the potential to target both nonspecific passive bulk cancerous tissue as well as specific or active/unique tumor cells (Ali et al., 2011a; Stanczyk et al., 2012; Wolinsky and Grinstaff, 2008). Delivery of anticancer drugs using dendrimers is further enhanced through the ability to colocate targeting moieties, and here successes have been reported with ligands ranging from PEG, folic acid, and carbohydrates to peptides and large proteins, this being elaborated on directly below.

13.3.2.1 PEGylation

Molecules modified through chemical conjugation with PEG can promote the delivery of high drug titers to their target cell population (Gajbhiye et al., 2009; Jain and Tekade, 2013; Medina and El-Sayed, 2009). The inclusion of PEG, or PEGylation as it is commonly known imparts additional features such as improved solubilization characteristics when compared with naked dendrimers or dendrimer–drug conjugates devoid of PEG (Bhadra et al., 2003; Kulhari et al., 2011; Medina and El-Sayed, 2009).

The incorporation of PEG provides a range of systemic benefits, too, including decreases in each of hemolytic toxicity, uncontrolled drug outflow, and macrophageal uptake (Bhadra et al., 2002; Gajbhiye et al., 2007; Jain et al., 2010; Zhu et al., 2010a, 2010b). PEGylation also significantly improves the kinetic stability of dendrimers and makes them candidates for the extended delivery of bioactives (Kulhari et al., 2011; Veronese and Pasut, 2005). Inclusion of PEG also provides a shielding effect, which reduces the immunogenicity and antigenicity of "coated" molecules such is the case with PEGylated dendrimers (Harris et al., 2001; Peracchia et al., 1999; Yang et al., 2008a).

Nonspecific or passive targeting of tumors is usually achieved by increasing the hydrodynamic diameter of the dendrimer, and this is very effectively achieved through PEGylation (Fig. 13.7) (Iyer et al., 2006; Medina and El-Sayed, 2009; Zhu et al., 2010a, 2010b). Extending the application of PEG further through the engineering of triblock systems, such as PAMAM-PEG-PAMAM, affords a highly water-soluble copolymer capable of forming polyplexes with pDNA that are devoid of toxicity despite their extended stability, this system achieving high transfection efficiencies on par with the inherently cytotoxic PEI (Kim et al., 2004).

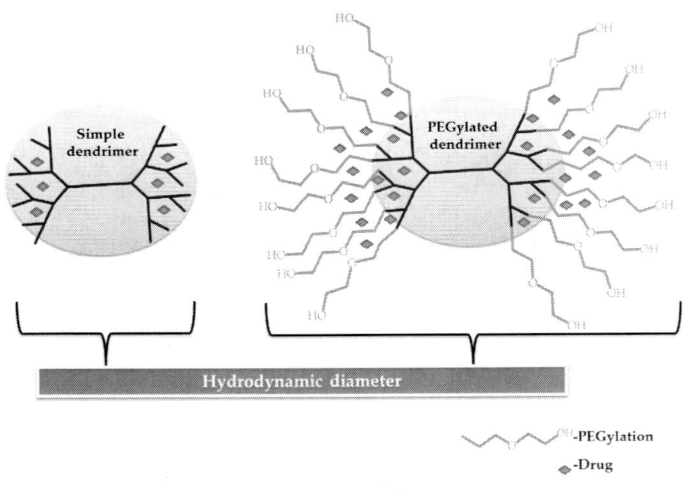

Figure 13.7 Hydrodynamic diameter comparison of a simple vs. a PEGylated dendrimer.

13.3.2.2 Folic acid conjugation

Folic acid (FA) has been extensively investigated in the field of chemotherapy, with the first documented evidence claiming a relationship between tumors and their absolute requirement for FA reported in 1944 (Leuchtenberger et al., 1944). Since then much has been elucidated about the health benefits of FA, and more recently this has extended into the realms of carrier systems. Specifically, FA-conjugated dendrimers have been proven to preferentially target tumor cells given their overexpression of folate receptors (Kono et al., 1999; Quintana et al., 2002; Wiener et al., 1997). Interestingly, FA receptor-binding efficiency increases with each additional FA molecule tethered to a dendrimer, this phenomenon peaking at five to six moieties per dendrimer. Although binding was found to increase, the rate of internalization did not follow similar trends, it being proposed that folate receptor clusters are typically preorganized on membranes of tumor cells, and that the key factor in reported tumor reduction is enhanced residence time on the cell rather than the rate of endocytosis (Hong et al., 2007). The antineoplastic agent MTX is an FA antagonist with a several 100-fold greater affinity than FA for the folate receptor (Antony, 1996). Hence, logical progression has led to the inclusion of MTX into dendrimeric systems, with elaborate systems comprising MTX, FA, and fluorescent markers (for tracking purposes) having been engineered and trialed with promising outcomes (Kukowska-Latallo et al., 2005; Majoros et al., 2005; Zong et al., 2012).

13.3.2.3 Monoclonal antibody conjugation

Tumor cells express a range of surface antigens, some that evolve and are cancer specific and others that are simply overexpressed when compared with normal cells. This phenomenon has led to an explosion in monoclonal antibody (mAb)-based approaches toward cancer therapy (Baselga et al., 2010; Bhatt et al., 2010; Nelson et al., 2010; Robert et al., 2011; Younes et al., 2012). If an overexpressed antigen is involved in cancer growth and proliferation, targeting it through a mAb may serve as a highly selective and potent therapeutic tool.

mAb-J591 is one such example, being specific toward prostate cancer–specific membrane antigen (PSMA), and it has been trialed as a G_5–PAMAM conjugate and then assessed for its binding affinity and internalization potential (Patri et al., 2004). The mAb–dendrimer conjugate was found to bind PMSA-positive LNCaP cells with exceptional specificity, this not being the case for PC-3 (PMSA-negative) cells. Furthermore, the conjugate was internalized as confirmed by confocal microscopy, while in the case of the native G_5–PAMAM, cells did not internalize the dendrimer to any appreciable extent. A similar outcome followed the application of anti-HER2–G_5–PAMAM for targeting of human growth factor receptor 2 (HER2), which is often overexpressed in breast and ovarian malignancies. The conjugates were highly effective at binding to and being internalized by HER2-expressing tumors both in vitro and in vivo (Shukla et al., 2006). A separate study investigated the efficacy of two different mAb–G_5–PAMAM conjugates, 60bca and J591, for the targeting of CD14 and PSMA, respectively with successful targeting achieved in vitro in both cases, which demonstrates the overwhelming promise of this approach (Thomas et al., 2004).

13.3.2.4 RGD–peptide conjugation

Arginylglycylaspartic acid (RGD) is a tripeptide with remarkable cell-binding tumor-targeting properties and its cyclized derivative, RGD-4C, has been conjugated to partially acetylated G_5–PAMAM fluorescently labeled (Alexa Fluor 488) for the targeting of tumor neovasculature, via uniquely expressed $\alpha_V\beta_3$ integrins (Shukla et al., 2005). Binding studies comparing free RGD-4C to dendrimer complexed-RGD-4C were performed on several cell lines with varying levels of integrin receptor expression, it being reported that while free RGD-4C bound much more rapidly than the RGD-4C-dendrimer complexes, the latter dissociated over 500 times slower, suggesting a synergistic effect of multiple peptide conjugation on binding avidity. Cyclic RGDs have also been attached to 1,4,7,10-tetraazacyclododecane-1,4,7,10-tetraacetic acid (DOTA)-conjugated mono-, di-, and tetravalent dendrimeric alkynes, again for $\alpha_V\beta_3$ integrin targeting (Dijkgraaf et al., 2007). Binding characteristics were evaluated in vitro and in vivo in mice bearing human kidney–derived tumors, and it was shown through biodistribution studies that the tetrameric RGD dendrimer demonstrated the highest levels of tumor targeting.

13.3.2.5 Glycosylation

Glycans and their interaction with cell surface carbohydrate/protein receptors signifies an important class of mechanism eliciting a number of biological and pathological processes, including cell recognition, signaling, trafficking, and the onset of infectious diseases (Varki et al., 2009). While monomeric glycans have a very low affinity for their receptor, recent advances have provided insights into the need for multivalent glycofunctionalities to be present in order to attain therapeutic relevance (Dam and Brewer, 2010; Marradi et al., 2013). As such, clustered carbohydrate structures comprising multivalent glycans attached to natural products or synthetic polymers, such as dendrimers, have been prepared to yield enhanced affinities to relevant targets (Kikkeri et al., 2011). G_5–PPI glycodendrimers with a dense maltose shell are internalized with greater efficiency by tumor cells than nontumor cells, demonstrating the potential of such systems in chemotherapy (Filimon et al., 2012). Glycodendrimers may also be utilized as biocompatible coatings to help lower the cytotoxicity and enhance cell adhesion properties of structures such as nanotubes (Chen et al., 2009; Wu et al., 2008). This class of dendrimers has also gained interest as therapeutic agents in their own right, given their ability to target and interact with biological membranes of a broad range of pathogens, including bacteria and viruses (Clayton et al., 2011; Garcia-Vallejo et al., 2013; Han et al., 2012; Heidecke and Lindhorst, 2007; Rojo and Delgado, 2004).

13.3.3 Theranostics

Applications of dendrimers and the multitude of conjugates in the literature are certainly vast in number, many extending beyond the scope of therapeutics into the realms of diagnostics or a hybrid of the two, as is expanded in next section.

13.3.3.1 Boron neutron capture therapy

Irradiation of ^{10}B (boron, nonradioactive stable isotope) with a low-energy neutron beam produces the unstable ^{11}B isotope, which is subsequently decomposed into He and Li; this process deposits large amounts of linear energy and causes high localized ionization, which promotes double-stranded DNA (dsDNA) breaks, damaging the mitotic potential of tumor cells (Barth, 2009; Wittig et al., 2008). To this end boronated mAbs have been used to facilitate accumulation of high ^{10}B titres at tumor sites (Barth, 2009; Yang et al., 2008b). Recently, alternate approaches to encourage boron accumulation have been investigated using L-4-dihydroxyborylphenylalanine and sodium mercaptoundecahydro-closo-dodecaborate (Barth et al., 2012), polyhedral boranes (Calabrese et al., 2012; Sivaev and Bregadze, 2009), and dendrimer-encapsulated icosahedral carborane cages (Spokoyny et al., 2011). Dendrimer encapsulation has been proven to enhance biocompatibility, water solubility, and biodistribution of icosahedral carborane cages, also muting immunogenic potential facing conventional therapy (Spokoyny et al., 2011; Tekade et al., 2009; Yinghuai et al., 2007). High amounts of ^{10}B have been loaded into various types of G_5 dendrimers and attached to the anti-epidermal growth factor receptor (EGFR) mAb cetuximab, these bioconjugates demonstrating a reasonable therapeutic activity against tumor proliferation (Barrett et al., 2009a, 2009b; Yang et al., 2008b). Furthermore, modification through PEGylation (Qualmann et al., 1996; Rolland et al., 2009b; Yinghuai et al., 2007), FA conjugation (Shukla et al., 2003), and dansyl-based fluorescent probe attachment (Bhattacharya et al., 2012; Qualmann et al., 1996; Rolland et al., 2009b) to a range of dendrimers (G_2–G_5 PAMAM and PLL) have enhanced tumor-specific uptake as well as therapeutic activity (Fig. 13.8) (Jang et al., 2009; Rolland et al., 2009b). Recently, peripheral attachment of carborane clusters on phenylene-cored dendrimers resulted in the accumulation of high amounts of carborane at tumor sites (Dash et al., 2012). Dendrimeric architecture has played a significant role in boron neutral capture therapy to enable maximum antitumor activity coupled with minimal toxicity (Cai et al., 2013; Majoros et al., 2008).

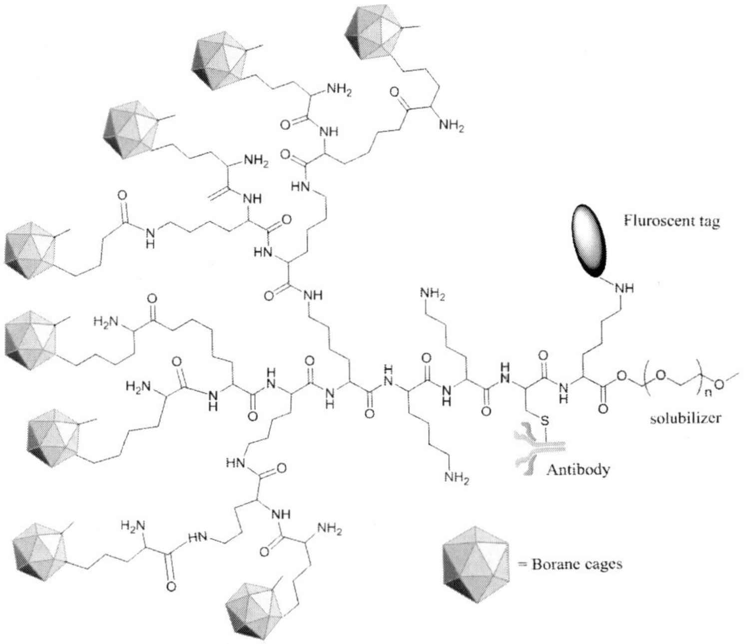

Figure 13.8 A multifunctional dendrimer used in boron neutral capture therapy. Reproduced with permission from Qualmann et al. (1996). Copyright Wiley-VCH Verlag GmBH & Co. KGaA.

13.3.3.2 Photodynamic therapy

Photodynamic therapy (PDT) is considered a less invasive method for the treatment of advanced stage tumors as compared to conventional surgical methods (Haasbeek et al., 2008; Huang et al., 2008a). Here, light irradiation of photosensitizers generates reactive oxygen species (ROS) that challenge tumor cell survival (Juarranz et al., 2008). However, tumor-selective accumulation of photosensitizers and reduction in peripheral toxicity have been challenging feats to overcome (Huang et al., 2008a; Juarranz et al., 2008). To address this, dendrimers have been investigated as potential carriers in PDT, this given that the dendrimeric architecture favors encapsulation of photosensitizers, while the attachment of targeting moieties can assist with preferential accumulation in tumor tissue (Bugaj, 2011; Klajnert et al., 2012; Wolinsky and Grinstaff, 2008).

Aluminium phthalocyanine tetrasulfonate ($AlPcS_4$) is a photosensitizing agent, having been successfully loaded into G_3–PAMAM dendrimers grafted onto the surface of porous Si nanoparticles (Tao et al., 2013). Inner voids within silica pores and amino functionalities facilitate high $AlPcS_4$ loading such that large amounts of the photosensitizer accumulated at cancerous sites, generating ROS that depleted tumor cell numbers. Furthermore, this dendrimer-based $AlPcS_4$ exhibited good biocompatibility and higher efficacy as compared to nondendrimeric $AlPcS_4$ (Tao et al., 2013).

13.3.3.3 Molecular probes

Combining a large molecular surface area and a high density of surface functionalities is essential for the development of molecular probes. Conjugation of sensing units on the dendrimer surface generates an integrated molecular probe, and when coordinated with the appropriate biomarker this serves as a very powerful (bio)medical tool (Avital-Shmilovici and Shabat, 2010; Jennings and Long, 2009).

Various nucleic acid-based dendrimeric probes have been developed and commercialized as "3DNA" technology (Caminade et al., 2006; Mora et al., 2008). Such probes are used for routine gene analysis applications such as rapid diagnosis of genetic and pathogenetic diseases and also for high-throughput functional genomic analysis (Caminade et al., 2006; Muyal et al., 2008). Fluorescent oligonucleotide dendrimer polylabeled DNA probes were developed for use as high-sensitivity contrast agents, while also serving as polymerase chain reaction (PCR) primers (Beier and Hoheisel, 1999; Caminade et al., 2006, 2008; Shchepinov et al., 1997). To demonstrate, an immobilized G_4 dendrimer armed with multiple single-stranded DNA copies was investigated for its efficacy as a probe. The dendrimer was able to hybridize with various DNA structures, with the interaction resulting in a large resonant frequency response, and such immobilized dendrimers demonstrated linear biosensibility and higher sensitivity as compared to free or unimmobilized dendrimer systems (Jang et al., 2009; Liu et al., 2010; Shinde et al., 2012). Separately, water-soluble, biocompatible, PEG-conjugated amino acid–based dendrimers comprising tetrabenzoporphyrin sensors have also been developed for in vivo oxygen imaging (Lebedev et al., 2009); these dendrimers emit high phosphorescence in the presence of oxygen at 800 nm, while tetrabenzoporphyrin is able to achieve pH-sensitive absorption and emission (Jang et al., 2009; Lebedev et al., 2009).

13.3.3.4 Contrast agents

Magnetic resonance imaging (MRI) contrast agents have been widely used to differentiate pathological and normal cells (Yan et al., 2007) with gadolinium (Gd) ions, in particular having been extensively investigated as clinical contrast agents (Kiessling et al., 2007; Kojima et al., 2011). However, low-molecular-weight Gd ions are somewhat toxic and they rapidly diffuse out of the systemic circulation, while also being rapidly excreted (Cheng et al., 2011b; Villaraza et al., 2010). To address these shortfalls, at least in part, complexation of Gd with albumin (Raatschen et al., 2008; Thurner et al., 2011) dextran (Battistini et al., 2008), and PLL (Li et al., 2012e; Shiraishi et al., 2009) has been employed, though low relaxivity (an indicator of MRI capability) and slow excretion remain problematic in MRI-based applications (Jang et al., 2009). To counteract these, Gd conjugated with dendrimers possessing various core functionalities has been investigated as a MRI contrast agent, with the most successful candidates reported to date being Gd–PAMAM-based dendrimers (Barrett et al., 2009a, 2009b). The dendritic structure of Gd–PAMAM allows changes in signal intensity and prolongs organ-specific imaging (Luo et al., 2011). A sixfold increase in relaxivity and prolonged blood circulation time was observed when using G_6–PAMAM (Jang et al., 2009; Langereis et al., 2007; Tóth et al., 1996; Wiener et al., 1994). Relaxivity was further improved with G_7 dendrimers; however, the effect plateaued as generations were further increased (Bryant et al., 1999; Langereis et al., 2007; Mohs and Lu, 2007). Specifically, diaminobutane (DAB) dendrimers have displayed faster excretion times when compared to PAMAM systems of comparable surface charge (Kim et al., 2007c; Kobayashi et al., 2003; Longmire et al., 2008a). To elaborate, the attachment of target specific moieties such as FA (Swanson et al., 2008; Yan et al., 2007), avidin (Majoros et al., 2008; Xu et al., 2007b), or mAbs (Wolinsky and Grinstaff, 2008; Xu et al., 2007a) to the dendrimer MRI agent enhances the pharmacokinetics and longitudinal relaxation rate of tumor cells. Given the promising role of dendrimers a PLL-based Gd dendrimer (Gadomer-17) proceeded to the clinical development phase as an MRI agent (Fig. 13.9) (Bumb et al., 2010; Misselwitz et al., 2001). Other MRI agents such as PPI dendrimers functionalized with 4 to 12 glucose moieties on the surface and Gd(III) ions coordinated to the core have also demonstrated potential with tunable molecular relaxivities (Langereis et al., 2006). Separately, Gd–DOTA conjugated onto a poly (L-glutamic acid) dendrimer through reversible disulfide bridges (via cystamine) enhanced complex stability, providing effective contrast diagnostic imaging (Xu et al., 2007c). On that basis a gallic acid–tetraethylene glycol (GATG) dendrimer was developed for use as an MRI contrast agent (Amaral et al., 2011; Fernández-Trillo et al., 2011). The molecular size (and volume) of dendrimers are key considerations impacting applicability, with hydrophilic small-sized dendrimers (<60 kDa) typically used as renal contrast agents, while larger hydrophilic dendrimers are used for imaging within the bloodstream and as lymphatic imaging agents and hydrophobic dendrimer complexes serve as liver contrasting agents (Fernández-Trillo et al., 2011; Longmire et al., 2008a; Menjoge et al., 2010; Svenson and Tomalia, 2012; Wijagkanalan et al., 2011).

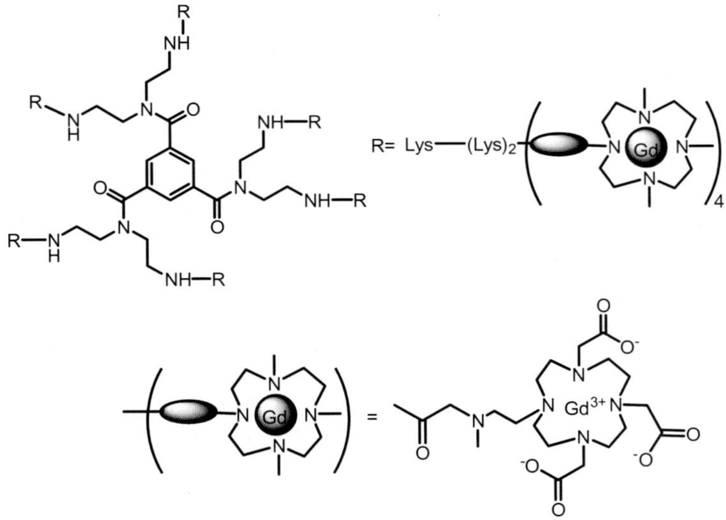

Figure 13.9 Gadomer 17.

In a similar fashion to MRI, high-molecular-weight X-ray contrast agents have also been developed with the aid of dendrimers (Longmire et al., 2008b). A computer tomography (CT) imaging contrast agent developed using a G_4 dendrimer with triiodobenzene derivatives resulted in high spatial resolution images (Yordanov et al., 2002). However, prolonged retention of these dendrimers proved problematic and the presence of iodine derivatives caused acute renal failure (Finn, 2006; Schild et al., 2006). With this in mind a PEG core containing PLL dendrimer–triiodobenzene derivatives were subsequently developed with high X-ray attenuation (Fu et al., 2006; Lee et al., 2013c). These types of X-ray contrast agents demonstrate good chemical stability along with superior properties such as high water solubility and low osmolality (Fu et al., 2006; Lee et al., 2013c; Pathak and Thigle, 2010).

Dendrimers have also been widely explored toward the development of optical imaging agents (Hilderbrand and Weissleder, 2010; Koyama et al., 2007). Some dendrimer types produce fluorescence under desirable conditions, with simple oxidation of OH-containing dendrimers able to generate blue fluorescence (Lee et al., 2004). PAMAM (G_3–G_6) dendrimers have been investigated along these lines, with conjugation of various fluorescent probes such as fluorescein isothiocyanate (FITC) (Saovapakhiran et al., 2009), Cy3-fluorophore (Striebel et al., 2004), and Alexa Fluor 594 (Ibey et al., 2005) and 488 (Shukla et al., 2006) achieved. These modified optical agents have improved biocompatibility and retain inherent fluorescence properties for use in in vitro and in vivo optical imaging techniques (Ibey et al., 2005; Leroueil et al., 2007; Saovapakhiran et al., 2009; Shukla et al., 2006; Striebel et al., 2004). To serve this purpose a G_6–PAMAM dendrimer complexed with Gd ions and Cy5.5 allowed dual mode imaging in response to MRI and near-infrared (NIR) fluorescence (Jennings and Long, 2009, Talanov et al., 2006). Similarly, five-color NIR fluorophores and radioisotope-conjugated dendrimers have also been developed for lymphatic imaging (Kobayashi et al., 2010; Longmire et al., 2008b). Dendrimers have also been introduced as agents for imaging in TEM, with dendrimer entrapped metals such as gold and silver further enhancing the quality of TEM images (Shi et al., 2007).

13.3.3.5 Dendrimers as drug molecules

Dendrimers have themselves shown promise as antimicrobial agents (Gajbhiye et al., 2009; Menjoge et al., 2010) and Vivagel® (Fig. 13.10), an anionic sulphonated PLL dendrimer is the best example of this to date, demonstrating activity against proliferation and infection by human immunodeficiency virus (HIV) and other sexually transmitted diseases (ClinicalTrials.gov, 2012; Greco and Vicent, 2008; Rupp et al., 2007).

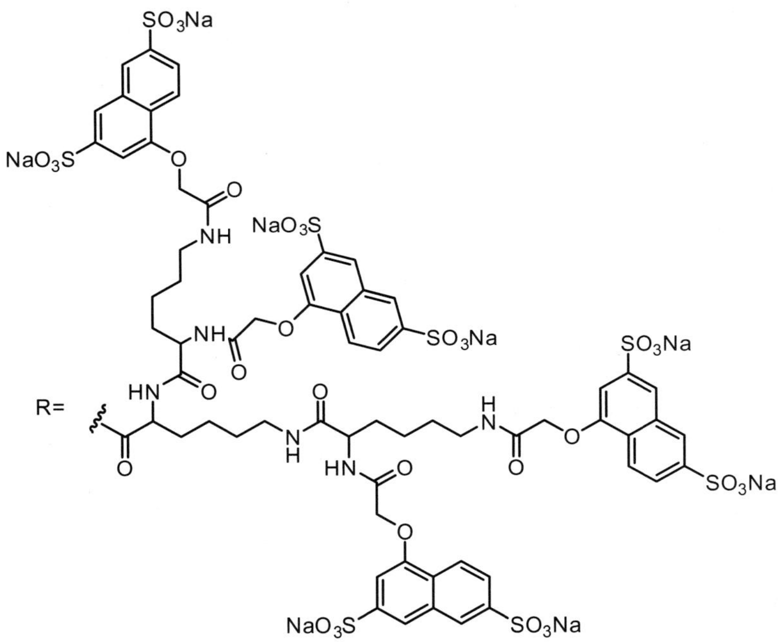

Figure 13.10 R group of Vivagel® (functionalized to a PLL core).

Although a range of popular dendrimers such as functionalized G_3–PAMAM, G_3–PPI, and G_3–PLL have been investigated for their anti-HIV activity, PLL dendrimers have been deemed the most suitable for further clinical investigation, this being a result of their ease of synthesis (Gajbhiye et al., 2009; McCarthy et al., 2005). Poly(phosphor-hydrazone) dendrimers (Pérez-Anes et al., 2009), PAMAM glycodendrimers (Clayton et al., 2011; Garcia-Vallejo et al., 2013), and galactosyl ceramide dendrimeric analogs (Jiménez et al., 2012; Rosa Borges et al., 2010) have also been evaluated for anti-HIV activity in different cell lines and with good biological activity. Separately, sialic acid–conjugated dendritic polyglycerol nanoparticles have demonstrated an ability to inhibit influenza A virus activity (Papp et al., 2011).

Various dendrimers have also displayed exceptional antibacterial properties, with PAMAM-, PPI-, and PLL-based dendrimers exhibiting activity against both gram-negative and gram-positive organisms (Castonguay et al., 2012; Tülü and Ertürk, 2012). Interestingly, PAMAM dendrimers functionalized with PEG and other terminal amine group modifications (e.g., carboxylic acid, hydroxyl and glycan termination) further enhanced bactericidal action (Branderhorst et al., 2008; Calabretta et al., 2007; Wang et al., 2010a). Carbosilane dendrimers (G_2 and G_3) with terminal quaternary ammonium groups showed a significantly potent bactericidal effect against gram-negative and gram-positive bacteria (Ortega et al., 2011; Rasines et al., 2009).

Functionalized PLL dendrimers of varying generation (G_0, G_1, and G_2) bearing linear tetra- and octapeptides (microbial surface recognition motifs) have also been developed to induce bactericidal activity (Niederhafner et al., 2005; Polcyn et al., 2009; Tam et al., 2002). Such dendrimers demonstrate antibacterial activity against gram-positive and gram-negative bacteria, with the findings of such studies being comparable to the protegrin and tachyplesin family of antimicrobial peptides (Niederhafner et al., 2005; Polcyn et al., 2009; Tam et al., 2002). Interestingly, PLL dendrimers terminated with modified mannose displayed antiadhesive properties, preventing the colonization of bacteria (Boas and Heegaard, 2004; Chabre and Roy, 2008).

The incorporation of inorganic species with dendrimers is also a promising approach with titanium-based materials conjugated to PAMAM also displaying bactericidal activity, particularly against highly resistant gram-negative strains such as *P. aeruginosa*, and with comparatively low cytotoxicity to human bone mesenchymal stem cells (Mintzer et al., 2011; Wang et al., 2011c). A strong antiprion activity has been shown when using cationic phosphorus-containing dendrimers with terminal tertiary amines, these dendrimers having also been investigated in immunological applications (Klajnert et al., 2007; Lim et al., 2010b). Here, aminobis-(methylene phosphonic acid) capping of phosphorus-containing dendrimers shows a remarkable dual effect of activating and facilitating the multiplication of human monocytes and natural killer cells, respectively (de Jong, 2010; Hayder et al., 2011; Rolland et al., 2009a). Similarly, a G_1 dendrimer having 12-aminobis-(methylene phosphonic acid) end groups derived from tyramine (G1–TamBP) (Fig. 13.11) demonstrates a promising ability to activate immune cells, while a glucosamine (or glucosamine-6-sulfate) partially functionalized 1,2-diaminoethane-centered $G_{4.5}$–PAMAM (carboxy-terminated) dendrimer displayed a marked anti-inflammatory effect (Griffe et al., 2007; Poupot et al., 2006; Quadir and Haag, 2012).

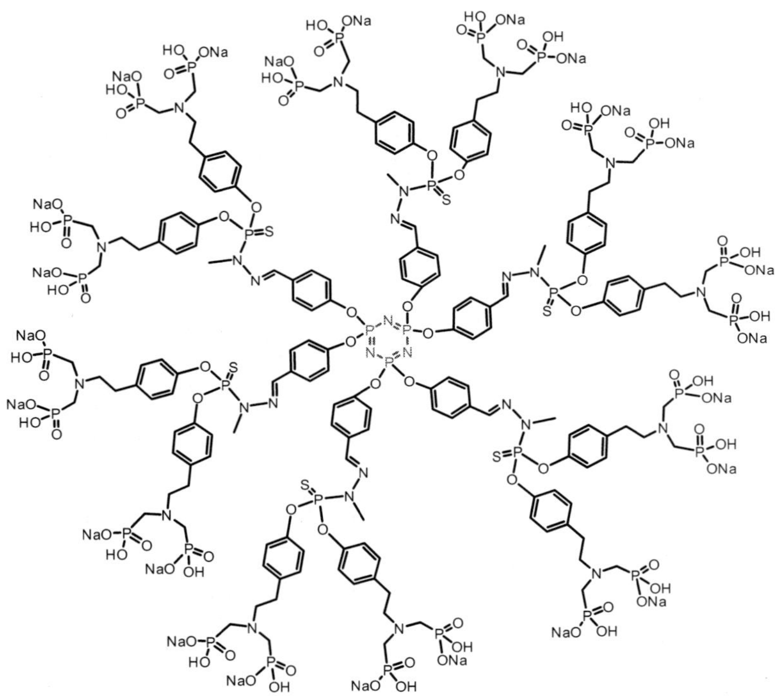

Figure 13.11 G1–TamBP dendrimer. Republished with permission of FASEB, from Poupot et al. (2006); permission conveyed through Copyright Clearance Center, Inc.

13.4 Concluding Remarks

Dendrimers possess a remarkably broad and diverse range of properties that have been explored in this chapter, leading to their calling in a broad range of disciplines across the fields of science and engineering. At the center of their appeal lies an unparalleled versatility, stemming from their architectural diversity, setting them apart from all other (bio)polymers. Given the rapid and ever-demanding advances in bio- and nanotechnology that this century has bought there remains little doubt that the once humble dendrimer, albeit in a morphed and more sophisticated form, has a pivotal role to play here. Dendrimers, with their innumerous and demonstrable (bio)chemical transformations, provide real, practical solutions toward multifunctional system design that will serve to meet the demands and challenges of translating basic concepts into advanced real-world applications for "generations" to come.

Chapter 14

Stimuli-Reactive Nanoparticles for Drug Targeting

Aditi Jhaveri, Tao Wang, and Vladimir Torchilin
Center for Pharmaceutical Biotechnology and Nanomedicine, Department of Pharmaceutical Sciences, Northeastern University, 140 The Fenway, Boston, MA 02115, USA
v.torchilin@neu.edu

14.1 Introduction

Advances in nanotechnology and nanomedicine over the past couple of decades have enabled the development of nanocarrier platforms for therapeutics that can potentially play a key role in improving disease management and treatment outcomes. A number of nanoparticulate carriers, including liposomes, polymeric micelles, dendrimers, nanoemulsions, polymeric nanoparticles, metallic nanoparticles, nanogels (NGs), nanocapsules, and solid lipid nanoparticles (SLNs), are currently under investigation at various levels of sophistication (Fleige et al., 2012; Wang et al., 2012a; Ferrari, 2005; Torchilin, 2005, 2006a). Free drugs

Controlled Release Systems: Advances in Nanobottles and Active Nanoparticles
Edited by Alexander van Herk, Jacqueline Forcada, and Giorgia Pastorin

ISBN 978-981-4613-21-7 (Hardcover), 978-981-4613-22-4 (eBook)
www.panstanford.com

and therapeutic molecules may exhibit several limitations including poor solubility, loss of activity in vivo, unfavorable pharmacokinetics, lack of selectivity, toxicity, and poor biodistribution that pose a significant barrier to effective therapy (Allen and Cullis, 2004). Nanoparticles can overcome these limitations due to a multitude of favorable properties that include the ability to solubilize hydrophilic as well as hydrophobic agents, increased stability, tunable release characteristics, favorable pharmacokinetics and biodistribution, longevity in circulation, targeted delivery of drugs to specific cells or tissues, and reduced side effects (Torchilin, 2006a; Peer et al., 2007; Farokhzad and Langer, 2009). Additionally, these nanoparticles can be engineered in a way that enables them to selectively release their cargo in response to specific internal and external triggers to give rise to stimuli-reactive or stimuli-responsive nanoparticles (Torchilin, 2009; Fleige et al., 2012; Shim and Kwon, 2012). Internal stimuli comprise differences in pH, temperature, redox potential, and the presence or level of enzymes, while external stimuli can be provided by triggers like light, ultrasound, heat, and magnetic fields, which can control the temporal and spatial release of therapeutic agents (Fleige et al., 2012; Shim and Kwon, 2012; Ganta et al., 2008; Zhu and Torchilin, 2013).

In this chapter, we highlight various stimuli-reactive nanopreparations used for delivery of therapeutics with selected recent examples from the literature. More specifically, we discuss recent advances in the field of stimuli-reactive drug delivery relevant to cancer in particular, which has seen the greatest surge in nanotechnology-based innovative drug delivery platforms over the past few years. Major concepts relevant to cancer physiology and drug targeting are also discussed with examples to aid in understanding the existence of various biological stimuli in the tumor microenvironment.

14.2 Tumor Microenvironment and Nanoparticles

Tumor vasculature is abnormal and exhibits a markedly different physiology from the normal vasculature (Jain, 1994, 2001). The blood vessels in tumors show aberrant branching, are dilated and tortuous, and possess leaky or defective walls. The endothelial cells lining blood vessels in tumors are disorganized, with large fenestrations and give rise to inter-endothelial pore sizes with diameters in the order of several hundred nanometers (Jain, 1987a; Hobbs et al., 1998; Roberts and Palade, 1997). Architectural abnormalities as well as production of vascular permeability mediators (vascular endothelial growth factor [VEGF], bradykinin, nitric oxide, matrix metalloproteinases [MMPs], etc.) make the tumor vasculature highly permeable (Wu et al., 1998; Maeda et al., 1996; Fang et al., 2011). Rapidly proliferating cancer cells in tumor tissues compress lymphatic vessels, especially in the center of the tumor, causing their collapse, which results in poor lymphatic drainage from tumors (Padera et al., 2004). This defective lymphatic drainage, coupled with the increased permeability of the tumor vasculature, results in extensive leakage of blood plasma components, such as macromolecules and nanoparticles into the tumor tissue (Iyer et al., 2006; Maeda et al., 2000). This phenomenon, known as the enhanced permeability and retention (EPR) effect, which is the basis for passive targeting, was first described by Matsumura and Maeda (1986). Nanoparticles circulating in the blood can extravasate into tumors via the leaky vessels and are retained inside owing to the EPR effect. Although a promising strategy for anticancer drug delivery, this effect is not without its limitations (Fang et al., 2011). One of the main concerns with EPR is the heterogeneity within and between tumor types. The vascular pore cut off sizes vary between tumors, and even within the same tumor the vasculature can exhibit varying porosities and hence different permeabilities (Monsky et al., 1999; Prabhakar et al., 2013). Large tumors possess pathophysiological diversity, wherein some regions of tumors, particularly the central area of metastatic cancers, do not show the EPR effect and thus have a lower accumulation of macromolecules than other parts (Nagamitsu et al., 2009; Fang et al., 2011). A frequently overlooked aspect is the fact that not all tumor vessels are leaky, which causes a heterogeneous distribution of pore sizes that results in heterogeneous extravasation and delivery (Yuan et al., 1995; Yuan et al., 1996). It is thus important that additional avenues are explored to target tumors beyond the EPR effect.

Cancer cells or cancer-associated tumor vasculature frequently show the presence of antigens or receptors, which are either uniquely expressed or overexpressed relative to normal cells or the surrounding normal tissues (Allen, 2002). Nanoparticles can be designed specifically to target these overexpressed markers on cancer cells to provide selective delivery to the target cells (Allen, 2002; Torchilin, 2006a; Brannon-Peppas and Blanchette, 2004). Some examples of markers that are specifically or abundantly expressed on cancer cells and tissues include receptors such as folate (Lee and Low, 1997; Leamon and Low, 1991), transferrin (Ishida et al., 2001), vasoactive intestinal peptide (VIP) (Gespach et al., 1988; Reubi, 1996), integrin receptors (Brooks et al., 1994), antigens like carcinoembryonic antigen (CEA) (Crawford et al., 2003; Tiernan et al., 2013), CD20 (Rossi et al., 2009; Borisch et al., 2001), and CD19 (Hammer, 2012) among others. Various targeting ligands have been employed for actively targeting nanocarriers to tumor cells, including antibodies and their fragments (Torchilin et al., 2003; Park et al., 2001; Jin et al., 2010), aptamers (Farokhzad et al., 2006), peptides (Hu et al., 2008; Biswas et al., 2013a), sugar moieties (Sun et al., 2011; Nagasaki et al., 2001), proteins (Sawant et al., 2013; Huang et al., 2013b; Yin et al., 2013a), and small molecules like folate (Yin et al., 2013a).

A further enhancement to the selective delivery of nanoparticles to tumor tissues can occur by utilizing certain cues intrinsically characteristic to the tumor microenvironment or by applying various stimuli to this zone from outside the body so as to favorably modify the properties of the drug-loaded nanocarrier (Torchilin, 2009). Both internal and external stimuli could serve to provide controlled drug release, enhance cellular uptake of drug, control its intracellular fate, and also lead to physical action on the surrounding pathological tissue (Torchilin, 2009). The pathophysiological changes in tumors that contribute to stimuli-responsive release from nanoparticles include lowered pH, elevated redox potential, and up-regulated protein or enzyme levels relative to the normal tissues (Zhu and Torchilin, 2013). In solid tumors, extensive cell death and hypoxia contribute to a pH value that is more acidic (~6.5) than the normal blood pH (~7.4) (Vaupel et al., 1989; Tannock and Rotin, 1989). This value is even lower (~5.5 or lower) in the endosomes and lysosomes within cells compared to the cytosolic pH (Gerweck and Seetharaman, 1996). Elevated redox potential in cancer cells is attributed to a significantly higher concentration of intracellular glutathione (GSH) (100-fold higher) compared with the extracellular GSH concentration (Saito et al., 2003). Temperature can serve as an internal stimulus (local hyperthermia due to inflammation) as well as an external stimulus resulting from application of local ultrasound or electromagnetic field to the tumors (Torchilin, 2009). A biological trigger that can be exploited successfully for drug delivery from nanoparticles includes the presence of certain enzymes that are usually overexpressed in cancer such as MMP-2 (Zhu et al., 2012), MMP-9 (Chau et al., 2006), and cathepsin B (Chu et al., 2012). Nanocarriers can be designed to incorporate scaffolds or linkers that are sensitive to enzymatic cleavage or degradation (Fleige et al., 2012). Light (ultraviolet [UV], visible, and near-infrared [NIR]), ultrasound (low or high frequency), and magnetic field, in conjunction with magnetic nanoparticles, can also be utilized as triggers for selectively enhancing drug delivery to pathological tissues (tumors) using nanoparticles. We discuss these stimuli in detail in later sections of this chapter, giving specific examples of various nanocarriers designed for targeted drug delivery in cancer.

14.3 Tumor-Targeting Strategies

14.3.1 Passive Targeting

As discussed in the previous section, passive targeting relies mainly on the tumor microenvironment to achieve accumulation of nanocarriers via the EPR effect (Maeda et al., 2000). However, the properties of nanocarriers themselves, including size and surface characteristics, also play a key role in determining the effectiveness of passive targeting (Cho et al., 2008; Ganta et al., 2008). One of the advantages with nanocarriers is the ability to tune their sizes. The nanocarriers should be small enough so as to evade capture by the mononuclear phagocyte system (MPS) but large enough to prevent rapid leakage into blood capillaries (Cho et al., 2008). In general, after systemic administration, particles with diameters less than 5–10 nm are rapidly removed through renal clearance and extravasation, whereas particles larger than 200 nm can be sequestered in the spleen by mechanical filtration and are ultimately removed by cells of the MPS (Vinogradov et al., 2002; Stolnik et al., 2012; Moghimi et al., 2001). To take advantage of the EPR effect, the size of nanoparticles should be in 10–200 nm range (Ishida et al., 1999; Petros and DeSimone, 2010). However, Liu et al. recently showed that even 2.5 nm renal-clearable GSH-coated luminescent gold nanoparticles (GS-AuNPs) could show passive targeting due to the EPR effect when compared to the IRDye 800CW (Liu et al., 2013c). The GS-AuNPs had a longer retention time in tumors and cleared more than thrice as fast as the dye molecules from normal tissues. This finding could potentially avoid the dilemma of different size requirements for minimizing opsonization and achieving the EPR effect (Liu et al., 2013c). Another critical requirement for passive targeting of nanoparticles is their longevity in circulation (Torchilin, 2006a; Torchilin and Trubetskoy, 1995; Petros and DeSimone, 2010). Long-circulating nanocarriers can accumulate slowly in pathological sites with leaky and abnormal vasculature via the EPR effect and enhance drug delivery to those sites (Torchilin, 2006a, 2009). This helps in sustaining the requisite level of pharmaceutical agent in the blood for extended intervals (Torchilin, 2009). Moreover, longevity of nanocarriers can produce a better targeting effect for targeted drugs and drug carriers by allowing more time for their interaction with the target because of a larger number of passages of these carriers through the target with blood (Torchilin, 1996).

The most widely used strategy to impart longevity to nanoparticles has been to adsorb or graft flexible hydrophilic polymeric coatings like poly(ethylene glycol) (PEG) to their surface (Klibanov et al., 1990; Otsuka et al., 2003; Torchilin and Trubetskoy, 1995). The hydrophilic polymers can act as a steric barrier to protect nanocarriers from interactions with blood components and reduce plasma protein adsorption (Torchilin, 2006a). Some other examples of hydrophilic and biocompatible polymers used for steric protection include poly *N*-vinyl pyrrolidone (Torchilin et al., 2001b), poly(vinyl) alcohol (Takeuchi et al., 2001), poloxamines (Moghimi, 1999), poly[*N*-(2-hydroxypropyl)methacrylamide (Whiteman et al., 2001), and poly(acryloyl morpholine) (Sartore et al., 1997). An important consequence of modification of nanocarriers with such hydrophilic polymers is decreased accumulation by the MPS (Torchilin, 2009).

A number of nanoparticles based on passive targeting have been validated clinically, the majority of which include liposomes, polymeric micelles, and polymeric nanoparticles (Zhang et al., 2008). Table 14.1 lists some examples of passively targeted drug delivery systems that are under clinical investigation or have been approved.

Table 14.1 Some examples of passively targeted delivery systems

Nanocarrier and composition	API/Product name	Medical indication	Trial phase	Company
Polymeric Micelles				
PEG-PLA copolymer	Paclitaxel/Genexol®PM	Recurrent breast cancer	IV	SanYang Co.
		Breast, lung, pancreatic cancer	II–III	
PEG-b-PAsp copolymer	Doxorubicin/NK911	Various cancers	I–II	Nippon Kayaku Co.
Pluronic micelles				
	Doxorubicin/SP1049C	Gastric cancer	III	Supratek Pharma Inc.
Nanoparticles				
Albumin	Paclitaxel/Abraxane®	Metastatic breast cancer	Approved	Celgene Co.
Proprietary polymer	Doxorubicin/Transdrug®	Hepatocarcinoma	III	BioAlliance Pharma
Liposomes				
SoyHPC/CHOL/DSPE-PEG	Doxorubicin/DOXIL®	Kaposi's sarcoma	Approved	Sequus Pharma
EPC/CHOL	Doxorubicin/ Myocet®	Metastatic breast cancer	Approved	Elan Pharma
DSPC/CHOL	Daunorubicin/DaunoXome®	Kaposi's sarcoma	Approved	Nexstar Pharma
DSPPC/CHOL/sphingosine	Vincristine/Marqibo®	Non-Hodgkin's lymphoma	Phase II/III	Inex Pharm
SoyPC/DPPG/CHOL	Cisplatin/Lipoplatin™	Several cancer types	Phase II/III	Regulon Inc.

Source: Danhier et al., 2010; Immordino et al., 2006.

Although clinically useful, passive targeting strategies suffer from several limitations. A vast majority of nanoparticles utilize a surface coating of PEG for biocompatibility; however, this hydrophilic surface hampers endosomal escape and intracellular uptake of the carriers into tumor cells creating the so-called "PEG dilemma"(Hatakeyama et al., 2011). Furthermore, it is not possible to ubiquitously target cells within a tumor due to limitations imposed by inefficient diffusion for certain drugs. This makes the passive targeting approach difficult to control, which could lead to multiple-drug resistance (MDR) (Peer et al., 2007). Overexpressed transporter proteins on the cancer cell surface expel drugs to lower their therapeutic effect, causing cancer cells to become resistant to a variety of drugs (Gottesman et al., 2002). Additionally, the inherent heterogeneity within tumors further complicates passive targeting (Jain, 1994).

Active targeting of nanocarriers can overcome the limitations imposed by passive targeting and can provide specific target recognition to nanocarriers (Peer et al., 2007; Torchilin, 2009).

14.3.2 Active Targeting

Active targeting relies on the modification of nanoparticles with specific ligands on their surface that will be recognized by molecules selectively or differentially expressed on the tumor cells (Torchilin, 2006a; Kamaly et al., 2012). It enables preferential accumulation of nanoparticles in the tumor tissue, individual tumor cells, or intracellular organelles inside the cells (Nie et al., 2007). The binding of ligands to receptors allows internalization of nanoparticles via receptor-mediated endocytosis (Chou et al., 2011). Active targeting can be combined with the ability of the nanocarrier to circulate for longer times by decorating the surface of PEGylated nanocarriers with targeting ligands, a strategy that can improve the intracellular uptake and selectivity compared to non-targeted PEGylated carriers (Hatakeyama et al., 2011; Torchilin, 2006a). Targeting ligands can be attached to nanocarriers via the PEG spacer arm so as to extend the ligand outside the dense PEG brush and avoid steric hindrance for its binding to target receptors (Torchilin et al., 2001a; Torchilin, 2009). A number of ligands have been investigated to date for active targeting of various nanocarriers, as mentioned in Section 14.2.

Antibodies are the most widely used targeting ligands for nanocarriers, and various antibody platforms, including murine, chimeric, humanized, and human monocolonal antibodies (mAbs), as well as antibody fragments, have been developed using antibody-engineering technologies (Reichert et al., 2005). Recently, for the first time, Han and Davis reported the in vivo efficacy of a single-antibody targeted nanoparticle delivery system (Han and Davis, 2013a). This nanoparticle system consisted of camptothecin (CPT) conjugated to a mucic acid–containing polymer (MAP) to which one Herceptin antibody on average was attached as the targeting agent using the boronic acid-diol complexation methodology (Han and Davis, 2013b). This system with just one targeting antibody per MAP-CPT nanoparticle resulted in complete tumor regression and superior antitumor activity compared to that of non-targeted MAP-CPT nanoparticles and Herceptin antibody alone groups in mice bearing human epidermal growth factor receptor 2 (HER-2) overexpressing BT-474 human breast tumors. Herceptin functioned to enhance the BT-474 cell uptake and also to provide a therapeutic effect (Han and Davis, 2013a). Our group has reported both liposomes and micelles targeted with antinuclear mAb 2C5, which possesses a nucleosome restricted specificity, for enhanced tumor accumulation and treatment in various cancers (Gupta and Torchilin, 2007; Perche et al., 2012; Koren et al., 2012; Sawant et al., 2013; Elbayoumi and Torchilin, 2007; Zhu et al., 2012). In another study, NIR-active gold nanorods (NRs) were coated with polydopamine (PD) and conjugated with anti–epidermal growth factor receptor (EGFR) antibodies for targeted photothermal therapy of cancer (Black et al., 2013). In vitro, these targeted NRs were significantly more toxic to breast and oral cancer cells overexpressing EGFR compared to non-targeted NRs when irradiated with a broadband light source (Black et al., 2013). Tivnan et al. demonstrated the first silica nanoparticle–based targeted delivery of a tumor-suppressive, pro-apoptotic micro-RNA, miR-34a, to neuroblastoma tumors overexpressing the cell surface antigen disialoganglioside GD2 (GD_2) in a murine orthotopic xenograft model (Tivnan et al., 2012). The targeted nanoparticles enabled tumor-specific delivery following systemic administration and resulted in a significant decrease in tumor growth, increased apoptosis, and reduced vascularization (Tivnan et al., 2012). Many nanocarrier platforms like dendrimers

(Shukla et al., 2008), iron oxide nanoparticles (Hsieh et al., 2012), NGs (Nukolova et al., 2011), polymeric nanoparticles (Liu and Feng, 2011, Ling et al., 2011), quantum dots (Lu et al., 2011), and gold nanoparticles (AuNPs) (Liu et al., 2013d) have been investigated to date for targeted delivery of drugs or contrast agents by conjugation with antibodies or antibody fragments.

Small molecules present an attractive option for targeting due to the possibility of achieving higher ligand densities on nanoparticle surfaces because of their small size, easy coupling chemistries for conjugation, less immunogenic effects compared to macromolecules in vivo, and reproducible and scalable manufacturing (Kamaly et al., 2012). Folate is widely used as a targeting ligand as a number of cancers overexpress folate receptors (Leamon and Low, 1991).

Zhang et al. developed folate-targeted polymeric micelles loaded with sorafenib and superparamagnetic iron oxide nanoparticles (SPIONs) against hepatic carcinoma (HepG2) cells in vitro (Zhang et al., 2013b). The targeted micelles achieved better cell growth inhibition and a stronger apoptotic effect in vitro, and their targeting effects could be monitored using a clinical magnetic resonance imaging (MRI) scanner. Folate has been used as a targeting ligand in a number of delivery vehicles, including liposomes (Liu et al., 2013f; Watanabe et al., 2012; Chaudhury et al., 2012), dendrimers (Thomas et al., 2012; Sunoqrot et al., 2012; Wang et al., 2011e), nanosuspensions (Talekar et al., 2013), polymeric nanoparticles (Liu et al., 2012c; Xu et al., 2013a), carbon nanotubes (Niu et al., 2013), and gold NRs (Huang et al., 2011a). Other small molecules such as biotin (Bu et al., 2013) and sugar moieties like mannose (Gajbhiye et al., 2013), galactose (Wang et al., 2013c; Zheng et al., 2012), and glucose (Xiong et al., 2012c) have been used successfully for targeting. Riboflavin was used as a targeting agent for the first time for breast cancer–directed therapeutics composed of mitomycin C (MMC) conjugated to *N*-(2- hydroxypropyl)methacrylamide (HPMA) (Bareford et al., 2013). The targeted drug–polymer conjugates resulted in an increase in MMC uptake and nuclear localization and thus a potent cytotoxic effect (Bareford et al., 2013).

Transferrin receptor (TfR) is overexpressed in many cancers and its endogenous ligand transferrin (Tf) or antibodies against Tf can be used to selectively target these receptors using nanoparticulate drug carriers (Prost et al., 1998; Huang et al.,

2013b; Chen et al., 2013b; Galaway and Stockley, 2013; Jiang et al., 2012). Camp et al. reported liposomal nanoparticles targeted by a single-chain antibody fragment to the TfR (TfRscFv) to deliver wild-type human p53 (SGT-53) for tumor-specific targeting in combination with gemcitabine in pancreatic cancer (Camp et al., 2013). The targeted liposomal formulation (SGT-53) sensitized pancreatic cancer to treatment with conventional chemotherapy and in combination with gemcitabine prolonged the median survival and reduced tumor burden compared with single drug treatment. Recently, a first-in-man phase I clinical trial with this targeted liposomal formulation showed that systemically delivered SGT-53 was well tolerated, exhibited anticancer activity, and allowed tumor-targeted delivery of the SGT-53 to metastatic lesions (Senzer et al., 2013). Growth factors like epidermal growth factor (EGF) (Cai et al., 2008; Jung et al., 2012; Chen et al., 2012) and fibroblast growth factor (FGF) are other protein ligands that can be attached to nanoparticles for cell-specific delivery (Szlachcic et al., 2012). Peptides are interesting as targeting ligands due to their small size and lower immunogenicity compared to proteins as well as ease of conjugation to nanoparticles. A number of peptides have been investigated extensively as targeting ligands for nanoparticles, including arginine-glycine-aspartic acid (RGD) tripeptide sequence, to target integrin ($\alpha_v\beta_5$, $\alpha_v\beta_3$) receptors (Desgrosellier and Cheresh, 2010; Brooks et al., 1994), VIP to target the VIP receptors (Gespach et al., 1988, Dagar et al., 2012), Lyp-1 (Cys-Gly-Asn-Lys-Arg-Thr-Arg-Gly-Cys) that binds to the stress-related protein p32 that is up-regulated on tumor cells (Park et al., 2010), and cell-penetrating peptides such as *trans*-activating transcriptional transactivator (TAT) from human immunodeficiency virus (HIV-1) (Koren et al., 2012; Sawant and Torchilin, 2009), transportan (Song et al., 2011), and octaarginine (Koshkaryev et al., 2012a; Biswas et al., 2013b).

Nucleic acid ligands or aptamers are oligonucleotides that can fold into well-defined 3D structures to bind with high affinity and specificity to their target molecules (proteins, peptides, or small molecules) (Zhang et al., 2011). Farokhzad et al. reported poly(D,L-lactic-*co*-glycolic acid)-block-poly(ethylene glycol) (PLGA-*b*-PEG) nanoparticles encapsulating docetaxel and surface-functionalized with A10 2'-fluoropyrimidine RNA aptamers, which recognize the prostate-specific membrane antigen (PSMA), expressed on the surface of prostate cancer cells (Farokhzad et al., 2006). These nanoparticles showed effective tumor regression in an LNCap xenograft mouse model of prostate cancer (Farokhzad and Langer, 2009). Yu et al. utilized MUC1 protein aptamers conjugated to the surface of paclitaxel-loaded PLGA nanoparticles using a DNA spacer as the targeting agents against MUC-1 overexpressing breast cancer cells MCF-7 (Yu et al., 2011a). The aptamer-conjugated nanoparticles increased nanoparticle uptake and enhanced in vitro drug delivery and cytotoxicity in MUC1+ cancer cells. Table 14.2 shows some examples of actively targeted nanoparticles.

14.3.3 Organelle-Specific Targeting

The delivery of therapeutic agents and drug-loaded nanocarriers to specific organ and tissues may need to be followed by their subsequent entry into the cells to exert their therapeutic effect in the cytoplasm, nucleus, or specific subcellular organelles such as the mitochondria, lysosomes, and endoplasmic reticulum (ER) (Torchilin, 2006b). Genes and antisense therapeutics, pro-apoptotic drugs, and lysosomal enzymes exert their action in the nucleus, mitochondria, and lysosomes, respectively (Torchilin, 2006b). Macromolecules and smaller particles may be taken up from extracellular space into the cells via the endocytic pathway (Varga et al., 2000). These molecules are enveloped in the endosome, which eventually fuses with lysosomes where an acidic environment and lysosomal enzymes may cause degradation of the cargo. As a result, a very small percentage of the intact material actually reaches the cytoplasm (Torchilin, 2006b; Ganta et al., 2008). Methods to deliver macromolecules and nanocarriers directly into the cytoplasm use either invasive (microinjection or electroporation) or non-invasive methods (pH-sensitive carriers that can destabilize the endosomal membrane to release the therapeutics into the cytoplasm and cell-penetrating peptides) (Torchilin, 2006b). However, getting into the cytoplasm does not ensure delivery of therapeutics to subcellular organelles by mere diffusion or random interaction with organelles in the cell (Boddapati et al., 2008). To ensure that cargoes loaded into nanocarriers are targeted to subcellular organelles, specific ligands may need to be attached to nanocarriers to direct them to the desired compartment (Kamaly et al., 2012; Ma et al., 2013;

Biswas et al., 2012b). This approach is beneficial as it avoids chemical modification of the therapeutic molecule itself (Boddapati et al., 2008). The accumulation of nanoparticles would then depend on the recognition of these targeting moieties by the endogenous intracellular trafficking mechanisms (Paulo et al., 2011).

Intracellular delivery of nanoparticles can be mediated using internalizable ligands (e.g., transferrin, folate) (Thomas et al., 2012; Jiang et al., 2012) or by cell-penetrating proteins and peptides (CPPs) (Zhu et al., 2012; Koren et al., 2012), which have the ability to translocate through cell membranes to deliver the cargo inside cells (Torchilin, 2006b). Proteins like TAT from HIV-1 (Frankel and Pabo, 1988), Antennapedia (Antp) from *Drosophila* (Derossi et al., 1994; Perez et al., 1992), and VP22 from the herpes virus (Elliott and O'Hare, 1997) among others have shown this translocation property. The transduction property of these proteins resides in short sequences of about 20 amino acids or less, which are rich in basic residues like lysine (Lys) and arginine (Arg) (Torchilin, 2006b, 2008). Various CPPs have already been used to modify liposomes, polymeric micelles and nanoparticles, dendrimers, AuNPs, silica nanoparticles, gold NRs, iron oxide nanoparticles, and SLNs. For example, methoxy poly(ethylene glycol) (MPEG)/polycaprolactone (PCL) diblock copolymer nanomicelles, conjugated to the TAT peptide via a disulfide linkage, were used to deliver anti-VEGF siRNA (MPEG-PCL-ss-Tat/anti-VEGF siRNA) for silencing genes in vitro and in vivo (Kanazawa et al., 2012). These complexes showed enhanced cellular uptake as well as antitumor effects in tumor-bearing mice when administered systemically. Similar results were also seen with another CPP, octa-arginine (R8), when used to modify liposomes (Biswas et al., 2013a). Although used widely, a major challenge with CPPs is their lack of specificity for neoplastic versus non-neoplastic cells. Efforts have been made to overcome this limitation of CPPs by designing multifunctional nanocarriers that bear tumor cell–specific targeting ligands, in addition to CPPs to selectively deliver their payload inside tumor cells (Koren et al., 2012). Among other approaches, Kondo et al. screened a peptide library using mRNA display technology and identified artificial CPPs that can selectively home to tumor cells and tissues according to their lineage specificities in vitro and in vivo, thereby minimizing transduction into normal tissues (Kondo et al., 2012).

Table 14.2 Some examples of active targeting using targeted nanocarriers

Targeting strategies	Targeted cells/ receptors	Tageted ligands	Nanocarriers	API	Refs.
Tumor vasculature	Endothelium/Integrin $\alpha_v\beta_3$	Cyclic RGDfK peptide	Liposome	Doxorubicin	(Murphy et al., 2008)
	HUVEC/Integrin $\alpha_v\beta_3$	GRGDS peptides	PLGA NP	Paclitaxel	(Danhier et al., 2009)
	Endothelium/Selectin	P-selectin glycoprotein ligand 1	Liposome		(Carlisle et al., 2013)
Tumor angiogenic inflamed vessels	VCAM-1 and ICAM-1	Anti-VCAM-1 Abs Anti-ICAM-1 Abs	Polyurethane–urea NPs	CBO-P11	(Morral-Ruiz et al., 2013)
Cancer cells	Folate receptor	Folate	Nanospheres	Doxorubicin	(Li et al., 2013a)
	Lung cancer cells/ EGFR	Cetuximab	PLGA NPs	Paclitaxel	(Karra et al., 2013)
	Breast MCF-7 cancer cells	DMPGTVLP peptide	Liposomemicelle	Doxorubicin	(Wang et al., 2010b, 2010c)

Targeting strategies	Targeted cells/ receptors	Tageted ligands	Nanocarriers	API	Refs.
	Death receptor 5 (DR5)	Abs-targeting DR5	Hydrogel NPs	Photosensitizer	(Abdelghany et al., 2013)
	Pancreatic cancers/ Integrin and neuropilin-1	CRGDKGPDC peptide	Polymeric NPs	Micro-RNA-34a	(Hu et al., 2013c)
	Prostate-specific membrane antigen	PSMA-specific aptamer	Dendrimer NPs	Doxorubicin/ CpG	(Lee et al., 2011a)
Cancer stem cells	CD133	Anti-CD133 Mabs	PLGA NPs	Paclitaxel	(Swaminathan et al., 2013)
Metastatic cancer	Metastasis cancer cells	GCGNVVRQGC peptide	Liposome	Doxorubicin	(Wang et al., 2012f)
Tumor-associated macrophages	CD206	Mannose	Polymeric NPs	SiRNA	(Yu et al., 2013b)

In many cases, the disease target may be located inside a particular cell organelle, or a drug may act only upon subcellular localization, which makes it essential to develop nanocarriers that target intracellular organelles (Heller et al., 2012). Mitochondria represent an interesting subcellular target in cancer as they are the site of action for many chemotherapeutics and play a critical role in apoptosis (Costantini et al., 2000; Preston et al., 2001). Liposomes loaded with paclitaxel and modified with triphenylphosphonium-polyethylene glycol-phosphatidylethanolamine (TPP-PEG-PE) targeted mitochondria, showed improved cytotoxicity against cancer cell lines, and enhanced tumor growth inhibition in 4T1 breast tumor–bearing mice compared to non-modified drug-loaded liposomes (Biswas et al., 2012a). Similarly, TPP-modified G5 poly(amido) amine (PAMAM) dendrimers and cetyltrimethylammonium bromide (CTAB)-functionalized gold NRs have also been reported for mitochondrial targeting (Biswas et al., 2012b; Wang et al., 2011d). Zhang et al. reported a mitochondria-targeted-, teniposide-loaded-, steryl-amine-based self-assembled nanocarrier that could reverse MDR in breast cancer cells in vitro as well as in mice bearing MCF-7/ADR xenografts (Zhang et al., 2013c). In a proof-of-concept study, Marrache and Dhar reported the development and optimization of a rationally designed nanoparticle platform based on polymer blending for delivering therapeutics to mitochondria (Marrache and Dhar, 2012). A PLGA-*b*-PEG copolymer containing a terminal TPP cation was blended with either PLGA-*b*-PEG-OH or PLGA-COOH, and variations in size and charge were studied for their effect on intracellular trafficking of these nanoparticles. The resultant blended nanoparticles could enter mitochondria and showed high loading efficiency of therapeutics as well as improvement in the drug therapeutic index over their non-targeted counterparts or free drugs (Marrache and Dhar, 2012).

The cytotoxic action of a number of anticancer drugs depends on their binding to DNA within the cells. Moreover, therapeutic genes also exert their effect inside the nucleus. However, a number of intra- and extracellular biological barriers (plasma membrane, efflux proteins, endosomal escape, and nuclear membrane) need to be surmounted before therapeutics can make their way into the nucleus (Sui et al., 2011). Nuclear targeting of nanocarriers thus affords an interesting approach to deliver therapeutics.

Wang et al. reported SLNs with dexamethasone-conjugated cationic lipid to promote nuclear delivery of enhanced green fluorescent protein plasmid (pEGFP) as a model gene and modified it further with transferrin-PEG-PE (Tf-PEG-PE) (Wang et al., 2012f). The resultant Tf- SLN/pEGFP showed better transfection in vitro (HepG2 cells) and in vivo (HepG2 tumors) compared with nonmodified SLNs/pEGFP or Tf-non-dexamethasone SLN/pEGFP (Wang et al., 2012f). For nuclear delivery, drug-loaded nanocarriers are often modified with nuclear localization signals (NLSs), which are peptide sequences that permit the transport of cargoes from the cytoplasm to the nucleus (Sui et al., 2011). Yu et al. reported preparation and in vitro characterization of doxorubicin (DOX)-loaded, cholesterol-modified glycol chitosan (CHGCS) polymeric micelles conjugated to an NLS (Ac-CGYGPKKKRKVGG) (DNCHGC) (Yu et al., 2012a). The DNCHGC micelles resulted in significantly higher toxicity in HeLa and HepG2 cells over free DOX and DOX-loaded non-modified micelles. Moreover the DNCHGC micelles also showed greater cellular uptake, transported into the nucleus, and accelerated the accumulation of DOX in HeLa cells (Yu et al., 2012a).

The ER is essential for cell survival and many cell functions like intracellular calcium homeostasis, lipid biosynthesis, and protein secretion (Kim et al., 2008). Wang et al. showed that PEG-PE micelles could accumulate in the ER to induce ER stress by disturbing lipid homeostasis (Wang et al., 2012e). Moreover, PEG-PE micelles could selectively induce apoptosis in A549 cancer cells by activating the pro-apoptotic unfolded protein response (UPR) without affecting normal cells (MRC-5 and 293T) that exhibit a pro-survival-signaling response to the ER stress. The different responses of cancer cells and normal cells to ER stress suggest the suitability of PEG-PE micelles as a nanocarrier for drug delivery in cancer with a potential to alter sensitivity of cancer cells to chemotherapeutic agents (Wang et al., 2012d). Investigating ER-stress-induced toxicity of nanocarriers could thus open up an interesting avenue for cancer research. Liposomes with a phospholipid composition resembling that of the ER membrane (phosphoethanolamine, phosphocholine, phosphotidyl inositol, and phosphotidyl serine) were shown to traffic to the ER and colocalize with the ER 30 min after cell uptake (Pollock et al., 2010).

Lysosomes are acidic organelles (pH 4.5–5) that are primarily responsible for intracellular degradation, subsequent recycling of cellular constituents, and digestion of extracellular materials that they receive via endocytosis or autophagy (Kallunki et al., 2013; Kirkegaard and Jaattela, 2009). The degradation or digestion in lysosomes is carried out by many hydrolases (phosphatases, proteases, lipases, glycosidases, and sulfatases) present in the lumen of lysosomes, and the products thereof are recycled back to the cytosol via diffusion or through specific transport channels and released into the extracellular space by exocytosis (Bainton, 1981; Kirkegaard and Jaattela, 2009). Cancer causes a number of changes in lysosomes and their function, which include enhanced lysosomal biogenesis and hydrolase activity, altered membrane composition, and increased secretion of lysosomal contents into the extracellular space (Kallunki et al., 2013). As a result, many hydrolases are overexpressed in tumors and contribute to poor prognosis and a greater risk of recurrence. Extracellularly released lysosomal enzymes promote cancer progression, invasiveness, and angiogenesis, whereas intracellularly released enzymes can trigger apoptosis and the nonapoptotic lysosomal cell death pathway (Guicciardi et al., 2004; Joyce et al., 2004). Thus, inhibition of extracellularly released lysosomal enzymes, and lysosomal membrane destabilization or lysosomal membrane permeabilization (LMP) affords attractive targets for cancer therapy (Kallunki et al., 2013; Fehrenbacher and Jaattela, 2005). Koshkaryev et al. reported transferrin modified liposomes loaded with ceramides (short, C6, or long, C16), which are known lysosome-destabilizing agents, to specifically target cancer cells, as well as for increased lysosomal delivery and initiation of LMP-induced apoptosis both in vitro and in vivo (Koshkaryev et al., 2012b). In vivo, in an A2780 ectopic ovarian carcinoma mouse model, C_6Cer-loaded transferrin liposomes (TLs) demonstrated a strong antitumor and pro-apoptotic effect over plain C_6Cer-loaded liposomes or TLs not loaded with C_6Cer (Koshkaryev et al., 2012b). Selective LMP was also achieved using iron oxide magnetic nanoparticles (MNPs) targeted to the EGFR in cancer cells overexpressing EGFR under the influence of an alternating magnetic field (AMF) (Domenech et al., 2013). MNPs caused disruption of phospholipid bilayers under an AMF by mechanical rotation or released heat, causing increased permeabilization of lysosomal membranes, which also resulted in increased reactive oxygen species (ROS) production and reduced cell viability (Domenech et al., 2013).

14.4 Stimuli-Responsive Drug Release in Tumors

To elicit the maximum benefit from nanocarriers, efforts are being directed toward developing sophisticated nanoplatforms that, in addition to carrying therapeutic and/or imaging payloads, combine the properties of long circulation as well as active targeting and achieve spatial and temporal control over release of a payload from the carrier (Torchilin, 2009; Cheng et al., 2013). Stimuli-sensitive nanocarriers are designed to release their cargo under the influence of an intra- or extracellular stimulus of either physical, chemical, or biochemical origin (Fleige et al., 2012). These "smart" nanocarriers can be tuned with regard to their chemistry and offer great versatility in terms of delivery vehicles used and therapeutic payloads that can be delivered.

In the following sections, we discuss various internal and external stimuli that are used commonly to trigger the release of therapeutic payloads from nanocarriers. A schematic of various stimuli used to trigger drug release from nanoparticles is presented in Fig. 14.1.

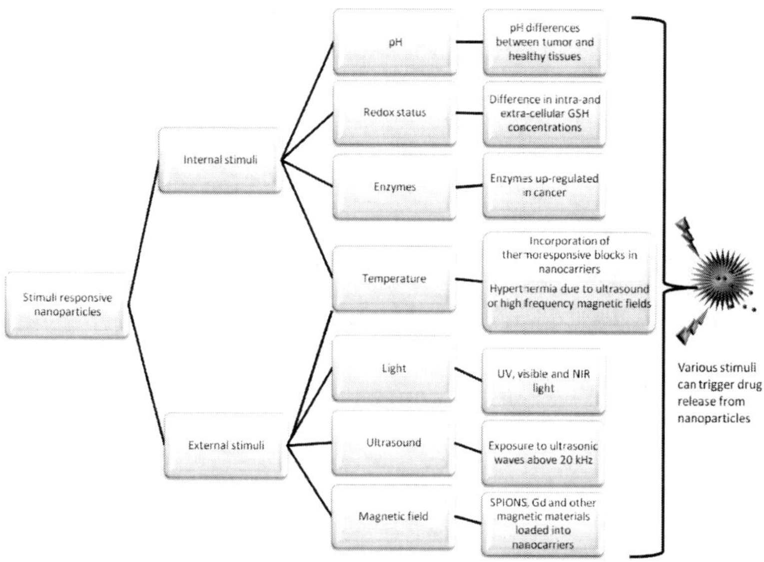

Figure 14.1 Schematic of various stimuli used to trigger drug release from nanoparticles.

14.4.1 Internal Stimuli

Internal stimuli used to trigger the release of therapeutics from nanoparticles include triggers intrinsically present in the tumor microenvironment, such as differences in pH, an altered redox potential, increased temperature (hyperthermia), and proteins or enzymes that are up-regulated during malignant transformation (Zhu and Torchilin, 2013; Torchilin, 2009).

14.4.1.1 pH

The nutritional and metabolic environment of tumors differs greatly from that of normal cells. The tumor vasculature is unable to meet the nutritional requirements of an ever-expanding population of cells, which results in an oxygen- and nutrient-deficient environment within tumors (Tannock and Rotin, 1989). Anaerobic conditions that lead to the production of lactic acid via the glycolytic pathway, and the hydrolysis of ATP in an energy-deficient environment, are responsible for the low pH environment found in many tumors (Tannock and Rotin, 1989). The low extracellular/interstitial pH in tumors also results from other mechanisms like glutaminolysis, carbon dioxide production, and biocarbonate depletion (Helmlinger et al., 2002). The extracellular pH value in solid tumors is lowered to ~6.5 compared to about 7.4, the normal blood supply's pH (Gerweck and Seetharaman, 1996; Vaupel et al., 1989). The pH values in the intracellular organelles are further lowered below that of the tumor pH. The pH is between 5 and 6 for early endosomes, whereas lysosomes are more acidic with a pH of about 4–5 (Casey et al., 2010). This pH gradient can be harnessed for drug delivery by designing nanocarriers that can selectively release their therapeutic payloads at specific sites, when they encounter a change in the pH of the environment, thus giving rise to pH-sensitive nanocarriers (Gao et al., 2010).

pH-sensitive nanocarriers have been investigated extensively and pH gradients probably account for the most widely used stimulus for designing responsive nanoparticles (Gao et al., 2010; Torchilin, 2009). Various strategies have been utilized for rendering nanoparticles pH responsive. One strategy is to incorporate ionizable functional groups like amines, carboxylic acids, and phosphoric acids in the polymer that can ionize at different pH values (Schmaljohann,

2006; Li et al., 2013e). The ionizable functional groups can reversibly change solubility upon ionization (Colson and Grinstaff, 2012; Li et al., 2013e). The pKa value of the polymer plays a critical role in design of a pH-sensitive nanocarrier. Cationic polymers deionize above their pKa value in contrast to anionic polymers, which undergo deionization below their pKa (Li et al., 2013e). Carboxylic acids carry a negative charge and are deprotonated at neutral pH, whereas they lose the charge and become undissociated at a lower pH. Amines on the contrary, are uncharged and hydrophobic when in neutral and alkaline solutions but are protonated and hydrophilic at mildly acidic pH (Colson and Grinstaff, 2012). Changes accompanying the protonation/deprotonation states of polymers transform a hydrophobic polymer into a hydrophilic one, which can then rapidly release its payload by swelling or due to an increase in its solubility (Colson and Grinstaff, 2012). Oh et al. utilized this strategy to design a pH-responsive DOX-loaded NG made of glycol chitosan (GCS) grafted with functional 3-diethylaminopropyl (DEAP) groups (GCS-g-DEAP) (Oh et al., 2010). At physiological pH, GCS-g-DEAP self-assembled with the hydrophilic GCS block and the hydrophobic DEAP block. At tumor extracellular pH (pHe), protonation of DEAP led to a decrease in hydrophobicity and caused the NG to destabilize. The subsequent release of DOX from the NGs was accelerated at lower pH values, leading to greater DOX-uptake by non–small cell lung carcinoma A549 cells under slightly acidic conditions, similar to those encountered in the tumor (Oh et al., 2010). Recently Yang et al. reported the development of pH-sensitive three-layer DOX-loaded micelles that self-assembled from amphiphilic four- and six-arm star triblock poly(ε-caprolactone)-b-poly(2-(diethylamino) ethyl methacrylate)-b-poly(poly(ethylene glycol) methyl ether methacrylate) (4/6AS-PCL-b-PDEAEMA-b-PPEGMA) (Yang et al., 2013c). The hydrophobic PCL layer was used to encapsulate DOX (base form); the middle layer of PDEAEMA was pH responsive, while the outer layer was formed by PEG. In vitro, DOX exhibited a pH-dependent release from the three-layer micelles. The release was increased significantly when the pH was decreased to 5.0 from 7.4 due to swelling of micelles upon protonation of the pendant tertiary amine groups in PDEAEMA in the middle layer of the micelles. The 4/6AS-PCL-b-PDEAEMA-b-PPEGMA micelles provided equivalent or enhanced anticancer activity relative to free DOX on HepG2 cells (Yang et al., 2013c).

Another strategy used to design pH-sensitive nanocarriers is the use of acid-labile functional groups that undergo cleavage at a specific pH. A variety of pH-labile linkers have been investigated to date, including benzoic imine (Guo et al., 2012b), hydrazone (Dong et al., 2013; Fang et al., 2012; Sawant et al., 2012; Koren et al., 2012; Huan et al., 2012), oxime (Jin et al., 2011), acetal (Gu et al., 2013), silyl ether (Parrott et al., 2010), ortho ester (Cheng et al., 2012b; Bruyere et al., 2010), Schiff base (Wang et al., 2008a), and vinyl ether (Xu et al., 2008b). The pH-sensitive linkers can be part of the polymeric components used to produce pH-responsive drug delivery systems that are stable in the normal tissues and circulation, but can release the entrapped drugs when they encounter low pH areas such as tumors (Bertrand et al., 2010; Torchilin, 2009). The acid-labile bonds may be present in the main chain, in side chains, or in the case of polymeric micelles as a linkage between the hydrophilic and hydrophobic blocks (Li et al., 2013e). Another important application of pH-cleavable linkers is in the design of long-circulating nanoparticles. Most nanocarriers are PEGylated to allow for increased stability, prevent protein binding, and extend the circulation time of nanoparticles in the body (Torchilin and Trubetskoy, 1995; Klibanov et al., 1990). However, PEGylation may not always be favorable from the drug delivery standpoint. It could inhibit interaction of the nanocarrier with the cell membrane and make the drug release from nanocarrier difficult. In cases where drug-loaded nanocarriers are taken up into cells via endocytosis, the PEG coat may prevent the contents from escaping the endosome and delivery to the cytoplasm (Torchilin, 2009). To overcome this limitation, a desirable option is to introduce a pH-sensitive linker between PEG and the nanoparticle surface. The pH-sensitive linkers are stable at physiological pH but are rapidly hydrolyzed and can release the PEG layer at low pH values found in tumors (Tiernan et al., 2013). Shedding of the PEG layer does not cause endosomal disruption per se but may help in the escape of nanoparticles by reducing the steric hindrance caused by PEG (Gao et al., 2010). Thambi et al. synthesized amphiphilic block copolymers (BCPs) from hydrophilic PEG and hydrophobic poly(γ-benzyl L-glutamate) (PBLG) with an acid-sensitive orthoester linkage between PEG and

PBLG (Thambi et al., 2011a). These copolymers formed nanosized micelles in aqueous conditions and were stable at physiological pH (pH 7.4). However, the micelles disintegrated readily at pH 5, due to cleavage of orthoester groups causing deshielding of PEG and aggregation of the hydrophobic PBLG core, leading to micron-sized particles. The release of DOX was significantly higher at acidic pH and the pH-sensitive micelles also showed greater toxicity in squamous carcinoma (SCC) cells relative to the non-pH-sensitive micelles (Thambi et al., 2011a). pH sensitivity of the PEG coats on nanoparticles also allows for designing multifunctional drug delivery systems: On the one hand PEG shielding can hide certain functions, while the nanocarrier is in circulation. On the other hand, it can expose those functions once the PEG coat is "deshielded" under the influence of environmental cues like lowered tumor pH (Koren et al., 2012).

Alternatively, drugs can be conjugated via acid-labile linkers to nanoparticles to control their release characteristics as well as for specific drug release in tumors. Kaminskas et al. reported generation 5 PEGylated polylysine dendrimers with an outer generation of L-lysine (LYS) or succinimyldipropyldiamine (SPN) which contained DOX linked through an acid-labile 4-(hydrazinosulfonyl) benzoic acid (HSBA) linker (Kaminskas et al., 2011). The acid-labile dendrimers showed preferential DOX release in vitro at acidic pH and retained their cytotoxic activity at levels similar to those of free DOX in MDA-MB-231 cells. In vivo, the dendrimers showed three to eight times higher accumulation in rat tumors over the control tissues (heart and muscle) (Kaminskas et al., 2011).

Among other pH-sensitive strategies, polycationic polymers such as polyethylenimine (PEI), which exhibit a substantial buffering capacity for the endosomal compartment, have been utilized extensively for gene delivery because of their ability to condense negatively charged nucleic acids (Boussif et al., 1995; Liu et al., 2011d). The membrane bound ATP-dependent proton pump acidifies the endosome to pH 5.0. Many primary and secondary amine groups with different pKa values (pKa 5.5–7.2) are present on the backbone of polymers like PEI, which protonate to different extents at a given pH. Upon endocytosis, the protonation of amines causes an increased influx of protons as well as counter ions such as chloride into the endosomal compartment, thus increasing its osmotic pressure (Grainger and El-Sayed, 2010). The osmotic gradient leads to water influx into the endosome, causing it to rupture and to release its contents into the cytoplasm in a process known as the "proton sponge" effect (Behr, 1997; Grainger and El-Sayed, 2010). Although the proton sponge mechanism has faced a lot of debate, it remains the most generally accepted mechanism for explaining polyplex escape from the endosome/lysosome pathway (Won et al., 2009; Benjaminsen et al., 2013). Shim et al. reported the development of an acid-degradable 25 kDa linear PEI, which was modified by incorporation of amino ketal branches to the secondary amines of linear PEI, to facilitate endosomal escape due to increased buffering capacity of amines (Shim and Kwon, 2009). The siRNA-ketalized linear PEI (KL-PEI) resulted in higher RNA interference (RNAi) efficiency independent of the presence of serum. Moreover, it was able to release the siRNA from polyplexes exclusively in the cytoplasm due to acid hydrolysis of amino ketal linkages versus naked siRNA or unmodified linear PEI, which localized in the cytoplasm as well as in the nucleus. The ketalization also reduced cytotoxicity of linear PEI without altering its serum-independent gene delivery efficiency (Shim and Kwon, 2009).

Like polyplexes, lipoplexes form complexes due to strong electrostatic interaction between cationic lipids and negatively charged nucleic acids (Felgner et al., 1987). pH sensitivity can be imparted to liposomes by utilizing fusogenic lipids like unsaturated dioleoylphosphatidyl ethanolamine (DOPE), which can adopt an inverted hexagonal phase (Munoz-Ubeda et al., 2010; Torchilin, 2006b). However, one of the major limitations of pH-sensitive liposomes is their limited gene transfection efficiency, as well as low binding affinity for plasmid DNA (pDNA) compared to cationic liposomes (Legendre and Szoka, 1992). To improve the transfection efficiency of pH-sensitive liposomes, Chen et al. prepared complexes containing cationic liposomes and pH-sensitive liposomes and studied the influence of various factors on pDNA transfection efficiency (Chen et al., 2013d). The complexes were formed from cationic 3β-[*N*-(*N*′, *N*′-dimethylaminoethane) carbamoyl] cholesterol (DC-CHOL) and dioleoylphosphatidyl ethanolamine (DOPE) liposomes, pDNA, and pH-sensitive PEGylated or non-PEGylated liposomes composed of cholesteryl hemisuccinate (CHEMS) and DOPE. Optimized complexes of DC-CHOL/DOPE liposomes/pDNA and pH-sensitive PEGylated liposomes exhibited potent pH sensitivity, showed high transfection efficiency in the presence of serum, and also showed enhanced accumulation in tumors (Chen et al., 2013d). Sawant et al. reported a pH-responsive micellar system for delivery of pDNA conjugated to low-molecular-weight PEI and modified by DOPE (PEI-PE/DNA) (Sawant et al., 2012). The PEI-PE/DNA complexes, when mixed with low-pH-degradable PEG-hydrazone-PE (PEG-Hz-PE), formed micelles that provided a tunable transfection efficiency and improved biocompatibility. The complexes exhibited higher transfection efficiency at pH 5.4 than at pH 7.4 due to the "deshielding" of PEI upon removal of PEG (Sawant et al., 2012).

Other strategies for pH-sensitive drug delivery make use of charge-conversional or charge-reversal nanoparticles (Lee et al., 2009c; Pittella et al., 2012). These nanoparticles are usually negatively charged at neutral pH and are capable of converting their surface charge from negative to positive in response to the tumor extracellular pH. This conversion takes place due to breakage of the acid-labile bonds (Li et al., 2013e). Pittella et al. reported calcium phosphate (CaP)/charge-conversional polymer hybrid nanoparticles carrying VEGF siRNA against pancreatic cancer (Pittella et al., 2012). These nanoparticles utilized a BCP of PEG and an endosome-disrupting polyanion, poly(*N'*-{*N'*-[(*N*-*cis*-aconityl)-2-aminoethyl]-2-aminoethyl}aspartamide) (PAsp(DET-Aco)), which is stable at extracellular pH (7.4) but degrades within the lysosome at pH 5.5 due to cleavage of *cis*-aconitic amide linkages, reverting back to the parent polycation PAsp (DET). The polymer was called a charge-conversional polymer (CCP) due to this anionic to cationic transition. The resulting hybrid CaP/PEG-CCP nanoparticles showed high gene-silencing efficiency in BxPC3 (pancreatic cancer) cells without any cytotoxicity. In vivo, upon systemic administration, nanoparticles containing siRNA against VEGF led to reduction in subcutaneous BxPC3 tumor growth and a significant VEGF gene silencing (68%) in the tumor (Pittella et al., 2012). Other groups have also reported such charge-conversional nanoparticles, including NGs (Du et al., 2010), a dual pH-sensitive system (pH-sensitive charge-conversion shielding and pH-sensitive drug release) for delivering DOX (Guan et al., 2013), micelles made from a charge-conversional poly(amino) acid derivative, which could respond to tumor pH to deliver DOX (Yoon et al., 2012) and a complex of zwitterionic chitosan (ZWC) and PAMAM dendrimers, where ZWC masked the PAMAM surface charge and was capable of charge conversion in the acidic tumor milieu to deshield PAMAM for pH-responsive drug delivery (Liu and Yeo, 2013).

To increase the accumulation of nanoparticles in tumors, one of the strategies suggested is to utilize aggregation of nanoparticles that is stimulated by the tumor microenvironment (Liu et al., 2013e). After extravasation into the tumor tissue, stimuli such as a reduced pH can cause the aggregation of nanoparticles to trap them at the tumor site, due to their increased size, and lead to increased nanoparticle retention in tumors (Liu et al., 2013e). The aggregates eventually sediment at the tumor site, and their interaction with cells increases their uptake by the cancer cells (Albanese and Chan, 2011; Cho et al., 2011). An interesting example of this was recently shown by Liu et al. (2013e). They designed a system based on AuNPs of ~16 nm prepared by surface modification with mixed-charge zwitterionic self-assembled monolayers. These AuNPs were stable in blood and normal tissues (pH 7.4) but aggregated in response to tumor extracellular pH (pH 6.5) in a fast, ultrasensitive, and reversible manner. This enabled the particles to have a good dispersibility at pH 7.4, as well as excellent stealth ability due to the non-fouling zwitterionic property. Moreover, in vivo, in BALB/c mice these particles showed superior blood half-life, tumor accumulation, retention, and cellular internalization over nonsensitive PEGylated AuNPs. The aggregation of AuNPs was also amenable to NIR photothermal therapy (Liu et al., 2013e). Wang et al. synthesized a novel amphoteric aliphatic polycarbonate via a one-step enzymatic copolymerization, which had both amine and carboxyl functionalities, to provide it with a pH-tunable self-aggregation feature (Wang et al., 2012c). The copolymer assumed various aggregation states in a controlled manner and had minimal toxicity on 293T and HeLa cells.

Figure 14.2 represents the structures of some of the commonly used pH-sensitive crosslinkers.

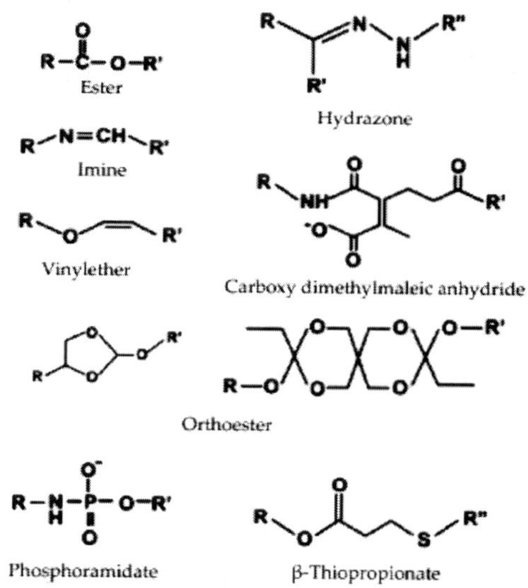

Figure 14.2 Commonly used pH-sensitive crosslinkers and their chemical structure. Modified with permission from Gao et al. (2010). Copyright 2010 American Chemical Society.

14.4.1.2 Redox potential

Nanoparticles that respond to changes in redox potential have emerged as a widely used class of stimuli-sensitive vehicles for delivery of small molecules as well as biologics like proteins, pDNA, siRNA, and antisense oligonucleotides. GSH tripeptide (γ-glutamyl-cysteinyl-glycine) is a reducing agent present inside the living cells in abundance (2–10 mM). In contrast, its extracellular concentration is almost 100–1000-fold lower (2–20 μM) (Schafer and Buettner, 2001). This disparity in GSH concentrations creates an oxidative environment in the extracellular space and a reductive environment inside the cell (Saito et al., 2003). Moreover, the cytosolic GSH concentration in tumor cells is elevated several times beyond that in normal cells (Russo et al., 1986). The difference in redox potential between intra- and extracellular environments, as well as that between tumor cells and normal cells provides an excellent opportunity to design nanocarriers that can selectively release their cargo in response to this altered GSH concentration (Zhu and Torchilin, 2013; Roy et al., 2010). The thiol–disulfide interconversion chemistry, which is a critical step in many biological processes, has been utilized synthetically for a number of bioconjugation protocols (Saito et al., 2003). Typically, redox-responsive nanocarriers contain disulfide bonds, which hold the cargo (DNA, siRNA, drugs, protein) within the nanocarrier under normal conditions. However, upon encountering reducing conditions within the cell, the carrier is destabililzed to release the entrapped cargo and facilitate intracellular accumulation (Torchilin, 2009; Wei et al., 2012).

To produce redox-responsive nanoparticles, various approaches have been utilized, including introduction of reduction-responsive linkers at the hydrophilic–hydrophobic junction in amphiphilic polymers (cleavable backbone), pendant groups, or side chains and redox-sensitive bonds as crosslinkers (Li et al., 2013e; Wei et al., 2012).

Polymeric micellar nanocarriers have received a lot of attention in recent years for the fabrication of redox-sensitive delivery systems. Redox-responsive micellar nanocarriers have been prepared from a variety of amphiphilic copolymers including linear (Li et al., 2011b; Khorsand Sourkohi et al., 2011), star (Ren et al., 2011), graft (Zhang et al., 2012d; Li et al., 2012a), and hyperbranched copolymers (Liu et al., 2011b). Thambi et al. reported polymeric micelles composed of PEG and poly(γ-benzyl-L-glutamate) bearing a disulfide bond (PEG-SS-PBLG) for the delivery of CPT (Thambi et al., 2011b). While only 40% CPT was released in the absence of GSH, complete release of CPT occurred in 20 hours in the presence of 10 mM GSH. Moreover, PEG-SS-PBLG micelles were more cytotoxic to SCC7 (squamous carcinoma) cells in vitro compared with PEG-*b*-PBLG and efficiently delivered CPT into the nuclei of these cells (Thambi et al., 2011b). A star-shaped, disulfide-linked copolymer of poly(ε-caprolactone) and PEG (6sPCL-SS-PEG) was used to prepare shell-sheddable micelles using DOX as a model drug (Ren et al., 2011). These micelles were stable under normal conditions but showed rapid destabilization under reducing conditions due to PEG shedding. GSH-dependent cytotoxicity for these DOX-loaded micelles was observed in MCF-7 cells in vitro. Redox-sensitive crosslinked micelles have been developed for the delivery of drugs as well as nucleic acids, which are core-crosslinked, shell-crosslinked, or crosslinked at the core–shell interface (Cajot et al., 2011; Matsumoto et al., 2009; Zhang et al., 2013a; Koo et al., 2012; Kim et al., 2010).

Cho et al. reported the development of a novel, biodegradable redox-sensitive polymer containing trimethyl-locked benzoquinone (TMBQ) as pendant groups (Cho et al., 2012a). Redox-sensitive nanoparticles loaded with paclitaxel (PTX) were prepared from this polymer using an emulsion method. The polymer was designed to release lactone and unmask free amino groups when reduced with a chemical agent, sodium dithionite. These nanoparticles released PTX upon conversion of the polymer from hydrophobic to hydrophilic with subsequent nanoparticle disassembly (swelling) resulting from protonation of the free amino groups. Reduction-responsive PEG-poly(aminoacid)/s NGs were developed to enable intracellular delivery of anticancer drugs in response to GSH (Ding et al., 2011). The NGs were prepared from PEG-poly(aminoacid)s which were synthesized through one-step ring-opening polymerization of L-phenylalanine *N*-carboxyanhydride (L-Phe NCA) and L-cystine (L-Cys NCA) with an amino group–terminated PEG monomethyl ether (mPEG-NH_2) as the macroinitiator, followed by their direct dispersion in a phosphate buffer solution at pH 7.4. The in vitro release of DOX, which was loaded into these NGs, could be controlled by altering the concentration of GSH as well as the concentration and content of poly(amino acid)s. These NGs also showed enhanced intracellular DOX release and greater inhibition of cell proliferation in HeLa cells that were pretreated with GSH over nonpretreated cells (Ding et al., 2011).

For successful gene delivery, it is essential to ensure the stability of the cargo in circulation and its efficient release once it reaches the cell milieu (Pack et al., 2005; Putnam, 2006). Redox-sensitive nanoparticles provide a promising strategy for intracellular delivery of siRNA and genes. To that end, bioreducible cationic lipids or polymers containing disulfide bonds such as cholesterol disulfide cationic lipids (Sheng et al., 2011), branched poly(disulfide amine) (Shim et al., 2012), hyperbranched PAMAM (Rahbek et al., 2010; Yu et al., 2012e) and disulfide-based polyethyleneimine (SS-PEI) (Xia et al., 2012) have been synthesized. Kataoka and colleagues have used redox-sensitive micelles for delivery of genes, anti-sense oligonucleotides and siRNA (Matsumoto et al., 2009; Oishi et al., 2005; Kumagai et al., 2012). For example, Kumagai et al. reported PEG-detachable polyplex micelles prepared from BCPs of PEG-b-poly{*N*-[*N*-2-aminoethyl)-2-aminoethyl]aspartamide} with a disulfide link between PEG and the cationic polymer (PEG-SS-P[Asp(DET)]), which could detach upon intracellular reduction (Kumagai et al., 2012). These micelles were evaluated in vivo in a mouse model of peritonially disseminated cancer. Upon intraperitoneal (i.p.) administration, the disulfide-linked micelles (PEG detachable) showed a significantly higher transgene expression in the tumor, spleen, and liver over the PEG-undetachable micelles (PEG-P[Asp(DET)]) without significant toxicity. Even with a similar distribution of both types of polyplex micelles, the disulfide-linked micelles encapsulating the human tumor necrosis factor α (hTNF-α) gene possessed a greater antitumor efficacy against disseminated cancer compared with the non-disulfide-linked micelles or saline controls (Kumagai et al., 2012). Xia et al. reported a bioreducible, nonviral carrier for intracellular delivery of siRNA against human telomerase reverse transcriptase (hTERT) (Xia et al., 2012). The authors used a biodegradable PEI of ~800 Da, containing multiple disulfide bonds (SS-PEI), which could strongly bind hTERT siRNA to form nanosized positively charged complexes that readily destabilize to release siRNA in a reducing environment both in vitro and in vivo. The SS-PEI/hTERT siRNA complexes transfected HepG2 cells in vitro and reduced the levels of hTERT mRNA and protein, decreased telomerase activity, caused cell growth inhibition, and showed significant apoptotic activity. Intratumoral administration of the complexes led to HepG2 tumor growth inhibition in a xenograft mouse model, and the complexes also showed low cytotoxicity in vitro, with no adverse effects on liver and kidney functions in vivo. However, a majority of the SS-PEI accumulated in the liver upon intravenous (i.v.) administration of ^{125}I-labeled SS-PEI/siRNA complexes, which warrants the inclusion of PEG for prolonged circulation in vivo. Other groups have reported redox-responsive PEGylated complexes with PEI for delivery of nucleic acids (Zhao et al., 2011a; Ping et al., 2013; Berthold et al., 2010). Lipid-based micelles and micelle-like nanoparticles are also being explored for redox-sensitive delivery of genes and siRNA. Musacchio et al. reported a nontoxic, redox-sensitive micellar system to stabilize and deliver siRNA in vitro (Musacchio et al., 2010). Double stranded GFP-siRNA was reversibly modified with phosphothioethanol (PE) via a reducible disulfide bond (siRNA-S-S-PE) and incorporated into PEG-PE micelles to form nanosized siRNA-

S-S-PE/PEG-PE mixed micelles. The mixed micelles protected siRNA against nuclease-mediated degradation for at least 24 hours and enabled the release of free siRNA under intracellularly relevant GSH concentrations. In vitro, in GFP-C166 endothelial cells, mixed micelles down-regulated GFP production 50-fold more effectively versus free siRNA and were devoid of the cytotoxic side effects typically seen with delivery systems in which siRNA is electrostatically associated with cationic carriers (Musacchio et al., 2010).

In the first attempt of its kind to polymerize siRNA in hydrogel particles, DeSimone and colleagues developed redox-responsive siRNA-conjugated hydrogel nanoparticles using the particle replication in nonwetting templates (PRINT) technology (Dunn et al., 2012). To incorporate siRNA covalently and reversibly into the PRINT hydrogel nanoparticles, it was derivatized with a photopolymeriazable acrylate possessing a degradable disulfide linkage. The siRNA cargo was retained in these "pro-siRNA hydrogels" until they reached the cell cytoplasm, where free siRNA was released under reducing conditions. By adding varying contents (5–50 wt.%) of an amine monomer 2- aminoethylmethacrylate hydrochloride (AEM), the PRINT particles were rendered cationic to facilitate cell internalization and endosomal escape. The gene-silencing efficiency and cytocompatibility of these particles were optimized by screening their amine content. PRINT hydrogel particles containing a 30% AEM-based hydrogel composition maintained the viability of HeLa cells above 80% after 4 hours treatment followed by 48 hours' incubation in media and exhibited dose-dependent silencing of luciferase expression versus controls, which did not cause a significant gene knockdown (Dunn et al., 2012).

In addition to disulfide linkages, diselenide bonds (SeSe) have also been investigated for redox-responsive gene delivery (Cheng et al., 2012a). Cheng et al. used diselenide bonds crosslinked to branhed oligoethylenimine (800 Da) OEI_{800} to develop novel reduction-sensitive gene carriers. Its potential for reduction sensitivity was also compared to disulfide-conjugated OEI_{800} in this study. Both SS- and SeSe-conjugated OEI_{800} nanoparticles demonstrated lower cytotoxicity and higher transfection efficiency when compared with nondegradable $PEI_{25k.}$ The diselenide-conjugated OEI_{800} showed higher transfection efficiency than disulfide-conjugated OEI_{800} at cationic polymer-to-plasmid DNA

(C/P) weight ratios ≥ 10 in B16F10 cells and at all C/P ratios in HeLa cells. Diselenide-conjugated OEI_{800} exhibited a cell-line-dependent reduction sensitivity and was shown to be as efficient as the disulfide-conjugated OEI_{800} in vitro (Cheng et al., 2012a).

Some commonly used redox-sensitive crosslinkers and their structures are shown in Fig. 14.3.

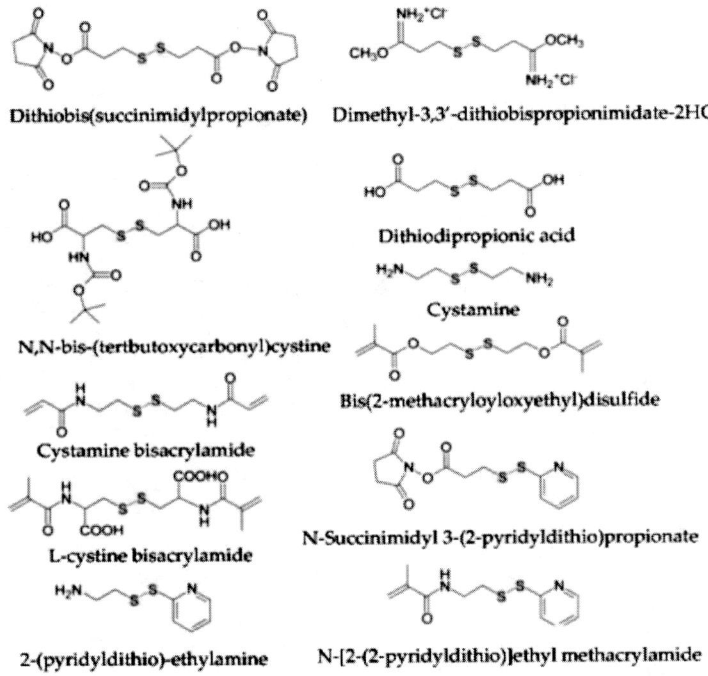

Figure 14.3 Commonly used redox-responsive crosslinkers and their chemical structures. Modified from Meng et al. (2009), Copyright 2009, with permission from Elsevier.

14.4.1.3 Tumor-expressed or overexpressed enzymes

Designing nanoparticles that can respond to the biocatalytic actions of enzymes is an emerging strategy in the field of bioresponsive delivery of therapeutics (Hu et al., 2012; de la Rica et al., 2012; Ghadiali and Stevens, 2008). Enzymes are selective biocatalysts and play a pivotal role in carrying out numerous biochemical interconversions that are critical to sustain life. Enzyme composition and expression levels are often altered in cancer and many other diseases, making them preferred targets for drug development and therapeutics. Often, the enzyme activity associated with a particular tissue or the enzyme concentration itself is found to be higher at the disease sites. In such cases, nanomaterials can be engineered to deliver therapeutics in a site-specific manner via enzymatic conversion of the carrier (de la Rica et al., 2012).

Enzyme-responsive delivery from nanoparticles can be brought about by incorporating enzyme reactive moieties in the main chain or side chain of the polymeric material, which is recognized by the enzyme, or a product of the enzymatic reaction, both of which result in enzymatic transformation of the nanoparticles (de la Rica et al., 2012). Alternately, when the nanomaterial itself is not sensitive to biocatalysis, its surface may be modified with molecules that elicit a change in the physical properties of the nanoparticle solution (sol-to-gel and gel-to-sol transition) upon enzymatic transformation (de la Rica et al., 2012; Hu et al., 2012). Enzyme-catalyzed reactions that cause the transformation of polymers can involve multiple non-covalent interactions (van der Waal's forces, hydrogen bonding, hydrophobic and electrostatic interactions), which may result in self-assembly, disintegration, and morphological transition of polymers or polymeric assemblies (Hu et al., 2012).

Some of the most commonly dysregulated enzymes in cancer include enzymes that belong to the class of hydrolases (proteases, lipases, and glycosidases), enzymes involved in cellular metabolic processes like glycolysis and fatty acid synthesis, and some oxidoreductases (Zhao et al., 2013; de la Rica et al., 2012). The MMP family of enzymes is predominantly involved in cancer progression, invasion, and metastasis (Nagase and Woessner, 1999; Gialeli et al., 2011; Overall and Kleifeld, 2006). Among the MMPs, MMP-2 and MMP-9 are the most important cancer-related MMPs (Li et al., 2013d). Polymeric nanoparticles can be covalently modified with drugs through MMP-cleavable linkers that can trigger drug delivery upon exposure to these enzymes in the target tissue. MMP-cleavable substrates (typically, cleavable peptides digested by proteases) have been used for delivery of drugs and imaging agents (Zhu et al., 2012; Chen et al., 2013c; Li et al., 2013d). Chen et al. recently reported a functionalized AuNP system for tumor imaging and targeting, wherein DOX was conjugated to the AuNPs by a thiol–Au bond with a MMP-2-cleavable peptide substrate CPLGLAGG (Chen et al., 2013c). The nanoparticles were decorated with PEG to allow for extended circulation. The presence of a thiol–Au bond led to accelerated DOX release once the particles were internalized via endocytosis due to the thiol group exchange reaction with GSH in the cytoplasm. Injection of the AuNP into SCC-7 tumor-bearing mice caused rapid release of DOX, which resulted in simultaneous tumor growth inhibition as well as fluorescent imaging. The functionalized AuNPs showed improved efficacy without side effects compared to free DOX (Chen et al., 2013c). Li et al. also reported a system to target MMP-2/MMP-9 (gelatinases A/B) using mPEG-PCL, with a gelatinase-cleavable peptide (PVGLIG) inserted between the mPEG and PCL blocks (Li et al., 2013d) (Fig. 14.4). The mPEG-peptide-PCL nanoparticles and positive control nanoparticles without the cleavable peptide were both loaded with docetaxel. The hollow structure of the gelatinase-responsive nanoparticles allowed for higher drug loading. The nanoparticles collapsed and aggregated in the presence of gelatinases and exhibited a gelatinase-dependent cellular uptake and in vitro cytotoxicty. The mPEG-peptide-PCL nanoparticles also showed better antitumor effects on primary cancer cells isolated from pericardial fluid of a late-stage lung cancer patient than the free docetaxel (Taxotere) (Li et al., 2013d).

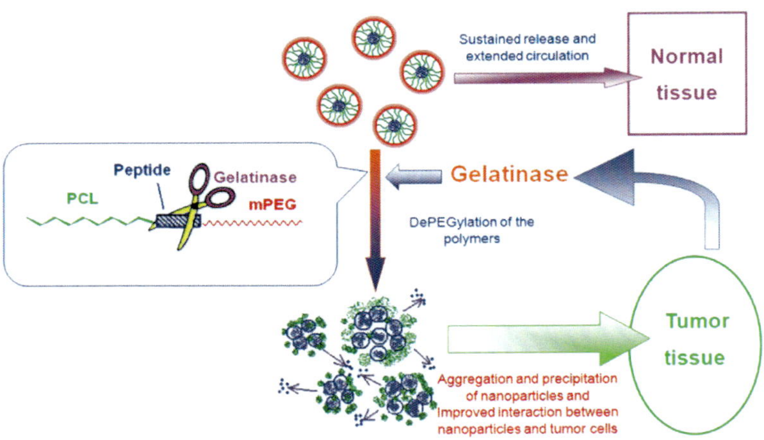

Figure 14.4 Gelatinase-responsive nanoparticles for enhanced interaction with cancer cells. Reprinted from *PLoS ONE*, Copyright Li et al. (2013d).

Among other enzymes, the cysteine protease cathepsin B is widely implicated in tumor invasion and metastasis and is overexpressed in many cancers at the mRNA, protein, and activity levels (Podgorski and Sloane, 2003; Elliott and Sloane, 1996). Cathepsin B responsive polymerosomes were prepared from a biodegradable BCP of mPEG and poly(d,l-lactide) (PDLLA) by incorporating a cathepsin-cleavable peptide sequence Gly-Phe-Leu-Gly-Phe (GFLGF) between the two blocks (mPEG-pep-PDLLA) (Lee et al., 2011b). These polymerosomes destabilized at pH 5.5 in the presence of the lysosomal enzyme cathepsin B and also showed rapid release of the model drug acridine orange under these conditions. The polymerosome surface was further modified with antiepidermal growth factor receptor antibody (abEGFR) to enhance cellular uptake. When tested in SKBR3 cells, the peptide-containing polymerosomes modified with abEGFR were endocytosed more rapidly and showed enhanced intracellular release of fluorescein isothiocyanate (FITC)-labeled dextran compared to control polymerosomes lacking EGFR or peptide or both (Lee et al., 2011b). CD44 is overexpressed in many solid tumors and is one of the major receptors for hyaluronic acid (HA) (Aruffo et al., 1990; Underhill, 1992). HA is a naturally occurring, nonimmunogenic, biocompatible, and biodegradable polysaccharide, which is degraded by hyaluronidase-I (Hyal-I), a lysosomal enzyme mainly found in the tumor microenvironment. HA is converted into low-molecular-weight fragments by this enzyme once it is endocytosed by the cancer cells and it has been used to target cancer cells overexpressing CD44 receptors (Toole, 2009; Choi et al., 2011). Chen et al. reported such a mesoporous silica nanoparticle (MSN)-based system, which utilized HA as a capping agent (MSN-HA) for targeted and enzyme-responsive drug delivery (Chen et al., 2013e). In vitro studies revealed that MSN-HA was nontoxic to cells even at high concentrations, whereas DOX-loaded MSN-HA (DOX-MSN-HA) showed dose-dependent cytotoxicity in MDA-MD-231 cells that was much higher than that for free DOX. In normal cells not expressing CD44 receptors (NIH 3T3), the DOX-MSN-HA exhibited only weak potency. To evaluate the targeting efficiency of these nanoparticles, FITC-doped MSNs (FITC-MSN-HA) were incubated with NIH 3T3- and CD44-overexpressing MDA-MB-231 cells. Stronger fluorescence signals were detected in the cancer cell line. HA-free MSNs used as a control showed a weak fluorescence when incubated with either of the cell lines, proving that HA-MSNs were more easily internalized by cancer cells than the HA-free MSNs. HA interacted with CD44 receptors on plasma membrane of MDA-MB-231 cells, resulting in the uptake of FITC-MSN-HA via endocytosis and their accumulation within endosomes and lysosomes. HA is degraded after endocytosis to lower-molecular-weight fragments by Hyal-I in the lysosomes (Chen et al., 2013e). HA-targeted nanocarriers have also been reported for delivery of drugs like SN-38 (Vangara et al., 2013), DOX, and CPT (Choi et al., 2011), as well as for gene delivery (Needham et al., 2012).

In addition to the examples mentioned here, proteases like plasmin have been explored for delivery of proteins using enzyme-responsive nanocapsules (Wen et al., 2011); gelatinase-responsive copper sulfide nanoparticles were reported for combined photoacoustic imaging, enzyme-responsive drug release, and photothermal therapy (Zha et al., 2013), and liposomal formulations of lipid pro-drugs of retinoids sensitive to degradation by secretory phospholipase A_2 (Pedersen et al., 2010a) have also been developed.

Another interesting class of materials comprises self-immolative materials. These materials are engineered such that a trigger initiates a cascade of intramolecular chemical reactions, which lead to complete disintegration of the material into small-molecule components (Wong et al., 2012). This trigger to "self-immolate" can be provided by an enzyme- sensitive linker. Dendrimers, oligomers, and linear polymers have been employed to design such self-immolative systems (Wong et al., 2012). Thiolated, hyperbranched polyglycerols were used to synthesize dendritic scaffolds, which were then conjugated with enzymatically cleavable maleimide-bearing pro-drugs of doxorubicin and methotrexate (Calderon et al., 2009). To attach these drugs to dendritic polyglycerol, Calderon et al. used either a self-immolative para-aminobenzyloxycarbonyl (PABC) spacer coupled to the dipeptide Phe-Lys or the tripeptide D-Ala-Phe-Lys as the protease substrate. These drug conjugates could effectively release drugs after cleavage by the enzyme cathepsin B and retained their activity (Calderon et al., 2009). Self-immolative systems have also been reported using *N*-(2-hyroxypropyl)-methacrylamide (HPMA) copolymer–drug conjugates with an AB_3 (i.e., three drug molecules attached to the head group "A" via linker) self-immolative dendritic linker (Erez et al., 2009). These self-immolative systems could be cleaved by cathepsin B to release a triple payload of pactitaxel and were more toxic to murine prostate carcinoma cells compared to a monomeric drug–polymer conjugate.

Other examples of enzyme-responsive delivery systems are listed in Table 14.3.

Table 14.3 Some examples of enzyme-responsive nanocarriers

Nanocarrier	Enzyme	Enzyme-cleavable linker	API	Results	Refs.
Micelle	MMP-2	GPLGIAGQ peptide	Paclitaxel	Enhanced antitumor activity	(Zhu et al., 2013)
Copper sulfide nanoparticles	Gelatinase	Gelatin	Doxorubicin	Enhanced cytotoxicity	(Zha et al., 2013)
Liposome	Phospholipase A(2)	Lipid C6	Retinoid	Enhanced cytotoxicity	(Pedersen et al., 2010b)
Core–shell nanoparticles	Esterase	Benzyl ester bond	Paclitaxel and photosensitizer	Enhanced antitumor activity	(Lee et al., 2013a)
Hydrogel	Thrombin	Phe-Pip-Arg-Ser	Heparin	Autoregulation of heparin release	(Maitz et al., 2013)
Hydrogel nanocapsules	Plasmin	KNRVK peptide	VEGF*	Blood vessel formation induced by enzyme-responsive release of VEGF	(Wen et al., 2011)

*VEGF: Vascular endothelial growth factor

14.4.1.4 Temperature (hyperthermia)

Temperature is one of the most widely investigated stimuli for drug delivery. Hyperthermia can be considered as both an internal as well as an external stimulus (Torchilin, 2009). Elevated temperature is commonly encountered in disease states such as fever or inflammation and can be applied from outside the body by local ultrasound or high-frequency AMF, which can cause oscillation of magnetic material accumulated at the target site with subsequent heat release (Torchilin, 2009). External control of local body temperature by physical means affords a more consistent signal and can be used as an extensive application for thermosensitive nanocarriers (Kang et al., 2008). An important characteristic of thermoresponsive polymers or carrier systems is their temperature-dependent change in the hydration state, and thus the hydrophilic–hydrophobic balance (Fleige et al., 2012). Thermo-sensitive polymers and nanosystems undergo coil-to-globule thermal-phase transitions (Kang et al., 2008). Most thermo-responsive polymers are soluble in water below a specific temperature but above that they are insoluble and cause phase separation. These polymers are said to have a lower critical solution temperature (LCST). The relation between temperature and solubility is the reverse for polymers that exhibit an upper critical solution temperature (UCST) (Li et al., 2013e). Polymers used in drug delivery applications majorly belong to the LCST type.

Among the thermoresponsive polymers, the most widely studied is poly(*N*-isopropylacrylamide) (pNIPAAm), which undergoes a sharp phase transition below its LCST at 32°C, changing from a hydrophilic "coil" to a hydrophobic "globule" above the LCST (Schild, 1992, Pelton, 2000, Wang et al., 1998). The LCST of pNIPAAm can be controlled by varying its hydrophilic or hydrophobic content. Its LCST increases when modified with a hydrophilic polymer, whereas copolymerization with hydrophobic polymers tends to decrease the LCST (Li et al., 2013e). A number of copolymers of pNIPAAm with various hydrophilic polymers have been developed to fine-tune its LCST closer to the body temperature for biomedical applications. Polymers like *N*-vinyl-2-pyrrolidone (Dincer et al., 2006), *N,N*-dimethylacrylamide (Qiu et al., 2009, Akimoto et al., 2010), acrylamide (AAm) (Yavuz et al., 2009), and PEG (Neradovic et al., 2004; Motokawa et al., 2005; Zhang et al., 2005), among others, have been utilized to that end.

Yang et al. reported thermo-sensitive comb-like copolymers of mPEG blocks and hydrophobic polyacrylate (PA) backbones with thermosensitive PNIPAM graft chains (mPEG-b-PA-g-PNIPAM) and optimized it for release of CPT by studying various parameters including copolymer composition, chain length of mPEG blocks and drug release behavior (Yang et al., 2013a). The copolymers spontaneously self-assembled to form nearly spherical core–shell nanomicelles (<200 nm) in aqueous solutions. The micelles also showed a thermoresponsive phase transition with a LCST from 40°C to 44.5°C and could stably incorporate CPT in the active lactone form within the micelle core. CPT could be delivered continuously from the micelles without the initial burst release, and the release was accelerated at temperatures above the LCST, confirming the thermotriggered drug release property of the micelles. Cytotoxicity studies showed that while the free drug was toxic even to normal cells (L929), the $mPEG_{45}$-b-PA-g-PNIPAM micelles showed toxicity only toward MDA-MB-231 breast cancer cells and could avoid the side effects associated with administering a free drug (Yang et al., 2013a). Li et al. recently reported thermosensitive polyelectrolyte complex (PEC) micelles for delivery of 5-fluorouracil (5-FU) using two biocompatible graft copolymers, chitosan-g-poly(*N*-isopropylacrylamide) (CS-g-PNIPAM) and carboxymethyl cellulose-g-poly(*N*-isopropylacrylamide) (CMC-g-PNIPAM), which self-assembled readily to form nanosized core–shell micelles (Li et al., 2014). PNIPAM formed the thermosensitive shell, whereas CS (positively charged) and CMC (negatively charged) formed the core of the PEC micelles. Gluteraldehyde (GA) was added to crosslink the PEC micelle core and decrease initial burst release. The LCST of these micelles was near 37°C. In vitro, on HeLa cells, the PEC micelles and PEC-GA micelles exhibited excellent temperature- and pH- controlled drug release properties (Li et al., 2014). pNIPAAm copolymer (pNIPAAm-co-Am-SH) was used to functionalize AuNPs (Salmaso et al., 2009). These nanoparticles had a transition temperature within the range of 37°C–40°C. Cell culture studies with MCF-7 cells showed that increase in temperature from 34°C to 40°C significantly increased the cellular uptake (~80-fold) of functionalized AuNPs, whereas no difference in the uptake was observed for nonmodified AuNPs at the same temperature. This increased uptake was due to phase transition of the micelle corona from a hydrophilic to hydrophobic state at 37°C (Salmaso et al., 2009). Such surface-modified thermoresponsive nanocarriers have also been reported by other groups (Akimoto et al., 2010; Zhao et al., 2011b).

A very interesting approach involved the use of a thermoresponsive polymer to regulate DNA interactions in a DNA-mediated assembly system as well as a DNA-encoded drug delivery system (Hamner et al., 2013). The authors functionalized AuNPs with thiol-modified single-stranded DNA (ssDNA) or oligonucleotides and a thiol modified copolymer of pNIPAAM-co-pAAm which had a transition temperature (Tc) of 51°C. The accessibility of sequence-specific hybridization between complementary DNA-functionalized AuNPs was dependent on the thermoresponsive behavior of the polymer. At $T < T_c$, the polymer was hydrophilic and in an extended conformation, blocking interaction between complementary DNA sequences at the periphery, whereas at $T > T_c$, a phase transition caused the polymer to convert from hydrophilic to hydrophobic, shrinking below the outer single-stranded DNA (ssDNA) and allowing hybridization to occur. This system was then utilized for delivery of DOX using a DNA-encoded carrier (Fig. 14.5). The AuNPs were functionalized with double-stranded DNA (dsDNA) whose sequences had a high affinity for DOX. At $T < T_c$, the drug release was decreased due to the polymer capping, whereas the reverse was true at $T > T_c$. This resulted in a fourfold improvement in the IC_{50} of the polymer+DNA-encoded nanocarrier over DOX alone (Hamner et al., 2013).

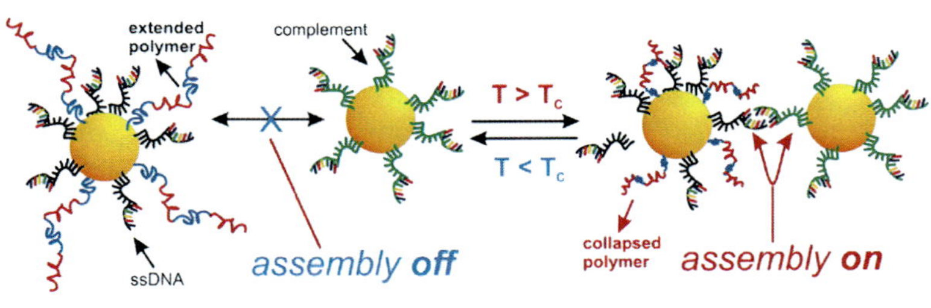

Figure 14.5 Using temperature-sensitive smart polymers to regulate DNA- mediated nanoassembly and encoded nanocarrier drug release. Reprinted with permission from Hamner et al. (2013). Copyright 2013 American Chemical Society.

Temperature-sensitive liposomes, in combination with hyperthermia produced by ultrasound or magnetic fluid hyperthermia, have also been used for stimuli-responsive release of drugs (Al-Jamal et al., 2012; Manzoor et al., 2012; Lapointe and Martel, 2009). Temperature-sensitive liposomes are made with lipids (typically dipalmitoylphosphatidylcholine, DPPC), which have transition temperatures in the hyperthermic range (39°C–42°C). At these temperatures, the lipid bilayers become leaky and release the entrapped cargo at a specific site only when the temperature is raised above 41°C, as in the case of DPPC (Bikram and West, 2008).

Elastin-like polypeptides (ELPs) have been extensively investigated as novel vehicles for thermoresponsive delivery by Chilkoti and colleagues (McDaniel et al., 2013b; McDaniel et al., 2013a; McDaniel et al., 2012). ELPs are genetically encoded biopolymers that possess repeats of a pentameric peptide sequence (Val-Pro-Gly-*X*aa-Gly) derived from tropoelastin, where *X* can be any amino acid except Pro (Gray et al., 1973; Tatham and Shewry, 2000). ELPs show a phase transition temperature similar to the LCST observed with some polymers, wherein they are soluble at low temperatures but can phase-separate into a gel-like phase above a critical transition temperature (McDaniel et al., 2013b). The transition temperature can be adjusted by changing the composition of the *X*aa residue, although there are also other factors determining it, including length of polypeptide, ionic strength and polarity of the solvent, and solution concentration. (Meyer and Chilkoti, 2004). McDaniels et al. reported a thermally responsive chimeric polypeptide (CP) derived from ELP for thermoresponsive delivery of DOX (McDaniel et al., 2012). CPs have hydrophilic ELPs as the N-terminal segment, while the C-terminal segment contains a variable number of periodically spaced Cys residues arranged in a Cys-(Gly-Gly-Cys)$_n$ motif, which is used to attach small-molecule chemotherapeutics. When multiple molecules of DOX were conjugated to the Cys end, the CP self-assembled into ~40 nm monodisperse nanoparticles, with DOX loaded in the core. The hydrophilic CP formed the nanoparticle corona with a phase transition temperature between 39°C and 42°C, suitable for mild hyperthermia of tumors. The CP-DOX nanoparticles remained stable upon dilution and were cytotoxic to C26 murine colon carcinoma cells at both 37°C and 42°C after one-hour treatment and released DOX into the nucleus. CPs without conjugated DOX showed no toxicity (McDaniel et al., 2012).

Other thermoresponsive polymers have been used to develop thermoresponsive vehicles, including poly(*N*-vinylcaprolactam) (PNVCL) (Prabaharan et al., 2009b), poly(ethylene oxide)/poly(propylene oxide) (PEO/PPO) BCPs (Bae et al., 2006), poly(2-isopropyl-2-oxazoline) (Park et al., 2007), and MPEG-cyclotriphosphazenes conjugated to hydrophobic oligopeptides (Toti et al., 2007). Additional examples of thermoresponsive nanocarriers are presented in Table 14.4.

14.4.2 External Stimuli

External stimuli that can be used to trigger drug release from nanoparticles include light, ultrasound, and magnetic field. These have the added advantage of providing more accurate temporal and spatial control over the release of therapeutics.

14.4.2.1 Light

Light is an attractive stimulus because of the flexibility it offers in terms of the spatiotemporal control over the release of cargo from nanocarriers (Fleige et al., 2012; Gohy and Zhao, 2013; Zhao, 2007). Moreover, the response to light can be tuned by varying the wavelength and intensity of applied light (Schumers et al., 2010). Ultraviolet (UV), visible, or NIR light can be used to trigger drug release from nanocarriers (Rijcken et al., 2007). UV light is absorbed by the skin as well as blood and tissues, which precludes its deep penetration in the body. NIR light, on the other hand, has a deep penetration (up to 10 cm) through tissues and water and avoids damage to cells (Zhao, 2007; Goodwin et al., 2005; Oerlemans et al., 2010).

To design light-responsive nanocarriers, photo-responsive groups (chromophores) can be incorporated within or conjugated to the nanocarriers (Oerlemans et al., 2010). All phototriggered reactions lead to a change in the structure or assembly of the nanocarrier either directly or indirectly, ultimately releasing the encapsulated cargo (Fomina et al., 2012). This change in the structure of the nanocarrier is due to structural transformations of specific functional groups either in the polymer backbone or in the side chains (Roy et al., 2010; Yan et al., 2013b; Zhao, 2007). A number of mechanisms can be used to induce light-triggered release of

Table 14.4 Examples of thermoresponsive nanocarriers

Thermoresponsive polymer	LCST *	Nanocarrier	Cargo	Refs.
Poly(*N*-vinylcaprolactam)	32°C	Micelle	5-FU	(Prabaharan et al., 2009b)
Poly(*N*-isopropylacrylamide-acrylamide-allylamine)	40°C	Iron oxide magnetic NPs**	–	(Wadajkar et al., 2013)
Poly(*N*-isopropylacrylamide-co-acrylamide)	43°C	Micelle	SPION and doxorubicin	(Kim et al., 2013a)
Leucine zipper peptides+lysolipid	42.5°C	Lipid-peptide vesicles	Doxorubicin	(Al-Ahmady et al., 2012)
PMEEECL-b-POCTCL***	38°C	Micelle	Nile Red and doxorubicin	(Cheng et al., 2012c)
Poly(ionic liquid-co-*N*-isopropylacrylamide)	pH 5.2: 36.3°C pH 7.4: 38.5°C	Ionically assembled NPs	Doxorubicin	(Cui et al., 2012)
Coiled-coil peptide motif	–	Mesoporous silica NPs	Fluorescein	(Martelli et al., 2013)

*LCST: Lower critical solution temperature

**NPs: Nanoparticles

***PMEEECL-b-POCTCL: Poly{γ-2-[2-(2-methoxyethoxy)ethoxy]ethoxy-ε-caprolactone}-b-poly(γ-octyloxy-ε-caprolactone) the cargo from nanocarriers, including photo-isomerization, photo-crosslinking, photosensitization-induced oxidation, light-triggered switch in polarity (reversible and irreversible), photo-decrosslinking, and photo-degradation of the polymer backbone (Fomina et al., 2012; Rijcken et al., 2007; Yan et al., 2013b). These mechanisms have been reviewed most recently by Fomina et al. (2012). Polymeric micelles and liposomes have been the two most widely utilized nanocarriers for the development of light-responsive vehicles (Alvarez-Lorenzo et al., 2009; Yavlovich et al., 2010; Zhao, 2007). However, other nanocarriers like dendrimers (Sun et al., 2012a), silica nanoparticles (Chang et al., 2012), AuNPs (Niikura et al., 2013), and NGs (Wang et al., 2013a) have also been employed.

Azobenzene and its derivatives account for the most widely studied reversible photo-responsive groups. Upon UV irradiation, azobenzenes undergo a reversible *trans* to *cis* photo-isomerization of their nitrogen double bond (N=N). This converts the apolar *trans* isomer into the polar *cis* isomer, which can be reconverted to the *trans* isomer by visible light (Rijcken et al., 2007; Yan et al., 2013b; Zhao, 2007). This isomerization causes the nanocarriers to disrupt and release the cargo. The reconversion to *trans* isomer is a useful feature for on-demand drug delivery (Fomina et al., 2012). A number of photo-responsive nanocarriers bearing azobenzene groups have been reported with the greater emphasis being on polymeric micelles (Tang et al., 2009; Boissiere et al., 2011; Jochum and Theato, 2010; Yuan et al., 2012; Xing et al., 2011a). In a novel proof-of-concept study, Yuan et al. reported the development of a photo-responsive, biocompatible DNA/mesoporous silica hybrid formed by introducing azo-benzene moieties into the DNA molecule (Yuan et al., 2012). Azobenzene was incorporated into DNA double strands, which were then immobilized at the pore mouth of MSNs, providing a novel photo-regulated release nanoplatform. The guest molecules (rhodamine 6G as model drug) were retained within nanopores of this platform under visible light (λ = 450 nm), with controlled release in the UV (λ = 365 nm) range due to release of the DNA linker into solution. Upon photo-isomerization of azobenzene, dehybridization/hybridization of complementary DNA caused opening/closing of pore gates of MSN. The authors tested the anticancer effects of DOX-loaded azo-DNA/MSNs on A549 (lung adenocarcinoma) as well as CCRF-CEM (acute lymphoblastic leukemia) cells. In the absence of UV light, the nanocarriers showed minimal toxicity to both cell lines even at high concentrations, but a dose-dependent cytotoxicity was evident upon UV irradiation for 30 min, confirming effective light-responsive behavior of these MSNs. Intracellular accumulation of DOX also increased with increase in the concentration of azo-DNA/MSN, suggesting efficient uptake and controlled release of DOX (Yuan et al., 2012) (Fig. 14.6).

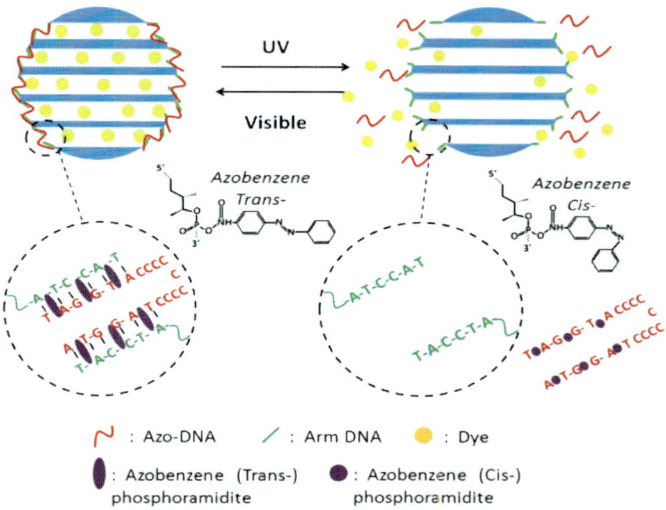

Figure 14.6 Example of a photo-responsive system based on mesoporous silica hybrid nanoparticles. Reprinted with permission from Yuan et al. (2012). Copyright 2012 American Chemical Society.

NIR has also been used as a trigger for drug delivery from nanocarriers containing azobenzene groups as chromophores (Liu et al., 2013a). Spiropyran (SP) derivatives exhibit a reversible photo-isomerization between a colorless SP and a colored zwitterionic merocyanine (ME) upon UV exposure (Rijcken et al., 2007; Roy et al., 2010). Such a change can affect the polarity or hydrophilicity, which, in turn, can alter microenvironments within polymers. Such photoswitching SP-containing nanoparticles were developed both for on-demand drug delivery and reversible changes in particle volume to enhance tissue penetration (Tong et al., 2012). Cinnamic acid and its derivatives are another UV-responsive class of chromophores, which undergo *trans–cis* photo-isomerization and either form cinnamate residues with an increased hydrophilicity or lead to reversible photo-dimerization (Rijcken et al., 2007). Recently, liposomes composed of dioleoylphosphatidyl ethanolamine (DOPE) and 2-(hexadecyloxy) cinnamic acid (HOCA) were reported where photodimerization of the cinnamic acid residue of HOCA destabilized the liposomal membrane and enabled significant release of the loaded cargo (Seo and Kim, 2013). A biodegradable copolymer was prepared by grafting dithiothreitol (DTT) into poly(3,4-dihydroxycinnamic acid) (PCA) to produce nanoparticles that could release a model protein, bovine serum albumin, upon shrinkage induced by UV-irradiation-triggered crosslinking (Shi et al., 2011). Dithienylethene (DTE) and stilbene also undergo reversible photo-isomerization upon UV or visible light irradiation, which causes structural deformation of nanocarriers (Chen et al., 2011b; Menon et al., 2011).

Another approach to designing photo-responsive nanocarriers is to incorporate chromophores that undergo irreversible photocleavage upon UV or NIR light exposure to change polarity of the nanocarriers, causing them to destabilize (Katz and Burdick, 2010; Gohy and Zhao, 2013; Oerlemans et al., 2010). Liu et al. employed NIR-light-sensitive polymeric micelles for delivery of DOX (Liu et al., 2012a). DOX-loaded micelles were formed from dextran-graft-(2-diazo-1,2-naphthoquinone) (Dex-DNQ), where DNQ was the hydrophobic portion and Dex was the hydrophilic part of the amphiphilic copolymers. Upon UV/NIR irradiation, DNQ, the phototrigger group, undergoes Wolff rearrangement to a hydrophilic 3-indene carboxylic acid (3-IC) molecule, causing micelle dissociation and release of encapsulated DOX. In vitro studies on HepG2 cells resulted in a significant growth inhibition of cells as a consequence of higher intracellular DOX release from micelles under NIR irradiation. Enhanced intracellular drug release was also confirmed by flow cytometry and fluorescence microscopy (Liu et al., 2012a). Yan et al. reported a novel strategy for NIR-light-triggered dissociation of BCP micelles using up-converting nanoparticles (UCNPs) (Yan et al., 2011). UCNPs can absorb NIR light and convert it to higher energy photons in the UV, visible, and NIR regions. UCNPs typically consist of transition metal, lanthanide, or actinide dopant ions embedded in the lattice of an inorganic crystalline host (Wang and Liu, 2009). $NaYF_4$:TmYb UCNPs were encapsulated within micelles of poly(ethylene oxide)-block-poly(4,5-dimethoxy-2-nitro-benzyl methacrylate) (PEO-b-PNBMA). Upon exposure of the micellar solutions to a continuous-wave diode NIR laser (980 nm), the UCNPs emit photons in the UV region, which are then absorbed by *o*-nitrobenzyl groups of the micelle core-forming block. The photocleavage of nitrobenzyl groups converted the polymethacrylate block to a hydrophilic block (methacrylic acid), which caused destabilization of BCP micelles and released the coloaded hydrophobic species (Nile Red dye). This approach avoids the need for UV or visible light excitation for biomedical applications (Yan et al., 2011).

Photo-crosslinking or photo-polymerization can also be used for light-triggered release. A polymerizable double bond irradiated directly or in presence of a radical sensitizer or initiator can lead to photo-crosslinking. Photo-polymerization of double bonds in the hydrophobic portion of a bilayer leads to shrinkage of some parts of the bilayer, disrupting the uniform packaging of molecules and creating pores in the bilayer to allow release (Fomina et al., 2012). A novel class of photo-triggerable liposomes made from diplamitoyl phosphatidylcholine (DPPC), a photo-polymerizable diacetylene phospholipid (1,2-*bis*(tricosa-10,12-diynoyl)-*sn*-glycero-3-phosphocholine) ($DC_{8,9}PC$) and DSPE-PEG2000 was reported by Yavlovich et al., which was the first study to show light-triggered release of an anticancer drug from photo-sensitive liposomes (Yavlovich et al., 2011). DOX was loaded into these liposomes and its release was triggered by irradiating them with visible light at 514 nm (166 mW/cm^2). When irradiated for 0-7 min, up to 22% higher release of DOX was seen compared to non-irradiated samples. The effects of released DOX on cellular toxicity were determined in vitro on a co-culture of MCF-7 and Raji cells with a two- to threefold improved cell killing over untreated samples (Yavlovich et al., 2011).

Gold nanostructures, including particles, rods, shells, and cages, possess strong surface plasmon resonance and show excellent NIR absorption as well as tunable optical resonances (Fomina et al., 2012). Gold nanoparticles can absorb the incident NIR light and convert it into heat in order to raise the temperature of tissues to cause photothermal ablation (PTA). You et al. reported the development of NIR-light-sensitive liposomes (DPPC, HSPC, CHOL, and DSPE-PEG2000) loaded with DOX with hydrophobically modified hollow gold nanospheres (HAuNS) attached to the liposome membrane (You et al., 2013). Rapid DOX release was observed form the liposomes upon NIR irradiation (808 nm). In vitro cytotoxicity studies with BEL-7402 human liver carcinoma cells showed significantly greater toxicity for the DOX-loaded formulation followed by NIR laser irradiation than for the formulation without irradiation or with DOX-loaded liposomes alone or with PTA-alone groups. This effect was also seen in vivo where the combination of DOX and HAuNS liposomes followed by NIR irradiation resulted in a significantly greater antitumor efficacy than with PTA alone (You et al., 2013).

Light can also be used to activate certain drugs (photosensitizers). Photodynamic therapy (PDT) combines a specific light source and the action of a photosensitizer to treat cancer (Dolmans et al., 2003). The photosensitizer is activated from the ground state to the excited state when exposed to light. When it returns to the ground state, it releases energy, which is transferred to oxygen to generate ROS such as singlet oxygen and free radicals that cause cellular toxicity. PDT can kill tumor cells directly, cause ROS-induced damage to the vascular system, or lead to activation of an immune response in the tumor (Dolmans et al., 2003). Many photosensitizers, including porphyrins, chlorophyll-based photosensitizers (chlorins, purpurins, bacteriochlorins), dyes (phthalocyanines, naphthalocyanins), and plant based photosensitizers like perylenequinones (hypericin) and α-terthienyl, have been investigated to date (Huang, 2005; Kiesslich et al., 2006; Ji and Shen, 2009). Peng et al. recently utilized an NIR combination of photodynamic and photothermal therapy for cancer, both in vitro and in vivo, using hollow silica nanoparticles (HSNs) loaded with the hydrophobic photosensitizer phthalocyanine (Pc) (Pc@HSNs) (Peng et al., 2013). NIR irradiation with a 730 nm laser caused the Pc@HSNs to convert the optical energy into heat and generate ROS for elimination of tumors. In vitro cytotototoxicity assessments in KB cells as well as in vivo studies after intratumoral injections of Pc@HSNs into S180 sarcoma tumor–bearing mice followed by NIR irradiation at 730 nm (1.5 W/cm^2 for 10 min) showed excellent treatment efficacy and tumor elimination without significant side effects, thus justifying the use of such a dual therapeutic strategy (Peng et al., 2013). Various other combinations of photosensitizers and nanocarriers have been reported for PDT (Skidan et al., 2008; Lima et al., 2013; Tanaka et al., 2011; Herlambang et al., 2011; Li et al., 2013f; Huang et al., 2011b; Gao et al., 2012; Wang et al., 2013b). Some examples of light-sensitive nanoparticles are given in Table 14.5.

Table 14.5 Examples of light-responsive nanocarriers

Nanocarrier	Light	Photo-responsive group	Cargo	Application	Refs.
Mesoporous silica nanoparticles	980 nm NIR irradiation	Theo-nitrobenzyl	Doxorubicin	Controllable drug delivery and fluorescence imaging	(Yang et al., 2013b)
Mesoporous silica nanoparticles	350 nm UV irradiation	Nitroveratryl carbamate-protected aminopropyl-functionality	Doxorubicin	Light- and pH-responsive release of doxorubicin	(Knezevic et al., 2011)
Dendritic micellar nanocarriers	808 nm NIR irradiation; 365nm UV irradiation	Diazonaphthoquinone	Doxorubicin	Light-sensitive release of Doxorubicin	(Sun et al., 2011)
Micelles	Light-emitting diodes with the peak intensity at about 660 nm	*Meta*-tetra (hydroxyphenyl)chlorine	*Meta*-tetra (hydroxyphenyl) chlorine	Photodynamic therapy	(Shieh et al., 2010)
Micelles	794 nm NIR irradiation	Coumarin groups	Nile Red	Controllable drug delivery	(Babin et al., 2009)
Single-walled carbon nanotubes	808 nm NIR irradiation	Single-walled carbon nanotubes	–	Cancer cell detection and NIR absorbance for selective photothermal ablation of tumors	(Xiao et al., 2009)

14.4.2.2 Ultrasound

Ultrasound refers the use of pressure waves with frequencies above 20 kHz (Husseini and Pitt, 2008). The utility of ultrasound stems from its non-invasiveness, ability to deeply penetrate tissues, and the ease with which ultrasonic waves can be controlled and focused at sites of tissues, including tumors (Husseini and Pitt, 2008, 2009; Gao et al., 2004c). Ultrasound can exert either thermal or non-thermal (mechanical) effects on biological tissues (Rapoport, 2007). Thermal effects are due to absorption of acoustic energy by fluids or tissues, while non-thermal effects are typically associated with oscillating or cavitating bubbles (Husseini and Pitt, 2008). Generation of thermal energy, perturbation of cell membranes due to cavitation, and enhanced permeability of blood capillaries constitute the main mechanisms of the biological effects of ultrasound (Gao et al., 2005). For the purpose of drug delivery, ultrasound-responsive systems can accumulate in desired areas and can be made leaky by locally applied ultrasound to liberate incorporated drugs and genes (Torchilin, 2009). Ultrasound-triggered drug release depends on a number of factors, including time of ultrasound application, frequency of ultrasound, power density, use of pulsed versus continuous-wave ultrasound, and drug lipophilicity (Husseini et al., 2000; Marin et al., 2001). Low-frequency ultrasound (20–100 kHz) can release the drug from nanocarriers more effectively and can penetrate deeper into the tissue than high-frequency ultrasound (1–3 MHz) (Rapoport, 2012; Marin et al., 2002). However, compared to high-frequency ultrasound, it does not allow sharp focusing (Rapoport, 2012). For in vivo applications, low-frequency ultrasound is more appropriate for larger and deep-seated tumors, while high-frequency ultrasound is used for small and superficial tumors (Rapoport, 2012; Marin et al., 2002).

Ultrasound-triggered drug release has been studied extensively in Pluronic-based polymeric micelles (Husseini et al., 2007, 2010, 2013; Ugarenko et al., 2009; Gao et al., 2005). Most recently, targeted, ultrasound-responsive Pluronic P105 micelles were developed by Husseini et al. (2013). P105 micelles with a folate-targeting moiety were loaded with DOX, and low-frequency ultrasound (70 kHz) was used as a stimulus to enable DOX release. The authors found an increase in drug release with increase in the power intensity of ultrasound, with maximum release (14%) reported at 5.4 W/cm^2. A power density threshold was observed at 0.55 W/ cm^2, below which significant drug release was not observed, suggesting a role for cavitation in triggering drug release from targeted micelles (Husseini et al., 2013). Xuan et al. reported a novel approach to amplify the effect of high-intensity focused ultrasound (HIFU) in disassembling amphiphilic BCP micelles in aqueous solution (Xuan et al., 2012). The diblock copolymer was made of a water-soluble PEO block and a block of poly(2-(2-methoxyethoxy)ethyl methacrylate) ($PMEO_2MA$) that is hydrophobic at temperatures above its LCST. A small amount of HIFU-labile 2-tetrahydropyranyl methacrylate (THPMA) was incorporated into $PMEO_2MA$ to form the micellar core above its LCST (Fig. 14.7). Under the effect of HIFU (1.1 MHz, 100 W, 20 min) the THPMA groups hydrolyzed and resulted in the conversion of hydrophobic THPMA comonomer units into hydrophilic methacrylic acid, and BCPs disassembled without a change in the solution temperature (Xuan et al., 2012).

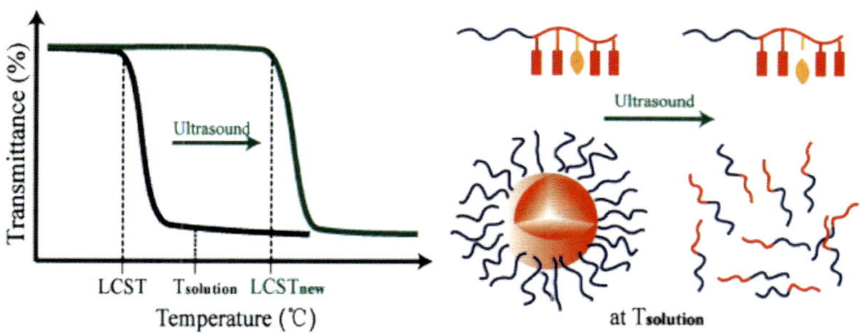

Figure 14.7 Schematic illustration of the amplification mechanism for ultrasound-disrupted block copolymer micelles based on an ultrasound-induced increase in the lower critical solution temperature (LCST) of the hydrophobic block. Upon application of ultrasound, THPMA groups are hydrolyzed and converted from hydrophobic THPMA comonomer units into hydrophilic methacrylic acid, disrupting the micelles. Reprinted with permission from Xuan et al. (2012). Copyright 2012 American Chemical Society.

HIFU-triggered drug release was also reported from poly(ethylene glycol)-b-poly[N-(2-hydroxypropyl) methacrylamide-lactate (mPEG-b-p(HPMAm-Lac_n) micelles (Deckers et al., 2013).

A very interesting example where whole-tumor ultrasound-mediated hyperthermia was used for triggering intravascular drug release from liposomes was recently reported by Kheirolomoom et al. (2013). The authors prepared temperature-sensitive liposomes that were stable in the circulation. A pH-sensitive complex was formed between DOX and copper (CuDOX) in the core of lysolipid-containing temperature-sensitive liposomes (LTSLs), such that it remained associated at neutral pH but dissociated into free DOX under low pH conditions. The CuDOX LTSLs were injected intravenously into *neu* deletion (NDL) metastatic mammary tumor–bearing mice (6 mg Dox/kg body weight) and ultrasound was used to trigger intravascular release of the drug. The tumor was insonified 5 min prior to drug administration and 20 min after drug injection. A single dose of the stabilized CuDOX-LTSLs combined with ultrasound was effective in significantly suppressing tumor growth compared to control groups of CuDOX-LTSLs or ultrasound alone. Continued treatment twice a week over 28 days regressed tumors to 30% of their initial volume. Upon termination of treatment, tumors in the CuDOX-LTSL+ultrasound group continued to regress and disappeared completely 53 days posttreatment, with no tumor recurrence for up to 8 months without any systemic toxicity (cardiac hypertrophy, leucopenia, weight and hair loss) detected for the CuDOX-LTSL group after therapy. None of the other control groups produced a complete response (Kheirolomoom et al., 2013).

Microbubbles have been used as ultrasound contrast agents (UCAs) in the clinic for a number of years. They usually incorporate a shell that is made up of protein, lipid, or polymer, which encloses a gas within it. The gas could be air, sulfur hexafluoride, or most commonly a perfluorocarbon gas (Nomikou and McHale, 2010). Stable cavitation and inertial cavitation of microbubbles in an ultrasonic field can cause transient, reversible cell membrane permeabilization by localized amplification of the effects of ultrasound on cellular targets (Geers et al., 2012). Microbubbles have also been investigated widely as drug carriers and enhancers of drug and gene delivery, as they perform the dual functions of target-specific delivery by responding to ultrasound and can deliver their payload intracellularly, resulting from transient membrane permeabilization (Lentacker et al., 2010; Nomikou and McHale, 2010; Fokong et al., 2012; Rapoport et al., 2011; Gao et al., 2008).

However, microbubbles pose inherent problems as drug carriers to solid tumors due to their large sizes (2–10 μm) and short circulation times (minutes), both of which preclude effective extravasation into tumors and prevent tumor targeting (Rapoport et al., 2011; Geers et al., 2012). A number of groups have reported the use of "nanobubbles" (NBs) to overcome this limitation (Huang et al., 2013a; Rapoport et al., 2011; Perera et al., 2013). Nanosized microbubble precursors can be developed, which accumulate in tumors via the EPR effect and, upon exposure to ultrasound, are converted to microbubbles (Rapoport et al., 2011). Alternately, NBs or "bubble liposomes" combining benefits of liposomes with ultrasound-responsiveness have been reported. They are generally prepared by sonicating liposomes in the presence of fluorinated gases, and they have been successfully utilized for drug and gene delivery (Wang et al., 2010d; Suzuki et al., 2010).

Yin et al. proposed the use of ultrasound-sensitive NBs for siRNA delivery to tumors (siRNA-NBs) using a heteroassembling strategy with SIRT-2 (against antiapoptosis gene sirtuin 2, overexpressed in gliomas) siRNA-complexed polymeric micelles (mPEG-b-PLLys) and gas-cored liposomes (Yin et al., 2013b). The exposure to low-frequency ultrasound enabled siRNA transfection, which effectively enhanced the gene-silencing ability of siRNA NBs, both in vitro and in vivo, causing elevated levels of cancer cell apoptosis. A significantly improved therapeutic effect was observed in a nude mouse glioma model using these siRNA NBs with the help of cavitation and sonoporation effects provided by ultrasound (Yin et al., 2013b). Perfluoropentane-encapsulated and superparamagnetic iron oxide (SPIO) nanoparticle-stabilized NBs were synthesized via a single-step emulsion process from thermosensitive F127 and poly(acrylic acid) (PAA) to serve as dual contrast agents for ultrasound and MRI (Huang et al., 2013a). These agents enabled enhanced passive targeting via increased permeability due to HIFU and magnetic targeting of chemotherapeutics to tumors, thus serving as tools to achieve combined in vivo tumor imaging and effective cancer therapy (Huang et al., 2013a). Novel nanocarriers known as eLiposomes have been described, which are essentially liposomes encapsulating an emulsion nanodroplet capable of phase change, as well as a therapeutic agent that can be used for drug delivery (Javadi et al., 2012; Lin et al., 2013). Typically, perfluorocarbon (PFC) liquids are used as nanoemulsions within the liposomes and are stabilized using phospholipids. Application of ultrasound can cause the emulsion droplet to change from liquid to gas, causing an increase in the volume within the vesicle, which results in rupture and release of the drug from the liposomes. The liposomes may also be modified with targeting ligands such as folate (Lin et al., 2013). Additional examples of ultrasound-sensitive nanocarriers are highlighted in Table 14.6.

14.4.2.3 Magnetic field

Magnetic field–responsive nanocarriers are generally loaded with the drug as well as a magnetically active component in to a pharmaceutically stable formulation, which can then be manipulated in an external magnetic field to control the rate of drug release by oscillating or heating the carrier (Torchilin, 2009; Lubbe et al., 2001). The most commonly employed magnetic particles for biomedical applications are iron oxides: magnetite (Fe_3O_4) or its oxidized form, maghemite (γ-Fe_2O_3), with a particle size of about 4–10 nm (Tartaj et al., 2003; Akbarzadeh et al., 2012b). These particles are superparamagnetic, which means that they are attracted by a magnetic field but do not retain residual magnetism once the field is removed (Šafařík and Šafaříková, 2002). They are referred to as superparamagnetic iron oxide nanoparticles (SPIONs). SPIONs have been used for a number of applications, including drug targeting, bioseparation (cell sorting), MRI, magnetic hyperthermia, and magnetic transfection (Torchilin, 2009). SPIONs that are not coated on their surface are prone to aggregation as they are hydrophobic and can be trapped by the immune system (Santhosh and Ulrih, 2013). Various coatings such as dextran, PEG, silica, PEI, poly(vinylalcohol), citric acid, pullulan, and folic acid can impart desired characteristics and also improve biocompatibility of the SPIONs (Santhosh and Ulrih, 2013). SPIONs can be engineered to perform multiple functions, such as carriers for drug, contrast enhancers for in vivo tumor imaging, hyperthermia, and magnetic drug targeting (Santhosh and Ulrih, 2013).

Magnetic drug targeting involves application of an external magnet at the intended pathological site in order to guide and concentrate magnetically susceptible delivery vehicles containing the drug at that site and to subsequently trigger the drug release.

Table 14.6 Examples of ultrasound-responsive nanocarriers

Nanocarrier	UCAs*	Cargo	Untrasonic exposure	Application	Refs.
Nanoemulsions	Perfluoropentane or perfluoro-15-crown-5-ether	Paxlitaxel	1 MHz or 5 MHz	Ultrasonic tumor imaging and targeted chemotherapy	(Rapoport et al., 2007; Rapoport et al., 2010; Rapoport et al., 2009) (Rapoport et al., 2011)
Liposomes	Perfluoropropane gas	pDNA	Frequency: 1 MHz; duty: 50%; burst rate: 2.0 Hz; intensity: 1.0 W/cm^2; time: 2 min	Ultrasound imaging and gene delivery	(Endo-Takahashi et al., 2013)
Heteroassembly of siRNA-loaded micelles and gas-cored liposomes	Octafluoropropane	siRNA	Frequency of 1 MHz, pulse repetition frequency of 1 kHz with 50% a duty cycle, intensity of 1.0 W/cm^2, and exposure time for 1 min	Tumor gene therapy	(Yin et al., 2013b)
Liposome–microbubble complexes	Perfluoropropane	Paxlitaxel	10 ms burst length, 1% duty cycle, 1 Hz pulse repetition frequency, and 10 min sonication duration	Ultrasound-triggered tumor-targeted chemotherapy	(Yan et al., 2013a)
Micelles	Perfluoropentane or perfluoro-15-crown-5-ether	Doxorubicin	MHz continuous-wave or pulsed ultrasound with 33% duty cycle at 3.4W/cm^2 nominal power density	Ultrasound-mediated intracellular and nuclear trafficking	(Mohan and Rapoport, 2010)

*UCAs: Ultrasound contrast agents

Dandamudi et al. showed that magnetic cationic liposomes (MCLs) loaded with the anticancer drug vinblastine significantly improved vascular uptake of MCLs in a murine melanoma model in the presence of an external magnet. A significant antitumor effect was evident with reduced tumor nodules in metastatic sites in the presence of an external magnet compared to the no-magnet treatment group (Dandamudi et al., 2009). Novel GoldMag (Fe_3O_4/Au) nanoparticles with Fe oxide as the core and a layer of Au deposited as the shell were modified with PEG and loaded with DOX. Biodistribution studies revealed a significantly higher DOX concentration in the liver exposed to a magnetic field versus no applied magnetic field, with higher particle aggregation in the targeted liver area (Chao et al., 2011). Hua et al. reported a carboxyl group containing a magnetically responsive core–shell nanocarrier with Fe_3O_4 MNPs and a water-soluble polyaniline derivative, poly[aniline-*co*-sodium *N*-(1-one-butyric acid) aniline] (SPAnNa) (Hua et al., 2010). PTX was immobilized on this nanocarrier to improve its thermal stability and water solubility. The PTX containing SPAnNa/MNP (bound-PTX) showed significantly improved cell killing under an external magnetic field. IC_{50} values for free PTX, bound PTX, and bound PTX under an external magnetic field were 7.1, 4.2, and 1.7 μg/mL, respectively, for CWR22R prostate cancer cells and 11.1, 9.7, and 4.6 μg/mL, respectively, for the PC3 prostate cancer cell line (Hua et al., 2010). Magnetic lipoplexes (ternary complexes of cationic lipids with plasmid DNA associated with SPIONs) can be concentrated onto target cells using magnetic fields. Wang et al. reported magnetic lipoplex delivery of short hairpin RNA (shRNA) to suppress Type 1 insulin-like growth factor receptor (IGF-1R) overexpression in lung cancer in vitro and in vivo (Wang et al., 2011a). Liposomal magnetofection of plasmids expressing GFP and shRNAs targeting IGF-1R (pGFPshIGF-1Rs) was studied in A549 cells and tumor-bearing mice and compared to lipofection by lipofectamine 2000. Liposomal magnetofection showed significantly improved transgene expression in vitro and better delivery efficiency in vivo than the lipofectamine group (Wang et al., 2011a).

SPIONs could also be used as MRI contrast agents for simultaneous imaging and stimuli-responsive drug delivery. Polymeric micelles encapsulating SPIONs and conjugated to antibodies have been used to facilitate precise tumor targeting and improve MRI contrast (T_2) (Liao et al., 2011; Sawant et al., 2009). PEG-PCL nanomicelles incorporating both SPIO and DOX and were targeted with subcutaneous anti-EGFR mAb (Cetuximab). Cetuximab immunomicelles bound to EGFR-overexpressing cells and delivered higher DOX and SPIO into the cells compared to non-targeted counterparts (Liao et al., 2011). Gultepe et al. demonstrated that MCLs loaded with SPIONs could be guided to the tumor site efficiently using an external magnet and their accumulation could be monitored non-invasively using MRI as well as biodistribution analysis. Quantitative MRI analysis and biodistribution studies in melanoma-bearing mice showed up to twofold-greater accumulation of the administered MCLs in the tumor mass by magnetic targeting (Gultepe et al., 2010). Other studies report tumor suppressor gene (*p53*) delivery using thermally crosslinked SPIONs functionalized by branched PEI, which can be tracked by MRI (Lee et al., 2012). A PEGylated, fullerene (C_{60})-based multifunctional magnetic nanocomposite with surface decorated iron oxide nanoparticles and a new PDT anticancer drug hematoporphyrin monomethyl ether (HMME) adsorbed to it was recently reported (C_{60}-IONP-PEG/HHMA) (Shi et al., 2013). This system was capable of excellent magnetic drug targeting, powerful PDT, and MRI both in vitro and in vivo, thus showing excellent potential as a theranostic agent.

Another interesting application of SPIONs, in addition to magnetically targeted drug delivery and MRI, involves the generation of magnetic fluid hyperthermia. Hyperthermia is the application of heat to tissues at temperatures slightly above those found in normal physiological conditions (May and Li, 2013). Cancer cells in particular are more sensitive to the effects of heat compared to the normal cells because of their tumor architecture (Moyer and Delman, 2008). Hyperthermia can thus be used as an adjunct to other strategies (surgery, radiation, and chemotherapy) to treat cancer. A major drawback in application of hyperthermia is achieving selectivity to heat only the tumor without damaging the surrounding normal tissue (van Vlerken and Amiji, 2006; Campbell, 2007). Iron oxide nanoparticles can be selectively localized to tumors, making them ideal candidates for local heat generation. Once the SPIONs are injected into the body in the form of ferrofluids, a strong alternating current (AC) magnetic field of sufficient strength and frequency can be applied. In the presence of this AMF, the magnetic moment inside the particles oscillates and magnetic energy is liberated in the form of heat through hysteresis losses or Néel relaxation and transferred to the tissue environment (Shinkai et al., 1995; Neuberger et al., 2005; Laurent et al., 2011). The degree of heat produced can be manipulated by adjusting the strength and duration of exposure of the magnetic field (Laurent et al., 2011). When a temperature of 41°C–43°C is reached, the tissue damage to normal tissues is reversible, while tumor cells are irreversibly damaged (Neuberger et al., 2005). Local hyperthermia, in addition to its direct effects on cells, can cause an increase in perfusion of systemically administered drugs into the core of the tumor and also lead to increased permeability (Kong and Dewhirst, 1999).

Pradhan et al. reported the development of folate-targeted, temperature-sensitive magnetic liposomes for magnetic hyperthermia triggered drug release (Pradhan et al., 2010). DOX-loaded liposomes were prepared using DPPC:cholesterol:DSPE-PEG_{2000}:DSPE-PEG_{2000}-folate at an 80:20:4.5:0.5 molar ratio (MagFolDox). FluidMag-HS, a commercially available aqueous iron oxide nanoparticle suspension, was used to prepare magnetic liposomes. The liposomes displayed temperature sensitivity with 52% DOX released in 50% FBS at 43°C after an incubation of one hour. Increased cellular uptake of DOX was observed from MagFolDox when it was targeted to tumor cells using a permanent magnetic field. This uptake was much higher than that observed in Caelyx®, a commercial liposomal doxorubicin formulation, nonmagnetic folate-targeted liposomes (FolDox) or free DOX, in folate receptor–expressing KB and HeLa cell lines. This also led to a corresponding increase in cytotoxicity of MagFolDox over Caelyx® and FolDox. Induction of magnetic hyperthermia at 42.5°C and 43.5°C caused a synergistic increase in the cytotoxicity of MagFolDox. Thus, the authors developed an optimized formulation that combined biological and physical drug targeting, triggered drug release, and AC magnetic field hyperthermia for thermochemotherapy of cancer (Pradhan et al., 2010). Cationic albumin-conjugated magnetite nanoparticles have been recently reported as novel candidates for hyperthermia in cancer therapy (Cheraghipour and Javadpour, 2013). Other nanoparticles capable of magnetic hyperthermia have also been reported (Bealle et al., 2012; Yoshida et al., 2010). Table 14.7 lists some examples of magnetic field–responsive nanoparticles.

Table 14.7 Examples of magnetic field–responsive nanoparticles

Nanocarrier	Contrast agents	Application	Refs.
Paramagnetic nanobeads	Mixed-metal oxo cluster, $Mn_8Fe_4O_{12}(O_2CCH_3)_{16}(H_2O)_4$ or Mn_8Fe_4 and styrene	T2-weighted magnetic resonance imaging	(Mh et al., 2013)
PEGylated fullerene/ iron oxide nanocomposites	Iron oxide	T2-weighted magnetic resonance imaging and photodynamic therapy, targeted drug delivery	(Shi et al., 2013)
Synthetic melanin-like nanoparticles	Melanins and Fe^{3+}	T1-weighted magnetic resonance imaging	(Ju et al., 2013)
Oil-in-water magnetic emulsion	Iron oxide	T2-weighted magnetic resonance imaging	(Ahmed et al., 2013)
Silica nanoparticles	$Mn(^{2+})$ ions	T1-weighted magnetic resonance imaging	(Kim et al., 2013c)
Gadolinium oxide (Gd_2O_3) nanoplates	Gd_2O_3	Multimodal imaging in biomedical applications	(Paik et al., 2013)
Long-circulating magnetoliposomes	Superparamagnetic iron oxide nanoparticles	T2-weighted magnetic resonance imaging	(Martins et al., 2013a)

14.5 Multifunctional Nanocarriers

The previous sections highlight several important features of nanocarriers, beginning from the very basic requirements for stability, longevity in circulation, and solubilization of drugs and contrast agents, to the need for selective targeting to cells that overexpress certain disease markers, complexation with nucleic acids, and internalization into various intracellular compartments, and, finally, to more complex features of nanocarriers such as responding to various internal and external environmental cues to release their payloads in a smart manner only at the pathological site. While all of these features are important individually, it has become increasingly evident that successful design of a nanocarrier may need a combination of two or more of the above-mentioned features. Such nanocarriers, which are engineered to combine multiple favorable functions so as to result in successful delivery of their payloads, constitute *multifunctional nanocarriers* (Fig. 14.8). To provide the desired combination of useful properties it is essential that various functionalities introduced in the nanocarriers perform their tasks in a coordinated manner (Torchilin, 2006a). A number of very interesting examples from recent literature report the development of such multifunctional nanocarriers for delivery of drugs as well as nucleic acids (Stevanovic et al., 2013; Meng et al., 2013; Kumagai et al., 2013; Kenny et al., 2013).

Zhu et al. reported a novel smart, multifunctional liposomal nanocarrier that could respond to up-regulated MMP-2 in the tumor microenvironment and also improve cancer cell–specific delivery of encapsulated drugs (Zhu et al., 2012). The key elements of this nanocarrier were two functional PEG-lipid conjugates, TATp-PEG(2000)-DSPE for enhanced intracellular uptake and MAL-PEG(3400)-peptide-DOPE for the stimulus response and to conjugate the liposome surface with tumor cell–specific antinucleosome monoclonal antibody (mAb 2C5). The resulting system had several drug delivery strategies combined within a single nanocarrier and coordinated in an optimal fashion. The features of the liposome included (i) long PEG chains for longevity and avoiding of non-specific interaction, (ii) nanosize to allow for passive targeting via the EPR effect, (iii) mAb 2C5 to specifically target tumor cells, (iv) MMP-2 sensitive bonds that could cleave in response to overexpressed MMP-2 to remove PEG chains, and

(v) cell-penetrating TATp to promote intracellular delivery after removal of protective PEG chains. This system could accumulate in tumors, target cancer cells, respond to overexpressed MMP-2, and provide enhanced cellular internalization via TATp (Zhu et al., 2012).

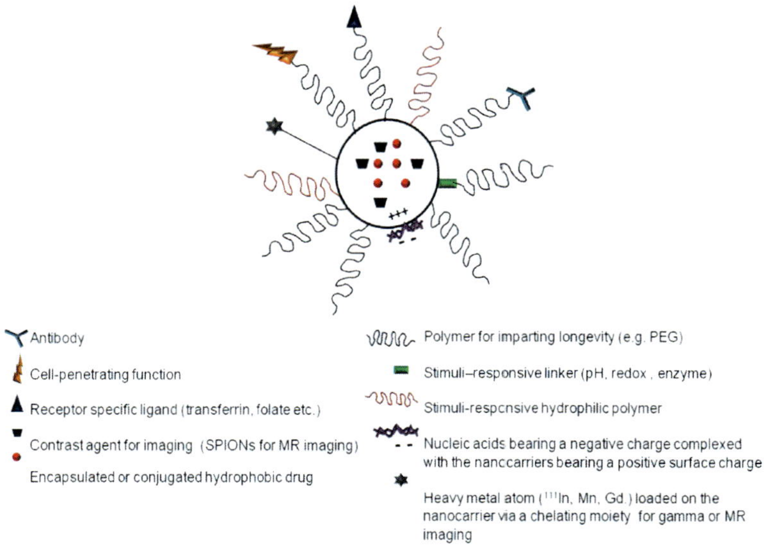

Figure 14.8 Schematic of a hypothetical multifunctional nanocarrier. Nanocarriers can be designed to have a combination of two or more of the above functions in order to generate multifunctional systems. Modified from Torchilin (2006a), Copyright 2006, with permission from Elsevier.

In another example, Kataoka and colleagues developed a smart multilayered assembly (SMA) fabricated by a layer-by-layer method using polyionic materials for multifunctional siRNA delivery (Suma et al., 2012). The SMA so designed had a siRNA-loaded core, a transiently core-stabilizing silica interlayer, an endosome disruption polycation interlayer, and a biocompatible PEG shell with reduction-responsive detachability. A unique feature of these SMAs was their construction from soft materials such as hydrophilic polymers and nonannealed silica that could be metabolized more rapidly from the body, providing a clear advantage over other materials for multiple-dose administration. The silica interlayer in SMA greatly improved the tolerance of siRNA polyion complexes (PICs) to dissociation triggered by anionic lipids, while significant siRNA release was observed after dialysis for removal of generated free silicates, suggesting the dissolvable nature of the silica interlayer. The endosome-disrupting polycation layer was a polyaspartamide derivative with two repeating aminoethylene units (PAsp(DET)) that was tethered with PEG through a disulfide bond (-SS-), which could be cleaved by reducing enzymes within endosomes/lysosome for efficient, less toxic endosomal escape. Compared to the nondisulfide controls, SMAs resulted in significantly better therapeutic gene (VEGF) silencing efficiency in cultured cells without associated cytotoxicity. Also, systemic administration of SMAs in subcutaneous tumor-bearing mice resulted in significant endogenous VEGF gene silencing in tumor tissue with no hematological toxicity (Suma et al., 2012). Some additional examples of multifunctional carriers are presented in Table 14.8.

14.6 Conclusions and Future Outlook

Advances in nanoparticle technologies have enabled the development of nanocarriers with specialized features and tunable surface chemistries to tailor them for delivery of a variety of different payloads from small molecules to complex biologics like DNA, siRNA, and proteins. From passively targeted nanocarriers to actively targeted nanocarriers bearing various functionalities (small molecules, antibodies, aptamers, and peptides), to nanocarriers with the ability to selectively internalize into intracellular structures (lysosomes and mitochondria), and finally to sophisticated smart nanocarriers, which can respond to subtle changes in the environment (pH, reductive conditions or overexpressed enzymes, or external stimuli like light, ultrasound, and magnetic fields), the levels of technological sophistication of these nanocarriers have increased. However, the main impetus for developing these nanocarrier technologies still focuses on the need to enhance the efficacy and reduce side effects of chemotherapeutics. As we strive to make treatments safer, it is important to ensure that the nanocarriers that are designed to deliver active molecules are safe and made from materials that are biocompatible and biodegradable. Decidedly, stimuli-responsive nanoparticles show a better control

Table 14.8 Some examples of multifunctional nanocarriers

Nanocarrier platform	Payload	Targeting agents	Imaging function	Stimuli-sensitive trigger	Refs.
Polymeric micelles					
P(NIPAAm-co-AAm-b-PCL	Doxorubicin	Integrin $\beta 4$ mAb recognizes A9 antigen	SPIONS for MRI	Thermoresponsive LCST: 43°C	(Kim et al., 2013b)
PEO-b-PCL	mdr1-siRNA and Doxorubicin	RGD4C (targets $\alpha_v\beta_3$) and TAT peptide	NA	pH	(Xiong and Lavasanifar, 2011)
TATp-PEG1000-PE/ PEG2000-peptide-PTX	Paclitaxel	TATp	NA	Enzyme MMP-2 sensitive peptide	(Zhu et al., 2013)
Nanoparticles					
Solid lipid nanoparticles	Bcl2-siRNA and paclitaxel	NA	Quantum dots	NA	(Bae et al., 2013)
Gold nanostars	Doxorubicin	cRGD	NIR light	NA	(Chen et al., 2013a)
Liposomes					
SPC:CHOL:DC-CHOL:DSPE-mPEG2000	Polo-like kinase 1 PLK1 siRNA	Folate and CPP	–	Enzyme Prostate-specific antigen (PSA-responsive peptide)	(Xiang et al., 2013)
Nanogels					
poly(NIPAM-co-AA) nanogel	Doxorubicin	iRGD	Fluorescent BSA-encapsulated gold nanoclusters	pH Thermoresponsive	(Su et al., 2013)

over the temporal and spatial release of drugs compared to conventional drug delivery systems. However, they are not devoid of off-target effects which could arise from non-specific stimuli such as up-regulated proteins in normal cells, higher concentrations of enzymes/reducing agents used in vitro compared to in vivo conditions, or lower pH conditions (pH ≤ 5) explored in some studies compared to the actual tumor pH (6–6.5) (Zhu and Torchilin, 2013). Hence, a careful understanding of not only the stimulus but also the extent of stimuli sensitivity of the designed nanomaterial is desired. Combining unique targeting ligands, along with the stimuli sensitivity feature has improved the capability of these systems for elimination of certain undesirable effects and offered better control.

In the future, the main challenge will be to develop nanocarriers that are responsive to biochemical signals or biomarkers typically present in nanomolar concentration ranges or even lower and enabling nanocarriers to incorporate mechanisms to amplify these signals. It will be useful to incorporate separate *sensing* and *responding* mechanisms within systems to preclude changes in the property of nanomaterials brought about by the stimulus (Stuart et al., 2010). With multiple functions incorporated into nanocarriers to make smart and versatile delivery systems, it is imperative that we not lose sight of critical features, including the biocompatibility and bioavailability of these nanocarriers. The toxicity profiles of novel nanomaterials and polymers should be thoroughly investigated before we think of integrating them into the current armamentarium of nanocarriers. Currently, the trend is toward incorporating multiple stimuli within nanocarriers to arrive at intelligent sensing systems capable of integrating detection, diagnosis, and therapy in a single carrier.

Although tremendous advancements have been made in the design of such nanocarriers, they have yet to fulfill the ultimate goal of clinical translation. Significant hurdles may include the scaling up of these complex systems on an industrial scale, as well as the process of approval by regulatory agencies. However, the future holds promise with the large number and diversity of nanocarriers currently under preclinical development. The examples mentioned in this chapter represent only a few of the many interesting possibilities that exist for the development of stimuli-responsive and multifunctional nanocarriers.

Chapter 15

Potential Applications of Nanotechnologies to Neurodegenerative Diseases (Alzheimer's and Parkinson's Diseases)

Joan Estelrich[a] and Gurutz Linazasoro[b]

[a]*Departament de Fisicoquímica, Facultat de Farmàcia, Universitat de Barcelona, Avda. Joan XXIII s/n, 08028-Barcelona, Spain*

[b]*Centro de Investigación Parkinson, Policlínica Gipuzkoa, Parque Tecnológico Miramón, 174, 20009 San Sebastián (Gipuzkoa), Spain*

joanestelrich@ub.edu; glinazasoro@inbiomed.org

15.1 Introduction

The number of different types of proteins in the human body approaches 100,000. Proteins are engaged in promoting and controlling virtually every event on which our lives depend. The amino acid sequence of a protein contains the information required for folding into a stable, functional, three-dimensional structure. It is becoming increasingly evident that biological systems have

Controlled Release Systems: Advances in Nanobottles and Active Nanoparticles
Edited by Alexander van Herk, Jacqueline Forcada, and Giorgia Pastorin

ISBN 978-981-4613-21-7 (Hardcover), 978-981-4613-22-4 (eBook)
www.panstanford.com

evolved elaborate procedures to ensure that proteins fold correctly or, if they do not, that they are detected and degraded before any serious harm can ensue to the host organism. One of these systems is formed by molecular chaperones. Chaperones are proteins that supervise protein folding so that, in most cases, mistakes are avoided and malfunctioning proteins are removed. Despite these systems of control, the likelihood of defects arising in the process of folding is significant, and a broad range of human diseases arises from the failure of a specific peptide or protein to adopt, or remain in, its native functional conformational state. These pathological conditions are generally referred to as protein conformational diseases. They include pathological states in which the impairment in the folding efficiency of a given protein results in a reduction in the quantity of the protein that is available to play its normal role. The reduction can arise as a result of several processes, but the largest group of misfolding diseases is associated with the conversion of specific peptides or proteins from their soluble functional states ultimately into highly organized fibrillar aggregates. These structures are generally described as amyloid fibrils, when they accumulated extracellularly, or intracellular inclusions, when the fibrils form inside the cell. In this chapter we use the term "amyloid" to refer also to these aggregates, without meaning that they are absolutely equivalents, and in consequence, the conformational diseases characterized by the presence of amyloid fibrils are called amyloidoses (Dobson, 1999; Thomas et al., 1995; Chiti and Dobson, 2006).

About 60 amyloid proteins have been identified. To date, there are 27 known extracellular fibril proteins in human, at least 9 of which have been studied in animal models. This large number of naturally occurring amyloid fibrils proteins underlies the need for a consistent use of standardized nomenclature (Sipe et al., 2010). Moreover, a large number of intracellular protein inclusions has been reported, although they have not been included in the nomenclature list. The established amyloid fibril nomenclature is based on the chemical nature of the fibril protein, which is designed protein A and followed by a suffix that is an abbreviated form of the parent or precursor protein name. Table 15.1 shows these 27 amyloidoses and their precursors in humans.

Table 15.1 Amyloid fibril proteins and their precursors in humans

Amyloid protein	Precursor	Syndrome or involved tissue
AL	Immunoglobulin light chain	Primary myeloma associated
AH	Immunoglobulin heavy chain	Primary myeloma associated
Ab_2M	b_2-microglobulin	Hemodyalisis associated
ATTR	Transthyretin	Familial, senile systemic
AA	(Apo)serum AA	Secondary, reactive
AApoAI	Apolipoprotein AI	Familial, aorta, meniscus
AApoAII	Apolipoprotein AII	Familial
AApoAIV	Apolipoprotein AIV	Sporadic, associated with aging
AGel	Gelsolin	Familial (Finnish)
ALys	Lysozyme	Familial
AFib	Fibrinogen a-chain	Familial
ACys	Cystatin C	Familial
ABri	ABriPP	Familial dementia, British
ALect2	Leukocyte chemotactic factor 2	Mainly kidney
ADan	ADanPP	Familial dementia, Danish
Ab	Ab protein precursor (AbPP)	Alzheimer's disease
APrP	Prion protein	Spongioform encephalopaties
ACal	(Pro)calcitonin	C cell thyroid tumours
AIAPP	Islet amyloid polypeptide	Insulinomas
AANF	Atrial natriuretic factor	Cardiac atria
APro	Prolactin	Aging pituitary, prolactinomas
AIns	Insulin	Iatrogenic
AMed	Lactadherin	Senile aortic, media
AKer	Kerato-epithelin	Cornea, familial
ALac	Lactoferrin	Cornea
AOaap	Odontogenic ameloblast-associated protein	Odontogenic tumours
ASemI	Semenogelin I	Vesicula seminalis

Source: According to Sipe et al., 2010.

Amyloids have a number of structural features in common, that is, the cross-β structure, the binding to characteristic dyes, like Congo Red (CR) and Thioflavine T (ThT), the fibrillar morphology observed by transmission electronic microscopy (TEM) or atomic force microscopy (AFM), and the green birefringence under polarized light upon staining with CR (Nilsson, 2004).

Table 15.2 shows the intracellular aggregates known as inclusion bodies. At least one of these aggregates, the neurofibrillary tangles, has the same structural features than the amyloids. The tangles are thus "intracellular amyloid." Other inclusions may have some, but not all, of these properties, for example, inclusion bodies within muscle fibers of inclusion body myositis, which are stained by CR, exhibit green birefringence and are immunoreactive with anti-β-amyloid.

Table 15.2 Intracellular inclusions with known biochemical composition, with or without amyloid properties (Sipe at al., 2010)

Inclusion name	Site	Protein nature	Examples of associate disease
Lewy bodies	Neurons intracytoplasmic	α-synuclein	Parkinson's disease
Huntington bodies	Neurons intranuclear	PolyQ expanded huntingtin	Huntington's disease
Hirano bodies	Neurons	Actin	Neurodegenerative disorders
Collins bodies	Neurons	Neuroserpin	Familial presenile dementia
Not specified	Neurons, many different cells	Ferritin	Familial neurodegenerative disorders

Most of conformational diseases share a striking number of common pathological features, such as evidence of membrane damage, oxidative stress, mitochondrial dysfunction, up-regulation of autophagy, and cell death (Glabe, 2006). In this way, amyloids from different diseases may share a common pathway for fibril formation. The initiating event is a protein misfolding, which results in the acquisition of the ability to aggregate in an infinitely propagating fashion. Quasistable intermediate aggregates ranging from dimers up to particles of a million Dalton or greater have been observed by a variety of methods. Soluble spherical aggregates have been observed for many different types of amyloids and it is assumed that these spherical oligomers, usually called amyloid-β-derived diffusible ligands (ADDLs) appear to represent intermediates in the pathway of fibril formation. Recent evidence suggest that amyloid oligomers, which represent intermediates in the fibril formation progress, may be primarily responsible for amyloid pathogenesis, rather than the mature fibrils that accumulate as large aggregates (Hardy and Selkoe, 2002; Lambert et al., 1998), even when the oligomers are formed from proteins that are not normally related to degenerative diseases (Bucciantini, 2002).

Amyloid oligomers and fibrils have in common the structural features indicated above. Amyloid oligomers are generically toxic to cells. This fact and the common structure shared by the amyloid oligomers imply that the primary mechanism of toxicity in these diseases must be the same or, said in other words, that the oligomers must act on the same primary target. As some amyloids arise from cytosolic proteins, while others are derived from extracellular proteins, the most obvious target that is accessible to both cytosolic and extracellular compartments is the plasma membrane that forms the interface between the two compartments. A growing body of evidence suggests that membrane permeabilization by amyloid oligomers may represent the common, primary mechanism of pathogenesis of amyloidosis (Glabe, 2006).

Concerning the neurodegenerative diseases, Parkinson's disease (PD) and Alzheimer's disease (AD) are the diseases with a greater incidence on the world population. Both diseases lack definite diagnostic approaches and effective cure at the present. Moreover, the currently available diagnostic tools are not sufficient for an early screening of AD/PD in order to start preventive approaches. However, the emerging field of nanotechnology has promised new techniques to solve some of the AD/PD challenges. Nanotechnology uses engineered nanomaterials or devices that can interact with biological systems at molecular levels with a high degree of specificity (Silva, 2006; Editorial, 2003). Thus they can stimulate, respond to, and interact with target cells and tissues in controlled ways to induce desired physiological responses, while minimizing undesirable effects. With all this potential, nanotechnology could have a revolutionary impact on diagnosis and therapy of PD and AD (Morris, 2004). In this chapter, we present the promises that nanotechnology brings in research on the PD/AD diagnosis and therapy.

15.2 Alzheimer's Disease

AD is one of a number of diseases in which proteins form amyloid aggregates. AD is a progressive, neurodegenerative disease, which inevitably leads to dementia and death. The brains of patients with AD contain a large number of amyloid deposits in the form of senile plaques. The amyloid core of these plaques is formed by interwoven fibrils that are composed by variants of the β-amyloid (Aβ) peptide. These fibrils are surrounded by dead neurones. Aβ varies from 39 to 43 amino acids in length, the most abundant forms being 40 and 42 amino acids (Glenner and Wong, 1984). Aβ is generated from the proteolytic processing of the amyloid precursor protein (APP) by β- and γ-secretases. Although a causal relationship between Aβ and the development of AD has not been conclusively demonstrated, considerable experimental data suggest that Aβ aggregates are important in the etiology of AD (Harper and Lanbury, 1997; Murphy, 2002; Soto, 2003; Gorman and Chakrabarty, 2001). Whereas early evidence suggested that Aβ fibrils initiate a cascade of events that result in neuronal cell death (Yankner, 1996), a number of investigators recently proposed that soluble aggregates of Aβ (also called oligomers or protofibrils), rather than monomers or insoluble amyloid fibrils, may be responsible for synaptic dysfunction in AD (Hardy and Selkoe, 2002; Hartley et al., 1999; Klein et al., 2001; Westerman et al., 2002; Kawarabayashi et al., 2004). Nevertheless, Aβ fibrillogenesis is still thought to play a critical role in the development of AD. On the other hand, the conversion of Aβ, which in native form is random, into fibrils rich in β-form, involves major structural changes, leading to the partial or complete disruption of the native fold (Klein et al., 2004).

In vitro studies have suggested that Aβ fibrillogenesis occurs by a multistep, nucleation-dependent process (Jarret and Lansbury, 1993). Formation of the nucleation seed is rate limiting, so in the absence of preformed seed fibrils, there is a significant lag period for the formation of the Aβ fibrils, followed by a rapid fibril elongation phase once seed fibrils, or nuclei, have been generated. Therefore, the fibrillogenesis can be considered to occur in two distinct stages, controlled by two key parameters, nucleation rate (k_n) and elongation rate (k_e) constants. This model is similar to a nucleation-dependent polymerization process that characterizes crystal growth. The kinetics of such a fibrillization process can be explained in terms of an autocatalytic reaction mediating the transition from the monomer to the aggregate species (Sabaté et al., 2003). Aβ fibrillogenesis can be completed by considering the presence of oligomers (Haass and Selkoe, 2007), as well as of micelles (Sabaté and Estelrich, 2005a).

As in other amyloidoses, one mechanism by which oligomers can damage the neurons in AD is the formation of pores or ion channels through the cell membrane. Early work in this area showed that Aβ can insert into planar lipid bilayers and allow a calcium current upon insertion and further that these channels can be blocked (Arispe et al., 1993), suggesting that the calcium current is really due to channel formation, not just bilayer permeabilization of the peptide. More recent work has been done using AFM, showing that peptide monomers oligomerize after insertion into the bilayer. Furthermore, in the presence of these oligomers, current can flow (Lin et al., 2001). However, other work has indicated that soluble oligomers specifically increase lipid bilayer conductance, while fibrils and soluble low-molecular-weight species have no effect (Kayed et al., 2004).

Microglia, the resident macrophage cells of the central nervous system (CNS), play a controversial role in AD: some types of reactive microglia seem detrimental, whereas others can actually be beneficial (Weitz and Town, 2012). A large part of microglial cell's role in the brain is maintaining homeostasis in noninfected regions and promoting inflammation in infected or damaged tissue. Microglia respond to different pathological agents with a reaction termed "microglia activation." Once activated, microglia release both neuroprotective and cytotoxic molecules. One of such molecules, interleukin-1β, when overexpressed, leads to the hyperphosphorilation of tau protein). Activated microglia synthesize APP and they are found to be associated with amyloid deposits in the brains of AD patients.

Although there is no cure, Alzheimer's medications can temporarily slow the worsening of symptoms and improve the quality of life for both patients and their caregivers. Of the five US Food and Drug Administration (FDA)-approved medications to treat the symptoms of AD, four are acetyl cholinesterase (AChE) inhibitors and the fifth an *N*-methyl-D-aspartate (NMDA) receptor antagonist. The AChE inhibitors work by slowing down the disease activity that breaks down a key neurotransmitter. Donepezil, galantamine, rivastigmine, and tacrine are cholinesterase inhibitors. Memantine is an NMDA and works by regulating the activity of glutamate, a chemical messenger involved in learning and memory. Memantine protects brain cells against excess glutamate, a chemical messenger released in large amounts by cells damaged by AD and other neurological disorders. Attachment of glutamate to cell surface "docking sites" called NMDA receptors permits calcium to flow freely into the cell. Over time, this leads to chromic overexposure to calcium, which can speed up cell damage. Memantine prevents this destructive chain of events by partially blocking the NMDA receptors.

Donepezil was approved for all stages of the disease, whereas galantamine, rivastigmine, and tacrine were for mild to moderate stages. Memantine is for moderate to severe stages. On average, the five cited AD drugs are effective for about 6 to 12 months in about half of the individuals who take them (http://www.alz.org).

Due to the limited success of the treatment, the pharmacotherapy of AD offers primarily symptomatic benefits, but with little or no evidence of slowing disease progression, researchers are looking for new ways to treat AD. There are several promising drugs in development (e.g., use of enzymes to decrease the Aβ production, substances that increase the Aβ clearance, immunotherapy, enzymatic degradation of Aβ, inhibition of Aβ aggregation, antioxidants, anti-inflammatory drugs, statins, etc. (for a review see Chopra et al., 2011; Corbett and Ballard, 2012). Among them, the inhibition of Aβ assembly has been considered as the primary therapeutic strategy for this neurodegenerative disease. Next, we review the use of soft and hard nanoparticles (NPs) to inhibit the amyloid fibrillation.

15.2.1 Inhibition of Aβ Assembly by Association Colloids (Micelles and Liposomes)

The effect of micelles on the inhibition of the aggregation is mainly due to the stabilization of nontoxic peptide conformations rather than to the disaggregation of the already formed fibrils. As a general rule, this effect is associated with the net charge of micelles. Micelles of sodium dodecyl sulphate (SDS), the most representative negatively charged surfactant, stabilize the peptide in an α-helical structure that does not aggregate (Marcinowski et al., 1998; Coles et al., 1998; Shao et al., 1999; Montserret et al., 2000; Pertinhez et al., 2002; Rangachari et al., 2006). Such stabilization may be critical to preventing the α-helix (soluble, monomeric) $\rightarrow$ β-sheet (insoluble amyloid deposit, toxic) conversion. The amount of the observed α-helical structure is pH dependent, suggesting that electrostatic interactions have an important role in the mechanism of SDS-inducing peptide conformation (Pertinhez et al., 2002). Such micelles present usually a diameter of 2–3 nm.

Similar is the effect of positively charged micelles (with a diameter slightly higher, 3–4 nm). The dodecyl trimethylammonium chloride produces, at high pH (9–12), a stabilization of the peptide in an α-helical structure while at pH 7.3, the α-helical structure is 35% (Marcinowski et al., 1998). Alkyl trimethylammonium bromides (dodecyl [12 C], tetradecyl [14 C], and hexadecyl [16 C]) at concentrations well above their critical micellar concentration (CMC) in water and at neutral pH stabilize the native conformation by delaying the aggregation (Sabaté and Estelrich, 2005b). The length of the aliphatic chain plays an important role in the antiaggregating effect. In this way, micelles of 12-carbon alkyl bromides are weaker delayers of the peptide aggregation than micelles with chains of 16 carbons. Micelles of cationic gemini surfactant hexamethylene-1,6-bis (dodecyl dimethylammonium bromide) ($C_{12}C_6C_{12}Br_2$) decrease the Aβ(1-40) nucleation rate and inhibit the lateral association between fibrils at pH 7.4 (Li et al., 2007). It has also been found that $C_{12}C_6C_{12}Br_2$ micelles can effectively disassemble mature Aβ(1-40) fibrils in vitro (Han et al., 2010). Aβ fibrils are disaggregated into soluble substances and simultaneously transported as small molecules to prevent further formation of Aβ fibrils. More curious is the effect of a tetrameric quaternary ammonium surfactant, N^1,N^{16}-didodecyl-7,10-bis(3-(2-dode-cyldimethylammonio) ethylamino)-3-oxopropyl)-N^1,N^1,N^{16},N^{16}-tetrame-thyl-4,13-dioxo-3,7,10,14-tetraazahexadecane-1,16-diaminium tetrabromide (PATC). It presents a low CMC (~0.08 mM) and a special aggregation behavior, since it forms large premicellar network-like aggregates far below the CMC (Hou et al., 2010). Micelles, as well as premicellar aggregates, of PATC are a powerful medium for disassembling the Aβ(1-40) fibrils even at 0.02 mM (He et al., 2011a).

Something different is the effect of neutral micelles. In this way, micelles formed by dodecylphosphocholine, have a different behavior since at neutral pH, only 5% of peptide structure is an α-helix, with the major conformation being a random coil (Marcinowski et al., 1998). Other micelles originated by the self-assembly of 1,2-distearoyl-*sn*-glycero-3-phosphoethanolamine-*N*-methoxy-poly(ethylene glycol 2000) (DSPE-PEG$_{2000}$) abrogate transformation of Aβ(1-42) to amyloidogenic β-sheet form (Pai et al., 2006).

The results obtained when considering the interaction of Aβ with liposomes are more controversial, although it is evident that the interaction results from two different scenarios: an electrostatic adsorption of Aβ to liposome surface, which is facilitated by the presence of negative surface potential in the membrane, and, second, an insertion of the Aβ peptide into membranes. As general rule, if Aβ inserted, then aggregation is inhibited, whereas binding of Aβ to membrane surfaces may accelerate fibrillogenesis (Bokvist et al., 2004). However, factors such as the Aβ variant (1-40 or 1-42), the lipid composition, physical state of the membrane, the lipid:protein ratio, vesicle size, and the medium in which the interaction occurs can modulate the interference of liposomes with Aβ. In this way, some studies have indicated that the binding of Aβ(1-40) to G$_{M1}$ ganglioside-containing sonicated liposomes (small unilamellar liposome [SUVs], with a diameter usually less than 100 nm) accelerated the rate at which soluble Aβ forms amyloid fibrils in vitro (Choo-Smith, et al., 1997; McLaurin and Chakrabartty, 1996; Kakio et al., 2002; Matsuzaki et al., 2007; Chi et al., 2007; Okada et al., 2008), and the same effect has been observed for the interaction at pH 7.4 of Aβ with positively charged large unilamellar liposomes (LUVs, with average diameters ranging from 100 to 200 nm) composed by of 1-palmitoyl-2-oleoly-*sn*-glycero-3-phosphocholine (POPC) and 1,2-dipalmitoyl-3-trimethyl-ammonium propane (DMTAP) at a 70:30 molar ratio (Chi et al., 2008). Moreover, cholesterol (CHOL) and G_{M1} play an important role in the release of Aβ from liposomal membranes of SUVs with a lipid composition similar to that found in the cerebral cortex membranes (Tashima et al., 2004). The effect of the surface charge is a debatable point, since some studies have found little or no interaction of Aβ with neutral liposomes (Bokvist et al., 2004; Choo-Smith, et al., 1997), but others have reported binding to phosphatidylcholine (PC) membranes of LUVs (Sabaté et al., 2005). Negatively charged liposomes (LUVs) usually delay the onset of Aβ fibril formation (Sabaté et al., 2012). The effect is the increase of lag time (by affecting the nucleation rate constant), but no clear trend could be detected in the elongation rates. This indicates that the mechanism behind the increased lag time must originate in interferences taking place during the build-up of small oligomers from monomers. Once nucleated, the aggregation process proceeds with the same rate in the absence and presence of liposomes. While negatively charged liposomes have shown to interact, increasing the ionic strength of the solution, this process did not abolish Aβ binding to PC/phosphoglycerol (PG) membranes of SUVs. This indicated that the peptide–lipid interaction is not purely electrostatic, but it also contains a hydrophobic component (Wong et al., 2009). Hellstrand et al. (2007) demonstrated that the phase behavior of the membrane in the LUVs has a large effect on the aggregation kinetics. By comparing vesicles with different membrane phase behavior they saw that the solid gel phase causes the largest retardation of Aβ fibril formation.

As a main conclusion, it is evident that micelles and liposomes provide an interface that can potentially act as a catalytic or inhibiting surface. Moreover, they reveal a structural approach of constructing efficiently disassembly agents to mature Aβ fibrils. However, the blood–brain barrier (BBB) permeability is the most important factor that limits the use of such substances. One requirement for any drug capable of interfering with the aggregation of β-peptide is that it should be able to cross the BBB. However, the BBB is a formidable gatekeeper in the body toward exogenous substances and is aimed to maintain the chemical composition of the neuronal "milieu" for proper functioning of neuronal circuits and synaptic transmission. This barrier results from the selectivity of the tight junctions between endothelial cells in CNS vessels that restricts the passage of solute. These tight junctions are composed of smaller subunits, frequently transmembrane proteins. Each of these transmembrane proteins is anchored into the endothelial cells by another protein complex.

15.2.2 Nanotechnologies for Alzheimer's Disease

Among the different attempts focused on the problem of the BBB, nanotechnology-based strategies have gained tremendous importance as some of them are capable of overcoming the limitations inherent to BBB passage. These include various types of lipidic, polymeric, inorganic, and other types of NPs for controlled drug delivery and release pertinent to various CNS conditions (Silva, 2010).

Advances in nanotechnology are beginning to exert a significant impact in neurology. These approaches, which are often based on the design and engineering of a plethora of nanoparticulate entities with high specificity for brain capillary endothelial cells, are currently being applied to early AD diagnosis and treatment (Brambilla et al., 2011a).

15.2.2.1 Nanotechnologies for AD diagnosis

Early diagnosis in AD before the onset of marked symptoms is critical in preventing the irreversible neuronal damage that eventually leads to dementia and ultimately death. The clinical development of positron emission tomography (PET) and single-photon emission computed tomography (SPECT) imaging agents that target deposits of aggregated Aβ peptides in vivo shows tremendous promise for detecting changes in the accumulation of senile plaques in living patients. Another important imaging technique with a potentially useful means to follow the pathological progression of AD is the magnetic resonance imaging (MRI). Since the formation of senile plaques precedes the neurofibrillar degeneration, the majority of efforts are concentrated either toward detection and identification of amyloid plaques by MRI using NPs doped with contrast agents or, alternatively, by NPs tagged with fluorescent probes.

Magnetic iron oxide NPs covalently tethered to the N-terminus of Aβ(1-40) through amide coupling were synthesized for targeting and imaging of senile plaques (Wadghiri et al., 2003). These NPs detected the senile plaques in transgenic mice using magnetic resonance microimaging when co-injected with mannitol used for permeating the BBB. Other selective labeling of Aβ(1-40) fibrils have been reported with three kinds of NPs: maghemite, maghemite containing the fluorescent probe rhodamine, and maghemite containing the fluorescent probe CR (Skaat and Margel, 2009).

The marker ThT has been encapsulated into NPs composed of a polystyrene (PS) core and a degradable poly(butyl-2-cyanoacrylate) (PBCA) shell. Its release into the brain after intracerebral injection has allowed visualizing amyloid aggregates in the hippocampus of transgenic mice (Siegemund et al., 2006).

Quantum dots (QDs) are inorganic conductor fluorophores composed of chemical elements belonging to transition groups in the periodic tables. NPs formed with QDs have been used as a tool to monitor and quantitatively describe the formation of fibrils and oligomers in solution and in a cellular system (Tokuraku et al., 2009). This approach allowed the study of the Aβ peptide aggregation kinetics but could also be used to follow the in vivo peptide aggregation. Regarding the latter task, the functionalization of these nanoassemblies with appropriate ligands such as transferrin for BBB-crossing purposes, have been considered (Xu et al., 2008a). On the other hand, it has been also observed that QDs surface-conjugated with lipopolysaccharide to form micelles induced significantly enhanced microglial activation in several microglial cells in vitro, as compared with lipopolysaccharide or QDs without lipopolysaccharide on the surface (Lalancette-Hébert et al., 2010). The authors suggested that microglia respond to NPs, even if they are in minute quantities, barely or not detectable by common biochemical or biophysical means.

Gold NPs also represent an interesting tool for studying Aβ aggregation kinetics. Choi et al. described the synthesis of heterodimeric NPs consisting of a cobalt(II) magnetic core and a platinum shell directly fused onto such NPs and stabilized by lipoic acid–PEG coating (Choi et al., 2008). The terminal carboxyl groups of the PEG chains enabled covalent binding with lysine residues of neutravidin at the surface of the NPs. The Co@Pt-Au nanoassemblies presented a high magnetization value, making them appropriate T_2-weighted spin echo MRI measurements. (Spin echo is one of the several methods to generate contrast images in MRI to visualize organs and tissues. Superparamagnetic materials produce images with negative contrast, the T_2-weighted images). The MRI measurements of Co@Pt-Au-neutravidin NP samples mixed with an increasing amount of biotinylated Aβ(1-40) peptides showed contrast agents governed by the peptide concentration. The results clearly showed that these NPs can be used in MRI to monitor key structural stages of Aβ self-assembly. In particular, a significant change in MRI signals during Aβ self-aggregation that corresponds to the detection of Aβ protofibrillar species in the early reversible stages of aggregation was observed.

In another approach, Gobbi et al. reported the preparation of two types of NPs (liposomes and solid lipid NPs) incorporating either phosphatidic acid or cardiolipin, to target with very high-affinity aggregated forms of Aβ(1-42) (fibrils and oligomers) (Gobbi et al., 2010).

Since methods to rapidly identify and quantify the association of small molecules to aggregated Aβ are critical, Capule and Yang developed an enzyme-linked immunoabsorbent assay (ELISA) for quantifying the binding of small molecules to aggregated Aβ peptides. This method can estimate binding constants of small molecules from the low nanomolar to the low micromolar range. A key step in this method is the air plasma treatment of the PS surface of the ELISA plates, which helps maintain the native β-sheet content of the amyloid peptides during analysis (Capule and Yang, 2012).

15.2.2.2 Nanotechnologies for detection of AD biomarkers in biological fluids

The main pathogenic biomarkers of AD are the soluble oligomers ADDLs and the hyperphosphorilated τ protein. The extremely low concentrations of these biomarkers in the cerebrospinal fluid (CFS) limit strongly their accurate identification by conventional ELISA.

To solve this problem, a strategy based on the termed biobarcode amplification has been used (Keating, 2005). Figure 15.1 illustrates this approach.

The key features of the system relied on the isolation of antigens by means of a *sandwich process* involving oligonucleotide (DNA barcode)-modified Au NPs and magnetic microparticles (MMPs), both functionalized with monoclonal or polyclonal antibodies specific to the ADDLs. The increased sensitivity derives mainly from the very effective sequestration of antigen and the amplification process that occurs as a result of the large number of barcode DNA strands released for each antigen recognition and binding event. The strands of a dehybridized double-stranded DNA were isolated and easily quantified by a scanometric method using DNA microarray (Georganopoulou et al., 2005). The system allows the identification of ADDLs at subfemtomolar concentrations, thus improving the ELISA test sensitivity by 6 orders of magnitude.

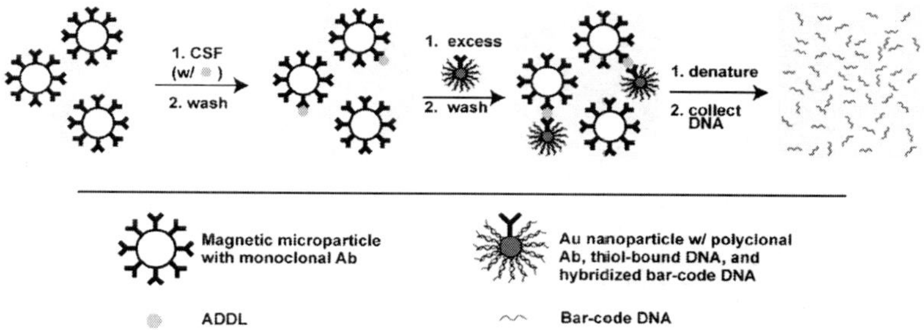

Figure 15.1 The biobarcode amplification assay. Reprinted with permission from Keating (2005). Copyright 2005 National Academy of Sciences, U.S.A.

Haes et al. designed a localized surface plasmon resonance (SPR) nanosensor to detect ADDLs (Haes et al., 2004, 2005). Sandwich architectures were prepared by synthesis and immobilization of surface-confined antigen nanotriangles onto a mica surface using nanosphere lithography. Then, a self-assembled monolayer consisting of a mixed monolayer of 1-octanethiol for passivation of the NPs toward nonspecific binding and 11-mercaptoundecanoic acid for covalent attachment of antibody to ADDLs was assembled on top of these Ag nanotriangles. The first antibody to ADDLs was covalently attached onto the surface of the NPs via the amidation reaction. The resulting biosensors were incubated with samples containing ADDLs, washed, and incubated with a polyclonal antibody solution specific to ADDLs to enhance the shift response for ADDLs. The localized SPR nanosensor allowed analysis of biological species in a surfactant-free environment and was demonstrated to be sensitive enough for the detection of ultralow concentrations of

ADDLs in biological samples. Moreover, this technique allowed the distinction between two different ADDLs species varying from their binding constant with the antibody specific to ADDLs.

Apolipoprotein E (ApoE) is another potential biomarker. Morales-Narvaez et al. explored the performance of CdSe@ZnS QDs for detecting ApoE. The assay is a sandwich immunocomplex microarray that functions via excitation by visible light. The QDs proved to be highly effective in the microarrays, although their performance strongly varied in function of the excitation wavelength. In the best conditions, the limit of detection was 62 pg mL^{-1} (Morales-Narváez et al., 2012).

The detection of τ protein and Aβ peptide in the CFS or plasma has been considered as a way to detect the AD in the earliest stages of the disease. However, both substances can be found in the biological fluids of healthy and AD subjects, and in consequence, the obtained results are inconclusive. Despite this hampering, several nanotechnology-based approaches have been developed for Aβ detection at submicromolar concentrations (Kolb et al., 2001; Chikae et al., 2008; Lee et al., 2009a; Kang et al., 2009b; Neely et al., 2009; Yang et al., 2011a; Wang et al., 2012b) or τ protein at concentrations greater than about 1 pg mL^{-1}. This concentration was about 2 orders of magnitude lower than typical τ protein concentration values (i.e., 195 pg mL^{-1}) in the CFS (Neely el at., 2010).

15.2.2.3 Nanotechnologies for AD treatment

NPs have been used for transporting active molecules across the BBB. Table 15.3 summarizes some of the approaches used to deliver bioactive molecules to the brain. These include the encapsulation of AChE inhibitors (i.e., tacrine and rivastigmine) (Joshi et al., 2010; Wilson et al., 2008a, 2008b). Results have shown an increase of their uptake within the brain compartment compared to free drug after intravenous injection into rats. Estrogen and an active antiprogesterone compound, mifepristone, slowed the progression of cognitive decline in AD patients. By encapsulating them, the bioavailability of these drugs after oral administration increased (Mittal et al., 2007; He et al. 2007). Another example is the poor stability of curcumin, which limits the exploitation of its numerous pharmacological properties (hypothetical, in the case of AD). The encapsulation into a nanoparticulate system dramatically increased curcumin half-life and concentration in the brain when compared to free curcumin (Sun et al., 2010; Mulik et al., 2010).

To avoid the severe complications found in the clinical trials of immunotherapy against Aβ(1-42) for AD treatment, nanocarriers functionalized with an antibody fragment specific to Aβ were prepared. Such nanocarriers were able to cross the BBB and to target the brain amyloid deposits.

Table 15.3 Bioactive molecules encapsulated or grafted in nanoparticles for AD treatment

Nanoparticles	Encapsulated or grafted bioactive molecules	Reference
Poly-sorbate-PBCA	Rivastigmine	Wilson et al. 2008a
PLGA-PBCA	Rivastigmine	Joshi et al., 2010
Poly-sorbate-PBCA	Tacrine	Wilson et al., 2008b
PBCA	Curcumin	Sun et al., 2010
PBCA decorated with ApoE3 ligands	Curcumin	Mulik et al., 2010
PLGA	Estradiol	Mittal et al., 2007
PLGA	Mifepristone	He et al., 2007
Lipidic	Gallic acid	Smith et al., 2010
Lipidic	D-penicillammine	Cui et al., 2009
NP	MAEHP	Liu et al., 2009b
Solid lipid	Ferulic acid	Picone et al., 2009
PEG	Glutation	Williams et al., 2009
Chitosan-based	Ab fragments	Agyare et al., 2008
PLGA	Superoxide dismutase	Reddy et al., 2008

Oxidation reactions, implied in the oxidative stress, are a risk factor of AD, can be catalyzed by metals such as iron (Fe^{2+}), copper (Cu^{2+}), aluminum (Al^{3+}), and zinc (Zn^{2+}). Thus, metal-chelating compounds have been used to improve the clinical conditions in AD patients. However, their poor brain uptake and toxic side effects have limited their use. To overcome this difficulty, a metal chelator (D-penicillamine) was grafted on lipidic NPs (Cui et al., 2005), and in another approach, carboxylic-functionalized PS NPs were grafted with the iron chelator 2-methyl-*N*-(2′-aminoethyl)-3-hydroxyl-4-

pyridinone (MAEHP) (Liu et al., 2009b). In both cases, the conjugate significantly protected cells from $A\beta$-associated cytotoxicity compared to cells treated with $A\beta$ alone. In another way of treatment, gallic acid, a promotor of the α-secretase, an enzyme involved in the nonamyloidogenic pathway of $A\beta$ peptides, has been encapsulated into lipidic NPs, demonstrating an increase of its oral bioavailability (Smith et al., 2010). Other strategy regarding the treatment of AD is directed toward the delivery of antioxidant species to the brain. In this way, two antioxidants, glutathione conjugated with PEG, and ferulic acid with solid lipid NPs, have been prepared with the aim of increasing levels of these drugs in the brain (Williams et al., 2009; Agyare et al., 2008). In another approach, a free-radical scavenger, the superoxide dismutase, has been conjugate with poly(lactic-co-glycolic acid) (PLGA) (Reddy et al., 2008).

Mourtas et al. (2011) reported the functionalization of azido-decorated liposomes with an alkyne-derivatized curcumin. Liposomes decorated with the planar curcumin had the highest affinity constant (in the 1–5 nM range) reported thus far for $A\beta$ fibrils, whereas nonplanar curcumin-decorated liposomes did not show any binding.

Special kinds of NPs are carboxifullerenes C_{60} derivatives. They are effective scavengers of oxygen radicals. One of them, fullerenol-1, also inhibits neuronal apoptosis induced by exposure to $A\beta$ peptide, which is associated with increased intracellular free-radical production. It has been demonstrated that this fullerene diminishes the influx of calcium ions caused by the creation of channels in the neuronal membrane (Cabaleiro-Lago et al., 2008).

As said previously, the inhibition of $A\beta$ assembly has been considered as the primary therapeutic strategy. Hence, the physical interaction of NPs with $A\beta$ will be able to interfere or delay the peptide aggregation process.

Different polymer NPs such as *N*-isopropylacrylamide: *N-tert*-butylacrylamide (PNIPAAM-co-PtBAM) (Rocha et al., 2008), sulfonated, sulfated, and fluorinate PS (Saraiva et al., 2010; Cabaleiro-Lago, 2010) have demonstrated to interfere with the aggregation process or suppress the $A\beta$ oligomerization. Interestingly, a dual effect of cationic amino-functionalized PS NPs toward the $A\beta$ aggregation process has been demonstrated (Brambilla et al., 2010). The possibility to alter the peptide aggregation simply by tuning the NP concentration was well described, hence highlighting the pivotal role of NP dosing on the aggregation behavior. Brambilla et al. have demonstrated that PEGylated, biodegradable poly(alkyl cyanoacrylate) (PAC) NPs could bind the soluble Aβ (1-42) peptide monomers and influence its aggregation kinetics, whereas non-PEGylated counterparts did not lead to detectable interaction (Brambilla et al., 2010). Later, this interaction was also observed with PEG-*block*-poly(lactic acid) (PEG-b-PLA) (Brambilla et al., 2011b). More recently, these authors have demonstrated that PEGylation of other NPs (copolymers formed by different families of PAC and PLA) can dramatically increase their affinity for Aβ(1-42) monomers and soluble oligomers both in solution and in the presence of serum proteins. As a result of collaborative hydrophilic and hydrophobic interactions between the PEG shell and Aβ (1-42), the latter could adsorb and aggregate at the surface of the NPs (Brambilla et al., 2012). PAC NPs coated with PEG and functionalized with either curcumin derivatives, or with a novel specific antibody, via the biotin-/streptavidin-binding strategy are able of binding not only the Aβ(1-42) monomer, but also Aβ(1-42) fibrillar aggregates. In the case of the anti Aβ(1-42) monoclonal antibody, dissociation rate constants in the picomolar range were obtained. With curcumin functionalization, the resulting NP offered a significant aggregation inhibition of Aβ(1-42), together with a marked toxicity rescue, both at low molar ratios (Le Droumaguet et al., 2012).

Skaat et al. (2012) designed amino-acid-based polymer NP containing two phenylalanines (F) in the polymer side chains. These residues FF are similar to the hydrophobic core sequence of Aβ (residues 16–20, KLVFF sequence). A significant inhibition of the Aβ(1-40) fibrillation process was observed in the presence of these NPs. This inhibition is probably due to the intermolecular attractive hydrophobic interactions between the pairs of FF residues of the NPs with the corresponding residues of the Aβ(1-40) prefibrillar aggregates, which disrupt the self-assembly of Aβ(1-40) into fibrils.

Inorganic NPs can also inhibit the nucleation and elongation processes of the amyloid fibrillation. Xiao et al. (2010) reported that CdTe QDs capped with *N*-acetyl-L-cysteine (NAC), drastically inhibited the Aβ fibrillation. Interestingly, the NPs concentration required to show inhibition effect was much lower than that of the amyloid peptide concentration. The authors postulated that the inhibition was due to the intermolecular attractive interactions such as the hydrogen bonding between NAC QDs and amyloid fibrils, resulting in the blockage of the active elongation sites on the fibrils. Moreover, these authors have explored the effect of particles size, surface charges, functionality and compositions of QDs and gold NPs on the elongation phase of Aβ fibrillogenesis. It was found that charge-to-surface area ratio of the NPs and the functional moiety and electrostatic charges of the conjugated ligands on the particle surfaces took a crucial regulatory role in fibrillogenesis (Chan et al., 2012). In another study, a strong inhibition of Aβ fibrillation has been observed with thioglycolid acid (TGA)-stabilized CdTe NPs (Yoo et al., 2011). The main disadvantage of such NPs is that CdTe are cytotoxic and cannot be used in vivo.

Gold NPs conjugate to an amphipatic peptide and a further irradiation with microwave produced dramatic effects on the Aβ aggregates. This approach offers a new strategy to inhibit, locally and remotely, the amyloidogenic process (Araya et al., 2008). The distribution in vivo of NPs of this kind has been studied by positron emission tomography (Guerrero et al., 2012).

Other NPs used to alter the conformation of the peptides and reduced the toxic activity are based on surface-modified nanogels (cholesterol-bearing pullulan nanogels) (Ikeda et al., 2006). Finally, dendrimers can be exploited as a powerful tool for reducing the peptide aggregation and investigating the formation mechanism of amyloid-like structures. Dendrimeric scaffolds functionalized with KLVFF served to inhibit the Aβ(1-42) aggregation and disassemble preexisting amyloid aggregates (Chafekar et al., 2007). Other dendrimers formed by polyamidoamine (PANAM) conjugated with sialic acid acted as Aβ-binding competing agents (Patel et al., 2007). Cationic phosphorus-containing dendrimers affect Aβ(1-28) peptide and τ protein aggregation (Wasiak et al., 2011).

To sum up, nanoparticulate systems can serve as efficient early diagnostic and/or therapeutic tools against AD. As indicated, these NPs make that a biological signal (a biomarker) transforms to a recordable signal and is amplified enough to be recorded. A common characteristic of them is the biodegradability. These NPs include magnetic particles, QDs, inorganic NPs (mainly, gold NPs), and organic NPs (especially lipidic and polymeric particles).

15.3 Parkinson's Disease

Although not unique, involvement of the dopamine system is the most prominent in PD. PD starts at a molecular level (a genetic defect and/or an environmental agent yield to a misfolding of proteins), with consequences at the cellular level (dysfunction of the ubiquitin–proteasome system and mitochondria with free-radical formation and protein aggregation, leading to the apoptotic death of dopaminergic cells), which in turn results in biochemical and neurophysiological disturbances (dopamine deficiency in the *striatum* and in other nuclei and increased firing rate with abnormal firing pattern in the *subthalamic nucleus* [STN] and internal segment of the *globus pallidus* [Gpi]) (Schapira, 2005). The clinical expression of all these disturbances is slowness of movement (bradykinesia), rigidity, postural abnormalities, and tremor.

There are several unmet needs, which require solutions to improve the care of patients with PD. Most of them are summarized in Table 15.4. Nanoscience could decisively contribute to the achievement of some major research objectives. Besides applications of interest for diagnostic purposes, nanotechnology could provide devices to limit and reverse neuropathological disease states, to support and promote functional regeneration of damaged neurons, to provide neuroprotection, and to facilitate the delivery of drugs and small molecules across the BBB (see Linazasoro, 2008; Pervaiz and Irshad, 2012, for a review). These and other potential applications of nanotechnology are analyzed in the following sections.

Table 15.4 Unmet needs and *ideal goals* in PD investigation

1. To find a cure: *To know the etiology of PD*
2. Early, presymptomatic diagnosis to apply neuroprotective treatments to stop disease progression: *To have a reliable biomarker and to know the mechanisms of cell death*
3. To improve current treatments (more efficacious and safer): *To better understand the physiopathology and the outcome of PD, to make a personalized treatment plan, and to improve the delivery of drugs to the brain*
4. To regenerate the damaged parts of the brain: *To improve the efficiency of cell sources and increase the survival of implanted cells and to improve the delivery of molecules to localized areas of the brain*

15.3.1 Nanotechnology and the Diagnosis of PD

NPs (gold and silica), QDs, nanodevices, and DNA, along with countless other forms of nanostructures, have been investigated regarding their usefulness in advancing molecular diagnostics (Kurek and Chandra, 2012). Other small-scaled materials like viruses and proteins also have potential for use as molecular diagnostic tools (Jalil et al., 2011).

The application of nanotechnology to imaging technology is paving the way for the future of personalized medicine. Using these diagnostics, individual therapies can be tailored to the very molecules that distinguish among patient's stages of the disease. One approach under development relies on the QD technique (Malvindi et al., 2011). These NPs use pieces of protein attached to their surface to lock on to the distinctive proteins on certain dysfunctional cells. Once injected into animals models, QDs produce high-resolution multicolored images of individual molecules moving within the animals' cells (Freundt et al., 2012). The photophysical properties of these inorganic nanocrystals make them unique for use in in vivo fluorescence imaging, as well as combining them with in vivo imaging. Thus, in PD, QDs can be "sent" to identify the impaired cells in the brain and a second collection of QDs loaded with therapeutic molecules can be discharged into the affected area.

One of the most recent and influential innovations in relation to diagnosis of PD is the discovery and creation of nanosized chips known as "lab on a chip" with the potential to diagnose disease rapidly. These devices have decreased drastically in size, up to one-millionth of their original size, and scientists can analyze many different things at once, for example, blood samples, tissues, etc., allowing the diagnosis of the disease (Johnston et al., 2005; Kumar et al., 2009). The diagnosis of PD with sensors designed to analyze the chemicals found in the breath is also being actively pursued (Righettoni et al., 2012; Tisch et al., 2013).

Aptamers represent another breakthrough in the domain of nanotechnology. Aptamers bind to specific target molecules and are made up of oligonucleic acids or peptide molecules (Nam et al., 2012; Javasena, 1999). Aptamers are classified into two types, nucleic acid aptamers and peptide aptamers. Recently, the first aptamer-based drug has been legalized by the US FDA, which could be used to treat age-related macular degeneration. Preliminary evidence suggests that PD is one of the many diseases that has the potential to be diagnosed by aptamers, particularly the RNA type (Nam et al., 2012), as well as immunosensors (Javasena, 1999), probes bound to QDs (Zhang et al., 2012c), and other assays with diagnostic and therapeutic applications (Qureshi et al., 2012; Urbán et al., 2012; Nair et al., 2010).

A PET scan is a valuable tool for the diagnosis of PD. It can be used to detect the deficiency of dopamine in the striatum and in other areas affected in very early phases of the disease (Bohnen and Müller, 2012).

α-synuclein has been shown to accumulate in the blood of patients with PD. α-synuclein is the basic constituent of Lewy bodies (Warr and Walker, 2012) and is considered as the major factor responsible of the damage and death of dopamine neurons. This key information could be used to carry out further research into new aptamers that can detect α-synuclein in blood, avoiding the need of more invasive tests such as the lumbar puncture (Tsukakoshi et al., 2012). The possibility of combining the use of a PET scan with RNA aptamers opens the door to new possibilities to accurately diagnose PD in very early phases.

Importantly, nanotechnology-based diagnostics, along with specialized and powerful bioinformatics software, can be used to acquire a much deeper understanding of the mechanisms associated with diseases like PD. This will result in far more effective treatments for such diseases than what is currently available (Cirillo et al., 2012a).

15.3.2 Therapeutic Applications of Nanotechnology

The potential therapeutic applications of nanotechnology in PD are showed in Table 15.5. Some of them are discussed in the following sections.

15.3.2.1 New routes of administration of drugs and other molecules

As it has been already mentioned, the nigrostriatal dopamine system is most affected in PD. In normal conditions, this system

Table 15.5 Potential role of nanotechnologies in PD and other neurodegenerative diseases

---

1. Early diagnosis:
   (a) To monitor the rate of cell loss in SNpc and other nuclei and alert the defense system and the compensatory mechanisms systems (neurogenesis included) when the rate is exceeding the normal expected decline
   (b) To monitor the state of defense and compensatory mechanisms
   (c) To create biosensors to monitor neurotransmitter levels in precise locations within the brain (dopamine and others), which can be reduced long before the appearance of motor and nonmotor symptoms
   (d) To improve image technologies (fluorophores and quantum dots), which may also facilitate surgical approaches
   (e) To detect abnormalities in plasma, thus avoiding the need of invasive tools
2. Study of pathogenic mechanisms: To increase our knowledge about the mechanisms of cell death, thereby opening the doors to new drug targets
3. Treatment:
   - To direct drugs to their targets in a very specific way through smart, targeted drug delivery systems. Drugs can act on signaling pathways required for neurotransmission (symptomatic effect) or in signaling pathways involved in neurodegeneration (neuroprotection and neurorescue) (e.g., nanoparticles, drug encapsulation strategies, multifunctional nanotherapeutics, ablation of areas with nanoparticles, DNAbots to identify and destroy pathogenic proteins, etc.).
   - To develop systems able to overcome the blood–brain barrier (e.g., barcodes).
   - To create biosensors to monitor neurotransmitter levels in precise locations within the brain (dopamine and others).
   - Intracellular manipulations and interventions: To repair DNA and other damages, cleaning of deposits of aggregated abnormal proteins, etc.
   - To introduce genes and proteins required for normal functioning in a highly controlled way (durable and controlled expression of the gene), avoiding the needs of viral vectors and complex control systems or of infusion pumps (e.g., organic silica particles).
   - To create bridges from SNpc to striatum and other basal ganglia nuclei and to favor their development by blocking the expression of antireparative signals (e.g., no-go, etc.).
   - To create media to push the development of functional dopaminergic neurons from stem cells.

---

works in a tonic way: nigral dopamine cells fire tonically with bursts of activity in special circumstances (e.g., excitation, fear, attention) (Grace, 1991). This phasic release is potent but has a brief and spatially limited effect as dopamine is rapidly removed from the synapsis. By contrast, tonic release of dopamine is mediated by the glutamatergic corticostriatal terminals (Floresco et al., 2003). The concentrations of tonically released dopamine are much lower than those achieved after the phasic release, but it is extended over a larger striatal volume, thus modulating striatal activity. Released dopamine exerts its biological effects by acting on dopamine receptors (Ruiz Ortega and Linazasoro, 2004). These are widely distributed in the brain and the final effect of a drug depends on the type and location of the dopamine receptor. Thus, a dopaminergic agent can elicit motor, cognitive, and limbic effects, which can be positive or negative for a particular patient. Alterations in basal ganglia dopamine receptors are well known in PD and they could contribute to the development of adverse effects, such as levodopa-induced dyskinesias (Jenner, 2000; Calon et al., 2000).

Although many antiparkinsonian drugs exist, levodopa continues to be the most potent and efficacious agent. Unfortunately, its initial excellent antiparkinsonian effects are lost after some years of treatment as side effects emerge limiting its use. Most of these long-term side effects are related to the short action of the drug, which results in a pulsatile stimulation of dopamine receptors (Linazasoro, 2007). This kind of stimulation is far from the normal physiology of the nigrostriatal pathway. Thus, continuous stimulation of dopamine receptors is highly desirable (Freundt et al., 2012; Linazasoro, 2007).

In PD the majority of drugs are administered orally. The journey from the mouth to the dopamine receptor is quite long and not without hurdles. Gastric pH and rate of emptying, dietary proteins, constipation, etc., determine the bioavailability of antiparkinsonian drugs (especially levodopa) and eventually the motor response. Other routes of administration such as intranasal, rectal, sublingual, or pulmonary are being explored, but presently none has shown to provide continuous dopaminergic stimulation. Therefore, alternative drug delivery systems, such as infusion pumps and skin patches, are now being used in PD. Infusion pumps allow the subcutaneous delivery of apomorphine (Katzenschlager et al., 2005) and intraduodenal delivery of levodopa (Nyholm et al., 2005). Their main objective is to avoid the oscillations of both plasma and striatal levels of these drugs to provide a more continuous and physiological stimulation of dopamine receptors. Each results in an excellent control of motor fluctuations and dyskinesias. The size of the infusion pump is considerable and difficult to cope with. Transdermal delivery by patches containing rotigotine is also available (Jenner, 2005). The drug contained in elastic vesicles is released to the silicon matrix following ionic gradients and finally crosses the stratum corneum to reach the epidermis, where it is eventually absorbed (Nugroho et al., 2004). Sustained delivery of apomorphine in rods of vinyl acetate applied subcutaneously and liposomes containing a precursor of levodopa is being investigated in animal models (Tsai et al., 2011). Miniaturization of drug delivery devices from the macroscale to the micro- or nanoscale is of considerable interest (Langer, 1998). Its ultimate objective is to integrate systems that combine device technology with therapeutic molecules to allow the creation of implantable devices that can provide prophylactic or therapeutic actions in situ. For this purpose, the development of new materials is essential (Shapira, 2005; Linasazoro, 2008; Pervaiz and Irshad, 2012). Polymers such as polycaprolactones, polylactide, PLGA, or poly(ethylene oxide) (PEO) (Sabaté et al., 2008; Soenen et al., 2009; Giordano et al., 2011; Bao et al., 2012; Simşek et al., 2013; Danhier et al., 2012), conducting polymers (Kammerich, 2011; Bhardwaj and Kundu, 2010), and polymerized ionic liquids (assembled into fibers, copolymers, etc.) begin to become a reality (Oliveira, 2012; Miklya, 2010; Prow et al., 2005). One of the main challenges is the need to balance the small scale of the devices with the quantities of drugs that are clinically necessary (Elele et al., 2012).

An ideal in vivo drug delivery system would be able to determine when and if a dose is needed and then deliver it automatically (Radadia et al., 2011; Yang et al., 2012a). These systems are based on biosensors, which are defined as devices giving information about the presence and quantity of a specific chemical compound in the studied media. Such a sensor detects the presence of a chemical reaction, which is a direct reflection of the analyte of interest (Lin et al., 2010). The adequate combination of a biosensor to a delivery system would give "smart" and "on demand" systems, allowing real-time control of drug dosage according to alterations from a physiological condition (LaVan et al., 2003). A limiting step in the creation of feedback-controlled drug delivery systems has been the development of stable sensors (Ochoteco et al., 2006; Tao and Desai, 2003). This is mainly due to the fact that biological compounds are used as active specific material, and the stability of these compounds over time used to be rather poor (Krol, 2012).

Besides these shortcomings, the advantages of these systems are obvious (Table 15.6).

Table 15.6 Advantages of improved drug delivery systems

(a) Continuous maintenance of drug levels in therapeutically desirable doses
(b) Reduction of harmful side effects due to targeted delivery to a particular cell-type tissue
(c) Potentially decreased amount of drug needed
(d) Decreased number of dosages, leading to improved patient compliance
(e) Facilitation of drug administration for pharmaceuticals with short half-lives

15.3.2.1.1 *Localized delivery of drugs*

These routes of administration mentioned above (intraduodenal, transdermal, respiratory, etc.) can guarantee a continuous stimulation of dopamine receptors, which is advantageous for PD. However, targeting and localized delivery are also key challenges in PD therapy. The undesirable distribution of any antiparkinsonian drug in healthy tissues may induce severe side effects, such as psychiatric disturbances and dyskinesias. This greatly limits the maximal allowable dose of the drug. Therefore, one of the main goals of drug delivery is to selectively attack the degenerative process and/or replenish the lacking neurotransmitters, while saving the normal tissue from excessive burdens due to drug toxicity (see previous section and Qiao et al., 2012).

Several approaches are being investigated. Dopamine-containing liposomes, after stereotactic implantation in the striatum, have shown to produce both behavioral recovery and enhanced striatal dopamine levels in a rat model of PD (Liu et al., 2012b; Carafa et al., 2010). Polymer NPs have an extraordinarily large surface area that presents diverse opportunities to place functional groups on the surface. These smart surfaces and materials could enable the development of better systems for in vivo drug delivery (see next section).

Many medical devices are implanted every year, and each of them could be viewed as a potential platform for providing local drug delivery. In the case of PD, deep brain stimulators are routinely implanted in the STN (Rodríguez et al., 2000), opening up the possibility of simultaneously delivering well-chosen drugs (glutamate antagonists, cannabinoids, and others) or other molecules (trophic factors, enzymes, etc.) that are able to modulate the activity of output structures of basal ganglia. Finally, implanting biosensors in the striatum and other brain nuclei to monitor the levels of dopamine and other neurotransmitters could be of high interest for other neurodegenerative diseases (Cass et al., 2012). This could be done by minimally invasive surgical techniques and the directing of the biocompatible sensor by magnetic NPs. The biocompatible sensor, together with the magnetic NPs, could be directed to the specific area of the brain by an external magnetic field. Although this guidance could be done by a permanent magnet, a more complex setup would be preferable for patients and, therefore, an external magnetic field could be applied by magnetic resonance-like devices.

15.3.2.1.2 *Overcoming the blood–brain barrier*

The BBB is an obstacle for many drugs, which otherwise could be highly useful for many neurological diseases. Therefore, the design of functionalized nanostructures that can be administered systemically and deliver drugs and small molecules across the BBB is an area of active research (Montenegro et al., 2012; Kortekaas et al., 2005; Lin et al., 2012a).

NPs designed to penetrate the BBB include solid colloidal matrix-like particles made of polymers with added properties such as magnetic particles, dendrimers, and lipids (Grabrucker et al., 2011; Wohlfart et al., 2012; Beg et al., 2011). Generally administered by the intravenous route, they have been developed for the targeted delivery of therapeutic or imaging agents (Vitaliano et al., 2012). They can be useful in brain drug delivery, accompanied by the local sustained release of peptides, proteins, genes, or antisense drugs. Compared to conventional drugs, they possess a high intrinsic pharmacological activity. However, NPs have poor stability in biological fluids, rapid enzymatic degradation, and unfavorable pharmacokinetic properties.

Moreover, the size of the nanostructures (generally between 50 and 300 nm) becomes a problem, because a size larger than 100 nm does not allow the nanostructure to cross the BBB. This is actually one of the topics under intensive research. A possible solution is their formulation in protective nanocontainers targeting the brain (Soppimath et al., 2001). These nanocarriers avoid the action of catabolic enzymes present in the BBB, thus allowing the passage of the molecule of interest, which is not degraded or destroyed once into the brain. This strategy might allow transvascular gene therapy, a procedure involving the global distribution of a therapeutic gene (tyrosine hydroxylase in the case of parkinsonism) to the brain after an intravenous administration and transport across the BBB (Pardridge, 2005). This is enabled with the development of a nonviral gene transfer technology that encapsulates plasmid DNA inside PEGylated immunoliposomes inside the nanocontainer.

15.3.2.1.3 *Localized delivery of genes and trophic factors, avoiding the use of viruses*

Safe and effective nanotechnology-based gene delivery vectors could potentially be used to counteract pathological processes in the brain, augment compensatory mechanisms, and influence the brain regenerative capacities. In vivo gene delivery is an area of considerable interest where genetic materials (e.g., DNA, RNA, and oligonucleotides) could be used to inhibit undesirable gene expression or to synthesize therapeutic proteins. Gene therapy for neurological diseases will become clinically relevant upon the development of effective and safe gene transfer vectors (During et al., 2001; Airaksinen and Saarma, 2002). The most commonly used vectors are of viral origin and raise safety concerns. That means that to be effective, they need to be inserted in the genome. This insertion process is quite random and occasionally it affects sites of the genome which can induce dangerous mutations. This is called insertional mutagenesis. Also, adenoviruses can induce severe hypersensitivity phenomena.

Most nonviral vectors are safer but they lack the high transfection efficiency obtained with viral vectors. For effective gene therapy, a genetic payload must be delivered to the targeted cell tissue, then enter such cell, and get transported to the nucleus to achieve expression. The initiating event, which is the entrance of the genetic material into the cell without the use of a viral vector, is limited because of the need to supply the DNA to the surface of the cell in sufficient concentration to effect entrance. Ultrafine silica NPs, functionalized with amino groups, have been shown to bind and protect plasmid DNA from enzymatic digestion and to effect cell transfection in vitro (Kneuer et al., 2000a). Recent work has established the feasibility of using organically modified silicate (ORMOSIL) NPs as a nonviral vector for in vitro gene transfection (Kneuer et al., 2000b). Moreover, researchers have delivered genes into the brains of living mice with an efficiency that is similar to, or better than, viral vectors and with no observable systemic or neurologic toxic effects. These results are highly promising but the long term toxicity of NPs has to be monitored. These gene–NP complexes might be applied to models of PD. Alternative delivery systems are being explored. Thus, genes can be delivered in liposomes administered intravenously and PEGylated immunoliposome delivery of the tyrosine hydroxylase gene has already been shown to alleviate behavioral deficits in rodent models of PD (Pardridge, 2005; Zhang et al., 2003c). Research using DNA NPs to deliver trophic factors in the striatum is producing promising results (Yurek et al., 2009; Fletcher et al., 2011).

15.3.2.1.4 *Regenerating the nigrostriatal pathway*

The prospect of being able to directly replace or stimulate re-growth indirectly of damaged and lost cells within the parkinsonian brain is highly attractive and may represent the best and most novel therapy for PD. After extensive studies in animal models demonstrating proof-of-principle (i.e., survival of implanted cells providing improvement of motor deficits without serious adverse effects), transplantation of fetal dopaminergic cells to parkinsonian patients was started in the late 1980s. Initial small open-label studies showed quite promising results that were not confirmed in double-blind clinical trials (Freed et al., 2001). Functional regeneration after injury to the mammalian CNS is extremely poor, largely because of the nonpermissive environment in the mature CNS. Several barriers to the repair process exist, including formation of scar tissue, gaps in nerve tissue formed during phagocytosis of dying cells, factors that inhibit axon growth, and the inability of many neurons to initiate axonal extension. Additionally, in the particular case of PD, the main objective of cell therapy should not be limited to replace the dopaminergic deficiency but to restore or reestablish the normal anatomy (connectivity) and physiology (appropriate synaptic contacts and functioning) of the striatum (Barker et al., 2013; Winkler et al., 2005).

The reason for the poor outcome in double-blind trials is probably multifactorial, but poor survival of cells after grafting seems to be one of the most influencing factors. Hence, stem cell–based replacement therapies have prompted high expectations, owing to their intrinsic capacity for generating millions of well characterized dopamine cells which, once implanted, could rewire the nigrostriatal system. Alternative ways to solve this problem are the use of cells as a vehicle of molecules essential for dopamine transmission and/or cell survival. Encapsulated cells are well suited to reach both aims: On the one hand, they can be genetically modified to produce molecules involved in dopamine transmission and survival; on the other hand, they are isolated from immunological attack (Orive et al., 2003). A major drawback of the encapsulation approach is that cells cannot make synaptic connections with the host neurons.

Would nanobiotechnology be capable of inducing the generation of lost neurons, when and where required and in the amount required? Conventional materials and materials with dimensions greater than 1 mm have not invoked proper cellular responses to regenerate tissue. In contrast, due to their ability to mimic the dimensions of constituent components of natural tissues, like proteins, nanomaterials may be an exciting successful alternative. Nanomaterials interact much more closely with cells, thereby inducing changes in cell functions. For instance, carbon nanofibers promote neural growth (Kabiri et al., 2012; Silva et al., 2004). Materials investigated to date include nanophase ceramics, metals, polymers, and composites. Nanoengineered scaffolds that support and promote neurite and axonal growth are highly promising (Scott, 2010). The use of rigid scaffolds combined with hydrogel (Mahoney and Saltzman, 2001) or extracellular matrix molecules are particularly interesting due to the possibility of complementation with embryonic tissues, cell lines, and neurotrophic growth factors that can enhance axonal regeneration. That is, embryonic tissues, cell lines, and neurotrophic growth factors can be added to these matrices, thus exerting a greater effect.

An unresolved problem of cell-based therapies is the unique combinations of matrix, growth factor, and cell adhesion cues that distinguish each microenvironment and that are critically important for controlling cell differentiation and histogenesis. Some investigators have created an approach in which cells, synthetic matrix elements, and controlled release technology are assembled and programmed, before transplantation, to mimic the chemical and physical microenvironment of developing tissue (Mahoney and Saltzman, 2001). Cells were assembled with cell adhesive/controlled release microparticles to form transplantable neotissues with an associated synthetic microenvironment. This is quite relevant since dose and distribution of protein have both been shown to influence the biological activity of cell populations. These authors used the technique with cholinergic neurons, but in theory, it could be applied to dopamine neurons.

Likewise these materials can be of great value in other aspects of regenerative therapies research. One of the main challenges in obtaining neurons with dopaminergic phenotype from embryonic stem cells concerns the differentiation process. Guiding differentiation is a key step once stem cells in the laboratory tend to differentiate in an uncontrolled manner, resulting in a mixture of cells of little medical use.

15.3.2.1.5 *Challenges of nanotechnology in Parkinson's disease*

A lot of experimental work is needed to know definitely the potential of these technologies. The challenges associated with nanotechnology applications in PD are numerous. Those related to the complexity of PD have been repeatedly noted throughout this article. Nanotechnology allows an intervention at a molecular level and any desired cellular signaling pathway can be targeted. One may believe that this extraordinary specificity is enough but there is a need for even greater specificity. Furthermore, there is a requirement for technologies that are able to concurrently carry out a diverse set of specific cellular and physiological functions, such as targeting multiple receptors or ligands. For instance, PD is the clinical expression resulting from numerous interdependent molecular and biochemical events affecting numerous and interdependent cell groups in numerous and interdependent basal ganglia and brainstem nuclei. A great cellular heterogeneity and multidimensional cellular interactions in the spatial and temporal domains underlie the basis of the nervous system's anatomical and functional wiring that is the basis of its extremely complex information processing. This process is disrupted in disease states and it is important not to cause further disruptions by the application of an interacting nanodevice, since the resulting side effects could be severe. Finally, long-term effects of nanotechnologies are unknown, raising concerns about their long-term safety.

Acknowledgments

Miryam Asunción contributed to the elaboration of the review.

Chapter 16

Nanotherapeutics for Cardiovascular, Pulmonary, and Hematologic Diseases

Denis B. Buxton
Division of Cardiovascular Sciences, National Heart, Lung, and Blood Institute, National Institutes of Health, Bethesda, MD 20817, USA
db225a@nih.gov

16.1 Introduction

Cardiovascular, pulmonary, and hematologic diseases exert a huge toll, both in terms of morbidity and mortality and in terms of economic impact. Globally, cardiovascular diseases (CVDs) account for 30% of deaths, and the global cost of CVDs for 2010 was estimated to be $863 billion (Bloom et al., 2011). These figures reflect worldwide incidences of more than 10 million for congestive heart failure (HF), 24 million for ischemic heart disease, and 28 million for stroke. By 2030 these incidences are predicted to increase by approximately 30%, reflecting a growing and aging population with increasingly sedentary lifestyles, resulting in an estimated global economic burden of $1044 billion (Bloom et al.,

Controlled Release Systems: Advances in Nanobottles and Active Nanoparticles
Edited by Alexander van Herk, Jacqueline Forcada, and Giorgia Pastorin

ISBN 978-981-4613-21-7 (Hardcover), 978-981-4613-22-4 (eBook)
www.panstanford.com

2011). Chronic respiratory diseases, such as chronic obstructive pulmonary disease (COPD) and asthma, account for an additional 7% of global deaths (Bloom et al., 2011). For COPD alone, the overall cost of the illness is estimated to be $2108 million for 2010, rising to $4708 million by 2030 (Bloom et al., 2011). Sepsis, infection of the bloodstream, remains a major cause of mortality, and abnormal regulation of blood clotting is a key component of a number of diseases.

There are thus compelling reasons to develop new therapeutics to reduce the huge burdens of heart, lung, and blood diseases both in the U.S. and globally. This chapter will focus on recent developments in the use of nanotechnology for the delivery of drugs and other therapeutics to the cardiovascular, pulmonary, and hematopoietic systems.

16.2 Cardiovascular Disease

CVD offers a broad range of opportunities for the application of nanotechnology, ranging from using targeted nanoparticles (NPs) for delivery of drugs and nucleic acid therapeutics to developing functionalized scaffolds to improve cardiovascular repair.

16.2.1 Atherosclerosis and Vascular Injury

The detection of atherosclerosis, and in particular of "vulnerable plaque" at high risk of rupture, has been a major focus of cardiovascular nanotechnology research (Lobatto et al., 2011). In contrast, the therapeutic use of nanotechnology against atherosclerotic plaque has been less well developed. One approach has been to target the *vasa vasorum*, the blood supply to vessel walls. Neoangiogenesis of the *vasa vasorum* accompanies plaque development and is a biomarker of unstable disease (Winter et al., 2008). Winter et al. used paramagnetic perfluorocarbon NPs targeted to the neovascular endothelium to deliver fumagillin, a mycotoxin produced by *Aspergillus fumigatus* that suppresses angiogenesis, to plaque. The NPs were targeted to the $\alpha_v\beta_3$ integrin using a quinolone $\alpha_v\beta_3$ antagonist. Hyperlipidemic rabbits were treated with atorvastatin alone or in combination with targeted fumagillin-containing NPs at 0 weeks only, or at 0 and 4 weeks. The investigators monitored the extent of neovascular endothelium serially using cardiac magnetic resonance (CMR) with $\alpha_v\beta_3$-targeted paramagnetic NPs for 8 weeks. Statin treatment alone had no antineovascular effect, while fumagillin treatment reduced the neovascular signal by 50%–75% at 1 week and maintained the effect for 3 weeks. A second fumagillin treatment at 4 weeks had a similar transient effect on neovascularization. However, in animals treated with statin and with fumagillin at 0 week and 4 weeks, a sustained decrease in neovascularization was demonstrated at the end of the 8-week study period (Fig. 16.1, lower panel, closed circles).

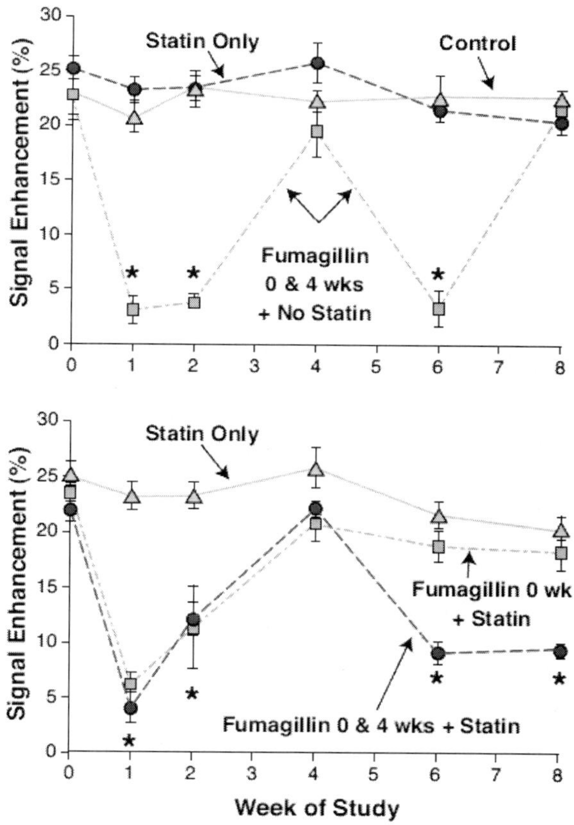

Figure 16.1 Cardiac magnetic resonance signal enhancement up to eight weeks after treatment with targeted fumagillin nanoparticles +/− oral atorvastatin. Reprinted from Winter et al. (2008), Copyright 2008, with permission from Elsevier.

Short interfering RNA (siRNA) modulation of monocyte-dependent inflammatory responses represents another promising avenue for the treatment of atherosclerotic plaque. Inflammatory monocytes (Ly-6C^{high}) depend on monocyte chemokine protein 1 (MCP1) binding to the chemokine receptor CCR2 to traffic to sites of injury, while noninflammatory monocytes (Ly-6C^{lo}) respond through fractalkine (in humans) protein and the chemokine receptor CX3CR1 (Leuschner et al., 2011). Increased invasion of Ly-6C^{high} monocytes promotes atherosclerosis, making the monocytes an attractive target for reducing atherosclerotic burden. Treating apoE$^{-/-}$ mice with established atherosclerosis for three weeks with lipid NPs carrying siCCR2 reduced monocyte/macrophage number in atherosclerotic plaques by 82%, including a marked reduction of inflammatory Ly-6C^{high} monocytes. Immunohistochemistry showed a 46% reduction in myeloid cells and a 38% reduction of lesion size in the aortic root (Leuschner et al., 2011).

Kamaly et al. designed polymeric NPs containing the anti-inflammatory peptide Ac2-26, an annexin A1/lipocortin 1-mimetic peptide, for use as therapeutics to dampen chronic inflammation (Kamaly et al., 2013). The NPs were built from biodegradable diblock poly(lactic-co-glycolic acid)-*block*-polyethyleneglycol (PLGA-b-PEG) and peptide-conjugated PLGA-PEG targeted using a collagen IV–targeted heptapeptide ligand identified by phage display biopanning. The targeting takes advantage of the exposure of collagen IV that occurs at sites of vascular inflammation and injury (Kamaly et al., 2013). In a hind-limb ischemia model, systemic injection of collagen-targeted Ac2-26 NPs reduced polymorphonuclear neutrophil (PMN) infiltration by 30% and decreased tissue damage. The NPs also reduced PMN infiltration and decreased the resolution interval in a zymosan-induced peritonitis model (Kamaly et al., 2013). The ability of the NPs to reduce inflammation in the chronic setting of atherosclerosis will be of great interest.

Another area of interest is the prevention of vascular smooth muscle proliferation following vascular injury, which can result, for example, in restenosis of vessels following angioplasty and stent placement. Cahn et al. developed multifunctional NPs, which they termed "nanoburrs," for controlled spatiotemporal delivery of the antiproliferative agent paclitaxel to injured vasculature (Chan et al., 2010). They designed 60 nm core–shell hybrid NPs composed of a paclitaxel–poly(lactic acid) (PLA)-conjugated core surrounded by a lipid interface with a PEG corona, and functionalized the NPs with the collagen IV targeting peptides described above. In vitro studies demonstrated slow release of paclitaxel over 12 days resulting from hydrolysis of the paclitaxel-PLA ester bond. The targeted NPs inhibited proliferation of human aortic smooth muscle cells in vitro, and showed greater retention in injured vasculature in rat balloon angioplasty models compared to nontargeted controls (Chan et al., 2010). In a subsequent study they used similar collagen-targeted nanoburrs but with a core of free paclitaxel in PLGA and a shell composed of soybean, lecithin, and peptide-conjugated distearoyl phosphatidylethanolamine–PEG. In a rat carotid injury model, systemic administration of targeted NPs reduced neointimal formation relative to saline, free paclitaxel or non-targeted paclitaxel-containing NPs (Fig. 16.2) (Chan et al., 2011).

The development of drug-eluting stents to replace bare metal stents for patients undergoing percutaneous transluminal coronary angioplasty has significantly decreased restenosis resulting from neointimal hyperplasia. However, lingering concerns remain about the long-term safety of drug-eluting stents, centered on delayed arterial healing, with persistent fibrin deposition and impaired re-endothelialization. Incomplete healing can lead to elevated levels of in-stent thrombosis and myocardial infarction (MI) late after stent placement (De Luca et al., 2012). This has led groups to explore the potential of nanotechnology for local drug delivery from stents. Tsukie et al. harnessed the antiproliferative pleiotropic actions of 3-hydroxy-3-methylglutaryl-coenzyme A (HMG-CoA) reductase inhibitors to improve healing. They coated metal stents with PLGA NPs containing pitavastatin and compared them to commercially available polymer-coated sirolimus-eluting stents in a pig coronary artery model. Both pitavastatin NP-coated and sirolimus-eluting stents reduced in-stent restenosis to a similar extent relative to bare metal stents. However, delayed endothelial healing was observed for sirolimus-eluting stents, but not for pitavastatin NP stents (Tsukie et al., 2013).

Gene delivery from the stent surface has also been explored as a tool for improving outcomes following stenting. For example, Brito et al. immobilized lipopolyplexes (LPPs) with plasmid DNA expressing human endothelial nitric oxide synthase (eNOS) on

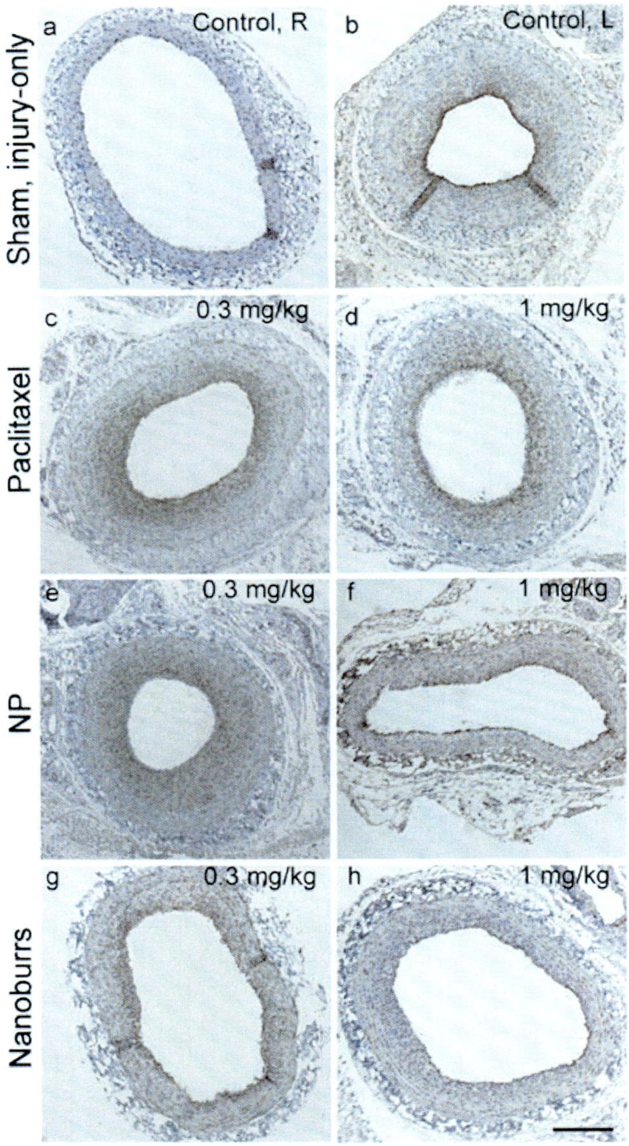

Figure 16.2 Immunohistochemical staining for α-smooth muscle actin in balloon-injured rat carotid arteries treated with paclitaxel (c, d), empty nanoparticles (e, f), or nanoburrs (g, h). Reprinted from Chan et al. (2011) with permission from the National Academy of Sciences.

stainless steel stents using PLGA and gelatin coatings. In a rabbit iliac artery restenosis model, human eNOS expression was detected in the artery within 24 hours. Local eNOS production suppressed smooth muscle cell proliferation (Fig. 16.3, left) and promoted arterial re-endothelialization (Fig. 16.3, right), resulting in a significant reduction in restenosis relative to stents with lipoplexes enclosing an empty vector (Brito et al., 2010).

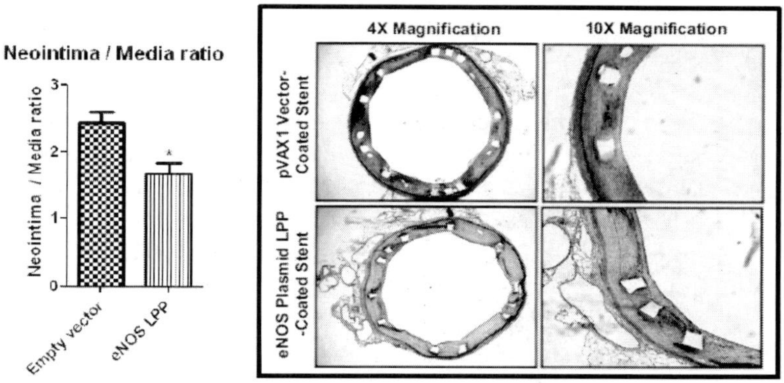

Figure 16.3 Inhibition of neointimal hyperplasia. Quantitative analysis from light microscopy images of polymethacrylate-embedded sections 14 days following implantation (left). Light microscopy images of rabbit iliac arterial tissue cross sections after two weeks of control and endothelial nitric oxide synthase (eNOS)-expressing plasmid DNA-loaded stent implantation (right). Reprinted from Brito et al. (2010) with permission from Biomed Central.

16.2.2 Myocardial Infarction and Heart Failure

For patients who suffer MIs, there has been a steady improvement in survival rates over the last 50 years, driven in part by the advent of revascularization, initially by thrombolysis and later by percutaneous angioplasty and stent placement (Parikh et al., 2009; Yeh et al., 2010). However, many patients who survive MI go on to suffer remodeling of the ventricle, resulting in decreased pumping and the development of HF. There are currently 5.7 million people in the U.S. living with HF, and this number is projected to increase to more than 8 million by 2030 (Heidenreich et al., 2013). Treatment options for HF are very limited, and so there is an urgent need both to decrease the number of MI patients who go on to develop HF, and to develop new HF treatments.

In the early stages following MI, the focus has been on improving blood flow to the damaged tissue and reducing oxidative damage. Adenosine is a potent vasodilator with multiple physiological effects through different receptor subtypes. It has been shown in clinical trials to have the potential to reduce myocardial damage during acute MI when given during reperfusion. A recent meta-analysis that included seven prospective clinical trials of intracoronary adenosine administration to acute MI patients during percutaneous coronary intervention demonstrated improvements in electrocardiographic endpoints (Singh et al., 2012a). However, while there was a trend toward improvement in clinical outcomes, including the incidence of major adverse cardiac events, HF, and cardiovascular mortality, these findings did not reach significance, possible reflecting suboptimal dosing and inadequate statistical power of the trials (Singh et al., 2012a). As a result of the failure to show clinical benefit, in addition to the short half-life of the drug in plasma and profound hypotensive and bradycardiac actions, adenosine has found limited use (Takahama et al., 2009). Takahama et al. encapsulated adenosine in liposomes, which resulted in prolonged circulation and accumulation of the liposomes preferentially in ischemic/reperfused tissue following intravenous injection. Administration of adenosine liposomes five minutes before reperfusion reduced infarct size without causing significant hemodynamic effects (Takahama et al., 2009).

Nitric oxide (NO), previously known as endothelial-derived relaxation factor, also causes vasodilation but has additional actions related to ischemia/reperfusion injury. NO modulates platelet adhesion, leukocyte adhesion, inflammation, and cardiomyocyte apoptosis. NO gas reduces infarct size in a porcine model of reperfusion injury but is impractical for clinical use, while many small-molecule donors release NO indiscriminately and may cause further damage by reacting with superoxide (Johnson et al., 2010). Johnston et al. developed S-nitroso-*N*-acetylpenicillamine-derivatized fourth-generation polyamidoamine dendrimers (G4-SNAP) that are triggered to release NO by physiological concentrations of glutathione. In ex vivo perfused rat hearts, 230 pM G4-SNAP (equivalent to 15 nM SNAP) in the presence of a physiological concentration of glutathione (500 μM) gave optimal reduction of infarct size, while higher or lower doses were less effective (Fig. 16.4) (Johnson et al., 2010).

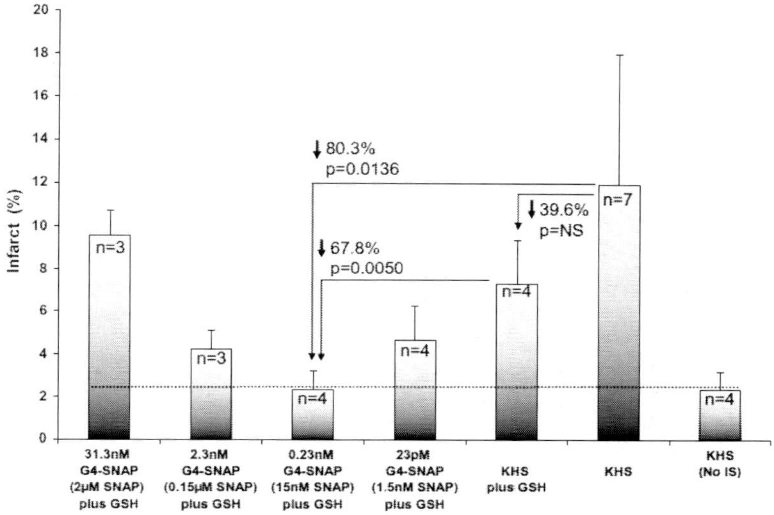

Figure 16.4 Infarct size in hearts subjected to 20 min global ischemia and reperfusion with KHS + GSH amended with G4-SNAP for 20 min followed by KHS + GSH alone for an additional 40 min. Reprinted from Johnson et al. (2010), Copyright 2010, with permission from Elsevier.

Modifying calcium uptake during reperfusion following ischemia represents another attractive target to reduce postischemic injury. Clemons et al. used a peptide derived from the alpha-interacting domain (AID) of the L-type Ca^{2+}channel to inhibit activation of the channel during reperfusion. They found that conjugation of the peptide to multifunctional polymeric NPs was more effective in promoting association of peptide with myocytes than delivering with a transactivator of transcription (TAT) cell-permeable peptide (Clemons et al., 2013). The NPs also distributed more evenly throughout perfused hearts than TAT-AID. Both formulations showed similar efficacy in decreasing tissue injury in reperfused guinea pig hearts, as indicated by release of creatine kinase and lactate dehydrogenase (Clemons et al., 2013).

Another approach to improving recovery post-MI has been to try to damp down protective responses that can be overactivated. For example, activation of the renin-angiotensin-aldosterone system (RAAS) by overexpression of angiotensin II (Ang II) following MI is a protective mechanism to maintain blood pressure. However, continuous stimulation can have deleterious effects through vasoconstriction and promotion of fibrosis, contributing to adverse ventricular remodeling and HF development. To address this, Liu et al. used a novel "tadpole" dendrimer constructed of cystamine core G4.0 poly(amido amine) (PAMAM) to deliver siRNA to down-regulate angiotensin II type 1 receptor (AT1R), the major receptor mediating most adverse effects of Ang II (Liu et al., 2013b). They tested two cell-penetrating peptides for efficacy in delivering dendrimers intracellularly, finding oligo-arginine to be superior to TAT peptide. Intramyocardial injection of oligo-arginine-conjugated dendrimer/siRNA complex into the ischemic border zone in a rat MI model resulted in down-regulation of AT1R gene expression in the left ventricle (LV) (Fig. 16.5, left), while AT2R expression was unaffected. Down-regulation of AT1R improved LV function three days post-MI (Fig. 16.5, right). Treatment with AT1R siRNA also reduced LV end systolic volume and decreased infarct size (Liu et al., 2013b).

Modulation of monocyte-dependent inflammatory responses, as discussed in Section 16.2.1, has also been tested as a therapy for MI. Delivery of CCR2 siRNA in mice subjected to myocardial ischemia/reperfusion resulted in silencing of CCR2, decreased accumulation of Ly-6C^{high} monocytes by 70%, and decreased infarct size by 40% (Leuschner et al., 2011). In proof-of-concept studies supporting the potential for clinical translation, monocyte gene silencing has been extended to non-human primates employing a similar approach with two novel lipid formulations; using CD45 siRNA, robust silencing of CD45 was observed in organ resident and blood circulating myeloid cells of cynomolgus macaques (Novobrantseva et al., 2012).

The post-MI myocardium represents a hostile environment in which a number of factors contribute to cell death and inhibit myocardial regeneration. Several groups have employed nanofibers to try to create a more benign environment that can stimulate endogenous and exogenous myocardial repair. Padin-Iruegas

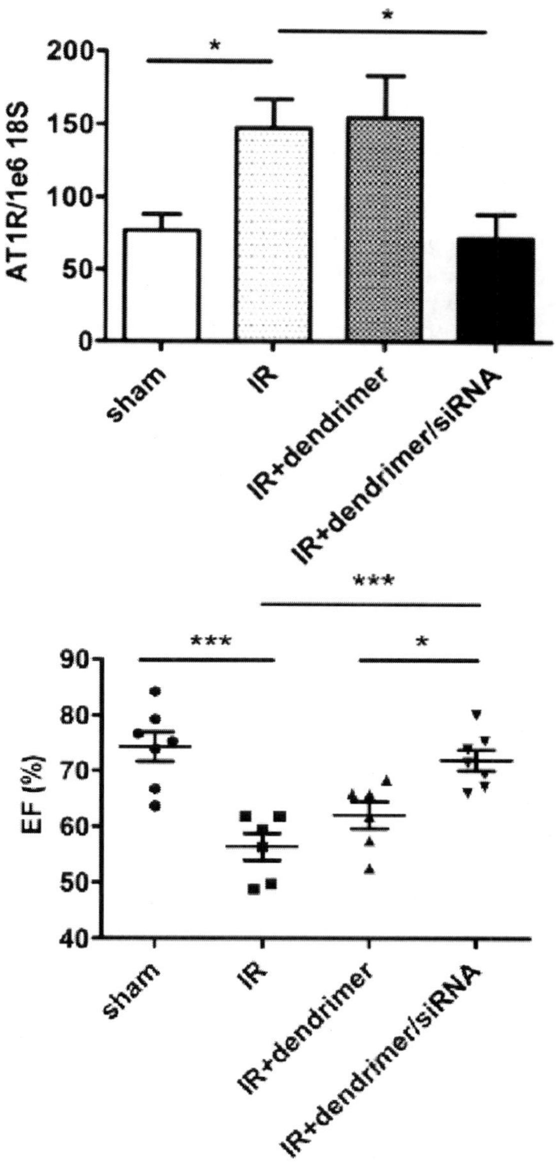

Figure 16.5 Angiotensin receptor expression and cardiac function in vivo. AT1R expression in LV tissue (left) and LV ejection fraction (right), three days' time after ischemia/reperfusion injury. Reprinted from Liu et al. (2013b), Copyright 2013, with permission from Elsevier.

et al. used a self-assembling peptide platform, consisting of AcN-RARADADARARADADA-CNH_2 16-mers (RAD16-II), to which biotinylated insulin-like growth factor I (IGF-1) was tethered through a biotin–streptavidin sandwich approach. Following injection into the ventricular wall, the peptides assemble into stable nanofibers on exposure to physiological pH and osmolarity. Injection of either RAD16-II or cardiac progenitor cells (CPCs) individually reduced infarct size relative to untreated hearts in a rat MI model, but in combination RAD16-II + CPCs were more effective than the individual therapies in improving cardiac function. The combination therapy also resulted in greater volume of regenerated myocytes increased length density of newly formed coronary arterioles relative to the individual therapies (Padin-Iruegas et al., 2009). Activation of resident CPCs by paracrine effects contributed to the stimulation of cardiomyogenesis and vasculogenesis (Padin-Iruegas et al., 2009).

Lin et al. used intramyocardial RAD16-II injection for sustained delivery of recombinant vascular endothelial growth factor (VEGF) in a rat MI model (Lin et al., 2012b). Again, combination therapy with both the nanofibers and VEGF was much more effective than nanofibers alone or free VEGF in improving angiogenesis, arteriogenesis, and cardiac performance 28 days post-MI. The engineered vascular niche promoted homing of a new population of cardiomyocyte-like cells to the injected sites. Preclinical studies in pigs also demonstrated functional improvements and enhanced revascularization in response to combination nanofiber/VEGF treatment, supporting the clinical potential of the therapy (Babar et al., 2012).

In a further development of the strategy, Guo et al. took the similar RADA16 peptide and modified it by addition of a heparin-binding domain, LRKKLGKA. The heparin-binding domain sequesters heparin in vivo, which is then available to bind a wide variety of heparin-binding growth factors such as VEGF. When VEGF was injected into the rat heart along with the RADA16-LRKKLGKA, 60% of the injected VEGF was retained at 1 day, and 15% remained at 28 days, much greater retention than observed with unmodified RADA16 (Guo et al., 2012a). In a rat MI model, this prolongation of the nanofiber-mediated sustained delivery resulted in better function in RADA16-LRKKLGKA/VEGF-treated hearts relative to RADA16/VEGF treatment, along with smaller infarcts and greater wall thickness (Fig. 16.6) (Guo et al., 2012a).

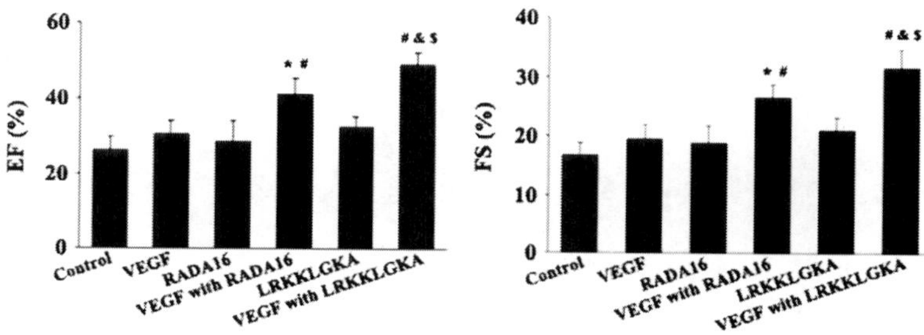

Figure 16.6 Improvement in LV contractile function and histological changes at four weeks after transplantation. The combination of VEGF plus peptide with heparin binding domain resulted in the greatest improvement in ejection fraction (EF) and fractional shortening (FS). Reprinted from Guo et al. (2012a), Copyright 2012, with permission from Elsevier.

Later after MI, the therapeutic focus switches to the prevention of cardiac remodeling and restoration of function to myocardial scar tissue. The development of tissue-engineered cardiac patches that can be implanted on the epicardial surface to prevent expansion of the infarct area, and potentially to integrate with host myocardium to improve cardiac function, has been an area of intense interest, and a wide range of micro- and nanomaterials have been tested as scaffold materials (Lakshmanan et al., 2012). A limitation of many of the matrices used for cardiac tissue engineering has been that the pore walls limit cell–cell interaction, delaying signal propagation (Dvir et al., 2011). To address this issue, Dvir et al. incorporated gold nanowires into alginate scaffolds to bridge the electrically-resistant pore walls and improve electrical communication. Tissues grown on the composite matrices were thicker and better aligned than tissues grown on alginate alone. When stimulated electrically, the cells in the nanowire constructs showed synchronous calcium transients and contraction (Dvir et al., 2011).

A related approach to the development of cardiac constructs with improved electrophysiological properties incorporated cardiac nanotubes (CNTs) into photo-crosslinked gelatin methacrylate (GelMA) hydrogels. Cardiac patches obtained by seeding neonatal rat cardiomyocytes onto hybrid CNT-GelMA hydrogel had threefold higher spontaneous beating rates and 85% lower excitation threshold relative to patches cultured on gelatin or GelMA (Shin et al., 2013).

16.2.3 Peripheral Arterial Disease

Peripheral arterial disease (PAD) results from atherosclerotic lesions in the extremities and primarily affects the legs. As the disease progresses, symptoms increase from pain when walking to rest pain to ulceration and eventually to gangrene. Risk factors include smoking, diabetes, dyslipidemia, and hypertension. Treatment options include lifestyle modification, endovascular approaches such as angioplasty and stenting, and surgical treatment to bypass blocked vessels using the saphenous vein or artificial grafts. However, PAD lesions are often diffuse, making endovascular and surgical approaches ineffective.

To overcome this problem, many groups have employed treatments with growth factors to try to grow new blood vessels in order to supply the affected limbs. Golub et al. used PLGA NPs to encapsulate VEGF, providing sustained release of VEGF over two to four days in vitro. In a mouse model of femoral artery occlusion, intramuscular injection of VEGF NPs resulted in significant increases in total blood vessel volume and connectivity, as measured by 3D micro–computed tomography (micro-CT). The VEGF NPs were an order of magnitude more effective than free VEGF (Golub et al., 2010).

Webber et al. investigated VEGF-mimetic peptide amphiphiles (PA) designed to assemble into cylindrical nanostructures. The VEGF PA forms nanofibers and entangled nanofiber gel networks that display the VEGF-mimetic peptide on the surface (Webber et al., 2011). Intramuscular injection into legs of mice in an ischemic hind-limb model demonstrated that the nanofibers increased tissue perfusion, functional recovery, limb salvage, and treadmill endurance compared to controls, which included the VEGF-mimetic peptide alone. Microcirculation density was also increased in nanofiber-treated limbs, consistent with a role for microcirculatory angiogenesis (Webber et al., 2011).

NPs have also been used to provide sustained local delivery of Food and Drug Administration (FDA)-approved drugs that stimulate vascularization through pleiotropic effects. HMG-

CoA reductase inhibitors, or statins, have a variety of pleiotropic vasculoprotective effects that are independent of their lipid-lowering activity, and have been shown to increase the angiogenic activity of mature endothelial cells and endothelial progenitor cells (EPCs) and augment collateral growth in ischemic heart and limb in experimental animals. Kubo et al. encapsulated the HMG-CoA reductase inhibitor pitavastatin in anionic PLGA NPs and found in in vitro experiments that they were taken up by endothelial cells but not smooth muscle cells (Kubo et al., 2009). In a murine model of hind-limb ischemia, a single injection of pitavastatin NPs improved blood flow at 7 and 14 days, while control NPs or free statin were ineffective. The pivastatin NPs enhanced both angiogenesis and arteriogenesis, and induced expression of VEGF and fibroblast growth factor (FGF). To move closer to clinical translation, the pitavastatin NPs were then tested in a rabbit model of chronic hind-limb ischemia (Oda et al., 2010). Treatment with pitavastatin NPs induced therapeutic arteriogenesis and ameliorated exercise-induced ischemia, consistent with the development of functional collateral arteries (Fig. 16.7). Fluorescently labeled PLGA NPs were found to localize to endothelial cells. Pretreatment with NPs loaded with vatalanib, a VEGF receptor tyrosine kinase inhibitor, abrogated the therapeutic effects of pitavastatin NPs (Oda et al., 2010).

Thiazolidines, peroxisome proliferator–activated receptor-γ (PPAR-γ) agonists, are used clinically to improve insulin sensitivity in skeletal muscles in type 2 diabetes mellitus. PPAR-γ is also expressed in vascular cells, and there is evidence that the thiazolidine pioglitazone stimulates therapeutic angiogenesis in vivo in hind-limb ischemia in a murine model. However, the systemic administration of pioglitazone is hampered by undesirable side effects, including edema and HF. Nagahama et al. packaged pioglitazone into PLGA NPs, which, as described above, localize to endothelial cells following intramuscular injection (Nagahama et al., 2012). A single injection of pioglitazone NPs improved blood flow through a PPAR-γ-dependent mechanism and stimulated angiogenesis and arteriogenesis, while free pioglitazone was ineffective. In animals treated with oral pioglitazone, a 1000-fold greater dose of drug was required to achieve the same beneficial effect on blood flow recovery (Nagahama et al., 2012).

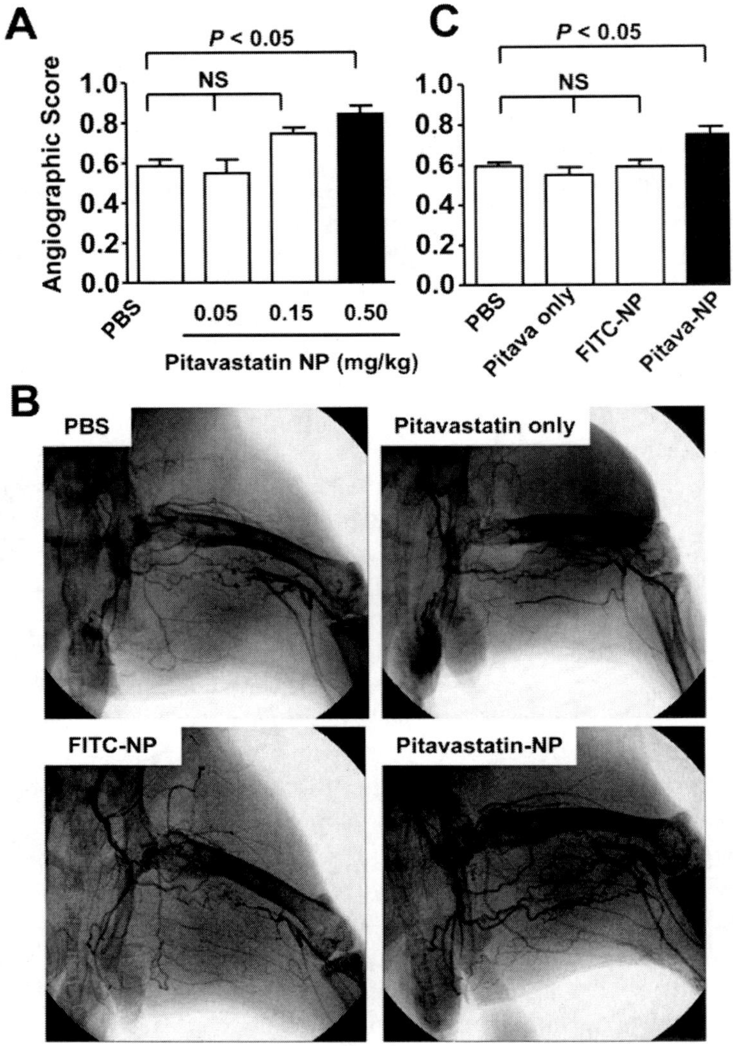

Figure 16.7 Effects of pitavastatin nanoparticles (NPs) on angiographically visible collateral arterial development 28 days after treatment. (A) Angiographic score. (B) Representative angiograms of PBS, pitavastatin-only, fluorescein isothiocyanate (FITC)–NP, and pitavastatin–NP groups at 28 days after treatment. Corkscrew-like collateral arteries were observed only in the pitavastatin–NP group. (C) Summary of the angiographic scores obtained in panel B. Reprinted with permission from Kubo et al. (2009), Copyright Wolters Kluwer Health.

16.2.4 Stroke

The brain extracellular space restricts penetration of particles into the brain, limiting the use of viruses or particles larger than 60 nm for delivery of therapeutics. Nance et al. found that the addition of a dense PEG coating facilitates diffusion, allowing NPs as large as 114 nm in diameter to move within the brain (Nance et al., 2012). The ability of NPs to penetrate brain tissue was correlated with a near-neutral surface charge; even small NPs with a negatively charged carboxylate surface failed to diffuse (Fig. 16.8). The extracellular space of human brain tissue has pores as large as 225 nm, with about 28% of the pores larger than 100 nm (right). Brain penetration with larger particles (>100 nm) will enable higher drug-loading efficiency and payload, greater drug dispersion, and extended periods of drug release for the treatment of stroke and other brain injuries. Systemic delivery of NP therapeutics may also be feasible in situations where the blood–brain barrier is compromised, for example, stroke and neuroinflammation (Nance et al., 2012).

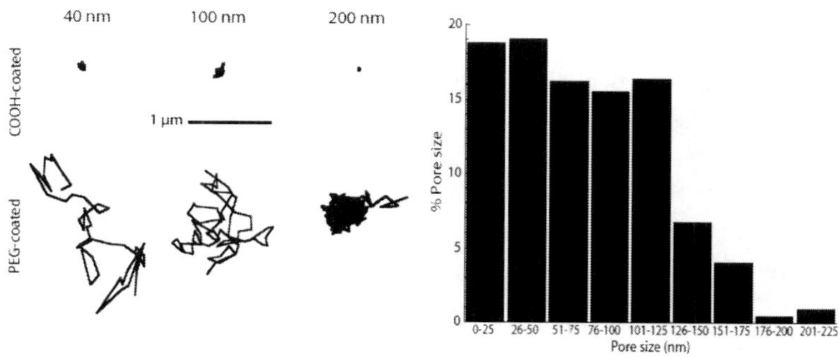

Figure 16.8 NP penetration of human brain tissue ex vivo. Representative particle trajectories for COOH- and PEG-coated NPs of various sizes in human brain tissue (left). Percentage of pores that fall in a dedicated size range in the extracellular space of fresh human brain tissue (right). Reprinted from Nance et al. (2012) with permission from AAAS.

16.2.5 Vascular Grafts

There is an urgent need for durable small-diameter synthetic grafts for coronary artery bypass grafting and replacement of lower leg vessels in adults and for congenital heart surgery in the pediatric population. Vascular prostheses made of expanded polytetrafluoroethylene (ePTFE) or polyethyleneterephthalate (Dacron), widely employed for the replacement of larger-diameter blood vessels, are generally unsuitable for small-diameter grafts with inner diameter <6 mm due to issues with thrombosis and intimal hyperplasia that often lead to complete graft occlusion (Ercolani et al., 2013). Nanotechnology provides opportunities to mimic the natural nanoscale topology encountered by cells in the vessel lumen and hence potentially to control endothelial coverage and neointimal growth, and a number of nanotechnology approaches have been undertaken (Hung et al., 2011; Rathore et al., 2012).

Electrospinning of micro- and nanofiber scaffolds for small-vessel grafts using a variety of polymers, including synthetic, natural, and composite, has been an area of particularly active research (Rathore et al., 2012; Ercolani et al., 2013). While there has been considerable progress in designing electrospun grafts with appropriate strength and compliance, efficient endothelialization and prevention of thrombosis remain as issues. Modification of the graft surface with heparin has been used as a tactic to decrease thrombus formation and stenosis. Janairo et al. modified 1 mm diameter electrospun poly(L-lactide) (PLLA) grafts with heparin using a di-amino-PEG linker (Janairo et al., 2012). In vitro experiments measuring inhibition of thrombin activity demonstrated that the immobilized heparin retained antithrombogenicity. Implantation of untreated, PEGylated, and heparin-modified grafts in a rat common carotid artery bypass model demonstrated that patency after one month increased from 43% for untreated grafts to 57% for PEGylated grafts and 86% for heparin-modified grafts (Janairo et al., 2012). Infiltration of cells into the heparin-modified grafts was also significantly higher, which may facilitate regeneration of ultimately self-supporting native vessels. The increased cell infiltration is likely to reflect the avid binding of growth factors to the heparin, promoting endothelial cell proliferation and migration (Janairo et al., 2012).

Du et al. developed a 3D gradient heparinized nanofibrous scaffold to promote endothelialization of the lumen of vessel and hence inhibit thrombosis (Du et al., 2012). Using coelectrospinning of chitosan and polycaprolactone (PCL), scaffolds were fabricated with a higher concentration of chitosan on the top surface and higher PCL on the basement. Heparinization of the scaffold by covalent attachment to chitosan-free amines resulted in a similar gradient of heparin from top to basement. Following loading of VEGF, which binds to heparin, onto gradient and uniform scaffolds, release of VEGF was more sustained for gradient scaffolds, with lower burst release. Adhesion and proliferation of human umbilical vein endothelial cells (HUVECs) were enhanced on the gradient scaffolds, with an intact monolayer growing on the top (luminal) side of the scaffold. Fluorescent staining following coculture with HUVECs and vascular smooth muscle cells (vSMCs) on the top and bottom surfaces of gradient chitosan/PCL showed that vSMC adhered and grew to a confluent cellular layer on the bottom surface, while HUVECs proliferated to form endothelium on the top surface, similar to the structure of luminal and adventitia in native blood vessels (Fig. 16.9). These in vitro studies demonstrate the potential of the 3D gradient scaffolds for creation of small-vein grafts (Du et al., 2012).

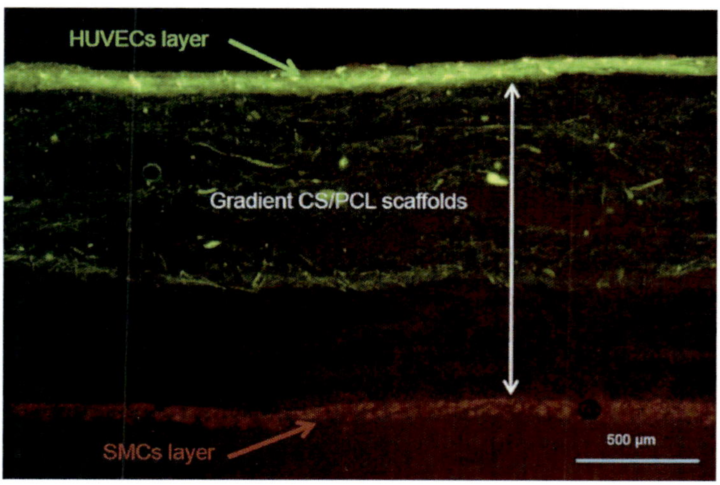

Figure 16.9 Fluorescent images of a cross section of sandwich cell–scaffold–cell structure. HUVECs on the lumen surface of Hep-g-chitosan/PCL were stained with antihuman vWF-FITC and vSMCs on the adventitia surface of Hep-g-chitosan/PCL stained with antihuman calponin-PE. Reprinted from Du et al. (2012), Copyright 2012, with permission from Elsevier.

16.3 Lung Disease

The lung offers unique opportunities for the delivery of therapeutics. To facilitate efficient gas exchange between the air and the bloodstream, the human lung has a very large absorptive area, 100–140 m^2 (Smola et al., 2008; Tam et al., 2008). This makes inhaled drug delivery an attractive route both for treatment of pulmonary diseases and for systemic drug delivery (Smola et al., 2008). The optimal particle diameter for deposition in the lung periphery is 1 to 5 μM, and a number of formulation methods are available for the generation of larger particles suitable for inhalational delivery of NPs. These include spray drying, NP flocculation, and supercritical fluid extraction; the details are beyond the scope of this chapter but have been reviewed extensively (Bailey and Berkland, 2009).

On the circulatory side, the lung also contains approximately half the body's total endothelial surface (Giordano et al., 2009). Since the lung is the first tissue bed encountered following intravenous delivery, and receives nearly all of the cardiac output, intravascular delivery can also be a very efficient route for lung uptake of therapeutics. However, the potential for systemic side effects is higher than for inhaled drug delivery.

16.3.1 Asthma and Chronic Obstructive Pulmonary Disease

Asthma and COPD, while distinct disease entities, share overlapping phenotypes. Both are characterized by chronic airway inflammation, and so therapeutic approaches targeting inflammation are common to both diseases. Asthma now represents one of the most common chronic illnesses, affecting as many as 200 million people globally. COPD is the third leading cause of death in the U.S. (Hoyert and Xu, 2012), affecting at least 15 million adults.

Atrial natriuretic peptide signaling through natriuretic peptide receptor A causes elevation of pro-inflammatory cytokines in the lung, and so Kandasamy et al. employed isatin, a naturally occurring heterocyclic indole-2,3-dione that inhibits signaling, to try to reduce pulmonary inflammation. They used chitosan, a derivative of the natural polysaccharide chitin, as an NP carrier for delivery because its mucoadhesiveness and positively charged surface groups facilitate uptake by lung cells (Yue et al., 2011). Intranasal delivery of isatin-chitosan NPs following ovalbumin (OVA) challenge in sensitized mice caused a similar down-regulation of natriuretic peptide receptor A to that resulting from administration of siRNA against the receptor. Isatin NPs improved lung histopathology, decreasing lung inflammation, goblet cell hyperplasia, and infiltration of inflammatory cells compared to the nontreated OVA group (Fig. 16.10, left) (Kandasamy et al., 2010). Isatin also reduced pulmonary production of inflammatory cytokines (right) and infiltration of inflammatory cells (not shown). Again, the actions of isatin were similar to those of the natriuretic peptide receptor A siRNA (Si9) (Kandasamy et al., 2010).

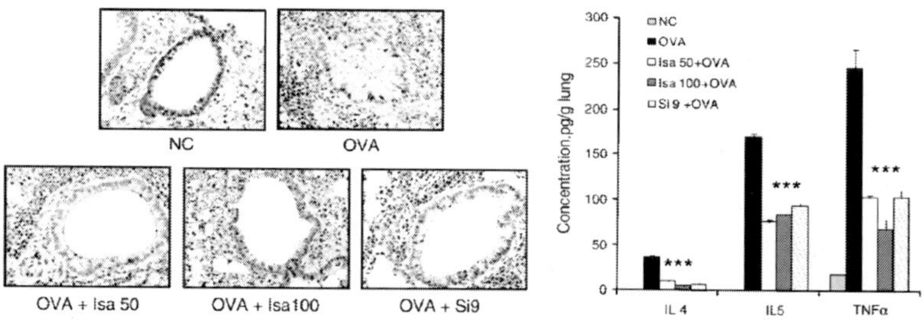

Figure 16.10 Therapeutic isatin treatment reduces allergen-induced lung pathology. Left: Isatin prevents OVA-induced lung histopathology. Lungs were fixed, sectioned, and stained with hematoxylin and eosin. Right: Isatin therapy reduces production of pro-inflammatory cytokines. Reprinted from Kandasamy et al. (2010), Copyright 2010, with permission from Elsevier.

Tranilast, *N*-(3,4-dimethoxycinnamoyl)-anthranilic acid, inhibits degranulation and histamine release from mast cells and also inhibits transforming growth factor beta (TGFβ) actions, making it potentially useful for the treatment of airway inflammatory diseases. However, poor solubility and systemic side effects have limited its use as an oral formulation (Onoue et al., 2011). Onoue et al. developed a nanocrystal solid dispersion of tranilast for inhalation therapy. Using wet-milling, they produced tranilast NPs of ~120 nm, which were mixed with lactose and micronized to give particles of ~5 μm diameter before mixing with 50 μm diameter carrier particles. The inhaled formulation attenuated antigen-induced inflammatory events in rats, as evidenced by histochemical analyses and inflammatory biomarkers such as lactate dehydrogenase, eosinophil peroxidase, and myeloperoxidase. When compared to oral administration, the systemic exposure of tranilast using a pharmacologically effective dose was fivefold lower than that following oral administration of a clinical dose of tranilast, which may make the inhaled formulation an attractive mode of delivery (Onoue et al., 2011).

16.3.2 Cystic Fibrosis

Cystic fibrosis (CF) is an autosomal recessive disorder, caused by mutations in the cystic fibrosis transmembrane conductance (CFTR) gene. The disease affects multiple organs, but pulmonary disease, involving decreased airway function and recurrent infections, is the major cause of morbidity and mortality. As a monogenic disease, gene therapy is an attractive option for treatment of CF, and since the pulmonary epithelium is the main target for gene therapy, inhalation has been the major route of administration. However, to reach the epithelium, therapeutics must cross a mucus layer, a viscoelastic gel that protects the epithelial layer from pathogens and particulates (Ensign et al., 2012). In CF, the mucus is thickened by dehydration and contains large amounts of DNA and filamentous actin, secreted by necrotic host cells and pathogens (Ibrahim et al., 2011). It has been demonstrated that common gene therapy viral vectors do not efficiently penetrate sputum from CF patients, which is likely to have been a significant contributor to the lack of efficacy of gene therapy in clinical trials (Hida et al., 2011).

To overcome the mucus barrier, Suk et al. employed the mucolytic *N*-acetyl cysteine (NAC), and showed in vitro using expectorated CF sputum that pretreatment with NAC increased the average spacing between sputum mesh elements. This resulted in greatly enhanced transport rates for 200 nm NPs coated with PEG (Suk et al., 2011). They then applied NAC pre-treatment in mice with *Pseudomonas aeruginosa* lipopolysaccharide (LPS)-induced mucus hypersecretion. Mice were pretreated intranasally prior to administration of a clinically tested nonviral gene carrier, poly-L-lysine (PLL) conjugated with a 10 kDa PEG segment ($CK_{30}PEG10k$).

NAC pretreatment increased gene delivery from CK_{30}PEG10k/DNA NPs up to 12-fold compared to saline control, similar to delivery in non-LPS treated (Suk et al., 2011).

Expression of wild-type CFTR at levels of 5%–10% of normal levels is expected to provide therapeutic efficacy (Padegimas et al., 2012). Padegimas et al. used the CK_{30}PEG10k compacted DNA NP platform formulated with PEG-substituted 30-mer lysine peptides to deliver an optimized human CFTR synthetic gene plasmid. Following intranasal delivery in mice, they were able to obtain long-term clinically meaningful expression, with 14% of normal expression at 60 days. Efficient delivery in the setting of mucus hypersecretion remains as a barrier to overcome.

Dehydration of the airway epithelial surface, which in part results from inappropriately raised sodium reabsorption through the epithelial sodium channel (ENaC), is a significant contributor to CF pathology (Clark et al., 2013). Clark et al. have identified an siRNA (GSK2225745) that provides long-term (>72 hours) suppression of ENaCα mRNA in human nasal epithelial cells. Following formulation into cationic lipid NP and intranasal administration in mice, the siRNA inhibited ENaCα mRNA expression in lung (Clark et al., 2013). Importantly, no stimulation of the innate immune system was found. However, the formulation has not yet been tested in a setting of mucus hypersecretion.

Arikace^{TM} is an inhaled liposomal formulation of the aminoglycoside antibiotic amikacin with high drug loading (drug-to-lipid ratio) and stability for administration via nebulization. Fluorescent liposomes penetrated readily into *P. aeruginosa* biofilms in vitro. Amikacin was released from liposomes through the action of rhamnolipid, a *P. aeruginosa* virulence factor associated with regulation of biofilm architecture and found in sputum of CF patients (Meers et al., 2008). Inhaled liposomal amikacin was released in a slow, sustained manner following inhalation in normal rat lung and was much more effective than inhaled free amikacin in treating chronic *P. aeruginosa* lung infections (Fig. 16.11, top). Daily dosing with liposomal amykacin was as effective as a clinically used treatment, twice daily free tobramycin (Fig. 16.11, bottom) (Meers et al., 2008).

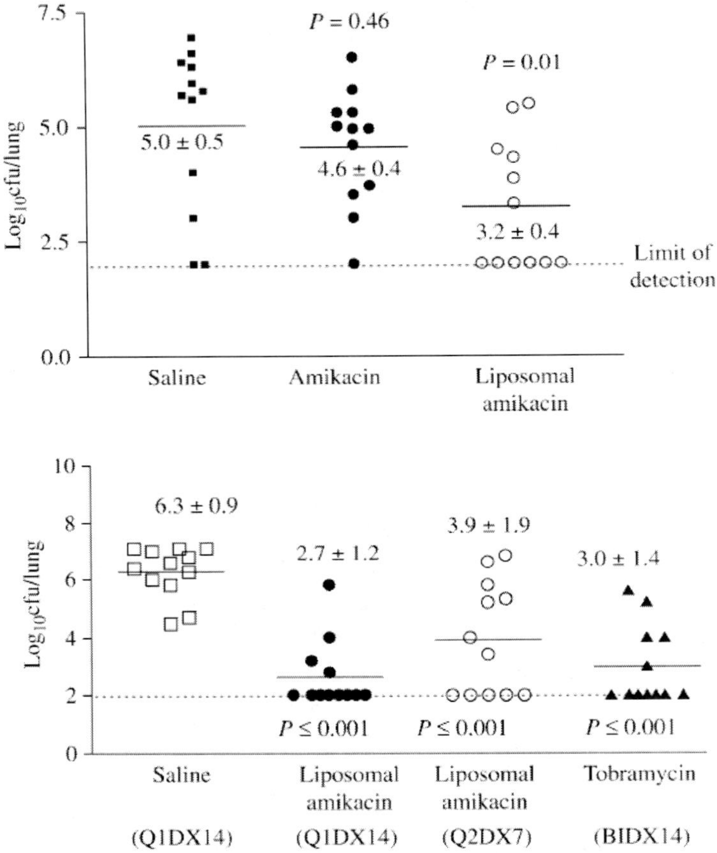

Figure 16.11 Efficacy of inhaled liposomal amikacin versus free amikacin or tobramycin. Left: Inhaled amikacin (free or liposomal) was delivered three times per week for two weeks. Right: Inhaled liposomal amikacin was delivered daily (Q1DX14) or every other day (Q2DX7), while tobramycin was delivered twice daily for two weeks. Reprinted from Meers et al. (2008) with permission from the Oxford University Press.

Liposomal amikacin has reached clinical trials for the treatment of *P. aeruginosa* treated with inhaled liposomal amikacin for 14 days, and pulmonary function measures were significantly improved (Okusanya et al., 2009). There was no change in sputum bacterial load overall, but improvements in pulmonary function correlated with decreases in sputum bacterial load, consistent with a causal link (Okusanya et al., 2009). In subsequent phase 2 dose escalation studies of 105 patients treated once daily, administration of the highest dose provided statistically significant improvements in pulmonary function and bacterial load and improved respiratory function (Clancy et al., 2010). Phase 3 trials are now underway, and Arikace^{TM} has also received Orphan Drug status from the FDA for treatment of nontuberculous mycobacteria.

16.3.3 Tuberculosis

Tuberculosis (TB) in humans is primarily a disease of the lungs, initiated by inhalation of infectious *Mycobacterium tuberculosis.* Following deposition in the alveolar space, the bacteria are taken up by alveolar macrophages into intracellular phagolysosomes. Some bacteria are also taken up by alveolar dendritic cells, which traffic to draining lymph nodes and prime naive T cells. The T cells traffic to the lung and contain the infection by the formation of granulomas, where the infected macrophages are contained within a core of foamy macrophages surrounded by T cells and a fibrous extracellular matrix laid down by fibroblasts. Patients can remain in this state of latency, in which the bacteria are nontransmissible for long periods, until external factors such as malnutrition, immunosuppression, or human immunodeficiency virus (HIV) infection stimulate bacterial activity, leading to active infection (Griffiths et al., 2010).

Many antitubercular drugs require daily dosing for extended periods, which can lead to patient noncompliance and subsequently to multidrug-resistant tuberculosis (MDR-TB) and extensively drug-resistant tuberculosis (XDR-TB). A major focus for the use of NPs for TB treatment has been to obtain extended release, reducing the dosing required and hence potentially increasing patient compliance (Buxton, 2009). In a recent example, Kumar et al. formulated the second line antitubercular ethionamide into PLGA NPs and compared plasma and organ distribution following oral administration of the free drug or NP formulation in mice (Kumar et al., 2011). Ethionamide-loaded PLGA NPs produced sustained plasma release of ethionamide for six days versus six hours for free ethionamide. Ethionamide was detected in organs (lung, liver, and spleen) for up to 5–7 days with encapsulated ethionamide, while free ethionamide was cleared within 12 hours. Ethionamide-loaded pharmacokinetic parameters were significantly improved by encapsulation of the drug into PLGA NPs (Kumar et al., 2011). Similar extensions in bioavailability were demonstrated for PLGA nano-formulation of the fluoroquinalone antibiotic levofloxacin for oral administration (Kumar et al., 2012a).

Encapsulation into NPs also shows promise for intravenous delivery of isoniazid. Saraogi et al. prepared mannosylated gelatin NPs, and demonstrated that retention in plasma was extended relative to administration of free drug. Isoniazid accumulation in the lung was higher for the NP formulation relative to free drug, probably reflecting uptake by alveolar macrophages (Saraogi et al., 2011). Antitubercular activity was higher for encapsulated isoniazid than for free drug, with a greater-than-twofold improvement in clearance of *M. tuberculosis* from lungs and spleens of infected mice. Hepatotoxicity was lower for the NP formulation despite higher liver uptake, possibly reflecting sequestration of drug in the Kupffer cells (Saraogi et al., 2011).

Clemens et al. used an innovative functionalized mesoporous silica NP design to obtain targeted intracellular delivery of antitubercular drugs (Clemens et al., 2012). Mesoporous silica NPs were efficiently taken up by human macrophages and deposited to lysosomes. Addition of a positively charged polyethyleneimine coating enhanced loading of increased efficacy against *M. tuberculosis*–infected macrophages. Mesoporous silica NPs were also equipped with cyclodextrin-based pH-sensitive valves that open at acidic pH for active release of internally-loaded isoniazid, a hydrophilic drug. Treatment of *M. tuberculosis*–infected macrophages with pH-gated mesoporous silica NPs resulted in a 1.5 log greater efficacy than treatment with the equivalent amount of free isoniazid (Clemens et al., 2012). Since eradication of TB requires long-term treatment, the translation of this platform will require demonstration of biocompatibility in vivo over an extended course of treatment.

Another novel approach to *M. tuberculosis* treatment employs Janus-faced liposomes to stimulate innate immune responses (Greco et al., 2012). The asymmetric liposomes have phosphatidylserine distributed to the outer surface and phosphatidic acid at the inner surface to resemble apoptotic bodies. The liposomes were demonstrated to promote phagolysosome killing and to promote mycobacterial killing through a reactive oxygen species (ROS)-

and phagolysosome-mediated mechanism (Greco et al., 2012). Stimulation of bronchoalveolar lavage cells from patients with active pulmonary TB or *Klebsiella pneumonia* with the asymmetric liposomes resulted in a strong eradication of intracellular bacteria (Fig. 16.12, left). In an experimental model of murine TB, with initiation of treatment two weeks after infection, intranasal treatment with the asymmetric liposomes three times per week for four weeks decreased lung bacteria 100-fold, while oral isoniazid decreased bacterial load only 2-fold. The liposomes also had more limited efficacy in the liver and spleen (Fig. 16.12, right). In parallel the liposomes reduced the inflammatory response to infection, decreasing inflammatory cytokines tenfold (Greco et al., 2012).

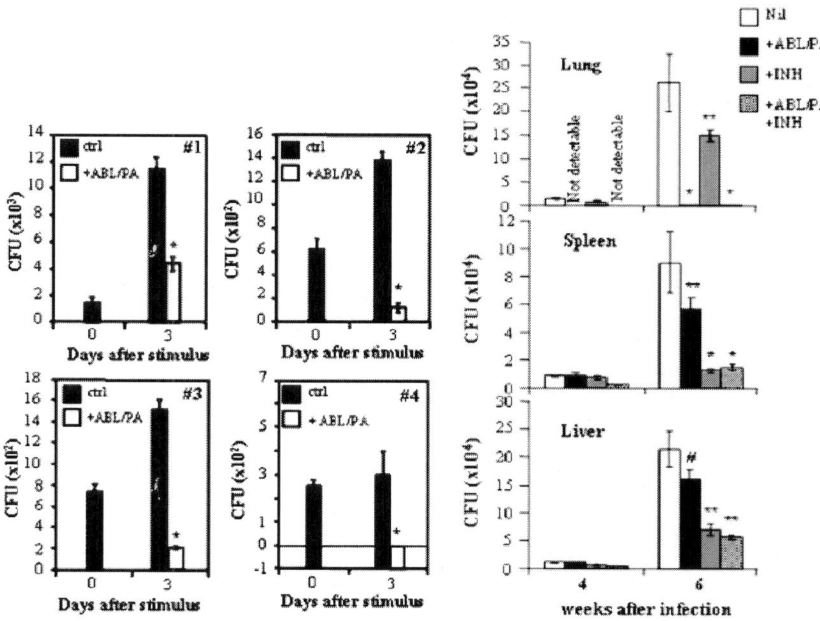

Figure 16.12 Reduction in bacterial load in vitro and in vivo. Left: Bacterial load in bronchioalveolar cells from patients with active pulmonary tuberculosis (patients 1–3) or *K. pneumonia* (patient 4) before and after 72 hours' stimulation with asymmetric liposomes. Right: Mycobacterial load after intranasal treatment with asymmetric liposomes three times weekly starting two weeks after infection, oral isoniazid, or a combination. Reprinted from Greco et al. (2012) with permission from the National Academy of Sciences.

16.3.4 Other Pulmonary Infections

Itraconazole is a widely used antifungal agent that is generally administered orally. However, bioabsorption of the drug is variable, resulting in a failure to reach therapeutic doses in many patients, and some patients are unable to swallow oral capsules. Intravenous administration of itraconazole has the advantage of providing rapid elevation of plasma itraconazole to high levels. However, due to poor solubility, itraconazole for intravenous administration is formulated commercially using hydroxypropyl-β-cyclodextrin, which causes nephrotoxicity in patients with renal impairment (Chen et al., 2008b). Nanoformulation offers the potential for solubilization of itraconazole without the use of hydroxypropyl-β-cyclodextrin, and one approach has been to bind the drug to albumin. Chen et al. demonstrated that itraconazole bound to albumin NPs caused less hemolysis than cyclodextrin-formulated drug and following intravenous administration in mice resulted in higher drug concentrations in the lung, liver, and spleen. Itraconazole has also been encapsulated into liposomes coated with carboxymethyl chitosan (CMC) (Wang and Huang, 2011). The CMC–liposome formulation of itraconazole demonstrated a longer plasma half-life relative to “naked” liposomal and cyclodextran formulations and also demonstrated improved lung uptake and decreased uptake by heart and kidney (Wang and Huang, 2011).

Voriconazole, a second-generation triazole drug, has been formulated into PLGA NPs for inhalational delivery using a multiple-emulsification technique (Sinha et al., 2013). NPs were prepared in the presence or absence of effervescence to give porous or solid particles, respectively. High loading efficiency (30%) and encapsulation efficiency (62%) could be achieved in both cases, with extended in vitro release up to 15 days. Porous NPs had smaller aerodynamic diameters and increased initial deposition of drug following delivery by inhalation chamber. Drug persistence in the murine lung was also longer for the porous NPs, seven days versus five days for solid particles (Sinha et al., 2013).

Silver carbene NPs have also been pursued as pulmonary antimicrobials for inhalation therapy. Hindi et al. loaded a novel silver *N*-heterocyclic carbene complex, SCC10, which possesses potent antimicrobial properties, into L-tyrosine polyphosphate

NPs (Hindi et al., 2009). Following intranasal administration of *P. aeruginosa,* delivery of nebulized SCC10 NPs at 1 hour and 24 hours provided a significant survival advantage at 72 hours relative to nebulized administration of a tenfold larger dose of free SCC10. The NP formulation also reduced lung bacterial burden relative to the free drug and decreased bacteremia, as demonstrated by dissemination of bacteria to the spleen (Hindi et al., 2009). In vitro experiments have demonstrated that silver carbene NPs formulated with L-tyrosine polyphosphate (Leid et al., 2012) or into degradable dextran NPs (Ornelas-Megiatto et al., 2012) had broad antimicrobial activities and were effective against bacterial strains, including those isolated from patients.

16.4 Hematologic Diseases

Diseases of the blood offer a diverse range of targets for novel nanobased therapeutics. At one end of the spectrum, sepsis represents an acute blood-based disease causing significant morbidity and mortality in the general population. At the other end, inherited coagulation disorders such as hemophilia affect a much smaller population but require lifelong management.

16.4.1 Thrombosis and Thrombolysis

The body must be able to respond rapidly to injury by clotting to prevent major blood loss, while avoiding the potentially lethal consequences of unwanted thrombosis. Nanotechnology offers a number of opportunities to influence this balance in controlled and/or localized ways.

16.4.1.1 Thrombolysis

Inappropriate thrombus formation can have devastating consequences, leading to MI, stroke, deep vein thrombosis, or pulmonary embolism. Thrombolytic therapy by systemic infusion of agents such as tissue plasminogen activator (tPA) can restore blood flow to occluded vessels, but is inefficient and also carries a substantial risk of bleeding complications (Lippi et al., 2013). This has led to a search for more targeted methods to deliver the therapeutic agent more specifically to the thrombus to increase thrombolytic efficiency and reduce systemic complications.

A number of laboratories have explored the potential use of magnetic targeting of NPs to improve the efficacy of thrombolysis. Bi et al. (2009) used urokinase conjugated to dextran-coated iron oxide NPs in a rat arteriovenous shunt thrombosis model. They found that targeting the NPs to the thrombus with a magnetic field was 5-fold more effective than urokinase alone and 2.6-fold more effective than the urokinase NPs without a magnetic field. Use of a magnetic field also prevented increases in tail bleeding time observed with free urokinase or NP-conjugate without a magnetic field (Bi et al., 2009). Ma et al. tested tPA conjugated to polyacrylate-coated iron oxide NPs in a rat iliac artery embolism model. Using an rtPA dose < 20% of that normally used clinically, they found that vessel blood flow and limb perfusion could be restored to ~80% of prethrombus levels within 75 min by moving the magnet back and forth along the iliac artery (Fig. 16.13); a stationary magnet was ineffective (Fig. 16.13, inset).

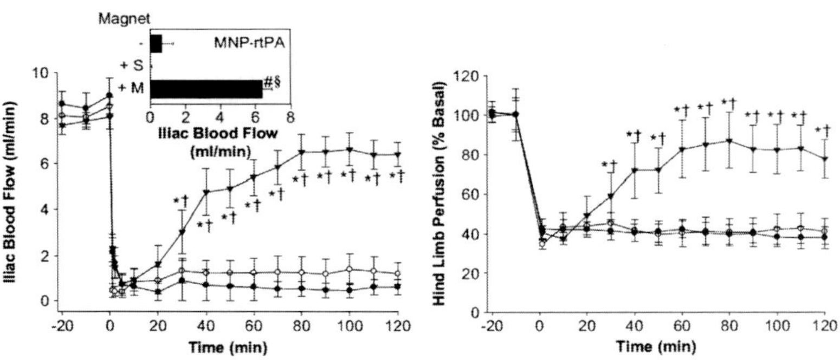

Figure 16.13 Thrombolytic effects of rtPA conjugated to magnetic nanoparticles in a rat embolic model. Magnetic targeting was achieved by a cyclic moving of external NdFeB magnet along the iliac artery. In some experiments, PAA–MNP–rtPA was administered with a moving (M) or stationary (S; $n = 3$) magnet and the iliac blood flow at 120 min was compared (left, inset). Reprinted from Ma et al. (2009), Copyright 2009, with permission from Elsevier.

The use of magnetic NPs with external magnetic fields in humans would probably be limited to superficial structures due to the difficulty in targeting magnetic fields to deeper structures. However, there may be a role for magnetic NPs as antithrombotic agents for intravascular stents, since the stent can act as a magnetizable implant to increase the magnetic force acting on the magnetic NPs. In proof-of-concept studies in pigs, Kempe et al. showed that administration of tPA-conjugated iron oxide NPs could restore patency to a magnetizable stent subjected to an external magnetic field at a much lower tPA dose than that used clinically (Kempe et al., 2010).

An alternate approach to targeted thrombolysis is to use biological targeting of multifunctional NPs to the thrombus. Marsh et al. developed NPs with up to 100 urokinase molecules and 40 antifibrin IgG-targeting moieties and found them to be as effective as a free enzyme in a lysing thrombus in vitro. They then tested them in a canine model of an electrically induced thrombus and found that the fibrin-targeted NPs, but not nontargeted particles, heavily decorated fibrin in the intra-arterial thrombus (Marsh et al., 2011). McCarthy et al. used a similar strategy employing an affinity peptide targeted to activated factor XII. Blood transglutaminase FXIIIa crosslinks fibrin strands to stabilize the nascent thrombus and also crosslinks a2antiplasmin into the clot, effectively increasing fibrinolytic resistance. The targeting peptide is based on the N-terminus of a2antiplasmin and so would be predicted to be specific for acute clots that require fibrinolysis rather than established thrombi that are protective (McCarthy et al., 2012). The peptide-targeted NPs, which carry tPA as a fibrinolytic agent, bind readily to thrombi in vitro and in vivo and had comparable thrombolytic activity to free tPA against murine pulmonary emboli. Determination of the safety profile of the nanoagent relative to free tPA will be the next step in order to address whether the nanoagent can reduce the rate of adverse events, including intracerebral hemorrhage (McCarthy et al., 2012).

A third approach to localized delivery of thrombolytic agent is to shield the thrombolytic activity while circulating and use either an external stimulus or a physiological activation mechanism to activate thrombolysis locally. Uesugi et al. employed PEG-coated gelatin tPA complexes to suppress tPA activity to 45% of the free enzyme; activity could be restored to normal levels in vitro by reactivation with ultrasound (Uesugi et al., 2010). Encapsulation of tPA in the gelatin complexes prolonged blood circulation threefold relative to free tPA. Administration of tPA–gelatin complexes followed by local ultrasound resulted in rapid and complete recanalization in a rabbit thrombosis model, faster and more complete than free tPA followed by ultrasound (Uesugi et al., 2010).

Korin et al. used an innovative biomimetic strategy to take advantage of the increased pathological shear stress caused by obstruction of blood vessels (Korin et al., 2012). They loaded tPA-coated PLGA NPs into microscale aggregates that break up when exposed to pathological shear stress, releasing the tPA NPs. Injection of the stress-activated nanotherapeutics (SA NTs) in a mouse arterial thrombus model resulted in accumulation of the NPs at the clot and rapid erosion of the thrombus, resulting in complete clearance of the occlusions (Fig. 16.14). In contrast, free tPA, predissociated tPA NPs, or fused aggregates that cannot disaggregate were ineffective (Fig. 16.14d). In a mouse pulmonary embolism model, six of seven animals treated with tPA NPs were still alive after one hour, while all control animals died. In an ex vivo pulmonary embolism model, tPA NPs were more effective than a tenfold higher dose of free tPA in reducing pulmonary artery pressure. This biomimetic targeting strategy thus looks very promising for maximizing thrombolytic efficacy, while minimizing side effects.

16.4.1.2 Hemostasis

Hemorrhage is a major cause of death in trauma victims and is particularly problematic in remote and rural areas and in combat casualties where access to platelet and plasma products may be limiting. The development of NP-based synthetic peptides with a long shelf life and reduced storage needs would have enormous potential, and several groups have pursued this goal. Okamura et al. developed liposomes decorated with the dodecapeptide HHLGGAKQAGDV (H12), corresponding to the fibrinogen γ-chain carboxy-terminal sequence. This represents a specific binding site of the ligand for the GPIIb/IIIa complex on activated platelets (Okamura et al., 2009). The liposomes contain the platelet agonist adenosine diphosphate (ADP) and release the ADP in an aggregation-dependent manner to augment agonist-induced platelet aggregation (Okamura et al., 2009). Pretreatment with the liposomes resulted in aggregation at sites of vascular injury in jugular veins treated with ferric

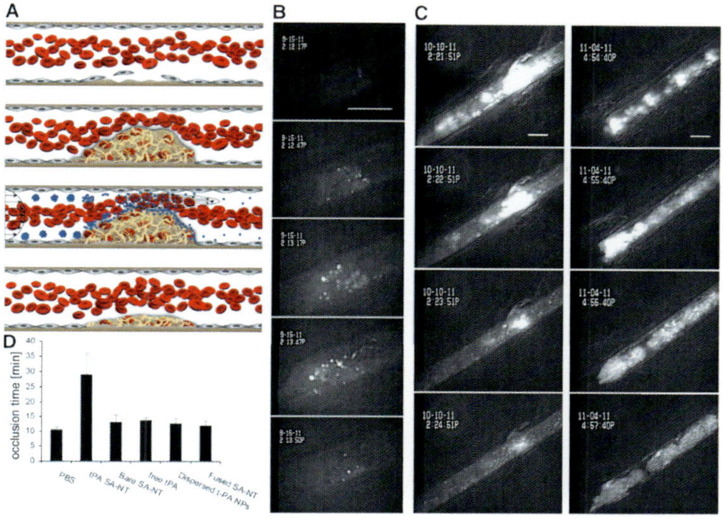

Figure 16.14 Shear targeting of a thrombolytic drug in a mouse arterial thrombosis model using SA-NTs. (A) Schematic of the experimental strategy. The ferric chloride–initiated thrombus (2nd panel) constricts the artery, increasing shear stress. Elevated shear stress causes the blue microscale aggregates to break down locally into nanoparticles (blue dots), which bind to the clot surface (3rd panel), breaking down the clot. (B) Sequential intravital fluorescence microscopic images of a thrombus in a partially occluded mesenteric artery recorded over a five-minute period beginning after bolus injection of fluorescent tPA-coated SA. (C) Intravital fluorescence microscopic images recorded over a five-minute period showing fluorescently labeled platelets accumulated within a forming thrombus. Thrombosis is then treated with injection of either tPA-carrying SA-NTs (50 ng tPA) (left) or PBS (right). (D) Bolus injection of SA-NTs carrying 50 ng tPA (tPA-SA-NT) significantly delayed the time to full vascular occlusion in FeCl-injured vessels. Reprinted from Korin et al. (2012) with permission from AAAS.

chloride (Okamura et al., 2010), and reduced bleeding in response to traumatic liver injury in thrombocytopenic rats (Nishikawa et al., 2012). Bertram et al. used Arg-Gly-Asp (RGD) as a targeting peptide for synthetic NPs; RGD interacts with the activated platelet receptors glycoprotein IIb-IIIa and integrin $\alpha_v\beta_3$. The synthetic platelets consisted of PLGA-PLL block copolymer cores to which conjugated PEG arms were conjugated and terminated with RGD functionalities, ~600 per particle (Bertram et al., 2009). In this case the efficacy results from multiple RGDs binding to multiple platelets, facilitating their aggregation. In rats with a femoral artery wound, pretreatment with synthetic platelets reduced bleeding times by ~50%, more effective than treatment with recombinant factor VIIa, the standard of care treatment (Fig. 16.15). Lyophilization and storage at room temperature for two weeks had no effect on the efficacy of the synthetic platelets. Treatment with the synthetic NPs postinjury also reduced bleeding time.

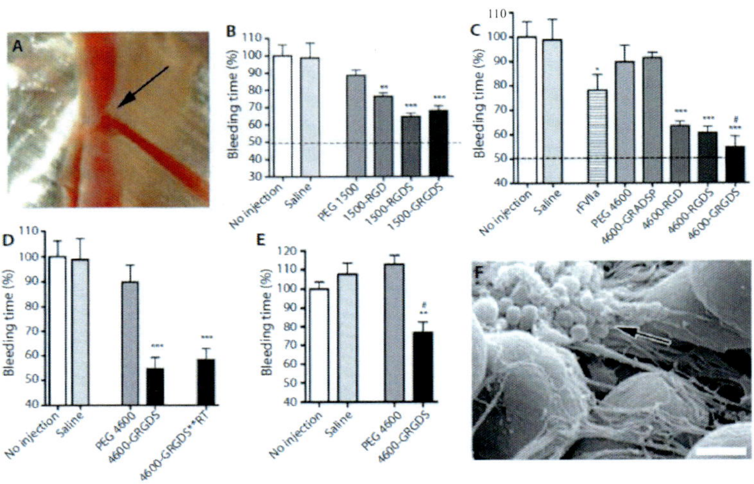

Figure 16.15 In vivo hemostatic efficacy of synthetic platelets. (A) Femoral artery injury model. Arrow: injury. (B) Bleeding times following intravenous administration of PEG 1500 synthetic platelets. (C) Bleeding times after intravenous administration of PEG 4600 synthetic platelets and rFVIIa. (D) Bleeding times comparing synthetic platelet administration, PEG-4600-GRGDS, to synthetic platelets stored in a lyophilized state at room temperature for two weeks, 4600-GRGDS**RT. (E) Bleeding times when treatments were administered after injury. Bleeding times represented as a percentage of no injection bleeding time values (255 ± 12 seconds). (F) SEM micrograph of clot excised from injured artery after synthetic platelet administration (4600-GRGDS). Synthetic platelets intimately associated with clot and connecting fibrin mesh (arrow). Reprinted from Bertram et al. (2009) with permission from AAAS.

More recently this group has shown that the RGD-functionalized synthetic NPs increase one-hour survival in a lethal rat liver resection model from 47% for saline-treated animals to 80% for RGD-NP treated animals (Shoffstall et al., 2012). Inactive NPs carrying a scrambled peptide had no effect on survival (40%).

The reversal of heparin anticoagulation represents another hemostasis-related area where NPs could play a role. In settings such as cardiopulmonary bypass, patients are treated with high doses of heparin to prevent thrombosis in the pump, and following disconnection from the pump the heparin antagonist protamine sulfate is given. However, protamine can cause serious adverse events, including severe anaphylactic reactions and arterial hypertension. Gale et al. have engineered virus-like particles based on the bacteriophage Qβ platform, displaying polyvalent cationic motifs that bind with high affinity to heparin (Gale et al., 2011). The virus-like particles consistently reversed heparin in plasma samples from patients treated with heparins, including patients who received high doses of heparin, with significantly less variability than protamine sulfate.

16.4.1.3 Inherited coagulation disorders

The availability of recombinant protein coagulation factors has had a significant impact on the treatment of patients with inherited coagulation disorders in the last two decades. The availability of longer-acting formulations of recombinant factors would benefit patients by reducing the frequency of injections. One approach to improving the plasma half-life has been to formulate the factors with liposomes. Recombinant activated factor VIIa (rFVIIa), used to treat bleeding episodes in hemophilia patients with inhibitors to factor VIII (FVIII) and factor IX, has a plasma half-life of 2.5 h, requiring multiple injections to treat bleeding episodes (Spira et al., 2010). Formulation of the protein with PEGylated liposomes significantly improved hemostatic activity in a phase I/II clinical trial, shortening clotting times and improving clot firmness. The authors speculate that the mechanism my involve binding of activated platelet to the liposomes; rFVIIa binds noncovalently to the PEGylated liposome surface, and so the proximity of rFVIIa to factor X may lead to faster initiation of the clotting process and greater initiation of the clotting cascade (Spira et al., 2010).

Recombinant FVIII has also been packaged as a PEGylated liposomal formulation (BAY 79-4980), with the goal of extending its effective half-life for use in prophylaxis to permit once weekly injection rather than the thrice-weekly injection needed for rFVIII. While preliminary dose-finding studies suggested efficacy in extending the time of protection from spontaneous bleeding, a larger phase II study was stopped because of unacceptably high rates of bleeding in the BAY 79-4980-treated patients (Powell et al., 2012). Once-weekly BAY 79-4980 resulted in a mean of 15 annualized bleeds, compared to >50 bleeds annually in patients without prophylactic treatment, and so the liposomal formulation could play a role in low-resource settings where thrice-weekly injections are not feasible (Powell et al., 2012).

16.4.2 Sepsis

Sepsis, an overwhelming infection of the bloodstream by toxin-producing bacteria, can lead rapidly to multiorgan failure and death. Sepsis is the leading cause of death in emergency rooms and ranks 11th on the list of cause of death overall in the U.S. and 7th for infant mortality (Murphy et al., 2012). Despite the use of antibiotics, fluid resuscitation and vasoactive support, mortality remains high, 28%–50%, which has led to searches for adjunct treatments such as extracorporeal removal of inflammatory mediators, but the outcomes have been disappointing (Lee et al., 2013b). Antibody-based methods for the removal of bacteria from blood have been developed, but the slow kinetics of antibody association with bacteria and the expense of antibody production are likely to limit clinical application. To address these issues, Lee et al. have developed a magnetic NP-based microfluidic technology using a synthetic ligand, zinc-coordinated bis(dipicolylamine) (*bis*-Zn-DPA). *Bis*-Zn-DPA forms coordination bonds with anionic phospholipids present at high density on the outer membrane of gram-positive and gram-negative bacterial cells but displays negligible binding to normal mammalian cells or negatively charged proteins such as albumin (White et al., 2012;

Lee et al., 2013b). Almost 100% of the gram-negative bacterium *E. coli* could be removed from whole blood by sequential microfluidic separations at flow rates as high as 60 mL/min (Fig. 16.16). *Bis*-Zn-DPA also binds specifically to endotoxin, a major contributor to septic shock, and *bis*-Zn-DPA magnetic NPs could also remove >90% of endotoxin in test-tube experiments (Lee et al., 2013b). Adaptation of the microfluidic device to extracorporeal use in a clinical setting could thus potentially remove both bacteria and endotoxin from the patient circulation simultaneously.

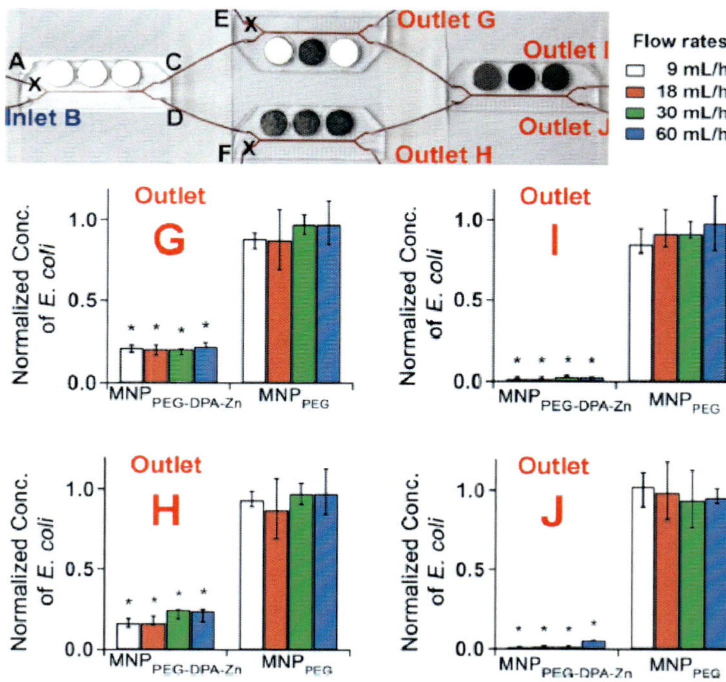

Figure 16.16 Effect of sequential and parallel microfluidic separations on normalized concentration of *E. coli* using MNPPEG-DPA-Zn in whole blood. Channels A, E, and F were closed. Blood was infused through inlet B and collected from outlets G, H, I, and J. *E. coli* concentrations were measured at various flow rates. Data are medians with 25th and 75th percentiles of the concentrations of *E. coli* at the outlets normalized to the initial *E. coli* concentration (5.0×10^6 CFU/mL). Reprinted from Lee et al. (2013b) with permission from the American Chemical Society.

A second approach focuses on pore-forming toxins such as staphylococcal α-hemolysin. The nanosponges consist of natural red blood cell (RBC) bilayer membranes fused on to PLGA NPs by extrusion (Hu et al., 2013a). The nanosponges absorb membrane-damaging nanotoxins, diverting them away from their cellular targets. Intravenous injection of nanosponges prior to injection of a lethal dose of staphylococcal α-hemolysin in mice resulted in 90% survival at 15 d, while mice treated with PEG-PLGA NPs or RBC vesicles all died within 6 h. Treatment of mice with nanosponges after α-hemolysin injection still provided a significant survival advantage, with more than 40% of mice surviving versus none in the control groups (Hu et al., 2013a).

Immune suppression is also a significant problem in sepsis, resulting in part from loss of immune effector cells, primarily dendritic cells and lymphocytes, through apoptosis. This leads to significant mortality late after sepsis due to inability to clear the primary infection or onset of secondary infections (Brahmamdam et al., 2009). Brahmamdan et al. used a novel cyclodextrin–polymer nanocomplex targeted to the transferrin receptor to deliver siRNA to lymphocytes, a cell type that has proven resistant to siRNA uptake. By down-regulating two pro-apoptotic cell death proteins, Bim and PUMA, they were able to decrease intracellular Bim and PUMA protein in lymphocytes following in vivo delivery. This resulted in decreased lymphocyte apoptosis and prevented the loss of splenic CD4 T and B cells (Brahmamdam et al., 2009).

16.4.3 Hematologic Malignancies

The treatment of hematologic malignancies such as leukemia, lymphoma, and multiple myeloma has been hampered by the lack of an active targeting strategy to direct therapy to the primary hematopoietic organ, the bone marrow (Mann et al., 2011). Mann et al. have developed E-selectin-targeted porous silica particles targeted E-selectin using a high-affinity thioaptamer ligand to deliver NP therapeutics to bone marrow. Injection of E-selectin-targeted particles carrying a payload of liposomes encapsulating paclitaxel resulted in approximately 20% of the injected dose reaching the bone marrow. In contrast, bone marrow uptake of liposomes alone or liposomes carried on nontargeted particles was negligible (Mann et al., 2011).

MicroRNA-155 (miR-155) has been implicated as a significant oncogenic factor in hematologic malignancies, making an attractive potential target for the treatment of lymphoma and leukemia (Babar et al., 2012). Babar et al. used PLGA NPs to deliver peptide nucleic acid anti-miRs to miR-155-dependent pre-B-cell tumors in a mouse model. Systemic delivery of the anti-miR-155 reduced miR-155 expression and slowed tumor growth, supporting the potential as a therapeutic option for leukemia and lymphoma (Babar et al., 2012).

The use of antisense oligonucleotides for the treatment of hematologic malignancies has met with limited success, due in part to immune-stimulating effects that can limit the tolerated dose (Yu et al., 2013a). Bcl-2 is a protein with antiapoptotic function that is dramatically overexpressed in chronic lymphocytic leukemia (CLL), contributing to drug resistance. An antisense oligodeoxynucleotide to Bcl-2, G3139, showed promise in preclinical CLL studies, but when tested in a phase I trial only a suboptimal dose could be tolerated due to cytokine release syndrome and immune activation (Yu et al., 2013a). To increase the effective dose achieved at the malignant cells, Yu et al. employed anti-CD20 antibody (rituximab) conjugation to target lipopolyplex NPs as a carrier for G3139. The selective B cell targeting afforded by rituxamib led to early endosomal compartmentalization of the NPs and resulted in robust Bcl-2 down-regulation and enhanced sensitivity to fludarabine-induced cytotoxicity (Yu et al., 2013a). In a xenograft CLL mouse model, survival following treatment with rituzamib-targeted G3130 NPs was significantly greater than for free G3139 or control G3139 targeted with Herceptin. Cytokine production was also lower for the CD20 targeted NPs, both in mice and in CLL patient blood, consistent with abrogation of the immunostimulatory effects of G3139 by selective targeting (Yu et al., 2013a).

16.5 Conclusions

There is an urgent need for new therapeutics for the treatment of heart, lung, and blood diseases, and in recent years there has been rapid growth in the development of nanotechnology-based approaches. As described above, this growth has included a number of innovative approaches to improved therapeutic efficacy. Biopanning approaches have identified improved targeting sequences to improve local delivery of therapeutics (Chan et al., 2010). Innovative strategies such as shear-activated release offer the opportunity to release nanotherapeutics specifically at the disease site (Korin et al., 2012). Surface modifications can enhance penetration of drug-bearing NPs into environments such as brain tissue and CF sputum (Suk et al., 2011; Nance et al., 2012). Targeted intracellular release using pH-gated mesoporous silica NPs also appears to be a promising approach for improving efficacy of drug delivery (Clemens et al., 2012). Janus-faced liposomes with an apoptotic-body-like exterior are efficiently taken up by macrophages, while phosphatidic acid on the interior stimulates the innate antimycobacterial response (Greco et al., 2012).

On the tissue-engineering front, addition of carbon nanotubes or nanowires to cardiac scaffolds improves electrical coupling between cells (Dvir et al., 2011; Shin et al., 2013), while 3D gradient nanoscaffolds help to optimize conditions for the growth of small-diameter vascular grafts with appropriate transmural orientation of endothelial cells and smooth muscle cells (Du et al., 2012). Nanosponges consisting of a toxin-absorbing nanocore surrounded by a natural red blood cell–like membrane show promise for the removal of toxins from the bloodstream in sepsis (Hu et al., 2013a). These types of innovative approaches to nanotherapy offer enormous promise for reducing the heavy burden of heart, lung, and blood diseases in coming years.

Chapter 17

Immunotherapy and Vaccines

Johanna Poecheim and Gerrit Borchard
School of Pharmaceutical Sciences, University of Geneva, University of Lausanne, 30, quai Ernest-Ansermet, 1211 Geneva, Switzerland
Gerrit.Borchard@unige.ch

17.1 Introduction

The human immune system has evolved to recognize antigenic properties of pathogens and initiate an immune response resulting in the restriction of the pathogen from entering the body or in the clearance of the pathogen after infection. Among the signals leading to an activation of the immune system are the pathogen's surface properties, its size, and its shape. It is therefore not surprising that the general paradigm of vaccine development demands the preparation of vaccines combining both antigen and adjuvant in the same—particulate—formulation. Vaccine development today, though, is fraught with the failure to achieve the required sufficient level of protection, possibly due to the lack of appropriate adjuvants—especially for mucosal vaccines—and delivery systems (Doroud and

Controlled Release Systems: Advances in Nanobottles and Active Nanoparticles
Edited by Alexander van Herk, Jacqueline Forcada, and Giorgia Pastorin

ISBN 978-981-4613-21-7 (Hardcover), 978-981-4613-22-4 (eBook)
www.panstanford.com

Rafati, 2011). However, and unfortunately, there is a bias between current knowledge on specific immune activation and its translation into real products. Whereas groundbreaking discoveries in the field of pathogenic pattern recognition by dedicated receptor families have been made over the last decades, only few ligands specifically interacting with these receptors have been introduced into clinical studies. One reason for this pause in vaccine development may have been the reluctance of manufacturers to introduce these novel adjuvants into the pipeline; another lies in the fact that they need to be presented to the cells of the immune system in an appropriate way to exploit their full potential. It may even be necessary to include combinations of novel adjuvants in one vaccine formulation to achieve synergistic effects.

Nanotechnology may offer the possibility to design novel, more effective vaccines. The nanoparticulate carrier systems should be of appropriate size, have favorable surface parameters for immune recognition, be targeted to antigen-presenting cells (APCs), and be suitable for the inclusion of the respective antigen. In addition, such novel vaccines may prove to have enhanced shelf-life stability, rendering refrigeration unnecessary, and to be applicable by mucosal pathways, allowing for the avoidance of needles for injection and the health risks related.

This chapter discusses the current influence of nanotechnology on the development of vaccines.

17.2 The Immune System

Vaccines are meant to elicit an immune response and to create long-lasting immune memory against a pathogen-specific antigen by staging an "artificial" infection. To rationally design novel vaccines it is essential to understand the function of the immune system itself. The vertebrate immune system consists of the innate and adaptive branches that cooperate to protect against infection and subsistence of pathogenic agents in the host. How both arms of the immune response are orchestrated is described in the following sections and illustrated in Fig. 17.1.

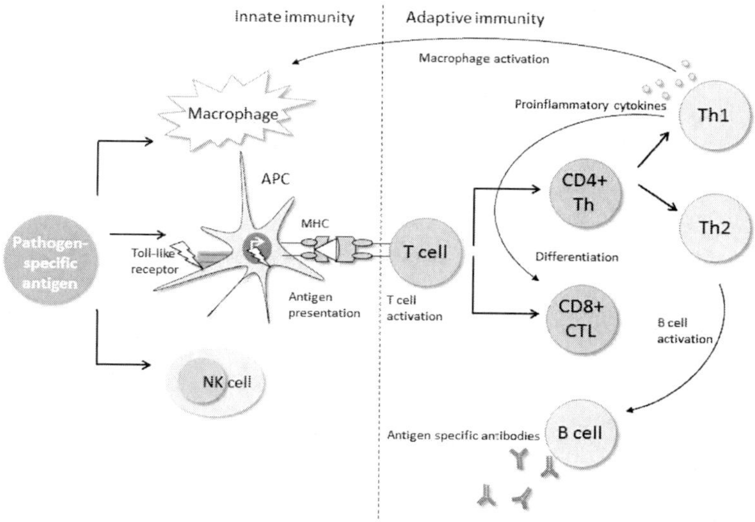

Figure 17.1 The role of the innate and adaptive immune responses following vaccination with a pathogen-specific antigen. Antigen-presenting cells (APCs), which activate naive T cells through antigen presentation, activate the adaptive immune system and induce T cell differentiation. Activated CD4+ T helper cells of type 1 activate macrophages and cytotoxic T lymphocytes (CTLs) through proinflammatory cytokine release. Type 2 helper T cells activate B cells to release specific antibodies against the antigen. Additionally to those antigen-specific B and T cells spread as effector cells, long-lived clones of memory cells are produced that form the immunological memory.

17.2.1 The Innate Immune System

When pathogens overcome the barriers formed by the skin and the mucosae and gain access to the body's soft tissues, the innate immune system detects the invading agent and is activated as a fast and early-stage immune response. Innate immunity provides immune surveillance and immediate defenses that are always available and do not improve with repeated exposure to the same antigen. The immune system has to identify danger signals related to the pathogen by detecting antigenic properties or patterns of a pathogen in the absence of prior immune recognition (first infection). Such danger signals are, for example, the recognition of pathogen-associated molecular patterns (PAMPs) by pattern recognition receptors (PRRs), missing components, or "missing self" signals distinguishing between bacterial cells and healthy human cells, tissue damage caused by pathogens, and alarm signals such as the release of certain cytokines. Upon recognition, complement activation is triggered, consisting in a system of plasma proteins and cell surface molecules that mark pathogens for destruction. These interactions assist resident macrophages to phagocytose the microbiological invaders, as well as to induce macrophages to secrete inflammatory cytokines that in turn attract neutrophils and natural killer (NK) cells to the site of infection (Parham, 2000) (Fig. 17.1).

17.2.2 The Adaptive Immune System

Under normal conditions, the adaptive immune system is silent and is only called into action directly when pathogens evade or overcome the innate immune defense. Thus, the adaptive immune response is initiated only if the innate immune system is activated and signals the presence of a pathogen. It "adapts" to the presence of a pathogen by activating, proliferating, and creating potent mechanisms to neutralize and/or eliminate pathogens.

Two types of adaptive immune responses are to be distinguished. Humoral immunity is mediated by antibodies, also known as immunoglobulins (Igs), which are produced by B lymphocytes. The most abundant isotypes are IgG, found in blood serum and lymph, and IgA, as the main effector of the mucosal immune system. Ratios of IgG2/IgG1 subclasses >1 are associated with type 1 helper (Th1) responses, whereas any ratio < 1 corresponds to type 2 helper (Th2)-biased immune responses. Simultaneously cell-mediated immunity is induced, mediated by T lymphocytes. The main difference between these lymphocyte subtypes is that Ig receptors of B cells bind whole molecules and intact pathogens, whereas T cells recognize only short peptide antigens (Kumar et al., 2012b), the so-called T cell epitopes. At the site of infection APCs internalize, process, and present T cell epitopes by the major histocompatibility complex (MHC or CD1) molecules expressed at the cell surface. These cells subsequently migrate to the draining lymph nodes (LNs), a collection point where

APCs interact with naive T cells. The T cell receptor (TCR) recognizes peptide antigens presented at the surface of these immune cells and, thus, activates the T cell (Von Andrian and Mackay, 2000). Antigen structures from intracellular infections are presented by MHC class I molecules and antigens from extracellular pathogens by MHC class II molecules. Depending on these two classes, T cells differentiate into one of two types of effector T cells: T helper cells ($CD4^+$ cells) to fight off extracellular pathogens and cytotoxic T cells ($CD8^+$ cells) to eliminate infected cells. The CD4+ T helper (Th) lymphocytes are commonly divided into Th1 and Th2 subtypes. Th1 cells release pro-inflammatory cytokines to activate macrophages and cytotoxic T lymphocytes (CTLs) that are a part of the cellular immune system. Th2 cells help B cells to maturate to antibody-producing plasma cells, thereby supporting humoral immunity (Fig. 17.1).

17.2.3 Immunological Memory

In the course of the adaptive immune response to infection, clones of pathogen-specific B and T cells spread as effector cells. Additionally, long-lived clones of memory T cells that form the immunological memory are produced. Subsequent immune responses to the same pathogen will be faster and stronger, since memory cells are more quickly activated than naive cells. Unlike naive T cells, memory T cells can patrol nonlymphoid tissues, such as mucosae, and detect infection at an earlier stage. The greater power of a secondary immune response supports the generation of vaccine-mediated protection. By applying antigenic structures of the pathogen to the body without inducing the disease, immunological memory is elicited. Additionally, Th1- and Th2-polarizing adjuvants may be introduced to direct the desired immune response (Parham, 2000; Plotkin et al., 2013).

17.3 Nanotechnology in Vaccines

17.3.1 Particle Characteristics Interacting with the Immune System

In the following section some of the key parameters of nanoparticles suitable for targeting to specific cells of the immune system will be discussed. Nanoparticles are of great scientific interest as they are effectively a bridge between bulk materials and atomic or molecular structures (Pignataro, 2010). To design an optimal vaccine carrier, physicochemical properties of nanoparticles such as size, surface charge, functional material(s) and their composition, hydrophilicity/hydrophobicity, and biodegradability have to be taken into consideration. Nanoparticles activate APCs, which phagocytose them and travel from the application site, that is, lung, skin, or nose, to the LNs where immune reactions are initiated.

17.3.1.1 Particle size

17.3.1.1.1 *Size-dependent uptake*

Particle size is a critical parameter that influences uptake by phagocytic immune cells via either direct penetration through the cell membrane (i.e., energy-independent mechanism) or by endocytosis (i.e., energy-dependent mechanism). Various types of endocytosis have been identified: virus-sized particles (20–200 nm) are taken up by clathrin-dependent endocytosis (particles < 150 nm), through caveolae-mediated endocytosis (particles within 50–80 nm), or via clathrin- and caveolae-independent pathways. Endocytosis of bacteria and larger-sized particles (>0.5 μm) occurs mainly via macropinocytosis and phagocytosis, effected only by macrophages and immature dendritic cells (DCs) (Xiang et al., 2006). It was verified in vitro that APCs are able to ingest particles up to 5 μm (Tabata and Ikada, 1988). Although nanoparticles' size seems to deeply influence the uptake mechanisms inside the cells, many reported data are still controversial. Some studies have shown that the most efficient uptake is achieved for particles in the nanosize range (Joshi et al., 2012; Yue et al., 2010), other studies have demonstrated a similar or even preferred internalization of microparticles of a size of up to 5 μm (Chua et al., 2011; Kobiasi et al., 2012).

It has been suggested that the immune system has evolved to react to particles on the scale of viruses (<0.5 μm) and bacteria (>0.5 μm) (Fifis et al., 2004). Uptake and immunostimulating mechanisms of nanoparticles and microparticles may be related to their similarity in size to these pathogens, and the immune system reacts to particles that fall within this size range (Chua et al., 2011).

In vivo, either the particles are phagocytosed by macrophages at the site of application and transported to the LNs or they diffuse to the LNs by interstitial flow where they become phagocytosed by LN-resident APCs. Studies have shown that only nanoparticles below 100 nm were taken up into the lymphatic system directly and activated LN-resident APCs more efficiently than nanoparticles of larger sizes (Reddy et al., 2006, 2007; Xie et al., 2009). This size dependency of absorption into the LNs is likely to be related to the process of particle transport through the interstitium. Large particles (>100 nm) will penetrate the interstitium less easily and will remain at the site of injection, targeting peripheral rather than LN-resident APCs (Oussoren and Storm, 2001). Manolova et al. reported on the impact of particle size on the specific targeting of distinct DC populations. Whereas small nanoparticles (20 nm) were found in LN-resident APCs, larger particles (500–2000 nm) were mainly associated with DC at the injection site. This size-dependent manner of transport to the LNs implicates a delayed appearance of nanoparticles > 500 nm in the LNs after subcutaneous injection and is consistent with the active transport of these particles by skin-resident DCs. Small particles of 20 nm were detected in the LNs within two hours after injection and taken up by LN-resident APCs, suggesting free drainage (Manolova et al., 2008). This explains the findings in other studies, where a more rapid appearance of chitosan nanoparticles in the LNs compared to chitosan microparticles after subcutaneous injection was observed (Chua et al., 2011, Kobiasi et al., 2012). From these findings it can be concluded that both nano- and microparticles are internalized by cells at comparable efficiency, but nanoparticles are transported to LNs much faster.

These results, however, do not necessarily implicate an increased immunogenicity for nanoparticles. Besides investigation of particle uptake into the LNs, their ability to elicit immune responses was examined. Both nano- and microparticles showed the ability to induce antigen-specific antibody responses. Polystyrene particles of 40 nm were found in more LN cells and induced higher levels of interferon gamma (IFN-γ) and antibody titers in mice than other smaller or larger-sized beads (Fifis et al., 2004). Particles of 1 μm elicited higher serum IgG levels than smaller particles (Gutierro et al., 2002). Some studies revealed no significant differences in IgG production between different-sized particles after parenteral administration (Chua et al., 2011; Gutierro et al., 2002; Nagamoto et al., 2004). However, it has to be mentioned that nanoparticles up to 1 μm applied intranasally elicited a significantly higher production of mucosal IgA when compared to microparticles (Nagamoto et al., 2004). Intranasal immunization with 200 nm nanoparticles enhanced CD4+ T cell responses in the lungs compared to 30 nm nanoparticles (Stano et al., 2012). Regarding the route of administration, following intranasal application, higher serum IgG2a/IgG1 ratios were found for 500 nm poly(lactic-co-glycolic) acid (PLGA) nanoparticles than for 200 nm particles, indicating a Th1 polarized response (Gutierro et al., 2002).

In general, these findings suggest that local immunization may be preferably induced by nanoparticles at a size range between 200 and 1000 nm. However, there is not always a clear size dependency with regard to particle uptake and immunogenicity. Extent and type of immune responses are presumably also associated with other physicochemical properties of the particles, the various materials used, the antigen type transferred, and the route of administration, as well as the vaccination regimen (Chua et al., 2011; Kobiasi et al., 2012).

17.3.1.1.2 *Nanoparticle size and Th response*

The concept of immune activation can be further extended by investigating the type of immune response induced. Particle size may influence Th cell differentiation, whereas Th1 cells are mainly developed following infections by intracellular bacteria and some viruses. Th1 cells produce cytokines that activate macrophages and are responsible for cell-mediated immunity and phagocyte-dependent protective responses. Th2 cells are predominant in responding to large extracellular pathogens and are responsible for antibody production and activation of eosinophils (Romagnani, 1999).

Seen the conflicting data available in the literature, it is difficult to accurately predict particle size ranges that will induce a Th1 or a mixed Th1/Th2 immune response outcome (Oyewumi et al., 2010). Some data are suggesting that nanoparticles promote cellular immune responses. For instance, codelivery of the hepatitis B viral protein HBcAg and monophosphoryl lipid A (MPLA) in the copolymer PLGA nanoparticles of around 300 nm promoted antigen-specific Th1 immune responses, including IFN-γ production (Chong et al., 2005). In a different study, mice vaccinated with 300 nm sized PLGA particles loaded with a model antigen (ovalbumin, OVA) generated the highest fraction of OVA-specific CTLs. They also induced more than a 50-fold increase in the IgG2a/IgG1 ratio compared to microparticles, suggesting polarization toward a Th1-type immune response (Joshi et al., 2012).

Unlike these observations, Henriksen-Lacey et al. described how DDA:TDB liposomes, regardless of their size, stimulated a characteristic Th1 immune response with production of IgG2 antibodies and IFN-γ (Henriksen-Lacey et al., 2011). Other reports showed that 200–600 nm particles favored induction of Th1 responses associated with higher levels of IFN-γ production, whereas microparticles of 2–8 μm in size promoted interleukin-4 (IL-4, a cytokine that induces differentiation of naive helper T cells (Th0 cells) to Th2 cells) secretion inducing humoral immunity (Kanchan and Panda, 2007).

It has also been reported that as the particle size increases from the nanometer to the micrometer range, antibody titers increase, as well. Gutierro et al. showed that 1 μm particles elicited higher total serum IgG levels than nanoparticles. IgG2a/IgG1 ratios typical for a Th1 response, on the other hand, were similar for all particle sizes (Gutierro et al., 2002). In addition, according to Katare et al., microparticles in the size range of 2–8 μm elicited the highest antibody titers, whereas decreasing the particles size resulted in lower-peak antibody titers (Katare et al., 2005).

By contrast, subcutaneous application of several particle types present in the environment led to different results. Diesel exhaust particles below a size of 500 nm increased the production of IgE and IgG1 antibodies as indicators for Th2-like responses, while IgG2a titers remained low. Carbon black particles, on the other hand, induced a mixed Th1/Th2 response, and the larger silica particles (1–5 μm) induced a Th1 bias (van Zijverden and Granum, 2000). In addition, data from Samuelsen et al. applying polystyrene particles of different diameters intratracheally, as well as from Mann et al. using orally applied bile salt stabilized vesicles (named bilosomes), showed that larger particles (≥1 μm) generated stronger Th1 responses than nanoparticles (Mann et al., 2009; Samuelsen et al., 2009).

Not only nanoparticle size as a parameter has been determined to initiate immunostimulatory reactions through mediating release of pro-inflammatory cytokines, but also the surface charge and the material used for nanoparticles production itself may modulate immune responses.

The studies described here are listed in Table 17.1 summarizing the effects of particle size on the resultant immune responses.

17.3.1.2 Charge

Charged nanoparticles have been shown to be more likely taken up by phagocytes than neutral particles of the same size. Polyethylene glycol (PEG)-modified neutral nanoparticles had the lowest percentage of uptake when compared to particles with cationic or anionic surface charges (Dobrovolskaia et al., 2008; Zahr et al., 2006). Nanoparticles rendered negatively charged, for example, by adsorption of mucus components after mucosal administration, again may interact to a much lesser extent with the negatively charged cell membrane surface. Consequently, it is more likely that positively charged particles are internalized by immune cells and induce mucosal immune responses, which was confirmed in a study showing a positive correlation between particle uptake and increasing zeta potential (Kwon et al., 2005). Usually particles displaying a cationic surface charge are thought to be more toxic, since they can penetrate into cells more easily.

However, some studies revealed a higher uptake of negatively charged particles compared to cationic ones, which was explained by nonspecific binding and clustering of the particles at cationic cell membrane domains and subsequent endocytosis (Verma and Stellacci, 2010). Another possibility in this case would be that these positively charged particles become neutralized by adsorption of negatively charged components of the mucus to their surface and thus reduced in their targeting ability (Rajapaksa and Lo, 2010).

Not only surface charge, but also surface charge density is crucial for interactions between nanoparticle surface and proteins, that is, opsonins. Nanoparticles have been shown to be covered with serum proteins in the bloodstream shortly after injection. This process is called opsonization and enables macrophages to recognize and subsequently phagocytose foreign particles. How different coating patterns of particles may change the pathway of opsonization

Table 17.1 List of representative studies demonstrating the effects of particle size on immune responses

Materials	Particle size (nm)	Route	Parameters measured	Comments
PS (Manolova et al., 2008), Chitosan (Chua et al., 2011; Kobiasi et al., 2012)	20/500/1000 150/1300, 163/2100	s.c.	Particle trafficking	Faster appearance of NPs in LNs than MPs.
PS (Fifis et al., 2004)	20/40/100/500/1000/2000	i.d.	IFN-γ, IgG	40 nm beads induced the highest levels of IFN-γ and IgG.
PLGA (Gutierro et al., 2002)	200/500/1000	s.c., p.o., i.n.	IgG	1 μm particles elicited higher total serum IgG levels than smaller particles for all tested routes.
Chitosan (Chua et al., 2011; Nagamoto et al., 2004)	163/2100 700/1300/3000	s.c., ip., i.n.	IgG, IgA	Similar ability in inducing IgG after parenteral, IgA levels higher for smaller particles after i.n. administration.
PPS sulfide (Stano et al., 2012), PLGA (Gutierro et al., 2002)	30/200, 200/500/1000	i.n.	IFN-γ, TNF-α, IL-2, IgG1, IgG2a	Higher Th1 responses were obtained with NPof 200–500 nm after i.n. administration.
PLGA (Chong et al., 2005)	633	s.c.	IFN-γ, IL-4, IgG	Only NPs induced Th1 responses with IFN-γ production.

(*Continued*)

Table 17.1 (*Continued*)

Materials	Particle size (nm)	Route	Parameters measured	Comments
Liposomes (Henriksen-Lacey et al., 2011)	<200/700/1500/ 2000–3000	i.m.	IL-1β, IL-2, IL-6, IFN-γ, IgG1, IgG2a, IL-10, IL-5	High levels of IFN-γ and IgG2a and low levels of IL-5 and IL-10 were noted regardless of the size. Particles of ~700–1000 nm produced the highest levels of IFN-γ and IL-1β.
PLA (Kanchan and Panda, 2007)	200–600/2000–8000	i.p.	IFN-γ, IL-4	NPs favored IFN-γ secretion,MPs IL-4 secretion and higher IgG1/IgG2a ratios.
PLA (Katare et al., 2005)	<2000/2–8 μm/>8 μm	i.m.	IgG	MPs of 2–8 μm elicited the highest antibody titers.
DEP, CBP, SIP (van Zijverden and Granum, 2000)	<500/1000–5000	s.c.	IgE, IgG1, IgG2a	NPs induced a strong Th2 response, MPsinduced Th1.
PS (Samuelsen et al., 2009)	64/202/1053/4646	i.t.	TNF-α, IL-1β	MPs induced higher cellular responses than NPs.
Bilosomes (Mann et al., 2009)	250/980	p.o.	IFN-γ, IgG1, IgG2a	Large vesicles are better inducers of Th1 responses, assessed by IFN-γ and IgG2a production.

Abbreviations: CBP: carbon black particle; DEP: diesel exhaust particle; i.d.: intradermal; i.m.: intramuscular; i.n.: intranasal; i.p.: intraperitoneal; i.t.: intratracheal; p.o.: per os; s.c.: subcutaneous; MP: microparticles; NP: nanoparticle; PLGA: poly(lactic-co-glycolic) acid; PPS: polypropylene sulfide; PS: polystyrene; SIP: silica particle.

(classic, alternative, lectin mediated) has recently been described by Moghimi et al. (2011, 2012). Particles of neutral surface potential may have slower opsonization kinetics, a much lower opsonization rate, and very possibly a different opsonization pattern than charged particles (Nie, 2010). Such particles avoiding opsonization are used when a "stealth effect" for prolonged circulation in the bloodstream and reduced immunogenicity is desired. Serum proteins bound to charged particles, on the other hand, have been shown to activate immune cell receptors upon unfolding of the protein (Deng et al.).

Last but not least, surface charge may also have an effect on the intracellular transport and processing following phagocytosis. As shown, for instance, by Harush-Frenkel et al. (2008), anionic nanoparticles are targeted to the degradative lysosomal route, whereas cationic particles, bound to the transcytotic pathway, are not found in lysosomes suggesting a longer intracellular residence time.

17.3.1.3 Immunogenicity of nanoparticle materials

Nanoparticles are classified as transport vehicles and their adjuvant activity is mainly related to enhanced delivery of antigens to APCs or to the LNs. However, mere transport function and interaction with target cells determining overall effect cannot be separated for these complex systems. Different types of bulk materials used for nanoparticle preparation may induce cytokine release, as has been shown in several studies (Plotkin et al., 2013). Materials used for nanoparticle preparation include (bio)polymers (e.g., PLGA and chitosan), lipids, and metals, of which certain substances have been shown to be immunogenic and/or have adjuvant properties themselves.

17.3.1.3.1 *Polymers*

Response to polymeric particles is characterized by mild inflammation, differentiation of immune cells, and degradation of the particles. (Chakravarthi et al., 2007). Biomaterials used in nanoparticulate vaccines have the potential to control the host response in order to reduce or enhance immune responses. Babensee et al. investigated the effect of biomaterial-induced DC phenotype changes and observed differential levels of DC maturation, depending on the biomaterial used. PLGA or chitosan films induced maturation of DCs, as seen in augmentation of pro-inflammatory cytokine production and expression of costimulatory molecules, such as CD86, providing increased interaction with T cells. By contrast, this effect was not observed when DCs were treated with alginate or agarose films, and was even inhibited by hyaluronic acid (Babensee, 2008; Rogers and Babensee, 2011; Yoshida et al., 2007).

In addition, a different effect on TNF-α secretion by the shape of biomaterials was observed, showing stronger effects for particles than for films of the same material (Yoshida et al., 2007). PLGA is considered to be safe and biocompatible at low immunogenicity, and is therefore approved by the US Food and Drug Administration (FDA). However, several studies exhibited the induction of weak inflammatory responses to PLGA, which in vaccine carriers could lead to an adjuvant effect enhancing the overall immunogenicity of the vaccine. After tracheal application of PLGA, an acute inflammatory response was demonstrated by the increase of neutrophils when compared with animals treated with saline solution. By contrast, the inflammatory response seen in animals instilled with saline and polystyrene was similar. This suggests that polystyrene under these conditions behaves as an inert material and does not induce a sustained inflammatory response in the airways (Avital et al., 2002; Springer et al., 2005).

Fiore et al. suggested that the degradation of PLGA into acidic by-products could be responsible for its potential to induce inflammatory responses. Having assessed macrophage and lymphocyte numbers, PLGA-treated lungs of mice displayed a slight increase in inflammation. By contrast, upon intratracheal administration of polyketal particles, which degrade into neutral compounds, no significant inflammatory reactions were detected (Fiore et al., 2010).

The adjuvant properties of chitosan, a biodegradable biopolymer, have also been subjected to investigations. Chitosan potently activated the NLRP3 (NOD-like receptor family, pyrin domain containing 3) inflammasome (a multiprotein oligomer that is a component of the innate immune system) in a phagocytosis-dependent manner. Chitosan is the deacetylated derivative of chitin, which was shown to be relatively inert, suggesting the influence of charge on the immunostimulatory properties of chitosan because of the presence of a secondary amino group (Bueter et al., 2011). Also Tokura et al., who studied the immunological aspects of chitin and chitin derivatives administered to animals, pointed out that chitosan itself, as an adjuvant, can induce polarized Th2 responses (Borchard et al., 2012; Tokura et al., 1999). Whether this activation is based on the interaction of chitosan with PRRs in a lectin-like fashion is still under discussion.

17.3.1.3.2 *Lipids*

Lipid-based delivery systems have been subjected to investigation as vaccine adjuvants, as well. It is generally assumed that hydrophobic surfaces of particles enhance their phagocytosis (Thiele et al., 2003). Among lipid-based vaccine deliver systems, such as solid lipid nanoparticles (SLNs), liposomes, polymerized and nonpolymerized liposomes, and new classes of lipid drug carriers such as immune-stimulating complexes (ISCOMs) and multilamellar vesicles (Alonso-Romanowski et al., 2003; Cai et al., 2011; Moon et al., 2011; Wilson et al., 2012), oil-in-water emulsions are the most extensively studied particulate vaccine carriers. In Europe, nanoscale emulsion-based adjuvants are licensed for influenza vaccines, containing MF59 (Novartis; 1997) or AS03 (GlaxoSmithKline; 2009). AF03 (Sanofi Pasteur) is another emulsion containing squalene, a naturally occurring intermediate metabolite of cholesterol. AF03 has been tested in humans, but is not yet licensed for use in humans (Montomoli et al., 2011; Salvador et al., 2011). The potential to improve vaccine performance has been shown, yet the mechanism of adjuvanticity of these oil-in-water emulsions remains unknown or undisclosed. However, as described above, several studies suggest that these nanoemulsions do not act only as antigen delivery systems but also as immune modulators.

In a study on the influence of squalene-based nanoemulsions on the activation of the innate immune system, it was hypothesized that changes in lipid metabolism and innate immunity are closely linked. Uptake of squalene-based nanoemulsions induced accumulation of neutral lipids that have been shown to have pro-inflammatory properties (Kalvodova, 2010; Lorentzen, 1999; Prieur et al., 2010). Other findings show that the nonionic surfactant polysorbate 80, used in the above-mentioned emulsions, may have an immunomodulatory effect, inducing Th2 responses (Kozutsumi et al., 2006). The systemic expression of IL-5 is in agreement with the Th2 immune response elicited by the polysorbate 80–containing nanoemulsion MF59 (Mosca et al., 2008). The contribution of vitamin E to the overall adjuvanticity in AS03 was investigated by Morel et al. (2011), where higher antibody responses with the inclusion of α-tocopherol than with the plain squalene-in-water emulsion were observed.

17.3.1.3.3 *Inorganic materials*

As an alternative to organic compounds, inorganic materials such as metals and metal oxides are also known to form nanoparticles. Due to their rigid nondeformable shape and much smaller size, they exhibit many distinct biological properties compared to polymeric or lipid nanoparticles (Huang et al., 2010). Being nonbiodegradable, these nanoparticles remain at the site of injection for extended periods of time, resulting in long-lasting presentation and thus enhanced immunogenicity (Maquieira et al., 2012). Among the most widely used metal nanoparticles, gold, silver, or copper and as magnetic metals iron or cobalt is taken into consideration for drug delivery; however, few attempts have been made to develop solid inorganic nanoparticles as vaccine platforms (Bhattacharyya et al., 2011; Pusic et al., 2012).

The main advantage of this type of nanoparticles is that they can be manufactured in fine-tunable sizes between 5 and 100 nm. In addition, they show a high affinity for sulfhydryl groups, which facilitates the binding of various biomolecules through chemisorption of thiol residues onto the metal nanoparticles (Algar et al., 2011; Cruz et al., 2012). Gold, silver and copper are known for being antimicrobial materials, probably due to their ability to react with the –SH groups of enzymes and hence inactivating these proteins (Bhattacharya and Mukherjee, 2008; Yoon et al., 2007). As an example, inhibitory effects on HIV-1 infection in CD4+MT-2 cells (human T cell leukemia [HTLV-1] virus carrier cell line) and cMAGI HIV-1 reporter cells (HeLa cell clone expressing human CD4 and HIV-*long terminal repeat*-coupled genes) due to the interaction of silver nanoparticles with the virus was reported by Elechiguerra et al. (2005).

Furthermore, the inflammatory potential of metals could be of interest in vaccine development. Goebel et al. (1995) showed that the T cell response in mice treated with gold(I) compounds for 12 weeks is directed against the oxidized gold(III), which is a metabolite generated in vivo by macrophages. In another study the inflammatory potential of colloidal silver on macrophages was evaluated through assessing the release of IL-8, cell viability, and induction of oxidative stress. While reactive oxygen species (ROS) are generated as part of the normal oxidative metabolism, their overproduction can lead to cell death. Early-stage oxidative stress, on the contrary, can stimulate inflammatory responses, and the generated ROS were found to trigger the release of IL-8 in macrophages (Park et al., 2011). While different types of metallic nanoparticles without any specific antigen loading reportedly exert antimicrobial activity, other studies revealed a compromise of innate immune response and clearance of bacterial pathogens after exposure. Shevedova et al. suggested that elimination of microbes depends to a larger extent on nitric oxide production than ROS from the oxidative burst (Nathan and Hibbs Jr., 1991; Shvedova et al., 2008). This could explain the controversial findings of decreased pulmonary bacterial clearance in mice after exposure to copper nanoparticles, despite induction of robust inflammatory responses (Kim et al., 2011).

Also metal oxide nanoparticles and their soluble ions were investigated with regard to their immunostimulatory/inflammatory effects. In vitro, nanoparticles made of nickel, copper, and zinc oxide all showed toxic effects, but only zinc and copper oxide nanoparticles induced IL-8 production and activation of transcription factors. Aqueous extracts of Zn and Cu oxide nanoparticles showed similar profiles of cytotoxicity and inflammatory effects, whereas nickel ions did not induce any cytotoxicity or immune responses. The inflammatory response assessed in rats given by nanoparticles was also different to the aqueous extracts, with zinc and copper oxide nanoparticles leading to the recruitment of eosinophils but not their soluble ions (Cho et al., 2012b).

17.3.2 Immune Receptor Targeting and Antigen Delivery

Currently, intense research efforts are aimed at the development of novel formulations and delivery systems for vaccines. By the use of nanoparticles, antigens are expected to be adequately presented to the immune system to elicit immune responses appropriate for sufficient protection against infectious diseases. Ideally, the nanoparticulate vaccine carrier system should have sufficient antigen- and adjuvant-loading capacity, through either incorporation or surface adsorption.

Formulation strategies can facilitate the capture by and the entry of the antigen into APCs. T cell antigens, in the form of peptides, proteins, plasmid DNA, or RNA formulated into nanoparticles appear to increase CTL responses in vivo. Such particulate formulations significantly increase efficiency as compared to the soluble antigen alone, as they are in general readily recognized as antigenic and ingested by immune cells, which subsequently leads to antigen processing and presentation to other cells of the immune system (Moingeon et al., 2002). Ligands of APC-specific PRRs, such as Toll-like receptors (TLRs), can be grafted onto the surface of particle-based vaccines to increase specific delivery to immune cells (Heuking and Borchard, 2012).

Application of vaccines in particulate form not only allows moving the antigen within the tissue to individual cells, it also facilitates its penetration through the cell membrane. This transfection process is a major requirement for making use of intracellular mechanisms, such as signaling pathways and antigen processing, to activate the immune system. Many biomolecules, used as antigens for vaccines, are not able to diffuse within tissues and through the plasma membrane, due to their size and other physicochemical properties. Particle-based strategies to overcome this problem have been developed for genetic vaccination relying on chemical materials used as carriers and physical treatment. In the "gene gun technique" DNA is coated on gold particles, energy is transferred to the DNA carriers to accelerate and propel the particles into the cells (Villemejane and Mir, 2009). This method is limited to low-depth penetration of particles in tissues but is sufficient for skin DNA vaccination (Luz Garcia-Hernandez et al., 2008). Magnetic nanoparticles have attracted attention because of their potential use for direct targeting to diseased tissues and organs by applying an external magnetic field. "Magnetofection" describes a technique of gene transfection involving naked plasmid DNA or DNA vectors associated to the surface of magnetic nanoparticles. Superparamagnetic iron oxide nanoparticles (SPIONs) are applied for magnetofection by delivering genes to the target cells, hence increasing their local concentration (Al-Deen et al., 2011; Vasir and Labhasetwar, 2007). Application of an external alternating magnetic field leads to the production of energy in the form of heat. Hyperthermia has been demonstrated to stimulate the innate immune response through the release of heat-shock proteins that activate neighboring immune cells (Baronzio et al., 2006; Colombo et al., 2012). As such, SPIONs may have adjuvant properties activating the immune system in a nonspecific manner. These two methods are still under investigation. Although a few clinical trials have already been performed or are in progress, long-term safety and potential risks have not been sufficiently evaluated (Villemejane and Mir, 2009). Another approach is impalefection by functionalizing nanoscale carbon fibers with DNA to impale cells. It is a useful tool for exploring gene delivery in vitro but has yet to be used in the therapeutic field (Pearce et al., 2013).

The types of effective immune responses against infectious diseases depend on the localization of the pathogens after infection. Generally, extracellular pathogens are combated by antibodies mediated by humoral immune responses, whereas protection against intracellular pathogens depends on cell-mediated immunity via CD4+ Th1 cells. In addition, CD8+ CTLs play an important role in the protection against these pathogens by directly killing infected cells (Nagata and Koide, 2010). Antigen recognition by T lymphocytes depends on the ability of APCs to process protein antigen into peptides and to present them via MHC molecules at the cell surface. The MHC I presentation pathway allows CD8+ T cells to identify and eliminate cells infected by intracellular pathogens. By contrast, MHC II and CD1 molecules present peptides derived from captured exogenous antigens to CD4+ T cells and NK cells (Paul, 2008) (Fig. 17.1).

17.3.3 Current Nanoparticle Vaccines on the Market and in Clinical Studies

Nanotechnology is enabling novel vaccines to enter clinical trials that often show superiority of the particulate systems over traditional vaccine types. We will discuss here several examples of nanoparticle-based vaccines.

On the basis of antigen presentation by DCs and innate immune system stimulation via TLRs, a vaccine against melanoma was designed. Melanoma-specific peptide Melan-A/Mart-1 was linked to virus-like nanoparticles (VLPs) loaded with CpG oligonucleotides as activators of TLR 9 located in the endosomal compartment of target cells (Speiser et al., 2005). Already vaccination of HLA-A2 transgenic mice in preclinical studies showed strong Melan-A-specific CD8 T cell responses. A phase I/II study in stage II–IV melanoma patients confirmed good tolerability of the vaccine and ex vivo T cell responses. Increased activated Melan-A-specific CD8 T cell populations were detected directly ex vivo as well as cytokine production of INF-γ, TNF-α, and IL-2, and LAMP-1 expression by vaccine-induced T cells. Furthermore, a central memory phenotype of those specific T cells was frequently observed and an enhancement of T cell responses was achieved by subsequent vaccination with the peptide emulsified in incomplete Freund's adjuvant (Speiser et al., 2010). Other vaccines using nanotechnology based on VLPs have already been licensed for the prevention of cervical and anogenital infections, that is, by GlaxoSmithKline (Engerix® and Cervarix®) and Merck and Co. (Recombivax HB and Gardasil®) (Buonaguro and Buonaguro, 2013). These products are noninfectious viral subunit vaccines with proteins of the respective pathogens assembled in nanosized VLPs, structures resembling the virus itself and thus recognized as antigens but lacking the genetic repertoire necessary for replication.

A novel liposomal influenza subunit vaccine is currently under clinical investigation. It consists of liposomes containing the viral surface proteins hemagglutinin (HA) and neuraminidase (NA) derived from various influenza strains (Babai et al., 2001). An IL-2-supplemented liposomal formulation injected intramuscularly has proven to be both safe and effective in inducing strong anti-HA and anti-NA antibody responses in mice and humans (Ben-Yehuda et al., 2003).

DNA vaccines with one specific or even several pathogenic antigenic epitopes encoded by the plasmid are known to drive cellular and humoral immune responses at least in the animal model. To increase transfection efficiency, and to enhance and direct immune responses to DNA vaccines, particle-based delivery systems have been explored (Bivas-Benita et al., 2004, 2009).

FDA-controlled phase I studies to assess safety and immunogenicity of particle-mediated DNA vaccines against influenza and hepatitis B viruses (HBVs) have been completed only for microparticles. These studies were a part of the clinical development of a particle-mediated epidermal delivery (PMED) DNA vaccine or "gene gun" based on a dry powder formulation with DNA plasmid adsorbed onto microscopic gold particles (PowderMed, 2007, 2008). Particle-mediated DNA administration permits the use of small quantities of DNA and application via the skin as an immunological active tissue increases the efficacy of the vaccine. HBV-specific protective antibody responses of at least 10 mIU/mL and antigen-specific CD8+ cells were detected. Hepatitis B surface antigen–specific IFN-γ-secreting Th cells were measured in the majority of individuals treated, indicating Th1-biased responses. Current data demonstrate that particle-mediated DNA administration with this needle-free method of administration was safe and well tolerated. Since Th1 and CD8+ T cell responses are associated with resolution of HBV and other chronic viral infections, the results reported suggest a potential for particle-mediated DNA vaccine delivery (Roy et al., 2000).

Another topical vaccine, DermaVir Patch for skin delivery of the DNA vaccine, has already entered clinical trials for the treatment of HIV infections (Immunity, 2013). Treatment of HIV-1/AIDS requires a vaccine to induce robust and long-lasting HIV-1-specific T cells to control viral replication. A plasmid DNA encoding the majority of HIV-1 genes has been chemically formulated into polyethylenimine mannose (PEIm), a cationic polymer that forms nanoparticles with the negatively charged plasmid through condensation. The mannose moiety targets the vaccine to receptors on the surface of APCs (Lisziewicz et al., 2007). On the basis of the potent induction of Gag, Tat, and Rev antigen-specific memory T cells, it is assumed that DermaVir boosts T cell responses specific to all the 15 HIV antigenic epitopes expressed by a single plasmid applied. Furthermore, a dose-dependent expansion of HIV-specific memory T cells with high proliferation capacity was detected (Lisziewicz et al., 2012).

Just recently a phase II clinical trial evaluated the safety and immunogenicity of a respiratory syncytial virus fusion (RSV F) protein vaccine, formulated as nanoparticles. The protein was extracted and purified from insect cell membranes and assembled into 40 nm nanoparticles composed of multiple RSV F oligomers arranged in the form of rosettes (Smith et al., 2012). The primary immune response measured confirmed the vaccine to be a potent antigen eliciting immune responses at levels that would be predicted to protect infants through maternal immunization (AZoNano, Novavax, 2013). A study using a polypeptide vaccine against malaria that self-assembles into spherical nanoparticles, displaying repetitive epitopes of the *Plasmodium falciparum* circumsporozoite protein, showed promising results in preclinical testing. The outcome was high-titer, long-lasting protective antibody and long-lived central memory CD8+ T cells and could provide sterile protection against a lethal challenge of the transgenic parasites in mice (Kaba et al., 2012).

From a soluble type of chitosan, Viscosan, a hydrogel was formed, which was further mechanically processed into gel particles. This so-called ViscoGel was mixed with a commercial vaccine against *Haemophilus influenza* type b (Act-HIB). ViscoGel-Act-HIB with a tenth of the vaccine dose is as efficient in eliciting a humoral response in mice as the high-dose vaccine alone (Neimert-Andersson et al., 2011).

Mixing saponin, cholesterol, and phospholipid under controlled conditions, 40 nm spherical negatively charged ISCOMs of the ISCOMATRIX® adjuvant are formed. A range of ISCOMATRIX® vaccines have been tested in clinical trials, generating both antibody and T cell responses. The adjuvant has been combined with influenza viral proteins or recombinant antigens for two cancer/neoplasia vaccines and one hepatitis C virus (HCV) vaccine to be tested in humans. Higher and faster antibody responses compared to conventional influenza vaccines and virus-specific CTL responses were observed. Furthermore, strong virus-specific humoral and cellular immune responses were detected in vaccinated subjects of the two types of human papilloma virus (HPV) and of the HCV core ISCOMATRIX® vaccines (Pearse and Drane, 2005).

17.4 Conclusions

Prevention is better than treatment. Preventing a pathogen from gaining access to the body is better than to fight it off after infection.

Antibiotics are failing, occurrence of multidrug resistance is on the rise, and new pathogens are emerging: we are in dire need of novel vaccines that elicit long-lasting and protective immune responses. By our growing understanding of how adjuvants may work, the "dirty little secrets" of vaccinologists may be no more, and through nanotechnology we may have the opportunity to design vaccine delivery systems as nature intended them to be.

In practical terms, which antigens would be good candidates to be used in a novel (nanoparticulate) formulation, possibly using novel adjuvants? Using known antigens, for which successful vaccines already exist, might not be a good choice: the danger of nonacceptance by regulatory authorities and/or the public may be too high. Therefore, one would choose novel antigens that are being identified through sequencing of pathogen genomes, or a combination thereof. The concept of "new formulations for new vaccines" might rightfully apply, as tolerance toward undesired side effects is certainly higher for first-in-man vaccines than for new formulations of vaccines that have been tried and tested. The use of adjuvants, on the other hand, is fraught with their sometimes difficult handling; the skepticism of regulatory authorities toward new, not well-defined excipients/active principles; and the ongoing discussion in a public that remains often closed to scientific argumentation.

The way out of this dilemma would, in our opinion, be to enhance the immunogenicity of protein and peptide vaccines themselves, to use particulate systems to better target and present the antigen to the immune system, and to finally use well-characterized small-molecular-weight entities of a defined mechanism of action and known toxicity. These avenues are followed in modern vaccine development and will result in highly effective vaccines against old and new threats to our health. Nanotechnology is playing a prominent part in this endeavor.

Bibliography

Abad, J.M., Mertens, S.F., Pita, M., Fernandez, V.M., Schiffrin, D.J. (2005). *J. Am. Chem. Soc.*, **15**, 5689–5694.

Abd El-Rehim, H.A., Swilem, A.E., Klingner, A., Hegazy, E.-S.A., Hamed, A.A. (2013). *Biomacromolecules*, **14**, 688–698.

Abdelghany, S.M., Schmid, D., Deacon, J., Jaworski, J., Fay, F., Mclaughlin, K.M., Gormley, J.A., Burrows, J.F., Longley, D.B., Donnelly, R.F., Scott, C.J. (2013). *Biomacromolecules*, **14**, 302–310.

Abidian, M.R., Kim, D.H., Martin, D.C. (2006). *Adv. Mater.*, **18**, 405–409.

Agarwal, R., Kaye, S.B. (2003). *Nat. Rev. Cancer*, **3**, 502.

Aguirre, G., Ramos, J., Forcada, J. (2013). *Soft Matter*, **9**(1), 261–270.

Aguirre, G., Ramos, J., Heuts, J.P.A., Forcada, J. (2014). *Polym. Chem.*, **5**(15), 4569–4579.

Agyare, E.K., Curran, G.L., Ramakrishnan, M., Yu, C.C., Poduslo, J.F., Kandimalla, K.K. (2008). *Pharm. Res.*, **25**, 2674–2684.

Ahmed, M., Narain, R. (2012a). *Mol. Pharm.*, **9**, 3160–3170.

Ahmed, N., Fessi, H., Elaissari, A. (2012b). *Drug Discovery Today*, **17**, 928–934.

Ahmed, N., Jaafar-Maalej, C., Eissa, M.M., Fessi, H., Elaissari, A. (2013). *J. Biomed. Nanotechnol.*, **9**, 1579–1585.

Ahmed, N., Mora-Huertas, C.E., Jaafar-Maalej, C., Fessi, H., Elaissari, A. (2012c). Polymeric drug delivery systems for encapsulation hydrophobic drug, In *Drug Delivery Strategies for Poorly Water-Soluble Drugs*, Eds. Douroumis, D., Fahr, A., Wiley, UK.

Airaksinen, M.A., Saarma, M. (2002). *Nat. Rev. Neurosci.*, **3**, 383–394.

Akbarzadeh, A., Mikaeili, H., Zarghami, N., Mohammad, R., Barkhordari, A., Davaran, S. (2012a). *Int. J. Nanomed.*, 511–526.

Akbarzadeh, A., Samiei, M., Davaran, S. (2012b). *Nanoscale Res. Lett.*, **7**, 144.

Akbarzadeh, A., Samiei, M., Davaran, S. (2012c). *Nanoscale Res. Lett.*, 1–13.

Akbarzadeh, A., Samiei, M., Joo, S.W., Anzaby, M., Hanifehpour, Y., Nasrabadi, H.T., Davaran, S. (2012d). *J. Nanobiotechnol.*, **46**, 1–13.

Akerman, M.E., Chan, W.C.W., Laakkonen, P., Bhatia, S.N., Ruoslahti, E. (2002). *Proc. Natl. Acad. Sci. U S A*, **99**, 12617–12621.

Akhtar, M.J., Ahamed, M., Kumar, S., Khan, M.M., Ahmad, J., Alrokayan, S.A. (2012). *Int. J. Nanomed.*, **7**, 845–857.

Akimoto, J., Nakayama, M., Sakai, K., Okano, T. (2010). *Mol. Pharm.*, **7**, 926–935.

Al-Ahmady, Z.S., Al-Jamal, W.T., Bossche, J.V., Bui, T.T., Drake, A.F., Mason, A.J., Kostarelos, K. (2012). *ACS Nano*, **6**, 9335–9346.

Albanese, A., Chan, W.C. (2011). *ACS Nano*, **5**, 5478–5489.

Alberts, B., Johnson, A., Lewis, J., Raff, M., Roberts, K., Walter, P. (2002). *Molecular Biology of the Cell*, 4th ed., Garland Science, New York.

Al-Deen, F.N., Ho, J., Selomulya, C., Ma, C., Coppel, R. (2011). *Langmuir*, **27**, 3703–3712.

Al-Dosari, M.S., Gao, X. (2009). *AAPS J.*, **11**, 671–681.

Alexander, J.W. (2009). *Surg. Infect.*, **3**, 289–292.

Algar, W.R., Prasuhn, D.E., Stewart, M.H., Jennings, T.L., Blanco-Canosa, J.B., Dawson, P.E., Medintz, I.L. (2011). *Bioconjug. Chem.*, **22**, 825–858.

Ali, I., Salim, K., Rather, M.A., Wani, W.A., Haque, A. (2011a). *Curr. Cancer Drug Targeting*, **11**, 135–146.

Ali, S.I., Heuts, J.P.A., van Herk, A.M. (2011b). *Soft Matter*, **7**(11), 5283–5390.

Ali-Boucetta, H., Nunes, A., Sainz, R., Herrero, M.A., Tian, B., Prato, M., Bianco, A., Kostarelos, K. (2013). *Angew. Chem., Int. Ed. Engl.*, **52**, 2274–2278.

Alivisatos, A.P., Gu, W.W., Larabell, C. (2005). *Annu. Rev. Biomed. Eng.*, **7**, 55–76.

Al-Jamal, W.T., Al-Ahmady, Z.S., Kostarelos, K. (2012). *Biomaterials*, **33**, 4608–4617.

Allain, C., Poon, L., Chan, C., Richmond, W., Fu, P. (1974). *Clin. Chem.*, **20**, 470–475.

Allen, T.M. (2002). *Nat. Rev. Cancer*, **2**, 750–763.

Allen, T.M., Cullis, P.R. (2004). *Science*, **303**, 1818–1822.

Alonso-Romanowski, S., Chiaramoni, N.S., Lioy, V.S., Gargini, R.A., Viera, L.I., Taira, M.C. (2003). *Chem. Phys. Lipids*, **122**, 191–203.

Alvarez-Lorenzo, C., Bromberg, L., Concheiro, A. (2009). *Photochem. Photobiol.*, **85**, 848–860.

Alvarez-Puebla, R.A., Liz-Marzan, L.M. (2010). *Small*, **5**, 604–610.

Amaral, S.P., Fernandez-Villamarin, M., Correa, J., Riguera, R., Fernandez-Megia, E. (2011). *Org. Lett.*, **13**, 4522–4525.

Amstad, E., Kohlbrecher, J., Müller, E., Schweizer, T., Textor, M., Reimhult, E. (2011). *Nano Lett.*, **11**, 1664–1670.

An, Y., Zhang, L., Xiong, S., Wu, S., Xu, M., Xu, Z. (2012). *Colloid Surf. B*, **92**, 246–253.

Ancla, C., Lapeyre, V., Gosse, I., Catargi, B., Ravaine, V. (2011). *Langmuir*, **27**, 12693–12701.

Angelatos, A.S., Radt, B., Caruso, F. (2005). *J. Phys. Chem. B*, **7**, 3071–3076.

Angelova, M.I., Soleau, S., Meleard, P., Faucon, J.F., Bothorel, P. (1992). *Prog. Colloid Polym. Sci.*, **89**, 127–131.

Antony, A.C. (1996). *Annu. Rev. Nutr.*, **16**, 501–521.

Araya, E., Olmedo, Y., Bastús, N.G., Guerrero, S., Puntes, V.F., Giralt, E., Kogan, M.J. (2008). *Nanoscale Res. Lett.*, **3**, 435–443.

Arbab, A.S., Frenkele, V., Panditb, S.D., Andersonc, S.A., Yocuma, G.T., Burb, M., Khuud, H.M., Readd, E.J., Franka, J.A. (2005). *Stem Cells*, **24**, 671–678.

Arbab, A.S., Yocum, G.T., Kalish, H., Jordan, E.K., Anderson, S.A., Khakoo, A.Y., Read, E.J., Frank, J.A. (2004). *Blood*, **104**, 1217–1223.

Arias, J.L., Reddy, L.H., Couvreur, P. (2011). *Biomacromolecules*, **12**, 97–104.

Arispe, E., Rojas, E., Pollard, B. (1993). *Proc. Natl. Acad. Sci. U S A*, **90**, 567–571.

Arlt, M., Haase, D., Hampel, S., Oswald, S., Bachmatiuk, A., Klingeler, R., Schulze, R., Ritschel, M., Leonhardt, A., Fuessel, S., Büchner, B., Kraemer, K., Wirth, M.P. (2010). *Nanotechnology*, **21**, 335101 (9 pp).

Arruebo, M., Galán, M., Navascués, N., Téllez, C., Marquina, C., Ibarra, M.R., Santamaría, J. (2006). *Chem. Mater.*, **18**, 1911–1919.

Aruffo, A., Stamenkovic, I., Melnick, M., Underhill, C.B., Seed, B. (1990). *Cell*, **61**, 1303–1313.

Arunraj, T.R., Sanoj Rejinold, N., Ashwin Kumar, N., Jayakumar, R. (2013). *Int. J. Biol. Macromol.*, **62**, 35–43.

Arvizo, R.R., Saha, S., Wang, E., Robertson, J.D., Bhattacharya, R., Mukherjee, P. (2013). *Proc. Natl. Acad. Sci. U S A*, **110**, 6700–6705.

Asgeirsson, D., Venturoli, D., Rippe, B., Rippe, C. (2006). *Am. J. Physiol. Renal Physiol.*, **291**, F1083–F1089.

Ashcroft, J.M., Hartman, K.B., Kissell, K.R., Mackeyev, Y., Pheasant, S., Young, S., Van der Heide, P.A.W., Mikos, A.G., Wilson, L.J. (2007). *Adv. Mater.*, **19**, 573–576.

Ashwanikumar, N., Kumar, N.A., Nair, S.A., Kumar, G.V. (2012). *Int. J. Nanomed.*, **7**, 5769–5779.

Aslan, K., Luhrs, C.C., Perez-Luna, V.H. (2004). *J. Phys. Chem. B*, **40**, 15631–15639.

Averick, S.E., Paredes, E., Irastorza, A., Shrivats, A.R., Srinivasan, A., Siegwart, D.J., Magenau, A.J., Cho, H.Y., Hsu, E., Averick, A.A., Kim, J., Liu, S., Hollinger, J.O., Das, S.R., Matyjaszewski, K. (2012). *Biomacromolecules*, **13**, 3445–3449.

Avital, A., Shapiro, E., Doviner, V., Sherman, Y., Margel, S., Tsuberi, M., Springer, C. (2002). *Am. J. Respir. Cell Mol. Biol.*, **27**, 511–514.

Avital-Shmilovici, M., Shabat, D. (2010). *Soft Matter*, **6**, 1073–1080.

AZoNano. Positive data from Novavax Phase II RSV F protein nanoparticle vaccine clinical trial. April 4, 2013 [cited May 8, 2013]; Available from: http://www.azonano.com/news.aspx?newsID=27000.

Babai, I., Barenholz, Y., Zakay-Rones, Z., Greenbaum, E., Samira, S., Hayon, I., Rochman, M., Kedar, E. (2001). *Vaccine*, **20**, 505–515.

Babar, I.A., Cheng, C.J., Booth, C.J., Liang, X., Weidhaas, J.B., Saltzman, W.M., Slack, F.J. (2012). *Proc. Natl. Acad. Sci. U S A*, **109**, E1695–E1704.

Babensee, J.E. (2008). *Semin. Immunol.*, **20**, 101–108.

Babin, J., Pelletier, M., Lepage, M., Allard, J.F., Morris, D., Zhao, Y. (2009). *Angew. Chem.*, **48**, 3329–3332.

Bae, K.H., Choi, S.H., Park, S.Y., Lee, Y., Park, T.G. (2006). *Langmuir*, **22**, 6380–6384.

Bae, K.H., Lee, J.Y., Lee, S.H., Park, T.G., Nam, Y.S. (2013). *Adv Healthc Mater.*, **2**, 576–584.

Bai, F., Yang, X.L., Huang, W.Q. (2004). *Macromolecules*, **37**, 9746–9752.

Baibich, M.N., Broto, J.M., Fert, A., Vandau, F.N., Petroff, F., Eitenne, P., Creuzet, G., Friederich, A., Chazelas, J. (1988). *Phys. Rev. Lett.*, **21**, 2472–2475.

Bailey, M.M., Berkland, C.J. (2009). *Med. Res. Rev.*, **29**, 196–212.

Bailey, R.E., Smith, A.M., Nie, S.M. (2004). *Physica E*, **1**, 1–12.

Bainton, D.F. (1981). *J. Cell Biol.*, **91**, 66s–76s.

Bakalova, R., Ohba, H., Zhelev, Z., Ishikawa, M., Baba, Y. (2004). *Nat. Biotechnol.*, **11**, 1360–1361.

Baker, M. (2010a). *Nat. Methods*, **12**, 957–962.

Baker, S.N., Baker, G.A. (2010b). *Angew. Chem., Int. Ed.*, **38**, 6726–6744.

Bangham, A.D., Standish, M.M., Watkins, J.C. (1965). *J. Mol. Biol*, **13**, 238–252.

Bao, H., Jin, X., Li, L., Lv, F., Liu, T. (2012). *J. Mater. Sci. Mater. Med.*, **8**, 1891–901.

Bareford, L.M., Avaritt, B.R., Ghandehari, H., Nan, A., Swaan, P.W. (2013). *Pharm. Res.*, **30**(7), 1799–1812.

Barenholz, Y., Amselem, S., Lichtenberg, D. (1979). *FEBS Lett.*, **99**, 210–214.

Barker, R.A., Barrett, J., Mason, S.L., Björklund, A. (2013). *Lancet Neurol.*, **12**, 84–91.

Barnard, A.S. (2009). *Analyst*, **9**, 1751–1764.

Baronzio, G., Gramaglia, A., Fiorentini, G. (2006). *In Vivo*, **20**, 689–695.

Barrett, T., Ravizzini, G., Choyke, P., Kobayashi, H. (2009). *IEEE Eng. Med. Biol.*, **28**, 12–22.

Barth, R.F. (2009). *Appl. Radiat. Isotopes*, **67**, S3–S6.

Barth, R.F., Vicente, M.G.H., Harling, O.K., Riley, K.J., Binns, P.J., Kiger III, W., Kawabata, S. (2012). In *Malignant Gliomas (Radiation Medicine Rounds Volume 3 Issue 2)*, p. 209, Eds. Chakravarti, A., Fuss, M., Demos Medical, 1 Edn., Vol. 3.

Bartlett, G. (1959). *J. Biol. Chem.*, **234**, 466–468.

Bartlett, R., Medow, M., Panitch, A., Seal, B. (2012a). *Biomacromolecules*, **13**, 1204–1211.

Bartlett, R.L.R., Panitch, A., Rush L. (2012b). *Biomacromolecules*, **13**, 2578–2584.

Baselga, J., Gelmon, K.A., Verma, S., Wardley, A., Conte, P., Miles, D., Bianchi, G., Cortes, J., McNally, V.A., Ross, G.A. (2010). *J. Clin. Oncol.*, **28**, 1138–1144.

Bastús, N.G., Comenge, J., Puntes, V. (2011). *Langmuir*, **17**, 11098–11105.

Battistini, E., Gianolio, E., Gref, R., Couvreur, P., Fuzerova, S., Othman, M., Aime, S., Badet, B., Durand, P. (2008). *Chem.: Eur. J.*, **14**, 4551–4561.

Batzri, S., Korn, E.D. (1973). *Biochim. Biophys. Acta*, **298**, 1015–1019.

Bealle, G., Di Corato, R., Kolosnjaj-Tabi, J., Dupuis, V., Clement, O., Gazeau, F., Wilhelm, C., Menager, C. (2012). *Langmuir*, **28**, 11834–11842.

Beg, S., Samad, A., Alam, M.I., Nazish, I. (2011). *CNS Neurol. Disord. Drug Targets*, **10**, 576–588.

Behr, J.-P. (1997). *CHIMIA Int. J. Chem.*, **51**, 34–36.

Beier, M., Hoheisel, J.D. (1999). *Nucleic Acids Res.*, **27**, 1970–1977.

Beija, M., Marty, J.-D., Destarac, M. (2011). *Chem. Commun.*, **47**, 2826–2828.

Benito, J.M., Gómez-García, M., Ortiz Mellet, C., Baussanne, I., Defaye, J., García Fernández, J.M. (2004). *J. Am. Chem. Soc.*, **126**, 10355–10363.

Benjaminsen, R.V., Mattebjerg, M.A., Henriksen, J.R., Moghimi, S.M., Andresen, T.L. (2013). *Mol. Ther.*, **21**, 149–157.

Ben-Yehuda, A., Joseph, A., Zeira, E., Even-Chen, S., Louria-Hayon, I., Babai, I., Zakay-Rones, Z., Greenbaum, E., Barenholz, Y., Kedar, E. (2003). *J. Med. Virol.*, **69**, 560–567.

Berciaud, S., Cognet, L., Blab, G.A., Lounis, B. (2004). *Phys. Rev. Lett.*, **25**, 1–4.

Berthold, P.R., Shiraishi, T., Nielsen, P.E. (2010). *Bioconjug. Chem.*, **21**, 1933–1938.

Bertram, J.P., Williams, C.A., Robinson, R., Segal, S.S., Flynn, N.T., Lavik, E.B. (2009). *Sci. Transl. Med.*, **1**, 11ra22.

Bertrand, N., Leroux, J.C. (2012). *J. Controlled Release*, **161**, 152–163.

Bertrand, N., Simard, P., Leroux, J.C. (2010). *Methods Mol. Biol.*, **605**, 545–558.

Betageri, G.V., Jenkins, S.A., Parsons, D.L. (1993). Liposome Drug Delivery Systems, Technomic, Lancaster, USA.

Bhadra, D., Bhadra, S., Jain, P., Jain, N. (2002). *Pharmazie*, **57**, 5–29.

Bhadra, D., Bhadra, S., Jain, S., Jain, N.K. (2003). *Int. J. Pharm.*, **257**, 111–124.

Bhardwaj, N., Kundu, S.C. (2010). *Biotechnol. Adv.*, **28**, 325–347.

Bhatt, A.N., Mathur, R., Farooque, A., Verma, A., Dwarakanath, B. (2010). *Ind. J. Med. Res.*, **132**, 129–149.

Bhattacharya, R., Mukherjee, P. (2008). *Adv. Drug Delivery Rev.*, **60**, 1289–1306.

Bhattacharya, S., Alkharfy, K.M., Janardhanan, R., Mukhopadhyay, D. (2012). *Nanotechnol. Rev.*, **1**, 235–253.

Bhattacharya, S., Eckert, F., Boyko, V., Pich, A. (2007). *Small*, **3**, 650–657.

Bhattacharyya, S., Kudgus, R., Bhattacharya, R., Mukherjee, P. (2011). *Pharm. Res.*, **28**, 237–259.

Bhuchar, N., Sunasee, R., Ishihara, K., Thundat, T., Narain, R. (2012). *Bioconjug. Chem.*, **23**, 75–83.

Bi, F., Zhang, J., Su, Y., Tang, Y.C., Liu, J.N. (2009). *Biomaterials*, **30**, 5125–5130.

Bikram, M., West, J.L. (2008). *Expert Opin. Drug Deliv.*, **5**, 1077–1091.

Biswas, S., Deshpande, P.P., Perche, F., Dodwadkar, N.S., Sane, S.D., Torchilin, V.P. (2013a). *Cancer Lett.*, **335**(1), 191–200.

Biswas, S., Dodwadkar, N.S., Deshpande, P.P., Parab, S., Torchilin, V.P. (2013b). *Eur. J. Pharm. Biopharm.*, **84**(3), 517–525.

Biswas, S., Dodwadkar, N.S., Deshpande, P.P., Torchilin, V.P. (2012a). *J. Controlled Release*, **159**, 393–402.

Biswas, S., Dodwadkar, N.S., Piroyan, A., Torchilin, V.P. (2012b). *Biomaterials*, **33**, 4773–4782.

Biswas, S., Torchilin, V.P. (2013). *Pharmaceuticals*, **6**, 161–183.

Bivas-Benita, M., Lin, M.Y., Bal, S.M., van Meijgaarden, K.E., Franken, K.L.M.C., Friggen, A.H., Junginger, H.E., Borchard, G., Klein, M.R., Ottenhoff, T.H.M. (2009). *Vaccine*, **27**, 4010–4017.

Bivas-Benita, M., van Meijgaarden, K.E., Franken, K.L.M.C., Junginger, H.E., Borchard, G., Ottenhoff, T.H.M., Geluk, A. (2004). *Vaccine*, **22**, 1609–1615.

Black, K.C., Yi, J., Rivera, J.G., Zelasko-Leon, D.C., Messersmith, P.B. (2013). *Nanomedicine (London)*, **8**, 17–28.

Blackburn, W.H., Dickerson, E.B., Smith, M.H., McDonald, J.F., Lyon, L.A. (2009). *Bioconjug. Chem.*, **20**, 960–968.

Blanco, M.D., Guerrero, S., Benito, M., Teijón, C., Olmo, R., Muñíz, E., Katime, I., Teijón, J.M. (2010). *J. Biomed. Mater. Res. A*, **95**, 1028–1040.

Bloom, D.E., Cafiero, E.T., Jané-Llopis, E., Abrahams-Gessel, S., Bloom, L.R., Fathima, S., Feigl, A.B., Gaziano, T., Mowafi, M., Pandya, A., Prettner, K., Rosenberg, L., Seligman, B., Stein, A. Z., Weinstein, C. (2011). The global economic burden of noncommunicable diseases, 1–46, World Economic Forum, Geneva.

Boas, U., Heegaard, P.M. (2004). *Chem. Soc. Rev.*, **33**, 43–63.

Boddapati, S.V., D'Souza, G.G., Erdogan, S., Torchilin, V.P., Weissig, V. (2008). *Nano Lett.*, **8**, 2559–2563.

Bohnen, N.I., Müller, M.L. (2012). *J. Neural Transm.*, Dec 21. [Epub ahead of print).

Boissiere, O., Han, D., Tremblay, L., Zhao, Y. (2011). *Soft Matter*, **7**, 9410–9415.

Boken, J., Dalela, S., Kumar, D. (2013). *J. Chem. Eng. Proc. Technol.*, **4**, 8.

Bokvist, M., Lindstrom, F., Watts, A., Grobner, G. (2004). *J. Mol. Biol.*, **335**, 1039–1049.

Bonasio, R., Carman, C.V., Kim, E., Sage, P.T., Love, K.R., Mempel, T.R., Springer, T.A., von Andrian, U.H. (2007). *Proc. Natl. Acad. Sci. U S A*, **37**, 14753–14758.

Bonifazi, D., Nacci, C., Marega, R., Campidelli, S., Ceballos, G., Modesti, S., Meneghetti, M., Prato, M. (2006). *Nano Lett.*, **6**, 1408–1414.

Borchard, G., Esmaeili, F., Heuking, S. (2012). Chitosan-based delivery systems for mucosal vaccination, In *Chitosan-Based Systems for*

Biopharmaceuticals, pp. 211–224, Eds. Sarmento, B., Neves, J.d., Wiley, UK.

Borisch, B., Semac, I., Soltermann, A., Palomba, C., Hoessli, D.C. (2001). *Verh. Dtsch. Ges. Pathol.*, **85**, 161–166.

Bourlinos, A.B., Stassinopoulos, A., Anglos, D., Zboril, R., Georgakilas, V., Giannelis, E.P. (2008). *Chem. Mater.*, **14**, 4539–4541.

Boussif, O., Lezoualc'h, F., Zanta, M.A., Mergny, M.D., Scherman, D., Demeneix, B., Behr, J.P. (1995). *Proc. Natl. Acad. Sci. U S A*, **92**, 7297–7301.

Boutry, S., Laurent, S., Elst, L.V., Muller, R.N. (2006). *Contrast Media Mol. Imaging*, **1**, 15–22.

Bouwstra, J.A., Gooris, G.S., Bras, W., Talsma, H. (1993). *Chem. Phys. Lipids*, **64**, 83–98.

Boyer, C., Whittaker, M.R., Bulmus, V., Liu, J., Davis, T.P. (2010). *NPG Asia Mater.*, **1**, 23–30.

Boyer, D., Tamarat, P., Maali, A., Lounis, B., Orrit, M. (2002). *Science*, **5584**, 1160–1163.

Bradley, M., Lazim, A.M., Eastoe, J. (2011). *Polymers*, **3**(3), 1036–1050.

Brahmamdam, P., Watanabe, E., Unsinger, J., Chang, K.C., Schierding, W., Hoekzema, A.S., Zhou, T.T., McDonough, J.S., Holemon, H., Heidel, J.D., Coopersmith, C.M., McDunn, J.E., Hotchkiss, R.S. (2009). *Shock*, **32**, 131–139.

Brambilla, D., Le Droumaguet, B., Nicolas, J., Hashemi, S.H., Wu, L.-P., Moghimi, S.M., Couvreur, P., Andrieux, K. (2011a). *Nanomed. Nanotechnol. Biol. Med.*, **7**, 521–540.

Brambilla, D., Souguir, H., Nicolas, J., Mackiewicz, N., Verpillot, R., Le Droumaguet, B., Taverna, M., Couvreur, P., Andrieux, K. (2011b). *J. Biotechnol.*, **156**, 338–340.

Brambilla, D., Verpillot, R., Le Droumaguet, B., Nicolas, J., Taverna, M., Koňa, J., Lettiero, B., Hashemi, S.H., De Kimpe, L., Canovi, M., Gobbi, M., Nicolas, V., Scheper, W., Moghimi, S. M., Tvaroŝka, I., Couvreur, P., Andrieux, K. (2012). *ACS Nano*, **6**, 5897–5908.

Brambilla, D., Verpillot, R., Taverna, M., De Kimpe, L., Le Droumaguet, B., Nicolas, J., Canovi, M., Gobbi, M., Mantegazza, F., Salmona, M., Nicolas, V., Scheper, W., Couvreur, P., Andrieux, K. (2010). *Anal. Chem.*, **82**, 10083–10089.

Branderhorst, H.M., Kooij, R., Salminen, A., Jongeneel, L.H., Arnusch, C.J., Liskamp, R.M., Finne, J., Pieters, R.J. (2008). *Org. Biomol. Chem.*, **6**, 1425–1434.

Brannon-Peppas, L., Blanchette, J.O. (2004). *Adv. Drug. Delivery Rev.*, **56**, 1649–1659.

Brigger, I., Dubernet, C., Couvreur, P. (2002). *Adv. Drug Delivery Rev.*, **54**, 631–651.

Brinkley, M. (1992). *Bioconj. Chem.*, **1**, 2–13.

Brito, L.A., Chandrasekhar, S., Little, S.R., Amiji, M.M. (2010). *Biomed. Eng. Online*, **9**, 56.

Brooks, P.C., Clark, R.A., Cheresh, D.A. (1994). *Science*, **264**, 569–571.

Bruchez, M., Moronne, M., Gin, P., Weiss, S., Alivisatos, A.P. (1998). *Science*, **5385**, 2013–2016.

Brunton, L., Chabner, B., Knollman, B. (2010). *Goodman and Gilman's The Pharmacological Basis of Therapeutics*, 12th ed., McGraw-Hill Professional.

Brust, M., Walker, M., Bethell, D., Schiffrin, D.J., Whyman, R. (1994). *J. Chem. Soc.: Chem. Commun.*, **7**, 801–802.

Bruyere, H., Westwell, A.D., Jones, A.T. (2010). *Bioorg. Med. Chem. Lett.*, **20**, 2200–2203.

Bryant, L.H., Brechbiel, M.W., Wu, C., Bulte, J.W., Herynek, V., Frank, J.A. (1999). *J. Magn. Reson. Imaging*, **9**, 348–352.

Bu, L., Gan, L.C., Guo, X.Q., Chen, F.Z., Song, Q., Qi, Z., Gou, X.J., Hou, S.X., Yao, Q. (2013). *Int. J. Pharm.*, **452**(1-2), 355–362.

Bucciantini, M., Giannoni, E., Chiti, F., Baroni, F., Formigli, L., Zurdo, J., Taddei, N., Ramponi, G., Dobson, C. M., Stefani, M. (2002). *Nature*, **416**, 507–511.

Bueter, C.L., Lee, C.K., Rathinam, V.A.K., Healy, G.J., Taron, C.H., Specht, C.A., Levitz, S.M. (2011). *J. Biol. Chem.*, **286**, 35447–35455.

Bugaj, A.M. (2011). *Photoc. Photobio. Sci.*, **10**, 1097–1109.

Buhleier, E., Wehner, W., Vögtle, F. (1978). *Synthesis*, **2**, 155–158.

Bulte, J.W.M., et al. (2001). *Nat. Biotechnol.*, **19**, 1141–1147.

Bumb, A., Brechbiel, M.W., Choyke, P. (2010). *Acta Radiol.*, **51**, 751–767.

Buonaguro, F., Buonaguro, L. (2013). *Expert Rev. Vaccines*, **12**, 99–99.

Burtea, C., Laurent, S., Mahieu, I., Larbanoix, L., Roch, A., Port, M., Rousseaux, O., Ballet, S., Murariu, O., Toubeau, G., Corot, C., Vander Elst, L., Muller, R.N. (2011). *Contrast Media Mol. Imag.*, **4**, 236–250.

Burtea, C., Laurent, S., Roch, A., Vander Elst, L., Muller, R.N. (2005). *J. Inorg. Biochem.*, **99**, 1135–1144.

Bütün, V., Atay, A., Tuncer, C., Baş, Y. (2011). *Langmuir*, **27**, 12657–12665.

Buxton, D.B. (2009). *Nanomedicine (London)*, **4**, 331–339.

Buzea, C., Pacheco, I.I., Robbie, K. (2008). *Biointerphases*, **2**, MR17–MR71.

Cabaleiro-Lago, C., Quinlan-Pluck, F., Lynch, I., Dawson, K.A., Linse, S. (2010). *ACS Chem. Neurosci.*, *1*, 279–287.

Cabaleiro-Lago, C., Quinlan-Pluck, F., Lynch, I., Lindman, S., Minogue, A.M., Thulin, E., Walsh, D.M., Dawson, K.A., Linse, S. (2008). *J. Am. Chem. Soc.*, **130**, 15437–15443.

Cai, S., Yang, Q., Bagby, T.R., Forrest, M.L. (2011). *Adv. Drug Delivery Rev.*, **63**, 901–908.

Cai, T., Marquez, M., Hu, Z. (2007). *Langmuir*, **23**, 8663–8666.

Cai, X., Hu, J., Xiao, J., Cheng, Y. (2013). *Expert. Opin. Ther. Pat.*, 1–15.

Cai, X.M., Xie, H.L., Liu, M.Z., Zha, X.L. (2008). *BMC Biotechnol.*, **8**, 74.

Cajot, S., Lautram, N., Passirani, C., Jerome, C. (2011). *J. Controlled Release*, **152**, 30–36.

Calabrese, G., Nesnas, J.J., Barbu, E., Fatouros, D., Tsibouklis, J. (2012). *Drug Discovery Today*, **17**, 153–159.

Calabretta, M.K., Kumar, A., McDermott, A.M., Cai, C. (2007). *Biomacromolecules*, **8**, 1807–1811.

Calderon, M., Graeser, R., Kratz, F., Haag, R. (2009). *Bioorg. Med. Chem. Lett.*, **19**, 3725–3728.

Callari, M., Aldrich-Wright, J., de Souza, P.L., Stenzel, M.H. (2014). *Prog. Polym. Sci.*, **39**, 1614–1643.

Calon, F., Tahar, A.H., Blanchet, P.J., Morisette, M., Grondin, R., Goulet, M. et al. (2000). *Trends Neurosci.*, **23**, S92–S100.

Caminade, A.-M., Padié, C., Laurent, R., Maraval, A., Majoral, J.-P. (2006). *Sensors*, **6**, 901–914.

Caminade, A.M., Turrin, C.O., Majoral, J.P. (2008). *Chem.: Eur. J.*, **14**, 7422–7432.

Camp, E.R., Wang, C., Little, E.C., Watson, P.M., Pirollo, K.F., Rait, A., Cole, D.J., Chang, E.H., Watson, D.K. (2013). *Cancer Gene Ther.*, **20**, 222–228.

Campbell, R.B. (2007). *Nanomedicine (London)*, **2**, 649–52.

Campbell, R.B., Fukumura, D., Brown, E.B., Mazzola, L.M., Izumi, Y., Jain, R.K., Torchilin, V.P., Munn, L.L. (2002). *Cancer Res.*, **62**, 6831–6836.

Cao, L., Wang, X., Meziani, M.J., Lu, F., Wang, H., Luo, P.G., Lin, Y., Harruff, B.A., Veca, L.M., Murray, D., Xie, S.-Y., Sun, Y.-P. (2007). *J. Am. Chem. Soc.*, **37**, 11318–11319.

Cao, Y.C. (2008). *Nanomedicine*, **3**, 467–469.

Capule, C.C., Yang, J. (2012). *Anal. Chem.*, **84**, 1786–1791.

Carafa, M., Marianecci, C., Di Marzio, L., De Caro, V., Giandalia, G., Giannola, L.I., Santucci, E. (2010). *J. Liposome Res.*, **20**, 250–257.

Carlisle, R., Seymour, L.W., Coussios, C.C. (2013). *Pharm. Res.*, **30**, 352–361.

Caruso, F., Trau, D., Möhwald, H., Renneberg, R. (2000). *Langmuir*, **4**, 1485–1488.

Casey, J.R., Grinstein, S., Orlowski, J. (2010). *Nat. Rev. Mol. Cell Biol.*, **11**, 50–61.

Cass, W.A., Peters, L.E., Fletcher, A.M., Yurek D.M. (2012). *Neurochem. Int.*, **60**, 186–191.

Castillejos, E., Debouttière, P.J., Roiban, L., Solhy, A., Martinez, V., Kihn, Y., Ersen, O., Philippot, K., Chaudret, B., Serp, P. (2009). *Angew. Chem., Int. Ed. Engl.*, **48**, 2529–2533.

Castonguay, A., Ladd, E., van de Ven, T.G., Kakkar, A. (2012). *New J. Chem.*, **36**, 199–204.

Casula, M.F., Floris, P., Innocenti, C., Lascialfari, A., Marinone, M., Corti, M., Sperling, R.A., Parak, W.J., Sangregorio, C. (2010). *Chem. Mater.*, **22**, 1739–1748.

Chaban, V.V., Savchenko, T.I., Kovalenko, S.M., Prezhdo, O.V. (2010). *J. Phys. Chem. B.*, **114**, 13481–13486.

Chabre, Y.M., Roy, R. (2008). *Curr. Top. Med. Chem.*, **8**, 1237–1285.

Chafekar, S.M., Malda, H., Merkx, M., Meijer, E.W., Viertl, D., Lashuel, H.A., Baas, F., Scheper, W. (2007). *ChemBioChem*, **8**, 1857–1864.

Chakravarthi, S.S., Robinson, D.H., De, S. (2007). Nanoparticles prepared using natural and synthetic polymers, In *Nanoparticulate Drug Delivery Systems*, Vol. 166, pp. 51–60, Eds. Thassu, D., Deleers, M., Pathak, Y., CRC Press, USA.

Chakravarty, P., Marches, R., Zimmerman, N.S., Swafford, A.D., Bajaj, P., Musselman, I.H., Pantano, P., Draper, R.K., Vitetta, E.S. (2008). *Proc. Natl. Acad. Sci. U S A*, **105**, 8697–8702.

Chan, H.-M., Xiao, L., Yeung, K.-M., Ho, S.-L., Zhao, D., Chan, W.-H., Li, H.-W. (2012). *Biomaterials*, **33**, 4443–4450.

Chan, J.M., Rhee, J.W., Drum, C.L., Bronson, R.T., Golomb, G., Langer, R., Farokhzad, O.C. (2011). *Proc. Natl. Acad. Sci. U S A*, **108**, 19347–19352.

Chan, J.M., Zhang, L., Tong, R., Ghosh, D., Gao, W., Liao, G., Yuet, K.P., Gray, D., Rhee, J.-W., Cheng, J., Golomb, G., Libby, P., Langer, R., Farokhzad, O.C. (2010). *Proc. Natl. Acad. Sci. U S A*, **107**, 2213–2218.

Chan, W.C.W., Nie, S. (1998). *Science*, **5385**, 2016–2018.

Chan, Y., Bulmus, V., Zareie, M.H., Byrne, F.L., Barner, L., Kavallaris, M. (2006). *J. Controlled Release*, **115**, 197–207.

Chandrawati, R., Caruso, F. (2012). *Langmuir*, **28**, 13798–13807.

Chang, B., Guo, J., Liu, C., Qian, J., Yang, W. (2010a). *J. Mater. Chem.*, **20**, 9941–9947.

Chang, B., Sha, X., Guo, J., Jiao, Y., Wang, C., Yang, W. (2011). *J. Mater. Chem.*, **21**, 9239–9247.

Chang, C., Wang, Z.-C., Quan, C.-Y., Cheng, H., Cheng, S.-X., Zhang, X.-Z., Zhuo, R.-X. (2007). *J. Biomater. Sci. Polym. Ed.*, **18**, 1591–1599.

Chang, P.V., Prescher, J.A., Sletten, E.M., Baskin, J.M., Miller, I.A., Agard, N.J., Lo, A., Bertozzi, C.R. (2010b). *Proc. Natl. Acad. Sci. U S A*, **5**, 1821–1826.

Chang, Y.-R., Lee, H.-Y., Chen, K., Chang, C.-C., Tsai, D.-S., Fu, C.-C., Lim, T.-S., Tzeng, Y.-K., Fang, C.-Y., Han, C.-C., Chang, H.-C., Fann, W. (2008). *Nat. Nanotechnol.*, **5**, 284–288.

Chang, Y.-T., Liao, P.-Y., Sheu, H.-S., Tseng, Y.-J., Cheng, F.-Y., Yeh, C.-S. (2012). *Adv. Mater.*, **24**, 3309–3314.

Chao, X., Guo, L., Zhao, Y., Hua, K., Peng, M., Chen, C., Cui, Y. (2011). *J. Drug Targeting*, **19**, 161–170.

Charleux, B., Delaittre, G., Rieger, J., D'Agosto, F. (2012). *Macromolecules*, **45**, 6753–6765.

Chau, Y., Padera, R.F., Dang, N.M., Langer, R. (2006). *Int. J. Cancer*, **118**, 1519–1526.

Chaudhuri, P., Soni S., Sengupta, S. (2010). *Nanotechnology*, **21**, 025102 (11 pp).

Chaudhury, A., Das, S., Bunte, R.M., Chiu, G.N. (2012). *Int. J. Nanomed.*, **7**, 739–751.

Chen, C., Xie, X.X., Zhou, Q., Zhang, F.Y., Wang, Q.L., Liu, Y.Q., Zou, Y., Tao, Q., Ji, X.M., Yu, S.Q. (2012a). *Nanotechnology*, **23**, 045104.

Chen, D., Jiang, M., Li, N., Gu, H., Xu, Q., Ge, J., Xia, X., Lu, J. (2010a). *J. Mater. Chem.*, **20**, 6422–6429.

Chen, G., He, Y., Wu, X., Zhang, Y., Luo C., Jing P. (2012b). *Braz. J. Med. Biol. Res.*, **45**, 771–776.

Chen, H., Maa, X., Li, Z., Shi, Q., Zheng, W., Liu, Y., Wang P. (2012c). *Biomed. Pharmacother.*, **66**, 334–338.

Chen, H., Zhang, X., Dai, S., Ma, Y., Cui, S., Achilefu, S., Gu, Y. (2013a). *Theranostics*, **3**, 633–649.

Chen, J., Patil, S., Seal, S., McGinnis, J. F. (2006). *Nat. Nanotechnol.*, **1**, 142–150.

Chen, J.Y., Chen, S.Y., Zhao, X.R., Kuznetsova, L.V., Wong, S.S., Ojima, I. (2008a). *J. Am. Chem. Soc.*, **130**, 16778–16785.

Chen, M.L., He, Y.J., Chen, X.W., Wang, J.H. (2012d). *Langmuir*, **28**, 16469–16476.

Chen, M.L., He, Y.J., Chen, X.W., Wang, J.H. (2013b). *Bioconjug. Chem.*, **24**, 387–397.

Chen, W., Gu, B., Wang, H., Pan, J., Lu, W., Hou, H. (2008b). *Int. J. Pharm.*, **362**, 133–140.

Chen, W.H., Xu, X.D., Jia, H.Z., Lei, Q., Luo, G.F., Cheng, S.X., Zhuo, R.X., Zhang, X.Z. (2013c). *Biomaterials*, **34**, 8798–8807.

Chen, X., Chen, H., Tripisciano, C., Jedrzejewska, A., Rümmeli, M.H., Klingeler, R., Kalenczuk, R.J., Chu, P.K., Borowiak-Palen, E. (2011a). *Chem. Eur. J.*, **17**, 4454–4459.

Chen, X., Wu, P., Rousseas, M., Okawa, D., Gartner, Z., Zettl, A., Bertozzi, C.R. (2009). *J. Am. Chem. Soc.*, **131**, 890–891.

Chen, Y., Sun, J., Lu, Y., Tao, C., Huang, J., Zhang, H., Yu, Y., Zou, H., Gao, J., Zhong, Y. (2013d). *Int. J. Nanomed.*, **8**, 1573–1593.

Chen, Y., Zheng, X., Qian, H., Mao, Z., Ding, D., Jiang, X. (2010b). *Appl. Mater. Interfaces*, **2**, 3532–3538.

Chen, Y.C., et al. (2012e). DOI:10.1016/j.biomaterials.2012.02.059.

Chen, Y.-C., Liao, L.-C., Lu, P.-L., Lo, C.-L., Tsai, H.-C., Huang, C.-Y., Wei, K.-C., Yen, T.-C., Hsiue, G.-H. (2012f). *Biomaterials*, 1–13.

Chen, Y.C., Liao, L.C., Lu, P.L., Lo, C.L., Tsai, H.C., Huang, C.Y., Wei, K.C., Yen, T.C., Hsiue, G.H. (2012g). *Biomaterials*, **33**, 4576–4588.

Chen, Z., He, Y., Wang, Y., Wang, X. (2011b). *Macromol. Rapid Commun.*, **32**, 977–982.

Chen, Z., Li, Z., Lin, Y., Yin, M., Ren, J., Qu, X. (2013e). *Chemistry*, **19**, 1778–1783.

Cheng, G., He, Y., Xie, L., Nie, Y., He, B., Zhang, Z., Gu, Z. (2012a). *Int. J. Nanomed.*, **7**, 3991–4006.

Cheng, J., Ji, R., Gao, S.J., Du, F.S., Li, Z.C. (2012b). *Biomacromolecules*, **13**, 173–179.

Cheng, R., Feng, F., Meng, F., Deng, C., Feijen, J., Zhong, Z. (2011a). *J. Controlled Release*, **152**, 2–12.

Cheng, R., Meng, F., Deng, C., Klok, H.A., Zhong, Z. (2013). *Biomaterials*, **34**, 3647–3657.

Cheng, Y., Gao, Y., Rao, T., Li, Y., Xu, T. (2007). *Comb. Chem. High. T. Scr.*, **10**, 336–349.

Cheng, Y., Hao, J., Lee, L.A., Biewer, M.C., Wang, Q., Stefan, M.C. (2012c). *Biomacromolecules*, **13**, 2163–2673.

Cheng, Y., Zhao, L., Li, Y., Xu, T. (2011b). *Chem. Soc. Rev.*, **40**, 2673–2703.

Cheraghipour, E., Javadpour, S. (2013). *Int. J. Hyperthermia*, **29**, 511–519.

Chernousova, S., Epple, M. (2013). *Angew. Chem., Int. Ed.*, **6**, 1636–1653.

Chi, C., Cai, T., Hu, Z. (2009). *Langmuir*, **25**, 3814–3819.

Chi, E.Y., Ege, C., Winans, A., Majewski, J., Wu, G., Kjaer, K., Lee, K.Y.C. (2008). *Proteins*, **72**, 1–24.

Chi, E.Y., Frey, S.L., Lee, K.Y.C. (2007). *Biochemistry*, **46**, 1913–1924.

Chi, X., Huang, D., Zhao, Z., Zhou, Z., Yin, Z., Gao, J. (2012). *Biomaterials*, **33**, 189–206.

Chiang, W.-H., Ho, V.T., Chen, H.-H., Huang, W.-C., Huang, Y.-F., Lin, S.-C., Chern, C.-S., Chiu, H.-C. (2013). *Langmuir*, **29**, 6434–6443.

Chigurupati, S., Mughal, M.R., Okun, E., Das, S., Kumar, A., McCaffery, M., Seal, S., Mattson, M.P. (2012). *Biomaterials*, **34**, 2194–2201.

Chikae, M., Fukuda, T., Kerman, K., Idegami, K., Miura, Y., Tamiya, E. (2008). *Bioelectrochemistry*, **74**, 118–123.

Chin, C.F., Yap, S.Q., Li, J., Pastorin, G., Ang, W.H. (2014). *Chem. Sci.*, **5**, 2265.

Chiti, F., Dobson, C.M. (2006). *Annu. Rev. Biochem.*, **75**, 333–366.

Chiu, H.-C., Lin, Y.-W., Huang, Y.-F., Chuang, K., Chern, C.S. (2008). *Angew. Chem., Int. Ed.*, **47**, 1875–1878.

Cho, E.C., Zhang, Q., Xia, Y. (2011). *Nat. Nanotechnol.*, **6**, 385–391.

Cho, H., Bae, J., Garripelli, V.K., Anderson, J.M., Jun, H.W., Jo, S. (2012a). *Chem. Commun. (Camb.)*, **48**(48), 6043–6045.

Cho, K., Wang, X., Nie, S., Chen, Z.G., Shin, D.M. (2008). *Clin. Cancer Res.*, **14**, 1310–1316.

Cho, W.-S., Duffin, R., Poland, C.A., Duschl, A., Oostingh, G.J., MacNee, W., Bradley, M., Megson, I.L., Donaldson, K. (2012b). *Nanotoxicology*, **6**, 22–35.

Choi, J.S., Choi, H.J., Jung, D.C., Lee, J.H., Cheon, J. (2008). *Chem. Commun.*, **19**, 2197–2199.

Choi, K.Y., Yoon, H.Y., Kim, J.H., Bae, S.M., Park, R.W., Kang, Y.M., Kim, I.S., Kwon, I.C., Choi, K., Jeong, S.Y., Kim, K., Park, J.H. (2011). *ACS Nano*, **5**, 8591–8599.

Choi, S.H., Yoon, J.J., Park, T.G. (2002). *J. Colloid Interface Sci.*, **251**, 57–63.

Choi, S.W., Kim, W.S., Kim. J.H. (2003). *J. Dispersion Sci. Technol.*, **24**, 475–487.

Chong, C.S.W., Cao, M., Wong, W.W., Fischer, K.P., Addison, W.R., Kwon, G.S., Tyrrell, D.L., Samuel, J. (2005). *J. Controlled Release*, **102**, 85–99.

Choo-Smith, L.-P., Garzon-Rodriguez, W., Glabe, C.G., Surewicz, W.K. (1997). *J. Biol. Chem.*, **272**, 22987–22990.

Chopra, K., Misra, S., Kuhad, A. (2011). *Expert Opin. Pharmacother.*, **12**, 335–350.

Chou, L.Y., Ming, K., Chan, W.C. (2011). *Chem. Soc. Rev.*, **40**, 233–245.

Chow, P.Y., Gan, L.M. (2005). *Adv. Polym. Sci.*, **175**, 257–298.

Chu, D.S., Johnson, R.N., Pun, S.H. (2012). *J. Controlled Release*, **157**, 445–454.

Chua, B.Y., Al Kobaisi, M., Zeng, W., Mainwaring, D., Jackson, D.C. (2011). *Mol. Pharm.*, **9**, 81–90.

Chujo, Y. (2007). KONA, 25, 255.

Chujo, Y., Sada, K., Kawasaki, T., Saegusa, T. (1992). *Polym. J.*, **24**, 1301–1306.

Cirillo, D., Agostini, F., Klus, P., Marchese, D., Rodriguez, S., Bolognesi, B., Tartaglia, G.G. (2012a). *RNA*, Dec 21. [Epub ahead of print].

Cirillo, G., Iemma, F., Spizzirri, G., Puoci, F., Curcio, M., Parisi, O.I., Picci N. (2012b). *J. Biomater. Sci. Polym. Ed.*, 823–844.

Clancy, J.P., Minic, P., Dupont, L.J., Goss, C.H., Quittner, A.L., Lymp, J.F., Burns, J.L., Govan, J.R., Ramsey, B., Gupta, R. (2010). *Pediatr. Pulmonol.*, **33**(45 Suppl), 299–300.

Clark, K.L., Hughes, S.A., Bulsara, P., Coates, J., Moores, K., Parry, J., Carr, M., Mayer, R.J., Wilson, P., Gruenloh, C., Levin, D., Darton, J., Weber, W.M., Sobczak, K., Gill, D.R., Hyde, S. C., Davies, L.A., Pringle, I.A., Sumner-Jones, S.G., Jadhav, V., Jamison, S., Strapps, W.R., Pickering, V., Edbrooke, M.R. (2013). *Mol. Ther. Nucleic Acids*, **2**, e65.

Clark, M.R., Aliyar, H.A., Lee, C., Jay, J.I., Gupta, K.M., Watson, K.M., Stewart, R.J., Buckheit, R.W., Kiser, P.F. (2011). *Int. J. Pharm.*, **413**, 10–18.

Clayton, R., Hardman, J., LaBranche, C.C., McReynolds, K.D. (2011). *Bioconjug. Chem.*, **22**, 2186–2197.

Clemens, D.L., Lee, B.Y., Xue, M., Thomas, C.R., Meng, H., Ferris, D., Nel, A.E., Zink, J.I., Horwitz, M.A. (2012). *Antimicrob. Agents Chemother.*, **56**, 2535–2545.

Clemons, T.D., Viola, H.M., House, M.J., Iyer, K.S., Hool, L.C. (2013). *ACS Nano*, **7**, 2212–2220.

ClinicalTrials.gov. (2012). A Phase 3 Study of SPL7013 Gel (VivaGel) for the Treatment of Bacterial Vaginosis, http://clinicaltrials.gov/ct2/show/NCT01577537.

Cobley, C.M., Skrabalak, S.E., Campbell, D.J., Xia, Y.N. (2009). *Plasmonics*, **2**, 171–179.

Cohen, S., Yoshioka, T., Lucarelli, M., Hwang, L.H., Langer, R. (1991). *Pharm. Res.*, **8**(6), 713–720.

Coles, M., Bicknell, W., Watson, A.A., Fairlie, D.P., Craik, D.J. (1998). *Biochemistry*, **37**, 11064–11077.

Colombo, M., Carregal-Romero, S., Casula, M.F., Gutierrez, L., Morales, M.P., Bohm, I.B., Heverhagen, J.T., Prosperi, D., Parak, W.J. (2012). *Chem. Soc. Rev.*, **41**, 4306–4334.

Colon, J., Herrera, L., Smith, J., Patil, S., Komanski, C., Kupelian, P., Seal, S., Jenkins, D. W, Baker, C. H. (2009). *Nanomedicine*, **5**, 225–231.

Colon, J., Hsieh, N., Ferguson, A., Kupelian, P., Seal, S., Jenkins, D.W, Baker, C.H. (2010). *Nanomedicine*, **6**, 698–705.

Colson, Y.L., Grinstaff, M.W. (2012). *Adv. Mater.*, **24**, 3878–3886.

Cooke, V.G., et al. (2012). *Cancer Cell*, **21**, 66–81.

Corbett, A., Ballard, C. (2012). *Expert Opin. Emerging Drugs*, **17**, 147–156.

Corot, C., Robert, P., Idee, J.-M., Port, M. (2006). *Adv. Drug Delivery Rev.*, **14**, 1471–1504.

Costantini, P., Jacotot, E., Decaudin, D., Kroemer, G. (2000). *J. Natl. Cancer Inst.*, **92**, 1042–1053.

Cozzoli, P.D., Manna, L., Curri, M.L., Kudera, S., Giannini, C., Striccoli, M., Agostiano, A. (2005). *Chem. Mater.*, **6**, 1296–1306.

Craig, D., Simpson, J., Faulds, K., Graham, D. (2013). *Chem. Commun.*, 1, 30–32.

Crawford, N.P., Colliver, D.W., Galandiuk, S. (2003). *J. Surg. Oncol.*, **84**, 239–248.

Cruz, L.J., Tacken, P.J., Rueda, F., Domingo, J.C., Albericio, F., Figdor, C.G. (2012). *Methods Enzymol.*, **509**, 143–163.

Cui, W., Lu, X., Cui, K., Niu, L., Wei, Y., Lu, Q. (2012). *Langmuir*, **28**, 9413–9420.

Cui, Z., Lockman, P.R., Atwood, C.S., Hsu, C.H., Gupte, A., Allen, D.D., Mumper, R.J. (2005). *Eur. J. Pharm. Biopharm.*, **59**, 263–272.

Cullis, P.R., Fenske, D.B., Hope, M.J. (1996). Physical properties and functional roles of lipid in membranes, In *Biochemistry of Lipids, Lipoproteins and Membranes*, Eds. Vance, D.E., Vance, J., Elsevier, Amsterdam.

Cuomo, F., Lopez, F., Miguel, M.G., Lindman, B. (2010). *Langmuir*, **26**, 10555–10560.

Da Ma. (2014). *Nanoscale*, **6**, 6415–6425.

Dabbousi, B.O., Rodriguez-Viejo, J., Mikulec, F.V., Heine, J.R., Mattoussi, H., Ober, R., Jensen, K.F., Bawendi, M.G. (1997). *J. Phys. Chem. B*, **46**, 9463–

9475.

Dadsetan, M., Taylor, K.E., Yong, C., Bajzer, Ž., Lu, L., Yaszemski, M.J. (2013). *Acta Biomater.*, **9**, 5438–5446.

Dagallier, C., Dietsch, H., Schurtenberger, P., Scheffold, F. (2010). *Soft Matter*, **6**, 2174–2177.

Dagar, A., Kuzmis, A., Rubinstein, I., Sekosan, M., Onyuksel, H. (2012). *Drug Delivery Transl. Res.*, **2**, 454–462.

Dahan, M., Laurence, T., Pinaud, F., Chemla, D.S., Alivisatos, A.P., Sauer, M., Weiss, S. (2001). *Opt. Lett.*, **11**, 825–827.

Daigle, J.C., Claverie, J.P. (2008). *J. Nanomater.*, Article ID 609184 (8 pages).

Dam, T.K., Brewer, C.F. (2010). *Adv. Carbohyd. Chem. Biochem.*, **63**, 139–164.

Dandamudi, S., Patil, V., Fowle, W., Khaw, B.A., Campbell, R.B. (2009). *Cancer Sci.*, **100**, 1537–1543.

Danhier, F., Ansorena, E., Silva, J.M., Coco, R., Le Breton, A., Préat, V. (2012). *J. Controlled Release*, **161**, 505–522.

Danhier, F., Feron, O., Preat, V. (2010). *J. Controlled Release*, **148**, 135–146.

Danhier, F., Vroman, B., Lecouturier, N., Crokart, N., Pourcelle, V., Freichels, H., Jerome, C., Marchand-Brynaert, J., Feron, O., Preat, V. (2009). *J. Controlled Release*, **140**, 166–173.

Daniel, M.-C., Astruc, D. (2004). *Chem. Rev.*, **1**, 293–346.

Das, M., Patil, S., Bhargava, N., Kang, J.-F, Riedel, L.M., Seal, S., Hickman, J. J. (2007a). *Biomaterials*, **28**, 1918–1925.

Das, M., Sanson, N., Fava, D., Kumacheva, E. (2007b). *Langmuir*, **23**, 196–201.

Dash, B.P., Satapathy, R., Bode, B.P., Reidl, C.T., Sawicki, J.W., Mason, A.J., Maguire, J.A., Hosmane, N.S. (2012). *Organometallics*, **31**, 2931–2935.

Dastidar, S.G., Sharma, S.K. (1989). *Exp. Cell Biol.*, **57**, 159–164.

Daum, N., Tscheka, C., Neumeyer, A., Schneider, M. (2012). *Wiley Interdiscipl. Rev.: Nanomed. Nanobiotechnol.*, **4**, 52–65.

de Jong, E.R. (2010). Johannes Gutenberg Universität, Mainz.

de Jong, W.H., Borm, P.J. (2008). *Int. J. Nanomed.*, **3**, 133–149.

de Jong, W.H., et al. (2008). *Biomaterials*, **29**, 1912–1919.

de La Zerda, A., Zavaleta, C., Keren, S., et al. (2008). *Nat. Nanotechnol.*, **3**, 557–562.

De Luca, G., Dirksen, M.T., Spaulding, C., Kelbaek, H., Schalij, M., Thuesen, L., van der Hoeven, B., Vink, M.A., Kaiser, C., Musto, C., Chechi, T., Spaziani, G., Diaz de la Llera, L.S., Pasceri, V., Di Lorenzo, E., Violini, R., Cortese, G., Suryapranata, H., Stone, G.W. (2012). *Arch. Intern. Med.*, **172**, 611–621.

de Meyer, F., Smit, R. (2009). *Proc. Natl. Acad. Sci. U S A*, **106**, 3654–3658.

De Volder, M.F.L., Tawfick, S.H., Baughman, R.H., Hart, A.J. (2013). *Science*, **339**, 535–539.

De la Rica, R., Aili, D., Stevens, M.M. (2012). *Adv. Drug. Delivery Rev.*, **64**, 967–978.

Decher, G., Schlenoff, J.B. (2006). *Multilayer Thin Films: Sequential Assembly of Nanocomposite Materials*, John Wiley and Sons.

Deckers, R., Paradissis, A., Oerlemans, C., Talelli, M., Storm, G., Hennink, W.E., Nijsen, J.F. (2013). *Langmuir*, **29**, 9483–9490.

Deka, S.R., Quarta, A., Corato, R. Di, Falqui, A., Manna, L., Cingolani, R., Pellegrino, T. (2010). *Langmuir*, **26**, 10315–10324.

Delcea, M. Möhwald, H., Skirtach, A.G. (2011). *Adv. Drug Delivery Rev.*, **63**, 730–747.

Delogu, L.G., Magrini, A., Bergamaschi, A., Rosato, N., Dawson, M.I., Bottini, N., Bottini, M. (2009). *Bioconjugate Chem.*, **20**, 427–431.

Delogu, L.G., Vidili, G., Venturelli, E., Ménard-Moyon, C., Zoroddu, M.A., Pilo, G., Nicolussi, P., Ligios, C., Bedognetti, D., Sgarrella, F., Manetti, R., Bianco, A. (2012). *Proc. Natl. Acad. Sci. U S A*, **109**, 16612–16617.

Delplace, V., Couvreur, P., Nicolas, J. (2014). *Polym. Chem.*, **5**, 1529–1544.

D'Emanuele, A., Attwood, D. (2005). *Adv. Drug Delivery Rev.*, **57**, 2147–2162.

Demers, L.M., Mirkin, C.A., Mucic, R.C., Reynolds, R.A., Letsinger, R.L., Elghanian, R., Viswanadham, G. (2000). *Anal. Chem.*, **22**, 5535–5541.

Demirel, G.B., von Klitzing, R. (2013). *ChemPhysChem*, **14**, 2833–2840.

Deng, Z.J., Liang, M., Toth, I., Monteiro, M., Minchin, R.F. *Nanotoxicology*, **0**, 1–9.

Derossi, D., Joliot, A.H., Chassaing, G., Prochiantz, A. (1994). *J. Biol. Chem.*, **269**, 10444–10450.

Desgrosellier, J.S., Cheresh, D.A. (2010). *Nat. Rev. Cancer*, **10**, 9–22.

Devalapally, H., Shenoy, D., Little, S., Langer, R., Amiji, M. (2007). *Cancer Chemother. Pharmacol.*, **59**, 477–484.

DeVita, V.T., Jr, Hellman, S., Rosenberg, S.A. (2000). *Cancer: Principles & Practice of Oncology*, 6 Cdr Sub., Lippincott Williams and Wilkins.

Dhanikula, A.B., Panchagnula, R., (2005). *Curr. Drug Delivery*, **2**, 75–91.

Dhanikula, R.S., Argaw, A., Bouchard, J.-F., Hildgen, P. (2008). *Mol. Pharm.*, **5**, 105–116.

Dhar, S., Liu, Z., Thomale, J., Dai, H., Lippard, S.J. (2008). *J. Am. Chem. Soc.*, **130**, 11467–11476.

Dickerson, E.B., Blackburn, W.H., Smith, M.H., Kapa, L.B., Lyon, L.A., McDonald, J.F. (2010). *BMC Cancer*, **10**, 10.

Dicko, A., Kwak, S., Frazier, A.A., Mayer, L.D., Liboiron, B.D. (2010). *Int. J. Pharm.*, **391**, 248–259.

Dieny, B., Speriosu, V.S., Metin, S., Parkin, S.S.P., Gurney, B.A., Baumgart, P., Wilhoit, D.R. (1991). *J. Appl. Phys.*, **8**, 4774–4779.

Dijkgraaf, I., Rijnders, A.Y., Soede, A., Dechesne, A.C., van Esse, G.W., Brouwer, A.J., Corstens, F.H., Boerman, O.C., Rijkers, D.T., Liskamp, R.M. (2007). *Org. Biomol. Chem.*, **5**, 935–944.

Dincer, S., Rzaev, Z.O., Piskin, E. (2006). *J. Polym. Res.*, **13**, 121–131.

Ding, J., Shi, F., Xiao, C., Lin, L., Chen, L., He, C., Zhuang, X., Chen, X. (2011). *Polym. Chem.*, **2**, 2857–2864.

Doane, T.L., Burda, C. (2012). *Chem. Soc. Rev.*, **7**, 2885–2911.

Dobrovolskaia, M.A., Aggarwal, P., Hall, J.B., McNeil, S.E. (2008). *Mol. Pharm.*, **5**, 487–495.

Dobson, C.M. (1999). *TIBS*, **24**, 329–332.

Dolmans, D.E., Fukumura, D., Jain, R.K. (2003). *Nat. Rev. Cancer*, **3**, 380–387.

Domenech, M., Marrero-Berrios, I., Torres-Lugo, M., Rinaldi, C. (2013). *ACS Nano.*, **7**(6), 5091–5101.

Domenech, R., Abian, O., Bocanegra, R., Correa, J., Sousa-Herves, A., Riguera, R., Mateu, M.G., Fernandez-Megia, E., Velazquez-Campoy, A., Neira, J.L. (2010). *Biomacromolecules*, **11**, 2069–2078.

Donath, E., Sukhorukov, G.B, Caruso,F., Davis, S.A., Möhwald, H. (1998). *Angew. Chem., Int. Ed.*, **37**, 2201–2205.

Dong, D.W., Xiang, B., Gao, W., Yang, Z.Z., Li, J.Q., Qi, X.R. (2013). *Biomaterials*, **34**, 4849–4859.

Dong, H., Mantha, V., Matyjaszewski, K. (2009). *Chem. Mater.*, **21**, 3965–3972.

Dong, H., Matyjaszewski, K. (2010). *Macromolecules*, **43**, 4623–4628.

Doroud, D., Rafati, S. (2011). *Expert Rev. Vaccines*, **11**, 69–86.

Douziech-Eyrolles, L., Marchais, H., Herve, K., Munnier, E., Souce, M., Linassier, C., Dubois, P., Chourpa, I. (2007). *Int. J. Nanomedicine*, **2**, 541–550.

Dreaden, E.C., Alkilany, A.M., Huang, X., Murphy, C.J., El-Sayed, M.A. (2012). *Chem. Soc. Rev.*, **7**, 2740–2779.

Dromi, S., et al. (2007). *Clin. Cancer Res.*, **13**, 2722–2727.

Drummond, D.C., Meyer, O., Hong, K., Kirpotin, D.B., Papahadjopoulos, D. (1999). *Pharmacol. Rev.*, **51**, 691–744.

Du, F., Wang, H., Zhao, W., Li, D., Kong, D., Yang, J., Zhang, Y. (2012). *Biomaterials*, **33**, 762–770.

Du, J.Z., Sun, T.M., Song, W.J., Wu, J., Wang, J. (2010). *Angew. Chem., Int. Ed. Engl.*, **49**, 3621–3626.

Duan, C., Gao, J., Zhang, D., Jia, L., Liu, Y., Zheng, D., Liu, G., Tian, X., Wang, F., Zhang, Q. (2011). *Biomacromolecules*, **12**, 4335–4343.

Duan, X., Li, Y. (2013). *Small*, **9**, 1521–1532.

Dubertret, B., Calame, M., Libchaber, A.J. (2001). *Nat. Biotechnol.*, **4**, 365–370.

Dubertret, B., Skourides, P., Norris, D.J., Noireaux, V., Brivanlou, A.H., Libchaber, A. (2002). *Science*, **5599**, 1759–1762.

Dulkeith, E., Morteani, A.C., Niedereichholz, T., Klar, T.A., Feldmann, J., Levi, S.A., van Veggel, F.C., Reinhoudt, D.N., Moller, M., Gittins, D.I. (2002). *Phys. Rev. Lett.*, **20**, 1–4.

Dulkeith, E., Ringler, M., Klar, T.A., Feldmann, J., Javier, A.M., Parak, W.J. (2005). *Nano Lett.*, **4**, 585–589.

Dumortier, H., Lacotte, S., Pastorin, G., Marega, R., Wu, W., Bonifazi, D., Briand, J.P., Prato, M., Muller, S., Bianco, A. (2006). *Nano Lett.*, **6**, 1522–1528.

Dunn, S.S., Tian, S., Blake, S., Wang, J., Galloway, A.L., Murphy, A., Pohlhaus, P.D., Rolland, J.P., Napier, M.E., Desimone, J.M. (2012). *J. Am. Chem. Soc.*, **134**, 7423–7430.

During, M.J., Kaplitt, M.G., Stern, M.B., Eidelberg, D. (2001). *Hum. Gene Ther.*, **12**, 1589–1591.

Dutta, T., Burgess, M., McMillan, N.A., Parekh, H.S. (2010). *Nanomed. Nanotech. Biol. Med.*, **6**, 463–470.

Dvir, T., Timko, B.P., Brigham, M.D., Naik, S.R., Karajanagi, S.S., Levy, O., Jin, H., Parker, K.K., Langer, R., Kohane, D.S. (2011). *Nat. Nanotechnol.*, **6**, 720–725.

Dykman, L., Khlebtsov, N. (2012). *Chem. Soc. Rev.*, **6**, 2256–2282.

Ebato, Y., Kato, Y., Onishi, H., Nagai, T., Machida, Y. (2003). *Drug Devp. Res.*, **58**, 253–257.

Eckhardt, S., Brunetto, P.S., Gagnon, J., Priebe, M., Giese, B., Fromm, K.M. (2013). *Chem. Rev.*, **7**, 4708–4754.

Editorial (2003). *Nat. Biotechnol.*, **21**, 1113.

Edwards, K.A., Baeumner, A.J. (2006). *Talanta*, **68**, 1432–1441.

Eichenbaum, G.M., Kiser, P.F., Dobrynin, A.V., Simon, S.A., Needham, D. (1999a). *Macromolecules*, **32**, 4867–4878.

Eichenbaum, G.M., Kiser, P.F., Shah, D., Simon, S.A., Needham, D. (1999b). *Macromolecules*, **32**, 8996–9006.

Elbakry, A., Zaky, A., Liebl, R., Rachel, R., Goepferich, A., Breunig, M. (2009). *Nano Lett.*, **5**, 2059–2064.

Elbayoumi, T.A., Torchilin, V.P. (2007). *Eur. J. Pharm. Sci.*, **32**, 159–168.

Elechiguerra, J., Burt, J., Morones, J., Camacho-Bragado, A., Gao, X., Lara, H., Yacaman, M. (2005). *J. Nanobiotechnol.*, **3**, 6.

Elele, E., Shen, Y., Susarla, R., Khusid, B., Keyvan, G., Michniak-Kohn, B. (2012). *J. Pharm. Sci.*, **101**, 2523–2533.

Elghanian, R., Storhoff, J.J., Mucic, R.C., Letsinger, R.L., Mirkin, C.A. (1997). *Science*, **5329**, 1078–1081.

Elliott, E., Sloane, B. (1996). *Perspect. Drug Discovery Des.*, **6**, 12–32.

Elliott, G., O'hare, P. (1997). *Cell*, **88**, 223–233.

El-Sayed, M.E.H., Hoffman, A.S., Stayton, P.S. (2005). *J. Controlled Release*, **1–3**, 47–58.

Endo-Takahashi, Y., Negishi, Y., Nakamura, A., Suzuki, D., Ukai, S., Sugimoto, K., Moriyasu, F., Takagi, N., Suzuki, R., Maruyama, K., Aramaki, Y. (2013). *Biomaterials*, **34**, 2807–2813.

Ensign, L.M., Schneider, C., Suk, J.S., Cone, R., Hanes, J. (2012). *Adv. Mater.*, **24**, 3887–3894.

Ercolani, E., Del Gaudio, C., Bianco, A. (2013). *J. Tissue Eng. Regen. Med.* doi: 10.1002/term.1697.

Erez, R., Segal, E., Miller, K., Satchi-Fainaro, R., Shabat, D. (2009). *Bioorg. Med. Chem.*, **17**, 4327–4335.

Erickson, B., DiMaggio, S.C., Mullen, D.G., Kelly, C.V., Leroueil, P.R., Berry, S.A., Baker, J.R., Jr., Orr, B.G., Banaszak Holl, M.M. (2008). *Langmuir*, **24**, 11003–11008.

Esfand, R., Tomalia, D.A. (2001). *Drug Discovery Today*, **6**, 427–436.

Espinosa, E., Zamora, P., Feliu, J., González Barón, M. (2003). *Cancer Treat. Rev.*, **29**, 515–523.

Esser-Kahn, A.P., Odom, S.A., Sottos, N.R., White, S.R., Moore, J.S. (2011). *Macromolecules*, **44**, 5539–5553.

Estrada, A.C., Daniel-da-Silva, A.L., Trindade, T. (2013). *RSC Adv.*, **27**, 10828–10836.

Etrych, T., Jelinkova, M., Rihova, B, Ulbrich, K. (2001). *J. Controlled Release*, **73**, 89–102.

Evans, D.F., Wennerström, H. (1994). *The Colloidal Domain: Where Physics, Chemistry, Biology and Technology Meet*, VCH, New York.

Fang, C., Kievit, F.M., Cho, Y.C., Mok, H., Press, O.W., Zhang, M. (2012). *Nanoscale*, **4**, 7012–7020.

Fang, J., Nakamura, H., Maeda, H. (2011). *Adv. Drug Delivery Rev.*, **63**, 136–151.

Faridi-Majidi, R., Sharifi-Sanjani, N., Agend, F. (2006). *Thin Solid Films*, **515**, 368–374.

Farokhzad, O.C., Cheng, J., Teply, B.A., Sherifi, I., Jon, S., Kantoff, P.W., Richie, J.P., Langer, R. (2006). *Proc. Natl. Acad. Sci. U S A*, **103**, 6315–6320.

Farokhzad, O.C., Langer, R. (2009). *ACS Nano*, **3**, 16–20.

Fatouros, P.P., Corwin, F.D., Chen, Z.J., Broaddus, W.C., Tatum, J.L., Kettenmann, B., Ge, Z., Gibson, H.W., Russ, J.L., Leonard, A.P., Duchamp, J.C., Dorn, H.C. (2006). *Radiology*, **240**, 756–764.

Fattahi, H., Laurent, S., Liu, F., Arsalani, N., Elst, L.V., Muller, R.N. (2011). *Nanomedicine*, **6**, 529–544.

Fayi, S., Warren, C.W.C. (2011). *Nanotechnology*, **49**, 1–7.

Fechheimer, M., Boylan, J.F., Parker, S., Sisken, J.E., Patel, G.L., Zimmer, S.G. (1987). *Proc. Natl. Acad. Sci. U S A*, **84**, 8463–8467.

Feczkó, T., Muskotál, A., Gál, L., Szépvölgyi, J., Sebestyén, A., Vonderviszt, F. (2008). *J. Nanopart. Res.*, **S1**, 227–232.

Fehrenbacher, N., Jaattela, M. (2005). *Cancer Res.*, **65**, 2993–2995.

Felgner, P.L., Gadek, T.R., Holm, M., Roman, R., Chan, H.W., Wenz, M., Northrop, J.P., Ringold, G.M., Danielsen, M. (1987). *Proc. Natl. Acad. Sci. U S A*, **84**, 7413–7417.

Fennimore, A.M., Yuzvinsky, T.D., Han, W.Q., Fuhrer, M.S., Cumings, J., Zettl, A. (2003). *Nature*, **424**, 408–410.

Fenske, D.B. (1993). *Chem. Phys. Lipids*, **64**, 143–162.

Ferguson, C.J., Hughes, R.J., Pham, B.T.T., Hawkett, B.S., Gilbert, R.G., Serelis, A.K., Such, C.H. (2002). *Macromolecules*, **35**, 9243–9245.

Fernández-Trillo, F., Pacheco-Torres, J., Correa, J., Ballesteros, P., Lopez-Larrubia, P., Cerdán, S., Riguera, R., Fernandez-Megia, E. (2011). *Biomacromolecules*, **12**, 2902–2907.

Ferrara, K.W., Borden, M.A., Zhang, H. (2009). *Acc. Chem. Res.*, **42**, 881–892.

Ferrari, M. (2005). *Nat. Rev. Cancer*, **5**, 161–171.

Fielding, L.A., Derry, M.J., Ladmiral, V., Rosselgong, J., Rodrigues, A.M., Ratcliffe, L.P.D., Sugihara, S., Armes, S.P. (2013). *Chem. Sci.*, **4**, 2081–2087.

Fifis, T., Gamvrellis, A., Crimeen-Irwin, B., Pietersz, G.A., Li, J., Mottram, P.L., McKenzie, I.F.C., Plebanski, M. (2004). *J. Immunol.*, **173**, 3148–3154.

Filimon, A., Sima, L., Appelhans, D., Voit, B., Negroiu, G. (2012). *Curr. Med. Chem.*, **19**, 4955–4968.

Finn, W.F. (2006). *Nephrol. Dial. Transpl.*, **21**, i2–i10.

Fiore, V.F., Lofton, M.C., Roser-Page, S., Yang, S.C., Roman, J., Murthy, N., Barker, T.H. (2010). *Biomaterials*, **31**, 810–817.

Fleige, E., Quadir, M.A., Haag, R. (2012). *Adv. Drug. Delivery Rev.*, **64**, 866–884.

Fleming, D.A., Thode, C.J., Williams, M.E. (2006). *Chem. Mater.*, **9**, 2327–2334.

Fletcher, A.M., Kowalczyk, T.H., Padegimas, L., Cooper, M.J., Yurek, D.M. (2011). *Neuroscience*, **194**, 220–226.

Floresco, S.B., West, A.R., Ash, B., Moore, H., Grace, A.A. (2003). *Nat. Neurosci.*, **6**, 968–973.

Fokong, S., Theek, B., Wu, Z., Koczera, P., Appold, L., Jorge, S., Resch-Genger, U., Van, Z.andvoort, M., Storm, G., Kiessling, F., Lammers, T. (2012). *J. Controlled Release*, **163**, 75–81.

Fomina, N., Sankaranarayanan, J., Almutairi, A. (2012). *Adv. Drug. Delivery Rev.*, **64**, 1005–1020.

Fortina, P., Kricka, L.J., Graves, D.J., Park, J., Hyslop, T., Tam, F., Halas, N., Surrey, S., Waldman, S.A. (2007). *Trends Biotechnol.*, **25**, 145–152.

Fox, M.E., Szoka, F.C., Fréchet, J.M. (2009). *Acc. Chem. Res.*, **42**, 1141–1151.

Frankel, A.D., Pabo, C.O. (1988). *Cell*, **55**, 1189–1193.

Frederiksen, L., Anton, K., van Hoogevest, P., Keller, H.R., Leuenberger, H. (1997). *J. Pharm. Sci.*, **86**, 921–927.

Freed, C.R., Greene, P.E., Breeze, R.E., Tsai, W.Y., Dumuchel, W., Kao, R., et al. (2001). *N. Engl. J. Med.*, **344**, 710–719.

Freitas, P.P., Silva, F., Oliveira, N.J., Melo, L.V., Costa, L., Almeida, N. (2000). *Sens. Actuators A*, **1–3**, 2–8.

Freundt, E.C., Maynard, N., Clancy, E.K., Roy, S., Bousset, L., Sourigues, Y., Covert, M., Melki, R., Kirkegaard, K., Brahic, M. (2012). *Ann. Neurol.*, **72**, 517–524.

Froehlich, E., Mandeville, J.S., Jennings, C.J., Sedaghat-Herati, R., Tajmir-Riahi, H.A. (2009). *J. Phys. Chem. B*, **113**, 6986–6993.

Fröhlich, E. (2012). *Int. J. Nanomed.*, 7, 5577.

Fu, Y., Nitecki, D.E., Maltby, D., Simon, G.H., Berejnoi, K., Raatschen, H.-J., Yeh, B.M., Shames, D.M., Brasch, R.C. (2006). *Bioconjug. Chem.*, **17**, 1043–1056.

Fu, Y.-S., Du, X.-W., Kulinich, S.A., Qiu, J.-S., Qin, W.-J., Li, R., Sun, J., Liu, J. (2007). *J. Am. Chem. Soc.*, **51**, 16029–16033.

Fujigaya, T., Morimoto, T., Nakashima, N. (2011). *Soft Matter*, **7**, 2647–2652.

Gabellieri, E., Strambini, G.B., Shcharbin, D., Klajnert, B., Bryszewska, M. (2006). *Biochim. Biophys. Acta*, **1764**, 1750–1756.

Gabizon, A., et al. (1999). *Bioconj. Chem.*, **10**, 289–298.

Gabizon, A.A. (1995). *Adv. Drug Delivery Rev.*, **16**, 285–294.

Gajbhiye, V., Ganesh, N., Barve, J., Jain, N.K. (2013). *Eur. J. Pharm. Sci.*, **48**, 668–679.

Gajbhiye, V., Palanirajan, V.K., Tekade, R.K., Jain, N.K. (2009). *J. Pharm. Pharmacol.*, **61**, 989–1003.

Gajbhiye, V., Vijayaraj Kumar, P., Kumar Tekade, R., Jain, N. (2007). *Curr. Pharm. Design*, **13**, 415–429.

Galaway, F.A., Stockley, P.G. (2013). *Mol. Pharm.*, **10**, 59–68.

Gale, A.J., Elias, D.J., Averell, P.M., Teirstein, P.S., Buck, M., Brown, S.D., Polonskaya, Z., Udit, A.K., Finn, M.G. (2011). *Thromb. Res.*, **128**, e9–e13.

Galow, T.H., Boal, A.K., Rotello, V.M. (2000). *Adv. Mater.*, **8**, 576–579.

Gan, D., Lyon, L.A. (2002). *Macromolecules*, **35**, 9634–9639.

Gan, Q., Lu, X., Yuan, Y., Qian, J., Zhou, H., Lu, X., Shi, J., Liu, C. (2011). *Biomaterials*, **32**, 1932–1942.

Gandubert, V.J., Lennox, R.B. (2005). *Langmuir*, **14**, 6532–6539.

Ganta, S., Devalapally, H., Shahiwala, A., Amiji, M. (2008). *J. Controlled Release*, **126**, 187–204.

Gao, C., Yan, D. (2004a) *Prog. Polym. Sci.*, **29**, 183–275.

Gao, J., Bender, C.M., Murphy, C.J. (2003). *Langmuir*, **21**, 9065–9070.

Gao, L., Fei, J., Zhao, J., Li, H., Cui, Y., Li, J. (2012). *ACS Nano*, **6**, 8030–8040.

Gao, W.-W., Chan, J.M., Farokhzad, O.C. (2010). *Mol. Pharm.*, **7**(6), 1913–1920.

Gao, X.H., Chan, W.C.W., Nie, S.M. (2002). *J. Biomed. Opt.*, **4**, 532–537.

Gao, X.H., Cui, Y.Y., Levenson, R.M., Chung, L.W.K., Nie, S.M. (2004b). *Nat. Biotechnol.*, **8**, 969–976.

Gao, Z., Fain, H.D., Rapoport, N. (2004c). *Mol. Pharm.*, **1**, 317–330.

Gao, Z., Kennedy, A.M., Christensen, D.A., Rapoport, N.Y. (2008). *Ultrasonics*, **48**(2008), 260–270.

Gao, Z.G., Fain, H.D., Rapoport, N. (2005). *J. Controlled Release*, **102**, 203–222.

Gaponik, N., Talapin, D.V., Rogach, A.L., Hoppe, K., Shevchenko, E.V., Kornowski, A., Eychmüller, A., Weller, H. (2002). *J. Phys. Chem. B*, **29**, 7177–7185.

Garcia, A., Marquez, M., Cai, T., Rosario, R., Hu, Z., Gust, D., Hayes, M., Vail, S.A., Park, C.-D. (2007). *Langmuir*, **23**, 224–229.

García-Jimeno, S., Escribano, E., Queralt, J., Estelrich, J. (2012). *Nanoscale Res. Lett.*, **7**, 452.

Garcia-Vallejo, J.J., Koning, N., Ambrosini, M., Kalay, H., Vuist, I., Sarrami-Forooshani, R., Geijtenbeek, T.B., van Kooyk, Y. (2013). *Int. Immunol.*, **25**, 221–233.

Gattás-Asfura, K.M., Weisman, E., Andreopoulos, F.M., Micic, M., Muller, B., Sirpal, S., Pham, S.M., Leblanc, R.M. (2005). *Biomacromolecules*, **6**, 1503–1509.

Gaulding, J.C., Smith, M.H., Hyatt, J.S., Fernandez-Nieves, A., Lyon, L.A. (2012a). *Macromolecules*, **45**, 39–45.

Gaulding, J.C., South, A.B., Lyon, L.A. (2012b). *Colloid Polym. Sci.*, **291**, 99–107.

Geers, B., Dewitte, H., De, S.medt, S.C., Lentacker, I. (2012). *J. Controlled Release*, **164**, 248–255.

Gentine, P., Bourel-Bonnet, L., Frisch, B. (2013). *J. Liposome Res.*, **23**, 11–19.

Gentine, P., Bubel, A., Crucifix, C., Bourel-Bonnet, L., Frisch, B. (2012). *J. Liposome Res.*, **22**, 18–30.

Georganopoulou, D.G., Chang, L., Nam, J.-M., Thaxton, C.S., Mufson, E.J., Klein, W.L., Mirkin, C.A. (2005). *Proc. Natl. Acad. Sci. U S A*, **102**, 2273–2276.

Germano, J., Martins, V.C., Cardoso, F.A., Almeida, T.M., Sousa, L., Freitas, P.P., Piedade, M.S. (2009). *Sensors*, **6**, 4119–4137.

Gerweck, L.E., Seetharaman, K. (1996). *Cancer Res.*, **56**, 1194–1198.

Gespach, C., Bawab, W., De, C.remoux, P., Calvo, F. (1988). *Cancer Res.*, **48**, 5079–5083.

Geszke-Moritz, M., Moritz, M. (2013). *Mater. Sci. Eng. C*, **3**, 1008–1021.

Geys, J., Nemmar, A., Verbeken, E., Smolders, E., Ratoi, M., Hoylaerts, M.F., Nemery, B., Hoet, P.H.M. (2008). *Environ. Health Persp.*, **12**, 1607–1613.

Ghadiali, J.E., Stevens, M.M. (2008). *Adv. Mater.*, **20**, 4359–4363.

Ghasemi, Y., Peymani, P., Afifi, S. (2009). *Acta Biomed.*, **2**, 156–165.

Ghosh, S., Dutta, S., Gomes, E., et al. (2009). *ACS Nano*, **3**, 2667–2673.

Gialeli, C., Theocharis, A.D., Karamanos, N.K. (2011). *FEBS J.*, **278**, 16–27.

Gibson, M.I., O'Reilly, R. (2013). *Chem. Soc. Rev.*, **42**, 7204–7213.

Giehm, L., Christensen, C., Boas, U., Heegaard, P.M., Otzen, D.E. (2008). *Biopolymers*, **89**, 522–529.

Gijs, M.A.M. (2004). *Microfluid. Nanofluid.*, **1**, 22–40.

Gilbert, R.G. (1995). *Emulsion Polymerization: A Mechanistic Approach*, Academic Press, London.

Giljohann, D.A., Seferos, D.S., Daniel, W.L., Massich, M.D., Patel, P.C., Mirkin, C.A. (2010) *Angew. Chem., Int. Ed.*, **19**, 3280–3294.

Gindy, M.E., Ji, S., Hoye, T.R., Panagiotopoulos, A.Z., Prud'homme, R.K. (2008). *Biomacromolecules*, **10**, 2705–2711.

Giordano, C., Albani, D., Gloria, A., Tunesi, M., Rodilossi, S., Russo, T., Forloni, G., Ambrosio, L., Cigada, A. (2011). *Int. J. Artif. Organs*, **34**, 1115–1127.

Giordano, R.J., Edwards, J.K., Tuder, R.M., Arap, W., Pasqualini, R. (2009). *Proc. Am. Thorac. Soc.*, **6**, 411–415.

Giraud, G., Schulze, H., Bachmann, T.T., Campbell, C.J., Mount, A.R., Ghazal, P., Khondoker, M.R., Ross, A.J., Ember, S.W.J., Ciani, I., Tlili, C., Walton, A.J., Terry, J.G., Crain, J. (2009). *Int. J. Mol. Sci.*, **4**, 1930–1941.

Giri, S., Karakoti, A., Graham, R.P., Maguire, J.L., Reilly, C.M., Seal, S., Rattan, R., Shridhar, V. (2013). *PLoS ONE*, **8**, e54578.

Gittins, D.I., Caruso, F. (2001). *Angew. Chem., Int. Ed.*, **16**, 3001–3004.

Gittins, D.I., Caruso, F. (2002). *ChemPhysChem*, **1**, 110–113.

Glabe, C.G. (2006). *Aging*, **27**, 570–575.

Glatzel, S., Laschewsky, A., Lutz, J.-F. (2011). *Macromolecules*, **44**, 413–415.

Glenner, G.G., Wong, C.W. (1984). *Biochem. Biophys. Res. Commun.*, **120**, 885–890.

Gobbi, M., Re, F., Canovi, M., Beeg, M., Gregori, M., Sesana, S., Sonnino, Broglioli, D., Musicanti, C., Gasco, P., Salmona, M., Masserini, M.E. (2010). *Biomaterials*, **31**, 6519–6529.

Goebel, C., Kubickamuranyi, M., Tonn, T., Gonzalez, J., Gleichmann, E. (1995). *Arch. Toxicol.*, **69**, 450–459.

Gohy, J.F., Zhao, Y. (2013). *Chem. Soc. Rev.*, **42**(17), 7117–7129.

Goldberg, M., Langer, R., Jia, X., (2007). *J. Biomater. Sci. Polym. Ed.*, **18**, 241–268.

Gole, A., Dash, C., Ramakrishnan, V., Sainkar, S.R., Mandale, A.B., Rao, M., Sastry, M. (2001). *Langmuir*, **5**, 1674–1679.

Gole, A., Murphy, C.J. (2004). *Chem. Mater.*, **19**, 3633–3640.

Golman, K., Zandt, R. in't, Lerche, M., Pehrson, R., Ardenkjaer-Larsen, J.H. (2006). *Cancer Res.*, **66**, 10855–10860.

Golub, J.S., Kim, Y.-t., Duvall, C.L., Bellamkonda, R.V., Gupta, D., Lin, A.S., Weiss, D., Robert Taylor, W., Guldberg, R.E. (2010). *Am. J. Physiol. Heart Circ. Physiol.*, **298**, H1959–H1965.

Gong, C., Wong, K.-L., Lam, M.H.W. (2008). *Chem. Mater.*, **20**, 1353–1358.

Gonzalez-Nilo, F., Perez-Acle, T., Guinez-Molinos, S., Geraldo, D.A., Sandoval, C., Yevenes, A., Santos, L.S., Laurie, V.F., Mendoza, H., Cachau, R.E. (2011). *Biol. Res.*, **44**, 43–51.

Gonzalez-Toro, D.C., Ryu, J.H., Chacko, R.T., Zhuang, J.M. Thayumanavan, S. (2012). *J. Am. Chem. Soc.*, **134**, 6964–6967.

Goodwin, A.P., Mynar, J.L., Ma, Y., Fleming, G.R., Frechet, J.M. (2005). *J. Am. Chem. Soc.*, **127**, 9952–9953.

Gorelikov, I., Field, L.M., Kumacheva, E. (2004). *J. Am. Chem. Soc.*, **126**, 15938–15939.

Gorman, P.M., Chakrabartty, A. (2001). *Biopolymers*, **60**, 381–394.

Gottesman, M.M., Fojo, T., Bates, S.E. (2002). *Nat. Rev. Cancer*, **2**, 48–58.

Grabolle, M., Kapusta, P., Nann, T., Shu, X., Ziegler, J., Resch-Genger, U. (2009). *Anal. Chem.*, **18**, 7807–7813.

Grabrucker, A.M., Garner, C.C., Boeckers, T.M., Bondioli, L., Ruozi, B., Forni, F., Vandelli, M.A., Tosi, G. (2011). *PLoS ONE*, **6**, e17851.

Grace, A.A. (1991). *Neuroscience*, **41**, 1–24.

Grainger, S.J., El-Sayed, M.E.H (2010). Stimuli-sensitive particles for drug delivery, In *Biologically-Responsive Hybrid Biomaterials*, pp. 171–190, Eds. Jabbari, E., Khademhosseini, A., World Scientific.

Grätzel, M., Thomas, J.K. (1973). *J. Am. Chem. Soc.*, **95**, 6885–6889.

Gray, W.R., Sandberg, L.B., Foster, J.A. (1973). *Nature*, **246**, 461–466.

Grecco, H.E., Lidke, K.A., Heintzmann, R., Lidke, D.S., Spagnuolo, C., Martinez, O.E., Jares-Erijman, E.A., Jovin, T.M. (2004). *Microsc. Res. Techniq.*, **4–5**, 169–179.

Greco, E., Quintiliani, G., Santucci, M.B., Serafino, A., Ciccaglione, A.R., Marcantonio, C., Papi, M., Maulucci, G., Delogu, G., Martino, A., Goletti, D., Sarmati, L., Andreoni, M., Altieri, A., Alma, M., Caccamo, N., Di Liberto, D., De Spirito, M., Savage, N.D., Nisini, R., Dieli, F., Ottenhoff, T.H., Fraziano, M. (2012). *Proc. Natl. Acad. Sci. U S A*, **109**, E1360–E1368.

Greco, F., Vicent, M.J. (2008). *Front. Biosci.*, **13**, 2744–2756.

Grieve, D.J., Byme, J.A., Cave, A.C., Shah, A.M. (2004). *Heart Lung Circ.*, **13**, 132–138.

Griffe, L., Poupot, M., Marchand, P., Maraval, A., Turrin, C.O., Rolland, O., Métivier, P., Bacquet, G., Fournié, J.J., Caminade, A.M. (2007). *Angew. Chem., Int. Ed.*, **46**, 2523–2526.

Griffiths, G., Nystrom, B., Sable, S.B., Khuller, G.K. (2010). *Nat. Rev. Microbiol.*, **8**, 827–834.

Grit, M., Crommelin, D.J.A. (1993). *Chem. Phys. Lipids*, **64**, 3–18.

Gruber, A., Drabenstedt, A., Tietz, C., Fleury, L., Wrachtrup, J., von Borczyskowski, C. (1997). *Science*, **5321**, 2012–2014.

Gu, Y., Zhong, Y., Meng, F., Cheng, R., Deng, C., Zhong, Z. (2013). *Biomacromolecules.*, **14**(8), 2772–2780.

Gu, Y.J., Cheng, J., Jin, J., Cheng, S.H., Wong, W.T. (2011). *Int. J. Nanomed.*, **6**, 2889–2898.

Guan, X., Li, Y., Jiao, Z., Chen, J., Guo, Z., Tian, H., Chen, X. (2013). *Acta Biomater.*, **9**, 7672–7678.

Gueroui, Z., Libchaber, A. (2004). *Phys. Rev. Lett.*, **16**, 1–4.

Guerrero, S., Herance, J.R., Rojas, S., Mena, J.F., Gispert, J.D., Acosta, G.A., Albericio, F., Kogan, M.J. (2012). *Bioconjug. Chem.*, **23**, 399–408.

Guicciardi, M.E., Leist, M., Gores, G.J. (2004). *Oncogene*, **23**, 2881–2890.

Gultepe, E., Reynoso, F.J., Jhaveri, A., Kulkarni, P., Nagesha, D., Ferris, C., Harisinghani, M., Campbell, R.B., Sridhar, S. (2010). *Nanomedicine (London)*, **5**, 1173–1182.

Guo, H.D., Cui, G.H., Yang, J.J., Wang, C., Zhu, J., Zhang, L.S., Jiang, J., Shao, S.J. (2012a). *Biochem. Biophys. Res. Commun.*, **424**, 105–111.

Guo, J., Cheng, W.P., Gu, J., Ding, C., Qu, X., Yang, Z., O'driscoll, C. (2012b). *Eur. J. Pharm. Sci.*, **45**, 521–532.

Guo, L., Yang, S., Yang, C., Yu, P., Wang, J., Ge, W., Wong, G.K.L. (2000). *Appl. Phys. Lett.*, **20**, 2901–2903.

Gupta, A.K., Berry, C., Gupta, M., Curtis, A. (2003). *IEEE Trans. Nanobiosci.*, **4**, 255–261.

Gupta, A.K., Gupta, M. (2005). *Biomaterials*, **18**, 3995–4021.

Gupta, B., Torchilin, V.P. (2007). *Cancer Immunol. Immunother.*, **56**, 1215–1223.

Gurdag, S., Khandare, J., Stapels, S., Matherly, L.H., Kannan, R.M. (2006). *Bioconjug. Chem.*, **17**, 275–283.

Gurunathan, S., Lee, K.-J., Kalishwaralal, K., Sheikpranbabu, S., Vaidyanathan, R., Eom, S.H. (2009). *Biomaterials*, **30**, 6341–6350.

Gutierro, I., Hernández, R.M., Igartua, M., Gascón, A.R., Pedraz, J.L. (2002). *Vaccine*, **21**, 67–77.

Guven, A., Rusakova, I.A., Lewis, M.T., Wilson, L.J. (2012). *Biomaterials*, **33**, 1455–1461.

Haasbeek, C.J., Senan, S., Smit, E.F., Paul, M.A., Slotman, B.J., Lagerwaard, F.J. (2008). *Oncologist*, **13**, 309–319.

Haass, C., Selkoe, D.J. (2007). *Nat. Rev. Mol. Cell. Biol.*, **8**, 101–112.

Hadidi, N., Kobarfard, F., Nafissi-Varcheh, N., Aboofazeli, R. (2013). *AAPS Pharm. Sci. Tech.*, **14**, 593–600.

Haes, A.J., Chang, L., Klein, W.L., Van Duyne, R.P. (2004). *Nano Lett.*, **4**, 1029–1034.

Haes, A.J., Hall, W.P., Chang, L., Klein, W.L., Van Duyne, R.P. (2005). *J. Am. Soc. Chem.*, **127**, 2264–2271.

Hainfeld, J.F., Slatkin, D.N., Focella, T.M., Smilowitz, H.M. (2006). Br. J. Radiol., 939, 248–253.

Hamano, N., Negishi, Y., Omata, D., Takahashi, Y., Manandhar, M., Suzuki, R., Maruyama, K., Nomizu, M., Aramaki, Y. (2013). *Mol. Pharm.*, **10**, 774–779.

Hammer, O. (2012). *mAbs*, **4**, 571–577.

Hamner, K.L., Alexander, C.M., Coopersmith, K., Reishofer, D., Provenza, C., Maye, M.M. (2013). *ACS Nano*, **7**, 7011–7020.

Hamoudeh, M., Kamleh, M.A., Diab, R., Fessi, H. (2008). *Adv. Drug Delivery Rev.*, **60**, 1329–1346.

Hampel, S., Kunze, D., Haase, D., Kramer, K., Rauschenbach, M., Ritschel, M., et al. (2008). *Nanomedicine (London)*, **3**, 175e82.

Han, G., You, C.C., Kim, B.J., Turingan, R.S., Forbes, N.S., Martin, C.T., Rotello, V.M. (2006). *Angew. Chem., Int. Ed.*, **19**, 3165–3169.

Han, H., Davis, M.E. (2013a). *Mol. Pharm.*, **10**(7), 2558–2567.

Han, H., Davis, M.E. (2013b). *Bioconjug. Chem.*, **24**, 669–677.

Han, H.D., Shin, B.C., Choi, H.S. (2006). *Eur. J. Pharm. Biopharm.*, **62**, 110–116.

Han, M.Y., Gao, X.H., Su, J.Z., Nie, S. (2001). *Nat. Biotechnol.*, **7**, 631–635.

Han, S., Kanamoto, T., Nakashima, H., Yoshida, T. (2012). *Carbohyd. Polym.*, **90**(2), 1061–1068.

Han, X., Zhang, X., Zhu, H., Yin, Q., Liu, H., Hu, Y. (2013). *Langmuir*, **29**, 1024–1034.

Han, Y.C., He, C.Q., Cao, M.W., Huang, X. Wang, Y.L., Li, Z.B. (2010). *Langmuir*, **26**, 1583–1587.

Häntzschel, N., Zhang, F., Eckert, F., Pich, A., Winnik, M.A. (2007). *Langmuir*, **23**, 10793–10800.

Hardman, R. (2006). *Environ. Health Persp.*, **2**, 165–172.

Hardy, J., Selkoe, D.J. (2002). *Science*, **297**, 353–356.

Harper, J.D., Lansbury Jr, P.T. (1997). *Annu. Rev. Biochem.*, **66**, 385–407.

Harris, J.M., Martin, N.E., Modi, M. (2001). *Clin. Pharmacokinet.*, **40**, 539–551.

Hartley, D.M., Walsh, D.M., Ye, C.P., Diehl, T., Vasques, S., Vassilev, P.M., Teplow, D.B., Selkoe, D.J. (1999). *J. Neurosci.*, **19**, 8876–8884.

Hartman, K.B., Hamlin, D.K., Wilbur, D.S., Wilson, L.J. (2007). *Small*, **3**, 1496–1499.

Harush-Frenkel, O., Rozentur, E., Benita, S., Altschuler, Y. (2008). *Biomacromolecules*, **9**, 435–443.

Harvey, R.A. (2011). *Pharmacology*, 5th ed., DeVita, Lippincott's Illustrated Reviews Series, USA.

Hatakeyama, H., Akita, H., Harashima, H. (2011). *Adv. Drug. Delivery Rev.*, **63**, 152–160.

Hatakeyama, H., Akita, H., Kogure, K., Oishi, M., Nagasaki, Y., Kihira, Y., Ueno, M., Kobayashi, H., Kikuchi, H., Harashima, H. (2007). *Gene Ther.*, **14**, 68–77.

Hawker, C.J., Frechet, J.M.J. (1990). *J. Am. Chem. Soc.*, **112**, 7638–7647.

Hayashi, K., Nakamura, M., Sakamoto, W., Yogo, T., Miki, H., Ozaki, S., Abe, M., Matsumoto, T., Ishimura, K. (2013). *Theranostics*, **6**, 366–376.

Hayder, M., Fruchon, S., Fournié, J.-J., Poupot, M., Poupot, R. (2011). *Sci. World J.*, **11**, 1367–1382.

He, C., Hou, Y., Han, Y., Wang, Y. (2011a). *Langmuir*, **27**, 4551–4556.

He, J., Tong, X., Zhao, Y. (2009). *Macromolecules*, **42**, 4845–4852.

He, J., Yan, B., Tremblay, L., Zhao, Y. (2011b). *Langmuir*, **27**, 436–444.

He, W., Horn, S.W., Hussain, M.D. (2007). *Int. J. Pharm.*, **334**, 173–178.

Heegaard, P.M., Boas, U., Otzen, D.E. (2007). *Macromol. Biosci.*, **7**, 1047–1059.

Heidecke, C.D., Lindhorst, T.K. (2007). *Chem.: Eur. J.*, **13**, 9056–9067.

Heidenreich, P.A., Albert, N.M., Allen, L.A., Bluemke, D.A., Butler, J., Fonarow, G.C., Ikonomidis, J.S., Khavjou, O., Konstam, M.A., Maddox, T.M., Nichol, G., Pham, M., Pina, I.L., Trogdon, J.G. (2013). *Circ. Heart Fail.*, **6**, 606–619.

Heister, E., Brunner, E.W., Dieckmann, G.R., Jurewicz, I., Dalton, A.B. (2013). *ACS Appl. Mater. Interfaces*, **5**, 1870–1891.

Heister, E., Neves, V., Lamprecht, C., Silva, S.R.P., Coley, H.M., McFadden, J. (2012). *Carbon*, **50**, 622–632.

Heister, E., Neves, V., Tîlmaciu, C., Lipert, K., Beltrán, V.S., Coley, H.M., Silva S.R.P., McFadden, J. (2009). *Carbon*, **47**, 2152–2160.

Heller, A., Brockhoff, G., Goepferich, A. (2012). *Eur. J. Pharm. Biopharm.*, **82**, 1–18.

Hellstrand, E., Sparr, E., Linse, S. (2010). *Biophys. J.*, **98**, 2206–2214.

Helmlinger, G., Sckell, A., Dellian, M., Forbes, N.S., Jain, R.K. (2002). *Clin. Cancer Res.*, **8**, 1284–1291.

Hendrickson, G.R., Lyon, L.A. (2010). *Angew. Chem., Int. Ed. Engl.*, **49**, 2193–2197.

Hendrickson, G.R., Smith, M.H., South, A.B., Lyon, L.A. (2010). *Adv. Funct. Mater.*, **20**, 1697–1712.

Henriksen-Lacey, M., Devitt, A., Perrie, Y. (2011). *J. Controlled Release*, **154**, 131–137.

Hentze, H.-P., Raghavan, S.R., McKelvey, C.A., Kaler, E.W. (2003). *Langmuir*, **19**, 1069–1074.

Herlambang, S., Kumagai, M., Nomoto, T., Horie, S., Fukushima, S., Oba, M., Miyazaki, K., Morimoto, Y., Nishiyama, N., Kataoka, K. (2011). *J. Controlled Release*, **155**, 449–457.

Heuking, S., Borchard, G. (2012). *J. Pharm. Sci.*, **101**, 1166–1177.

Heyden, K.V., Babooram, K., Ahmed, M., Narain, R. (2009). *Eur. Polym. J.*, **45**, 1689–1697.

Hida, K., Lai, S.K., Suk, J.S., Won, S.Y., Boyle, M.P., Hanes, J. (2011). *PLoS ONE*, **6**, e19919.

Hiemenz, P.C., Rajagopalan, R. (1997). *Principles of Colloid and Surface Chemistry*, **3rd ed.**, Marcel Dekker, New York.

Hilder, T.A., Hill, J.M. (2009). *Small*, **5**, 300–308.

Hilderbrand, S.A., Weissleder, R. (2010). *Curr. Opin. Chem. Biol.*, **14**, 71–79.

Hilliard, L.R., Zhao, X.J., Tan, W.H. (2002). *Anal. Chim. Acta*, **1**, 51–56.

Hindi, K.M., Ditto, A.J., Panzner, M.J., Medvetz, D.A., Han, D.S., Hovis, C.E., Hilliard, J.K., Taylor, J.B., Yun, Y.H., Cannon, C.L., Youngs, W.J. (2009). *Biomaterials*, **30**, 3771–3779.

Hines, M.A., Guyot-Sionnest, P. (1996). *J. Phys. Chem.*, **2**, 468–471.

Ho, K.C., Tsai, P.J., Lin, Y.S., Chen, Y.C. (2004). *Anal. Chem.*, 24, 7162–7168.

Hoare, T., McLean, D. (2006). *J. Phys. Chem. B*, **110**, 20327–20336.

Hoare, T., Pelton, R. (2004). *Macromolecules*, **37**, 2544–2550.

Hoare, T., Pelton, R. (2007). *Macromolecules*, **40**, 670–678.

Hoare, T., Pelton, R. (2008a). *Biomacromolecules*, **9**, 733–740.

Hoare, T., Pelton, R. (2008b). *Langmuir*, **24**, 1005–1012.

Hoare, T., Santamaria, J., Goya, G., Irusta, S., Lau, S., Stefanescu, C., Lin, D., Langer, R., Kohane, D.S. (2009). *Nano Lett.*, **9**, 3651–3657.

Hoare, T., Sivakumaran, D., Stefanescu, C.F., Lawlor, M.W., Kohane, D.S. (2012a). *Acta Biomater.*, **8**, 1450–1458.

Hoare, T., Timko, B.P., Santamaria, J., Goya, G.F., Irusta, S., Lau, S., Stefanescu, C.F., Lin, D., Langer, R., Kohane, D.S. (2011). *Nano Lett.*, **11**, 1395–1400.

Hoare, T., Young, S., Lawlor, M.W., Kohane, D.S. (2012b). *Acta Biomater.*, **8**, 3596–3605.

Hobbs, S.K., Monsky, W.L., Yuan, F., Roberts, W.G., Griffith, L., Torchilin, V.P., Jain, R.K. (1998). *Proc. Natl. Acad. Sci. U S A*, **95**, 4607–4612.

Högemann, D., Josephson, L., Weissleder, R., Basilion, J.P. (2000). *Bioconjug. Chem.*, **11**, 941–946.

Hong, H., Gao, T., Cai, W. (2009a). *Nano Today*, **4**, 252–261.

Hong, R.Y., Feng, B., Chen, L.L., Liu, G.H., Li, H.Z., Zheng, Y., Wei, D.G. (2008). *Biochem. Eng. J.*, **42**, 290–300.

Hong, S., Leroueil, P.R., Majoros, I.J., Orr, B.G., Baker Jr, J.R., Banaszak Holl, M.M. (2007). *Chem. Biol.*, **14**, 107–115.

Hong, S., Rattan, R., Majoros, I.J., Mullen, D.G., Peters, J.L., Shi, X., Bielinska, A.U., Blanco, L., Orr, B.G., Baker, J.R., Jr., Holl, M.M. (2009b). *Bioconjug. Chem.*, **20**, 1503–1513.

Hong, S.W., Kim, K.H., Huh, J., Ahn, C.-H., Jo, W.H. (2005). *Macromol. Res.*, **13**, 397–402.

Hong, S.Y., Tobias, G., Al-Jamal, K.T., Ballesteros, B., Ali-Boucetta, H., Lozano-Perez, S., Nellist, P.D., Sim, R.B., Finucane, C., Mather, S.J., Green, M.L., Kostarelos, K., Davis, B.G. (2010). *Nat. Mater.*, **9**, 485–490.

Hope, M.J., Bally, M.B., Webb, G., Cullis, P.R. (1985). *Biochim. Biophys. Acta*, **812**, 55–65.

Hoshino, Y., Imamura, K., Yue, M., Inoue, G., Miura, Y. (2012). *J. Am. Chem. Soc.*, **134**, 18177–18180.

Hou, Y., Han, Y.C., Deng, M., Xiang, J., Wang, Y.L. (2010). *Langmuir*, **26**, 28–33.

Hoyert, D.L., Xu, J.Q. (2012). *Natl. Vital Stat. Rep.*, **61**, 1–65.

Hsieh, W.J., Liang, C.J., Chieh, J.J., Wang, S.H., Lai, I.R., Chen, J.H., Chang, F.H., Tseng, W.K., Yang, S.Y., Wu, C.C., Chen, Y.L. (2012). *Int. J. Nanomed.*, **7**, 2833–2842.

Hu, C.M., Fang, R.H., Copp, J., Luk, B.T., Zhang, L. (2013a). *Nat. Nanotechnol.*, **8**, 336–340.

Hu, J., Zhang, G., Liu, S. (2012). *Chem. Soc. Rev.*, **41**, 5933–5949.

Hu, L., Sarker, A.K., Islam, M.R., Li, X., Lu, Z., Serpe, M.J. (2013b). *J. Polym. Sci. Part A Polym. Chem.*, **51**, 3004–3020.

Hu, Q.L., Jiang, Q.Y., Jin, X., Shen, J., Wang, K., Li, Y.B., Xu, F.J., Tang, G.P., Li, Z.H. (2013c). *Biomaterials*, **34**, 2265–2276.

Hu, S.-H., Liu, T.-Y., Huang, H.-Y., Liu, D.-M., Chen, S.-Y. (2007a). *Langmuir*, **24**, 239–244.

Hu, S.H., Liu, T.Y., Liu, D.M., Chen, S.Y. (2007b). *Macromolecules*, **40**, 6786–6788.

Hu, X., Tong, Z., Lyon, L.A. (2010a). *Colloid Polym. Sci.*, **289**, 333–339.

Hu, Z., Cai, T., Chi, C. (2010b). *Soft Matter*, **6**, 2115–2123.

Hu, Z., Luo, F., Pan, Y., Hou, C., Ren, L., Chen, J., Wang, J., Zhang, Y. (2008). *J. Biomed. Mater. Res. A*, **85**, 797–807.

Hua, M.Y., Yang, H.W., Chuang, C.K., Tsai, R.Y., Chen, W.J., Chuang, K.L., Chang, Y.H., Chuang, H.C., Pang, S.T. (2010). *Biomaterials*, **31**, 7355–7363.

Huan, M., Zhang, B., Teng, Z., Cui, H., Wang, J., Liu, X., Xia, H., Zhou, S., Mei, Q. (2012). *PLoS ONE*, **7**, e44116.

Huang, C.-H. (1969). *Biochemistry*, **8**, 344–352.

Huang, C.-Q., Pan, C.-Y. (2010). *Polymer*, **51**, 5115–5121.

Huang, H.M., Ou, H.C., Hsieh, S.J., Chiang, L.Y. (2000). *Life Sci.*, **66**, 1525–1533.

Huang, H.Y., Hu, S.H., Hung, S.Y., Chiang, C.S., Liu, H.L., Chiu, T.L., Lai, H.Y., Chen, Y.Y., Chen, S.Y. (2013a). *J. Controlled Release*, **172**, 118–127.

Huang, L.C.L., Chang, H.C. (2004). *Langmuir*, **14**, 5879–5884.

Huang, P., Bao, L., Zhang, C., Lin, J., Luo, T., Yang, D., He, M., Li, Z., Gao, G., Gao, B., Fu, S., Cui, D. (2011a). *Biomaterials*, **32**, 9796–9809.

Huang, P., Lin, J., Yang, D., Zhang, C., Li, Z., Cui, D. (2011b). *J. Controlled Release*, **152**(Suppl 1), e33–e34.

Huang, S., Fan, Y., Cheng, Z., Kong, D., Yang, P., Quan, Z., Zhang, C., Lin, J. (2009). *J. Phys. Chem. C*, **113**, 1775–1784.

Huang, X., Jain, P.K., El-Sayed, I.H., El-Sayed, M.A. (2007). *Nanomedicine*, **5**, 681–693.

Huang, X., Schwind, S., Yu, B., Santhanam, R., Wang, H., Hoellerbauer, P., Mims, A., Klisovic, R., Walker, A.R., Chan, K.K., Blum, W., Perrotti, D., Byrd, J.C., Bloomfield, C.D., Caligiuri, M.A., Lee, R.J., Garzon, R., Muthusamy, N., Lee, L.J., Marcucci, G. (2013b). *Clin. Cancer Res.*, **19**, 2355–2367.

Huang, X., Voit, B., (2013). *Polym. Chem.*, **4**, 435–443.

Huang, Y., Yu, F., Park, Y.-S., Wang, J., Shin, M.-C., Chung, H.S., Yang, V.C. (2010). *Biomaterials*, **31**, 9086–9091.

Huang, Z. (2005). *Technol. Cancer Res. Treat.*, **4**, 283–293.

Huang, Z., Xu, H., Meyers, A.D., Musani, A.I., Wang, L., Tagg, R., Barqawi, A.B., Chen, Y.K. (2008a). *Technol. Cancer. Res. T.*, **7**, 309.

Huang, Z., Zheng, X., Yan, D., Yin, G., Liao, X., Kang, Y., Yao, Y., Huang, D., Hao, B. (2008b). *Langmuir*, **24**, 4140–4144.

Huh, Y.M., Jun, Y.W., Song, H.T., Kim, S., Choi, J.S., Lee, J.H., Yoon, S., Kim, K.S., Shin, J.S., Suh, J.S., Cheon, J. (2005). *J. Am. Chem. Soc.*, **35**, 12387–12391.

Hung, H.S., Chen, H.C., Tsai, C.H., Lin, S.Z. (2011). *Cell Transplant.*, **20**, 63–70.

Hunter, R.J. (1987). *Foundations of Colloid Science*, Oxford University Press, New York.

Hunter, R.J. (1988). *Zeta Potential in Colloid Science. Principles and Applications*, Academic Press, London.

Huo, D., Li, Y., Qian, Q., Kobayashi, T. (2006). *Colloid Surf. B*, **50**, 36–42.

Hussain, I., Graham, S., Wang, Z., Tan, B., Sherrington, D.C., Rannard, S.P., Cooper, A.I., Brust, M. (2005). *J. Am. Chem. Soc.*, **47**, 16398–16399.

Hussain, M.M., Samir, T.M., Azzazy, H.M. (2013). *Clin. Biochem.*, **7–8**, 633–637.

Husseini, G.A., Diaz, D.e La, R.osa, M.A., Gabuji, T., Zeng, Y., Christensen, D.A., Pitt, W.G. (2007). *J. Nanosci. Nanotechnol.*, **7**, 1028–1033.

Husseini, G.A., Myrup, G.D., Pitt, W.G., Christensen, D.A., Rapoport, N.Y. (2000). *J. Controlled Release*, **69**, 43–52.

Husseini, G.A., Pitt, W.G. (2008). *Adv. Drug. Delivery Rev.*, **60**, 1137–1152.

Husseini, G.A., Pitt, W.G. (2009). *J. Pharm. Sci.*, **98**, 795–811.

Husseini, G.A., Stevenson-Abouelnasr, D., Pitt, W.G., Assaleh, K.T., Farahat, L.O., Fahadi, J. (2010). *Colloids Surf. A Physicochem. Eng. Asp.*, **359**, 18–24.

Husseini, G.A., Velluto, D., Kherbeck, L., Pitt, W.G., Hubbell, J.A., Christensen, D.A. (2013). *Colloids Surf. B Biointerfaces*, **101**, 153–155.

Huyen, D.C., Lu, P., Choi, S.-I., Jeong, U., Xia, Y. (2013). *Angew. Chem., Int. Ed.*, **52**, 1–5.

Hyeon, T., Lee, S.S., Park, J., Chung, Y., Na, H.B. (2001). *J. Am. Chem. Soc.*, **51**, 12798–12801.

Iancu, C., Mocan, L. (2011). *Int. J. Nanomed.*, **6**, 1675–1684.

Ibey, B.L., Beier, H.T., Rounds, R.M., Coté, G.L., Yadavalli, V.K., Pishko, M.V. (2005). *Anal. Chem.*, **77**, 7039–7046.

Ibrahim, B.M., Park, S., Han, B., Yeo, Y. (2011). *J. Controlled Release*, **155**, 289–295.

Iijima, S. (1991). *Nature*, **354**, 56–58.

Ikeda, K., Sawada, S., Akiyoshi, K., Matsuzaki, K. (2006). *Pept. Sci.*, **2005**, 313–316.

Imaz, A., Forcada, J. (2008a). *J. Polym. Sci. Part A Polym. Chem.*, **46**, 2766–2775.

Imaz, A., Forcada, J. (2008b). *J. Polym. Sci. Part Polym. Chem.*, **46**, 2510–2524.

Imaz, A., Forcada, J. (2009a). *Macromol. Symp.*, **281**, 85–88.

Imaz, A., Forcada, J. (2009b). *Eur. Polym. J.*, **45**, 3164–3175.

Imaz, A., Forcada, J. (2010). *J. Polym. Sci. Part A Polym. Chem.*, **48**, 1173–1181.

Imaz, A., Forcada, J. (2011). *J. Polym. Sci. Part A Polym. Chem.*, **49**, 3218–3227.

Imaz, A., Miranda, J.I., Ramos, J., Forcada, J. (2008). *Eur. Polym. J.*, **44**(12), 4002–4011.

Imlay, J.A. (2003). *Annu. Rev. Microbiol.*, **57**, 395–418.

Immordino, M.L., Dosio, F., Cattel, L. (2006). *Int. J. Nanomed.*, **1**, 297–315.

Immunity, G. (2013). Single DermaVir Immunization in HIV-1 Infected Patients on HAART (GIHU004), NCT00712530, May 6, 2013, Available from http://clinicaltrials.gov.

Imura, T., Gotoh, T., Otake, K., Yoda, S., Takebayashi, S., Yokoyama, S., Takebayashi, S., Sakai, H., Yuasa, M., Abe, M. (2003). *Langmuir*, **19**, 2021–2025.

Insin, N., Tracy, J.B., Lee, H., Zimmer, J.P., Westervelt, R.M., Bawendi, M.G. (2008). *ACS Nano*, **2**, 197–202.

Ishida, O., Maruyama, K., Sasaki, K., Iwatsuru, M. (1999). *Int. J. Pharm.*, **190**, 49–56.

Ishida, O., Maruyama, K., Tanahashi, H., Iwatsuru, M., Sasaki, K., Eriguchi, M., Yanagie, H. (2001). *Pharm. Res.*, **18**, 1042–1048.

Ishida, T., et al. (2006). *Int. J. Pharm.*, **309**, 94–100.

Ishida, T., Ichihara, M., Wang, X., Yamamoto, K., Kimura, J., Majima, E., Kiwada, H. (2006). *J. Controlled Release*, **112**, 15–25.

Israelachvili, J. (1991). *Intermolecular and Surface Forces*, **2nd ed.**, Academic Press, San Diego.

Israelachvili, J., Marcelja, S., Horn, R. (1980). *Rev. Biophys.*, **13**, 121–200.

Iyer, A.K., Khaled, G., Fang, J., Maeda, H. (2006). *Drug Discovery Today*, **11**, 812–818.

Jahn, A., Vreeland, W.N., DeVoe, D.L., Locascio, L.E., Gaitan, M. (2007). *Langmuir*, **23**, 6289–6293.

Jain, K., Kesharwani, P., Gupta, U., Jain, N. (2010). *Int. J. Pharm.*, **394**, 122–142.

Jain, N.K., Tekade, R.K. (2013). *Drug Delivery Strategies for Poorly Water-Soluble Drugs*, 373–409, John Wiley and Sons.

Jain, P.K., Eustis, S., El-Sayed, M.A. (2006). *J. Phys. Chem. B*, **37**, 18243–18253.

Jain, R.K. (1987a). *Cancer Metastasis Rev.*, **6**, 559–593.

Jain, R.K. (1987b). *Cancer Res.*, **47**, 3039–3051.

Jain, R.K. (1994). *Sci. Am.*, **271**, 58–65.

Jain, R.K. (2001). *Nat. Med.*, **7**, 987–989.

Jain, T.K., Morales, M.A., Sahoo, S.K., Leslie-Pelecky, D.L., Labhasetwar, V. (2005). *Mol. Pharm.*, **2**, 194–205.

Jaiswal, M.K., Banerjee, R., Pradhan, P., Bahadur, D. (2010). *Colloid Surf. B*, **81**, 185–194.

Jalil, M.A., Innate, K., Suwanpayak, N., Yupapin, P.P., Ali, J. (2011). *Artif. Cells BloodSubstit. Immobil. Biotechnol.*, **39**, 357–365.

Jana, N.R., Chen, Y., Peng, X. (2004). *Chem. Mater.*, **20**, 3931–3935.

Jana, N.R., Gearheart, L., Murphy, C.J. (2001a). *Langmuir*, **22**, 6782–6786.

Jana, N.R., Gearheart, L., Murphy, C.J. (2001b). *Adv. Mater.*, **18**, 1389–1393.

Jana, N.R., Gearheart, L., Murphy, C.J. (2001c). *J. Phys. Chem. B*, **19**, 4065–4067.

Jana, N.R., Yu, H.-h., Ali, E.M., Zheng, Y., Ying, J.Y. (2007). *Chem. Commun.*, **14**, 1406–1408.

Janairo, R.R., Henry, J.J., Lee, B.L., Hashi, C.K., Derugin, N., Lee, R., Li, S. (2012). *IEEE Trans. Nanobioscience*, **11**, 22–27.

Jang, W.-D., Kamruzzaman Selim, K., Lee, C.-H., Kang, I.-K. (2009). *Prog. Polym. Sci.*, **34**, 1–23.

Janib, S.M., Moses, A.S., MacKay, J.A. (2010). *Adv. Drug Delivery Rev.*, **62**, 1052–1063.

Jans, H., Huo, Q. (2012). *Chem. Soc. Rev.*, **7**, 2849–2866.

Jansen, T.G.T. (June 2014). *Miniemulsion Polymerization of Very Hydrophobic Monomers, A Reaction Engineering Approach*, PhD thesis, Eindhoven.

Jarrett, J.T., Lansbury, P.T. Jr. (1993). *Cell*, **73**, 1055–1058.

Javadi, M., Pitt, W.G., Belnap, D.M., Tsosie, N.H., Hartley, J.M. (2012). *Langmuir*, **28**, 14720–14729.

Jayasena, S.D. (1999). *Clin. Chem.*, **25**, 1628–1650.

Jenner, P. (2000). *Ann. Neurol.*, **47**(Suppl 1), 90–104.

Jenner, P. (2005). *Neurology*, **65**(Suppl 1), S3–S5.

Jennings, L.E., Long, N.J. (2009). *Chem. Commun.*, **28**, 3511–3524.

Jennings, T.L., Singh, M.P., Strouse, G.F. (2006). *J. Am. Chem. Soc.*, **16**, 5462–5467.

Jensen, L.B., Pavan, G.M., Kasimova, M.R., Rutherford, S., Danani, A., Nielsen, H.M., Foged, C. (2011). *Int. J. Pharm.*, **416**, 410–418.

Jeong, U., Teng, X., Wang, Y., Yang, H., Xia, Y. (2007). *Adv. Mater.*, **1**, 33–60.

Jesorka, A., Orwar, O. (2008). *Annu. Rev. Anal. Chem.*, **1**, 801–832.

Jeyamohan, P., et al. (2013). *Int. J. Nanomed.*, **8**, 2653–2667.

Ji, H.F., Shen, L. (2009). *J. Photochem. Photobiol. B*, **94**, 51–53.

Ji, Z., Lin, G., Lu, Q., Meng, L., Shen, X., Dong, L., Fu, C., Zhang, X. (2012). *J. Colloid Interface Sci.*, **365**, 143–149.

Jiang, T., Zukoski, C.F. (2012). *Langmuir*, **28**, 6760–6768.

Jiang, W., Lai, K.-L., Hu, H., Zeng, X.-B., Lan, F., Liu, K.-X., Wu, Y., Gu, Z.-W. (2011). *J. Nanoparticle Res.*, **13**, 5135–5145.

Jiang, W., Xie, H., Ghoorah, D., Shang, Y., Shi, H., Liu, F., Yang, X., Xu, H. (2012). *PLoS ONE*, **7**, e37376.

Jiménez, J.L., Pion, M., de la Mata, F.J., Gomez, R., Muñoz, E., Leal, M., Muñoz-Fernandez, M.A. (2012). *New J. Chem.*, **36**, 299–309.

Jin, C., Qian, N., Zhao, W., Yang, W., Bai, L., Wu, H., Wang, M., Song, W., Dou, K. (2010). *Biomacromolecules*, **11**, 2422–2431.

Jin, Y., Song, L., Su, Y., Zhu, L., Pang, Y., Qiu, F., Tong, G., Yan, D., Zhu, B., Zhu, X. (2011). *Biomacromolecules*, **12**, 3460–3468.

Jin, Z., Hildebrandt, N. (2012). *Trends Biotechnol.*, **7**, 394–403.

Jochum, F.D., Theato, P. (2010). *Chem. Commun.*, **46**, 6717–6719.

Jochum, F.D., Theato, P. (2013). *Chem. Soc. Rev.*, **42**, 7468–7483.

Johannsen, M., Thiesen, B., Wust, P., Jordan, A. (2010). *Int. J. Hyperth. Off. J. Eur. Soc. Hyperthermic Oncol. North Am. Hyperth. Group*, **26**, 790–795.

Johnson, T.A., Stasko, N., Matthews, J.L., Cascio, W.E., Holmuhamedov, E.L., Johnson, C.B., Schoenfisch, M.H. (2010). *Nitric Oxide*, **22**, 30–36.

Johnston, T.H., Fox, S.H., Brotchie, J.M. (2005). *Expert Opin. Drug Delivery*, **2**, 1059–1073.

Jones, C.D., Lyon, L.A. (2000). *Macromolecules*, **33**, 8301–8306.

Joshi, S.A., Chavran, S.S., Sawant, K.K. (2010). *Eur. J. Pharm. Biopharm.*, **76**, 189–199.

Joshi, V., Geary, S., Salem, A. (2012). *AAPS J.*, **15**, 85–94.

Joyce, J.A., Baruch, A., Chehade, K., Meyer-Morse, N., Giraudo, E., Tsai, F.Y., Greenbaum, D.C., Hager, J.H., Bogyo, M., Hanahan, D. (2004). *Cancer Cell*, **5**, 443–453.

Ju, K.Y., Lee, J.W., Im, G.H., Lee, S., Pyo, J., Park, S.B., Lee, J.H., Lee, J.K. (2013). *Biomacromolecules*, **14**(10), 3491–3497.

Juarranz, Á., Jaén, P., Sanz-Rodríguez, F., Cuevas, J., González, S. (2008). *Clin. Transl. Oncol.*, **10**, 148–154.

Juliano, R. (2006). *Ann. N. Y. Acad. Sci.*, **1082**, 18–26.

Jung, K.H., Park, J.W., Paik, J.Y., Quach, C.H., Choe, Y.S., Lee, K.H. (2012). *Nucl. Med. Biol.*, **39**, 1122–1127.

Jung, M., Hubert, D.H.W., Bomans, P., Frederik, P.M., van Herk, A.M., German, A.L. (2000a). *Adv. Mater.*, **12**, 210–213.

Jung, M., Hubert, D.H.W., van Veldhoven, E., Frederik, P., van Herk, A.M., German, A.L. (2000b). *Langmuir*, **16**, 3165–3174.

Justo, O.R., Moraes, A.M. (2011). *Chem. Eng. Res. Des.*, **89**, 785–792.

Kaba, S.A., McCoy, M.E., Doll, T.A.P.F., Brando, C., Guo, Q., Dasgupta, D., Yang, Y., Mittelholzer, C., Spaccapelo, R., Crisanti, A., Burkhard, P., Lanar, D.E. (2012). *PLoS ONE*, **7**, 1–11.

Kabiri, M., Soleimani, M., Shabani, I., Futrega, K., Ghaemi, N., Ahvaz, H.H., Elahi, E., Doran, M.R. (2012). *Biotechnol. Lett.*, **34**, 1357–1365.

Kakio, A., Nishimoto, S., Yanagisawa, K., Kozutsumi, Y., Matsuzaki, K. (2002). *Biochemistry*, **41**, 7385–7390.

Kallunki, T., Olsen, O.D., Jaattela, M. (2013). *Oncogene*, **32**, 1995–2004.

Kalvodova, L. (2010). *Biochem. Biophys. Res. Commun.*, **393**, 350–355.

Kam, N.W., Liu, Z., Dai, H. (2005a). *J. Am. Chem. Soc.*, **127**, 12492–12493.

Kam, N.W., O'Connell, M., Wisdom, J.A., Dai, H. (2005b). *Proc. Natl. Acad. Sci. U S A*, **102**, 11600–11605.

Kamaly, N., Fredman, G., Subramanian, M., Gadde, S., Pesic, A., Cheung, L., Fayad, Z.A., Langer, R., Tabas, I., Farokhzad, O. (2013). *Proc. Natl. Acad. Sci. U S A*, **110**, 6506–6511.

Kamaly, N., Xiao, Z., Valencia, P.M., Radovic-Moreno, A.F., Farokhzad, O.C. (2012). *Chem. Soc. Rev.*, **41**, 2971–3010.

Kamei, K., et al. (2009). *Biomaterials*, 30, 1809–1814.

Kaminskas, L.M., Kelly, B.D., Mcleod, V.M., Sberna, G., Owen, D.J., Boyd, B.J., Porter, C.J. (2011). *J. Controlled Release*, **152**, 241–248.

Kammerich, A.D. (2011). *Development and Characterization of Conducting Polymer Modified Electrodes for Neurological Applications*. Doctoral thesis, Georgetown University, 1–99.

Kanazawa, T., Sugawara, K., Tanaka, K., Horiuchi, S., Takashima, Y., Okada, H. (2012). *Eur. J. Pharm. Biopharm.*, **81**, 470–477.

Kanchan, V., Panda, A.K. (2007). *Biomaterials*, **28**, 5344–5357.

Kandasamy, R., Park, S.J., Boyapalle, S., Mohapatra, S., Hellermann, G.R., Lockey, R.F., Mohapatra, S.S. (2010). *Int. Immunopharmacol.*, **10**, 218–225.

Kang, B., Li, J., Chang, S., Dai, M., Ren, C., Dai, Y., Chen, D. (2012). *Small*, **8**, 777–782.

Kang, B., Yu, D., Dai, Y., Chang, S., Chen, D., Ding, Y. (2009a). *Small*, **11**, 1292–1301.

Kang, D.Y., Lee, J.H., Oh, B.K., Choi, J.W. (2009b). *Biosens. Bioelectron.*, **24**, 1431–1436.

Kang, H., Lee, E., Na, K., Bae, Y. (2008). Stimuli-sensitive nanosystems: for drug and gene delivery, In *Multifunctional Pharmaceutical Nanocarriers*, pp. 161–199, Ed. Torchilin, V., Springer, New York.

Kang, Y., Liu, Y.C., Wang, Q., Shen, J.W., Wu, T., Guan, W.J. (2009c). *Biomaterials*, **30**, 2807–2815.

Kang, Y.S., Risbud, S., Rabolt, J.F., Stroeve, P. (1996). *Chem. Mater.*, **9**, 2209–2211.

Karagoz, B., Eser, L., Duong, H.T., Basuki, J.S., Boyer, C., Davis, T.P. (2014). *Polym. Chem.*, **5**, 350–355.

Karakoti, A., Singh, S., Dowding, J.M., Seal, S., Self, W. T. (2010). *Chem. Soc. Rev.*, **39**, 4422–4432.

Karatasos, K., Posocco, P., Laurini, E., Pricl, S. (2012). *Macromol. Biosci.*, **12**, 225–240.

Karg, M., Hellweg, T. (2009). *Curr. Opin. Colloid Interface Sci.*, **14**, 438–450.

Karoonuthaisiri, N., Titiyevskiy, K., Thomas, J.L. (2003). *Colloid Surf. B*, **27**, 11.

Karousis, N., Tagmatarchis, N., Tasis, D. (2010). *Chem. Rev.*, **110**, 5366–5397.

Karra, N., Nassar, T., Ripin, A.N., Schwob, O., Borlak, J., Benita, S. (2013). *Small*, **9**(24), 4221–4236.

Kataoka, K., Harada, A., Nagasaki Y. (2012). *Adv. Drug Delivery Rev.*, **64**, 37–48.

Katare, Y.K., Muthukumaran, T., Panda, A.K. (2005). *Int. J. Pharm.*, **301**, 149–160.

Kateb, B., Chiu, K., Black, K.L., Yamamoto, V., Khalsa, B., Ljubimova, J.Y., Ding, H., Patil, R., Portilla-Arias, J.A., Modo, M., Moore, D.F., Farahani, K., Okun, M.S., Prakash, N., Neman, J., Ahdoot, D., Grundfest, W., Nikzad, S., Heiss, J.D. (2011). *NeuroImage*, **54**, S106–S124.

Kato, H., Kanazawa, Y., Okumura, M., et al. (2003). *J. Am. Chem. Soc.*, **125**, 4391–4397.

Kattumuri, V., Katti, K., Bhaskaran, S., Boote, E.J., Casteel, S.W., Fent, G.M., Robertson, D.J., Chandrasekhar, M., Kannan, R., Katti, K.V. (2007). *Small*, **3**, 333–341.

Katz, J.S., Burdick, J.A. (2010). *Macromol. Biosci.*, **10**, 339–348.

Katzenschlager, R., Hughes, A., Evans, A., Manson, A.J., Hoffman, M., Swinn, L., et al. (2005). *Mov. Disord.*, **20**, 151–157.

Kawamura, A., Hata, Y., Miyata, T., Uragami, T. (2012). *Colloid Surf. B*, **99**, 74–81.

Kawano, T., Niidome, Y., Mori, T., Katayama, Y., Niidome, T. (2009). *Bioconjug. Chem.*, **20**, 209–212.

Kawarabayashi, T., Shoji, M., Younkin, L. H., Lin, W. L., Dickson, D. W., Murakami, T., Matsubara, E., Abe, K., Ashe, K.H., Younkin, S.G. (2004). *J. Neurosci.*, **24**, 3801–3809.

Kayed, R., Sokolov, Y., Edmonds, B., McIntire, T.M., Milton, S.C., Hall, J.E., Glabe, C.G. (2004). *J. Biol. Chem.*, **279**, 46363–46366.

Keating, C.D. (2005). *Proc. Natl. Acad. Sci. U S A*, **102**, 2263–2264.

Kelly, C.V., Leroueil, P.R., Nett, E.K., Wereszczynski, J.M., Baker, J.R., Jr., Orr, B.G., Banaszak Holl, M.M., Andricioaei, I. (2008a). *J. Phys. Chem. B*, **112**, 9337–9345.

Kelly, C.V., Leroueil, P.R., Orr, B.G., Banaszak Holl, M.M., Andricioaei, I. (2008b). *J. Phys. Chem. B*, **112**, 9346–9353.

Kelly, C.V., Liroff, M.G., Triplett, L.D., Leroueil, P.R., Mullen, D.G., Wallace, J.M., Meshinchi, S., Baker, J.R., Orr, B.G., Banaszak Holl, M.M. (2009). *ACS Nano*, **3**, 1886–1896.

Kempe, M., Kempe, H., Snowball, I., Wallen, R., Arza, C.R., Gotberg, M., Olsson, T. (2010). *Biomaterials*, **31**, 9499–9510.

Kenny, G.D., Bienemann, A.S., Tagalakis, A.D., Pugh, J.A., Welser, K., Campbell, F., Tabor, A.B., Hailes, H.C., Gill, S.S., Lythgoe, M.F., Mcleod, C.W., White, E.A., Hart, S.L. (2013). *Biomaterials*, **34**, 9190–9200.

Kettel, M.J., Dierkes, F., Schaefer, K., Moeller, M., Pich, A. (2011). *Polymer*, **52**, 1917–1924.

Keyler, D.E., Pentel, P.R., Haughey, D.B. (2006). *J. Pharm. Sci.*, **76**, 101–104.

Khazaei, A., Soltani Rad, M.N., Borazjani, M.K. (2010). *Int. J. Nanomed.*, **5**, 639–645.

Kheirolomoom, A., Lai, C.Y., Tam, S.M., Mahakian, L.M., Ingham, E.S., Watson, K.D., Ferrara, K.W. (2013). *J. Controlled Release*, **172**(1), 266–273.

Khlobystov, A.N., Britz, D.A., Briggs, G.A. (2005). *Acc. Chem. Res.*, **38**, 901–909.

Khondee, S., Yakovleva, T., Berkland, C. (2010). *J. Appl. Polym. Sci.*, **118**, 1921–1932.

Khorsand, S.ourkohi, B., Cunningham, A., Zhang, Q., Oh, J.K. (2011). *Biomacromolecules*, **12**, 3819–3825.

Kiesslich, T., Krammer, B., Plaetzer, K. (2006). *Curr. Med. Chem.*, **13**, 2189–2204.

Kiessling, F., Morgenstern, B., Zhang, C. (2007). *Curr. Med. Chem.*, **14**, 77–91.

Kikkeri, R., Hong, S.Y., Grünstein, D., Laurino, P., Seeberger, P.H. (2011). Beilstein-Institut.

Kim, D.H., Vitol, E.A., Liu, J., Balasubramanian, S., Gosztola, D.J., Cohen, E.E., Novosad, V., Rozhkova, E.A. (2013a). *Langmuir*, **29**, 7425–7432.

Kim, D.H., Vitol, E.A., Liu, J., Balasubramanian, S., Gosztola, D.J., Cohen, E.E., Novosad, V., Rozhkova, E.A. (2013b). *Langmuir*, **29**(24), 7425–7432.

Kim, D.W., Kim, S.Y., Kim, H.K., Kim, S.W., Shin, S.W., Kim, J.S., Park, K., Lee, M.Y., Heo, D.S. (2007a). *Ann. Oncol.*, **18**, 2009–2014.

Kim, I., Xu, W., Reed, J.C. (2008). *Nat. Rev. Drug Discovery*, **7**, 1013–1030.

Kim, J., Dang, C.V. (2006). *Cancer Res.*, **66**, 8927–8930.

Kim, J., Nayak, S., Lyon, L.A. (2005). *J. Am. Chem. Soc.*, **127**, 9588–9592.

Kim, J., Singh, N., Lyon, L.A. (2007b). *Biomacromolecules*, **8**, 1157–1161.

Kim, J.E., Cha, E.-J., Ahn, C.-H. (2010). *Macromol. Chem. Phys.*, **211**, 956–961.

Kim, J.-H., Park, K., Nam, H.Y., Lee, S., Kim, K., Kwon, I.C. (2007c). *Prog. Polym. Sci.*, **32**, 1031–1053.

Kim, J.S., Adamcakova-Dodd, A., O'Shaughnessy, P., Grassian, V., Thorne, P. (2011). *Part. Fibre Toxicol.*, **8**, 29.

Kim, K., Huang, S.-W., Ashkenazi, S., O'Donnell, M., Agarwal, A., Kotov, N.A., Denny, M.F., Kaplan, M.J. (2007d). *Appl. Phys. Lett.*, **22**, 1–3.

Kim, S.M., Im, G.H., Lee, D.G., Lee, J.H., Lee, W.J., Lee, I.S. (2013c). *Biomaterials*, **34**, 8941–8.

Kim, T., Hyeon, T. (2014). *Nanotechnology*, **25**, 012001 (14 pp).

Kim, T.I., Seo, H.J., Choi, J.S., Jang, H.S., Baek, J.U., Kim, K., Park, J.S. (2004). *Biomacromolecules*, **5**, 2487–2492.

Kirkegaard, T., Jaattela, M. (2009). Lysosomal involvement in cell death and cancer. *Biochim. Biophys. Acta*, **1793**, 746–754.

Kirpotin, D., et al. (1986). *FEBS Lett.*, **388**, 115–118.

Kisak, E.T., Coldren, B., Zasadzinski, J.A. (2004). *Curr. Med. Chem.*, **11**, 199–219.

Klajnert, B., Cortijo-Arellano, M., Cladera, J., Majoral, J.-P., Caminade, A.-M., Bryszewska, M. (2007). *Biochem. Bioph. Res. Co.*, **364**, 20–25.

Klajnert, B., Rozanek, M., Bryszewska, M. (2012). *Curr. Med. Chem.*, **19**, 4903–4912.

Klein, W.L., Krafft, G.A., Finch, C.E. (2001). *Trends Neurosci.*, **24**, 219–224.

Klein, W.L., Stine, W.B., Teplow, D.B. (2004). *Neurobiol. Aging*, **25**, 569–580.

Klibanov, A.L., Maruyama, K., Torchilin, V.P., Huang, L. (1990). *FEBS Lett.*, **268**, 235–237.

Klinger, D., Landfester, K. (2012). *Polymer*, **53**, 5209–5231.

Kluza, E., van der Schaft, D.W., Hautvast, P.A., Mulder, W.J., Mayo, K.H., Griffioen, A.W., Strijkers, G.J., Nicolay, K. (2010). *Nano Lett.*, **10**, 52–58.

Kneipp, K., Kneipp, H., Kneipp, J. (2006). *Acc. Chem. Res.*, **7**, 443–450.

Kneuer, C., Sameti, M., Bakowsky, U., Schiestel, T., Schirra, H., Schmidt, H., Lehr C.M. (2000a). *Bioconjug. Chem.*, **11**, 926–932.

Kneuer, C., Sameti, M., Haltner, E.G., Schiestel, T., Schirra, H., Schmidt, H., Lehr, C.M. (2000b). *Int. J. Pharm.*, **196**, 257–261.

Knezevic, N.Z., Trewyn, B.G., Lin, V.S. (2011). *Chemistry*, **17**, 3338–3342.

Kobayashi, H., Kawamoto, S., Jo, S.-K., Bryant, H.L., Brechbiel, M.W., Star, R.A. (2003). *Bioconjug. Chem.*, **14**, 388–394.

Kobayashi, H., Ogawa, M., Alford, R., Choyke, P.L., Urano, Y. (2010). *Chem. Rev.*, **110**, 2620.

Kobiasi, M.A., Chua, B.Y., Tonkin, D., Jackson, D.C., Mainwaring, D.E. (2012). *J. Biomed. Mater. Res. Part A*, **100A**, 1859–1867.

Kojima, C., Turkbey, B., Ogawa, M., Bernardo, M., Regino, C.A., Bryant, L.H., Choyke, P.L., Kono, K., Kobayashi, H. (2011). *Nanomed. Nanotech. Biol. Med.*, **7**, 1001–1008.

Kolb, H.C., Finn, M.G., Sharpless, K.B. (2001). *Angew. Chem., Int. Ed. Engl.*, **40**, 2004–2021.

Kondo, E., Saito, K., Tashiro, Y., Kamide, K., Uno, S., Furuya, T., Mashita, M., Nakajima, K., Tsumuraya, T., Kobayashi, N., Nishibori, M., Tanimoto, M., Matsushita, M. (2012). *Nat. Commun.*, **3**, 951.

Kondratyuk, P., Yates, J.T. Jr. (2007). *Acc. Chem. Res.*, **40**, 995–1004.

Koneracka, M., Kopcansky, P., Antalik, M., Timko, M., Ramchand, C.N., Lobo, D., Mehta, R.V., Upadhyay, R.V. (1999). *J. Magn. Magn. Mater.*, **1–3**, 427–430.

Kong, G., Dewhirst, M.W. (1999). *Int. J. Hyperthermia*, **15**, 345–370.

Kono, K., Igawa, T., Takagishi, T. (1997). *Biochim. Biophys. Acta*, **1325**, 143–154.

Kono, K., Liu, M., Frechet, J.M. (1999). *Bioconjug. Chem.*, **10**, 1115–1121.

Koo, A.N., Min, K.H., Lee, H.J., Lee, S.U., Kim, K., Kwon, I.C., Cho, S.H., Jeong, S.Y., Lee, S.C. (2012). *Biomaterials*, **33**, 1489–1499.

Koren, E., Apte, A., Jani, A., Torchilin, V.P. (2012). *J. Controlled Release*, **160**, 264–273.

Korin, N., Kanapathipillai, M., Matthews, B.D., Crescente, M., Brill, A., Mammoto, T., Ghosh, K., Jurek, S., Bencherif, S.A., Bhatta, D., Coskun, A.U., Feldman, C.L., Wagner, D.D., Ingber, D.E. (2012). *Science*, **337**, 738–742.

Korsvik, C., Patil, S., Seal, S., Self, W.T. (2007). *Chem. Commun.*, 1056–1058.

Kortekaas, R., Leenders, K.L., Van Oostrom, J.C.H., Vaalburg, W., Bart, J., Willemsem, A. T.M., Hendrike, N.H. (2005). *Ann. Neurol.*, **57**, 176–179.

Koshino, M., Tanaka, T., Solin, N., Suenaga, K., Isobe, H., Nakamura, E. (2007). *Science*, **316**, 853.

Koshkaryev, A., Piroyan, A., Torchilin, V.P. (2012a). *Cancer Lett.*, **334**(2), 293–301.

Koshkaryev, A., Piroyan, A., Torchilin, V.P. (2012b). *Cancer Biol. Ther.*, **13**, 50–60.

Kostarelos, K., Bianco, A., Prato, M. (2009). *Nat. Nanotechnol.*, **4**, 627–633.

Kostarelos, K., Lacerda, L., Pastorin, G., Wu, W., Wieckowski, S., Luangsivilay, J., Godefroy, S., Pantarotto, D., Briand, J.P., Muller, S., Prato, M., Bianco, A. (2007). *Nat. Nanotechnol.*, **2**, 108–113.

Kotagiri, N., Kim, J.-W. (2014). *Int. J. Nanomed.*, **9**(Suppl 1), 85–105.

Koyama, Y., Talanov, V.S., Bernardo, M., Hama, Y., Regino, C.A., Brechbiel, M.W., Choyke, P.L., Kobayashi, H. (2007). *J. Magn. Reson. Imaging*, **25**, 866–871.

Kozutsumi, D., Tsunematsu, M., Yamaji, T., Murakami, R., Yokoyama, M., Kino, K. (2006). *Immunology*, **118**, 392–401.

Kremer, J.M.H., Esker, M.W., Pathmamanoharan, C., Wiersema, P.H. (1977). *Biochemistry*, **16**, 3932–3935.

Krishnan, K.M. (2010). *IEEE Trans. Magn.*, **7**, 2523–2558.

Kriz, C.B., Radevik, K., Kriz, D. (1996). *Anal. Chem.*, **11**, 1966–1970.

Krol, S. (2012). *J. Controlled Release*, **164**, 145–155.

Kroto, H.W., Heath, J.R., O'Brien, S.C., Curl, R.F., Smalley, R.E. (1985). *Nature*, **318**, 162–163.

Kruger, A., Liang, Y.J., Jarre, G., Stegk, J. (2006). *J. Mater. Chem.*, **24**, 2322–2328.

Krupka, T.M., Solorio, L., Wilson, R.E., Wu, H., Azar, N., Exner, A.A. (2010). *Mol. Pharm.*, **7**, 49–59.

Kubo, M., Egashira, K., Inoue, T., Koga, J., Oda, S., Chen, L., Nakano, K., Matoba, T., Kawashima, Y., Hara, K., Tsujimoto, H., Sueishi, K., Tominaga, R., Sunagawa, K. (2009). *Arterioscler. Thromb. Vasc. Biol.*, **29**, 796–801.

Kuckling, D., Vo, C.D., Adler, H.-J.P., Völkel, A., Cölfen, H. (2006). *Macromolecules*, **39**, 1585–1591.

Kukowska-Latallo, J.F., Candido, K.A., Cao, Z., Nigavekar, S.S., Majoros, I.J., Thomas, T.P., Balogh, L.P., Khan, M.K., Baker, J.R. (2005). *Cancer Res.*, **65**, 5317–5324.

Kularatne, B.Y., Lorigan, P., Browne, S., Suvarna, S.K., Smith, N.O., Lawry, J. (2002). *Cytometry*, **3**, 160–167.

Kulhari, H., Pooja, D., Prajapati, S.K., Chauhan, A.S. (2011). *Int. J. Pharm.*, **405**, 203–209.

Kumagai, H., Pham, W., Kataoka, M., Hiwatari, K., Mcbride, J., Wilson, K.J., Tachikawa, H., Kimura, R., Nakamura, K., Liu, E.H., Gore, J.C., Sakuma, S. (2013). *Int. J. Cancer*, **132**, 2107–2117.

Kumagai, M., Shimoda, S., Wakabayashi, R., Kunisawa, Y., Ishii, T., Osada, K., Itaka, K., Nishiyama, N., Kataoka, K., Nakano, K. (2012). *J. Controlled Release*, **160**, 542–551.

Kumar, A., Singh, D., Goyal, R.N., Kaur, D. (2009). *Talanta*, **78**, 964–969.

Kumar, A., Zhang, X., Liang, X.-J. (2013). *Biotechnol. Adv.*, **5**, 593–606.

Kumar, G., Sharma, S., Shafiq, N., Khuller, G.K., Malhotra, S. (2012a). *J. Pharm. Sci.*, **101**, 2165–2176.

Kumar, G., Sharma, S., Shafiq, N., Pandhi, P., Khuller, G.K., Malhotra, S. (2011). *Drug Delivery*, **18**, 65–73.

Kumar, V., Abbas, A.K., Fausto, N., Mitchell, R. (2012b). *Robbins Basic Pathology*, 9th Ed., Elsevier Saunders, USA.

Kurek, N.S., Chandra, S.B. (2012). *ARBS Ann. Rev. Biomed. Sci.*, **14**, 1–15.

Kurtoglu, Y.E., Mishra, M.K., Kannan, S., Kannan, R.M. (2010). *Int. J. Pharm.*, **384**, 189–194.

Kuszyk, B.S., Corl, F.M., Franano, F.N., Bluemke, D.A., Hofmann, L.V., Fortman, B.J., Fishman, E.K. (2001). *AJR Am. J. Roentgenol.*, **177**, 747–753.

Kuznetsova, A., Mawhinney, D.B., Naumenko, V., Yates Jr., J.T., Liu, J., Smalley, R.E. (2000). *Chem. Phys. Lett.*, **321**, 292–296.

Kwon, Y.J., Standley, S.M., Goh, S.L., Fréchet, J.M.J. (2005). *J. Controlled Release*, **105**, 199–212.

Laginha, K., Mumbengegwi, D., Allen, T. (2005). *Biochim. Biophys. Acta, Biomembr.*, **1711**, 25–32.

Lakowicz, J.R. (2006). *Plasmonics*, **1**, 5–33.

Lakshmanan, R., Krishnan, U.M., Sethuraman, S. (2012). *Expert Opin. Biol. Ther.*, **12**, 1623–1640.

Lalancette-Hébert, M., Moquin, A., Choi, A.O., Kriz, J., Maysinger, D. (2010). *Mol. Pharm.*, **7**, 1183–1194.

Lambert, M.P., Barlow A.K., Chromy, B.A., Edwards, C., Freed, R., Liosatos, M., Morgan, T. E., Rozovsky, I., Trommer, B., Viola, K. L., Wals, P., Zhang, C., Finchs, C.E., Krafft, G.A., Klein, W.L. (1998). *Proc. Natl. Acad. Sci. U S A*, **95**, 6448–6453.

Landini, L., Santarelli, M.F., Positano, V. (2008). *Curr. Drug Discovery Technol.*, **5**, 328–332.

Landmark, K.J., et al. (2008). *ACS Nano*, **2**, 773–783.

Langer, R. (1998). *Nature*, **392**, S5–S10.

Langereis, S., de Lussanet, Q.G., van Genderen, M.H., Meijer, E., Beets-Tan, R.G., Griffioen, A.W., van Engelshoven, J., Backes, W.H. (2006). *NMR Biomed.*, **19**, 133–141.

Langereis, S., Dirksen, A., Hackeng, T.M., van Genderen, M.H., Meijer, E. (2007). *New J. Chem.*, **31**, 1152–1160.

Lapeyre, V., Ancla, C., Catargi, B., Ravaine, V. (2008). *J. Colloid Interface Sci.*, **327**, 316–323.

Lapeyre, V., Gosse, I., Chevreux, S., Ravaine, V. (2006). *Biomacromolecules*, **7**, 3356–3363.

Lapinski, M.M., Castro-Forero, A., Greiner, A.J., Ofoli, R.Y., Blanchard, G.J. (2007). *Langmuir*, **23**, 11677–11683.

Lapointe, J., Martel, S. (2009). *Conf. Proc. IEEE Eng. Med. Biol. Soc.*, **2009**, 4246–4249.

Larsson, K., Kriz, K., Kriz, D. (1999). *Analysis*, **7**, 617–621.

Lasic, D. (1993). *Liposomes: From Physics to Applications*, Elsevier, Amsterdam.

Lasic, D.D. (1997). *J. Controlled Release*, 48, 203–222.

Laurent, S., Dutz, S., Hafeli, U.O., Mahmoudi, M. (2011). *Adv. Colloid Interface Sci.*, **166**, 8–23.

Laurent, S., Forge, D., Port, M., Roch, A., Robic, C., Elst, L.V., Muller, R.N. (2008). *Chem. Rev.*, **6**, 2064–2110.

Lauterbur, P.C. (1973). *Nature*, **242**, 190–191.

LaVan, D.A., McGuire, T., Langer, R. (2003). *Nat. Biotechnol.*, **10**, 1184–1191.

Le Droumaguet, B., Nicolas, J., Brambilla, D., Mura, S., Maksimenko, A., De Kimpe, L., Salvati, E., Zona, C., Airoldi, C., Canoci, M., Gobbi, M, Noiray, M.,

La Ferla, B., Nicotra, F., Scheper, W., Flores, O., Masserini, M., Andrieux, K., Couvreur, P. (2012). *ACS Nano*, **6**, 5866–5879.

Leamon, C.P., Low, P.S. (1991). *Proc. Natl. Acad. Sci. U S A*, **88**, 5572–5576.

Lebedev, A.Y., Cheprakov, A.V., Sakadžić, S., Boas, D.A., Wilson, D.F., Vinogradov, S.A. (2009). *ACS Appl. Mater. Interfaces*, **1**, 1292–1304.

Lechardeur, D., Lukacs, G.L. (2002). *Curr. Gene Ther.*, **2**, 183–194.

Lee, B.S., Fujita, M., Khazenzon, N.M., Wawrowsky, K.A., Wachsmann-Hogiu, S., Farkas, D.L., Black, K.L., Ljubimova, J.Y., Holler, E., (2006a). *Bioconj. Chem.*, **17**, 317–326.

Lee, C.C., Gillies, E.R., Fox, M.E., Guillaudeu, S.J., Fréchet, J.M., Dy, E.E., Szoka, F.C. (2006b). *Proc. Natl. Acad. Sci. U S A*, **103**, 16649–16654.

Lee, C.S., Park, W., Park, S.J., Na, K. (2013a). *Biomaterials*, **34**, 9227–9236.

Lee, H. (2012). *Mol. Simulat.*, **38**, 6.

Lee, H., Larson, R.G. (2011). *J. Phys. Chem. C*, **115**, 7.

Lee, H.J., Nguyen, Y.T., Muthiah, M., Vu-Quang, H., Namgung, R., Kim, W.J., Yu, M.K., Jon, S., Lee, I.K., Jeong, Y.Y., Park, I.K. (2012). *J. Biomed. Nanotechnol.*, **8**, 361–371.

Lee, I.H., An, S., Yu, M.K., Kwon, H.K., Im, S.H., Jon, S. (2011a). *J. Controlled Release*, **155**, 435–441.

Lee, J., et al. (2008a). *Adv. Funct. Mater.*, **18**, 258–264.

Lee, J.E., et al. (2010). *J. Am. Chem. Soc.*, **132**, 552–557.

Lee, J.-H., Jun, Y.-W., Yeon, S.-I., Shin, J.-S., Cheon, J. (2006c). *Angew. Chem., Int. Ed.*, **48**, 8160–8162.

Lee, J.H., Kang, D.Y., Lee, T., Kim, S.U., Oh, B.K., Choi, J.W. (2009a). *J. Nanosci. Nanotechnol.*, **9**, 7155–7160.

Lee, J.-H., Lee, K., Moon, S.H., Lee, Y., Park, T.G., Cheon, J. (2009b). *Angew. Chem., Int. Ed.*, **23**, 4174–4179.

Lee, J.J., Jeong, K.J., Hashimoto, M., Kwon, A.H., Rwei, A., Shankarappa, S.A., Tsui, J.H., Kohane, D.S. (2013b). *Nano Lett.* DOI:10.1021/nl3047305.

Lee, J.S., Groothuis, T., Cusan, C., Mink, D., Feijen, J. (2011b). *Biomaterials*, **32**, 9144–9153.

Lee, K.S., Chung, H.C., Im, S.A., Park, Y.H., Kim, C.S., Kim, S.B., Rha, S.Y., Lee, M.Y., Ro, J. (2008b). *Breast Cancer Res. Treat.*, **108**, 241–250.

Lee, M.S., Kim, J.-C. (2012). *J. Appl. Polym. Sci.*, **124**, 4339–4345.

Lee, N., Choi, S.H., Hyeon, T. (2013c). *Adv. Mater.*, **25**, 2641–2660.

Lee, N., Hyeon, T. (2012). *Chem. Soc. Rev.*, **7**, 2575–2589.

Lee, R.J., Low, P.S. (1997). *J. Liposome Res.*, **7**, 455–466.

Lee, S.K., Han, M.S., Asokan, S., Tung, C.-H. (2011c). *Small*, **3**, 364–370.

Lee, W.I., Bae, Y., Bard, A.J. (2004). *J. Am. Chem. Soc.*, **126**, 8358–8359.

Lee, Y., Ishii, T., Cabral, H., Kim, H.J., Seo, J.H., Nishiyama, N., Oshima, H., Osada, K., Kataoka, K. (2009c). *Angew. Chem., Int. Ed. Engl.*, **48**, 5309–5312.

Legendre, J.Y., Szoka, F.C., Jr. (1992). *Pharm. Res.*, **9**, 1235–1242.

Leid, J.G., Ditto, A.J., Knapp, A., Shah, P.N., Wright, B.D., Blust, R., Christensen, L., Clemons, C.B., Wilber, J.P., Young, G.W., Kang, A.G., Panzner, M.J., Cannon, C.L., Yun, Y.H., Youngs, W.J., Seckinger, N.M., Cope, E.K. (2012). *J. Antimicrob. Chemother.*, **67**, 138–148.

Lentacker, I., Geers, B., Demeester, J., De, S.medt, S.C., Sanders, N.N. (2010). *Mol. Ther.*, **18**, 101–108.

Leroueil, P.R., Hong, S., Mecke, A., Baker Jr, J.R., Orr, B.G., Banaszak Holl, M.M. (2007). *Acc. Chem. Res.*, **40**, 335–342.

Leswin, J.S.K., Meuldijk, J., Gilbert, R.G., van Herk, A.M. (2009). *Macromol. Symp.*, **275–276**(1), 24–34.

Leuchtenberger, C., Lewisohn, R., Laszlo, D., Leuchtenberger, R. (1944). *P. Soc. Exp. Biol. Med.*, **55**, 2.

Leuschner, F., Dutta, P., Gorbatov, R., Novobrantseva, T.I., Donahoe, J.S., Courties, G., Lee, K.M., Kim, J.I., Markmann, J.F., Marinelli, B., Panizzi, P., Lee, W.W., Iwamoto, Y., Milstein, S., Epstein-Barash, H., Cantley, W., Wong, J., Cortez-Retamozo, V., Newton, A., Love, K., Libby, P., Pittet, M.J., Swirski, F.K., Koteliansky, V., Langer, R., Weissleder, R., Anderson, D.G., Nahrendorf, M. (2011). *Nat. Biotechnol.*, **29**, 1005–1010.

Levchenko, T.S., Rammohan, R., Lukyanov, A.N., Whiteman, K.R., Torchilin, V.P. (2002). *Int. J. Pharm.*, **240**, 95–102.

Levin, C.S., Hofmann, C., Ali, T.A., Kelly, A.T., Morosan, E., Nordlander, P., Whitmire, K.H., Halas, N.J. (2009). *ACS Nano*, **3**, 1379–1388.

Levy, M., Luciani, N., Alloyeau, D., Elgrabli, D., Deveaux, V., Pechoux, C., Chat, S., Wang, G., Vats, N., Gendron, F., Factor, C., Lotersztajn, S., Luciani, A., Wilhelm, C., Gazeau, F. (2011). *Biomaterials*, **32**, 3988–3999.

Li, D., Li, G., Guo, W., Li, P., Wang, E., Wang, J. (2008a). *Biomaterials*, **18**, 2776–2782.

Li, D., Zhang, Y.T., Yu, M., Guo, J., Chaudhary, D., Wang, C.C. (2013a). *Biomaterials*, **34**, 7913–7922.

Li, F., Bi, W.T., Liu, L.Y., Li, Z., Huang, X.T. (2009a). *Colloid Surf. A: Physicochem. Eng. Asp.*, **1–3**, 160–164.

Li, G., Meng, Y., Guo, L., Zhang, T., Liu, J. (2014). *J. Biomed. Mater. Res. A*, **102**(7), 2163–2172.

Li, G.L., Möhwald, H., Shchukim, D.G. (2013b). *Chem. Soc. Rev.*, **42**, 3628.

Li, G.X., Wang, S.X. (2003). *IEEE Trans. Magn.*, **5**, 3313–3315.

Li, G.Y., Guo, L., Ma, S.M. (2009b). *J. Appl. Polym. Sci.*, 113, 1364–1368.

Li, J., Chen, Y.-C., Tseng, Y.-C., Mozumdar, S., Huang, L. (2010a). *J. Controlled Release*, 142, 416–421.

Li, J., Huo, M., Wang, J., Zhou, J., Mohammad, J.M., Zhang, Y., Zhu, Q., Waddad, A.Y., Zhang, Q. (2012a). *Biomaterials*, **33**, 2310–2320.

Li, J., Ng, H.T., Chen, H. (2005). *Methods Mol. Biol.*, **300**, 191–123.

Li, J., Yang, Y., Huang, L. (2012b). *J. Controlled Release*, **158**, 108–114.

Li, J., Yap, S.Q., Chin, C.F., Tian, Q., Yoong, S.L., Pastorin, G., Ang, W.H. (2012c). *Chem. Sci.*, **3**, 2083–2087.

Li, J., Yap, S.Q., Yoong, S.L., Nayak, T.R., Chandra, G.W., Ang, W.H., Panczyk, T., Ramaprabhu, S., Vashist, S.K., Sheu, F.S., Tan, A., Pastorin, G. (2012d). *Carbon*, **50**, 1625–1634.

Li, J.Y., Zu, B.Y., Zhang, Y., Guo, X.Z., Zhang, H.Q. (2010b). *J. Polym. Sci. Polym. Chem. Ed.*, **48**, 3217–3228.

Li, L., Tong, J., Wu, J., Du, J., Wang, J. (2009c). *Acta Polym. Sin.*, **9**, 257–263.

Li, M.-H., Keller, P. (2009). *Soft Matter*, **5**, 927–937.

Li, P., Xu, R., Wang, W., Li, X., Xu, Z., Yeung, K.W.K., Chu, P.K. (2013c). *Colloid Surf. B*, **101**, 251–255.

Li, P., Zhu, A.M., Liu, Q.L., Zhang, Q.G. (2008b). *Ind. Eng. Chem. Res.*, **47**, 7700–7706.

Li, R., Wu, A., Zhao, L., Hu, Z., Guo, S., Pan, X., Zou, H. (2011a). *Carbon*, **49**, 1797–1805.

Li, R., Wu, R., Zhao, L., Wu, M., Yang, L., Zou, H. (2010c). *ACS Nano*, **4**, 1399–1408.

Li, R., Wu, W., Liu, Q., Wu, P., Xie, L., Zhu, Z., Yang, M., Qian, X., Ding, Y., Yu, L., Jiang, X., Guan, W., Liu, B. (2013d). *PLoS ONE*, **8**, e69643.

Li, S.D., Huang, L. (2008). *Mol. Pharm.*, **5**, 496–504.

Li, W.H., Stöver, H.D.H. (2000). *Macromolecules*, **33**, 4354–4360.

Li, X.Q., Wen, H.Y., Dong, H.Q., Xue, W.M., Pauletti, G.M., Cai, X.J., Xia, W.J., Shi, D., Li, Y.Y. (2011b). *Chem. Commun. (Camb.)*, **47**, 8647–8649.

Li, Y., Armes, S. (2010). *Angew. Chem.*, **122**, 4136–4140.

Li, Y., Beija, M., Laurent, S., Elst, L.v., Muller, R.N., Duong, H.T., Lowe, A.B., Davis, T.P., Boyer, C. (2012e). *Macromolecules*, **45**, 4196–4204.

Li, Y., Cao, M., Wang, Y. (2007). *J. Phys. Chem. B*, **111**, 13436–13443.

Li, Y., Gao, G.H., Lee, D.S. (2013e). *Adv Healthc Mater.*, **2**, 388–417.

Li, Y., Wang, T., Wang, J., Jiang, T., Cheng, G., Wang, S. (2011c). *Appl. Surf. Sci.*, **257**, 5663–5670.

Li, Z., Chen, H., Bao, H., Gao, M. (2004). *Chem. Mater.*, **16**, 1391–1393.

Li, Z., Wang, C., Cheng, L., Gong, H., Yin, S., Gong, Q., Li, Y., Liu, Z. (2013f). *Biomaterials*, **34**, 9160–9170.

Liao, C., Sun, Q., Liang, B., Shen, J., Shuai, X. (2011). *Eur. J. Radiol.*, **80**, 699–705.

Liberti, P.A., Rao, C.G., Terstappen, L. (2001). *J. Magn. Magn. Mater.*, **1–2**, 301–307.

Lim, J., Majetich, S.A. (2013). *Nano Today*, **1**, 98–113.

Lim, J.K., Majetich, S.A., Tilton, R.D. (2009). *Langmuir*, **25**, 13384–13393.

Lim, W.T., Tan, E.H., Toh, C.K., Hee, S.W., Leong, S.S., Ang, P.C., Wong, N.S., Chowbay, B. (2010a). *Ann. Oncol.*, **21**, 382–388.

Lim, Y.-b., Mays, C.E., Kim, Y., Titlow, W.B., Ryou, C. (2010b). *Biomaterials*, **31**, 2025–2033.

Lima, A.M., Pizzol, C.D., Monteiro, F.B., Creczynski-Pasa, T.B., Andrade, G.P., Ribeiro, A.O., Perussi, J.R. (2013). *J. Photochem. Photobiol. B*, **125**, 146–154.

Lin, C.Y., Javadi, M., Belnap, D.M., Barrow, J.R., Pitt, W.G. (2013). *Nanomedicine*, **10**(1), 67–76.

Lin, H., Bhatia, R., Lal, R. (2001). *FASEB J.*, **15**, 2433–2444.

Lin, P., Chen, J.-W., Chang, L.W., Wu, J.-P., Redding, L., Chang, H., Yeh, T.-K., Yang, C.S., Tsai, M.-H., Wang, H.-J., Kuo, Y.-C., Yang, R.S.H. (2008). *Environ. Sci. Technol.*, **16**, 6264–6270.

Lin, P.C., Chou, P.H., Chen, S.H., Liao, H.K., Wang, K.Y., Chen, Y.J., Lin, C.C. (2006). *Small*, **4**, 485–489.

Lin, Y., Pan, Y., Shi, Y., Huang, X., Jia, N., Jiang, J.Y. (2012a). *Nanotechnol.*, **23**, 165101.

Lin, Y., Zhang, Z., Zhao, L., Wang, X., Yu, P., Su, L., Mao, L. (2010). *Biosens. Bioelectron.*, **25**, 1350–1355.

Lin, Y.D., Luo, C.Y., Hu, Y.N., Yeh, M.L., Hsueh, Y.C., Chang, M.Y., Tsai, D.C., Wang, J.N., Tang, M.J., Wei, E.I., Springer, M.L., Hsieh, P.C. (2012b). *Sci. Transl. Med.*, **4**, 146ra109.

Lin, Y.S., Tsai, P.J., Weng, M.F., Chen, Y.C. (2005). *Anal. Chem.*, **6**, 1753–1760.

Linazasoro, G. (2007). *Arch. Neurol.*, **64**, 137–140.

Linazasoro, G. (2008). *Parkinsonism Relat. D*, **14**, 383–392.

Ling, D., Hyeon, T. (2013). *Small*, **9–10**, 1450–1466.

Ling, Y., Wei, K., Luo, Y., Gao, X., Zhong, S. (2011). *Biomaterials*, **32**, 7139–7150.

Liong, M., Lu, J., Kovochich, M., Xia, T., Ruehm, S.G., Nel, A.E., Tamanoi, F., Zink, J.I. (2008). *ACS Nano*, **2**, 889–896.

Lippi, G., Mattiuzzi, C., Favaloro, E.J. (2013). *Semin. Thromb. Hemost.*, **39**, 48–58.

Lisziewicz, J., Bakare, N., Calarota, S.A., Bánhegyi, D., Szlávik, J., Újhelyi, E., Tőke, E.R., Molnár, L., Lisziewicz, Z., Autran, B., Lori, F. (2012). *PLoS ONE*, **7**, e35416.

Lisziewicz, J., Calarota, S.A., Lori, F. (2007). *Expert Opin. Biol. Ther.*, **7**, 1563–1574.

Liu, C., Guo, J., Yang, W., Hu, J., Wang, C., Fu, S. (2009a). *J. Mater. Chem.*, **19**, 4764–4770.

Liu, F., Urban, M.W. (2010). *Prog. Polym. Sci.*, **35**, 3–23.

Liu, G., An, Z. (2014). *Polym. Chem.*, **5**, 1559–15565.

Liu, G., Li, X., Xiong, S., Li, L., Chu, P.K., Yeung, K.W.K., Wu, S., Xu, Z. (2011a). *Colloid Polym. Sci.*, **290**, 349–357.

Liu, G., Men, P., Kudo, W., Perry, G., Smith, M.A. (2009b). *Neurosci. Lett.*, **455**, 187–190.

Liu, G.-Y., Chen, C.-J., Li, D.-D., Wang, S.-S., Ji, J. (2012a). *J. Mater. Chem.*, **22**, 16865–16871.

Liu, H., Tørring, T., Dong, M., Rosen, C.B., Besenbacher, F., Gothelf, K.V. (2010). *J. Am. Chem. Soc.*, **132**, 18054–18056.

Liu, J., Bu, W., Pan, L., Shi, J. (2013a). *Angew. Chem., Int. Ed. Engl.*, **52**, 4375–4379.

Liu, J., Gu, C., Cabigas, E.B., Pendergrass, K.D., Brown, M.E., Luo, Y., Davis, M.E. (2013b). *Biomaterials*, **34**, 3729–3736.

Liu, J., Lu, Y. (2007). *Chem. Commun.*, **46**, 4872–4874.

Liu, J., Pang, Y., Huang, W., Huang, X., Meng, L., Zhu, X., Zhou, Y., Yan, D. (2011b). *Biomacromolecules*, **12**, 1567–1577.

Liu, J., Qu, Y., Yang, K., Wu, Q., Shan, Y., Zhang, L., Liang, Z., Zhang, Y. (2014). *ACS Appl. Mater. Interfaces*.

Liu, J., Sun, Z., Deng, Y., Zou, Y., Li, C., Guo, X., Xiong, L., Gao, Y., Li, F., Zhao, D. (2009c). *Angew. Chem., Int. Ed.*, **32**, 5875–5879.

Liu, J., Yu, M., Zhou, C., Yang, S., Ning, X., Zheng, J. (2013c). *J. Am. Chem. Soc.*, **135**(13), 4978–4981.

Liu, K.C., Yeo, Y. (2013). *Mol. Pharm.*, **10**, 1695–1704.

Liu, K.S., Wen, C.J., Yen, T.C., Sung, K.C., Ku, M.C., Wang, J.J, Fang, J.Y. (2012b). *Nanotechnology*, **23**, 095103.

Liu, L., Patra, A., Scherf, U., Kissel, T. (2012c). *Macromol. Biosci.*, **12**, 1384–1390.

Liu, T., Cousins, A., Chien, C.C., Kempson, I., Thompson, S., Hwu, Y., Thierry, B. (2013d). *Cancer Lett.*, **328**, 271–277.

Liu, X., Chen, Y., Li, H., Huang, N., Jin, Q., Ren, K., Ji, J. (2013e). *ACS Nano.*, **7**(7), 6244–6257.

Liu, X., Ma, S., Dai, C., Cai, F., Yao, Y., Yang, Y., Feng, M., Deng, K., Li, G., Ma, W., Xin, B., Lian, W., Xiang, G., Zhang, B., Wang, R. (2013f). *Endocrinology*, **154**, 1414–1423.

Liu, Y., Feng, S.S. (2011). *J. Controlled Release*, **152**(Suppl 1), e64–e65.

Liu, Y., Miyoshi, H., Nakamura, M. (2006). *J. Controlled Release*, **114**, 89–99.

Liu, Y.-Y., Yu, Y., Tian, W., Sun, L., Fan, X.-D. (2009d). *Macromol. Biosci.*, **9**, 525–534.

Liu, Z., Chen, K., Davis, C., Sherlock, S., Cao, Q., Chen, X., Dai, H. (2008a) *Cancer Res.*, **68**, 6652–6660.

Liu, Z., Davis, C., Cai, W., He, L., Chen, X., Dai, H. (2008b). *Proc. Natl Acad. Sci. U S A*, **105**, 1410–1415.

Liu, Z., Fan, A.C., Rakhra, K., Sherlock, S., Goodwin, A., Chen, X., Yang, Q., Felsher, D.W., Dai, H. (2009e). *Angew. Chem., Int. Ed. Engl.*, **48**, 7668–7672.

Liu, Z., Sun, X., Nakayama-Ratchford, N., Dai, H. (2007a). *ACS Nano*, **1**, 50–56.

Liu, Z., Yanagi, K., Suenaga, K., Kataura, H., Iijima, S. (2007b). *Nat. Nanotechnol.*, **2**, 422–425.

Liu, Z., Yanga, K., Lee, S.T. (2011c). *J. Mater. Chem.*, **21**, 586–598.

Liu, Z., Zheng, M., Meng, F., Zhong, Z. (2011d). *Biomaterials*, **32**, 9109–9119.

Liz-Marzán, L.M. (2004). *Mater. Today*, **2**, 26–31.

Liz-Marzan, L.M., Giersig, M., Mulvaney, P. (1996a). *Chem. Commun.*, **6**, 731–732.

Liz-Marzán, L.M., Giersig, M., Mulvaney, P. (1996b). *Langmuir*, **18**, 4329–4335.

Ljubimova, J.Y., Fujita, M., Khazenzon, N.M., Lee, B.S., Wachsmann-Hogiu, S., Farkas, D.L., Black, K.L., Holler, E. (2008). *Chem. Biol. Interact.*, **171**, 195–203.

Lo, C-L., Lin, S-J., Tsai, H-C., Chan, W-H., Tsai, C-H., Chen, C-H., Hsiue, G-H. (2009). *Biomaterials*, **30**, 1–13.

Lobatto, M.E., Fuster, V., Fayad, Z.A., Mulder, W.J. (2011). *Nat. Rev. Drug Discovery*, **10**, 835–852.

Lodhi, N., Mehra, N. K., Jain, N.K. (2013). *J. Drug Targeting*, **21**, 67–76.

Longmire, M., Choyke, P.L., Kobayashi, H. (2008a). *Nanomedicine*, **3**, 703–717.

Longmire, M., Choyke, P.L., Kobayashi, H. (2008b). *Curr. Top. Med. Chem.*, **8**, 1180.

Loomis, K., McNeeley, K., Bellamkonda, R.V. (2011). *Soft Matter*, **7**, 839–856.

Lorentzen (1999). *Scand. J. Immunol.*, **49**, 45–50.

Lowery, A.R., et al. (2006). *Int. J. Nanomed.*, **1**, 149–154.

Lu, F., Gu, L., Meziani, M.J., Wang, X., Luo, P.G., Veca, L.M., Cao, L., Sun, Y.P. (2009a). *Adv. Mater.*, **21**, 139–152.

Lu, J., Yang, J.-x., Wang, J., Lim, A., Wang, S., Loh, K.P. (2009b). *ACS Nano*, **8**, 2367–2375.

Lu, R.M., Chang, Y.L., Chen, M.S., Wu, H.C. (2011). *Biomaterials*, **32**, 3265–3274.

Lubbe, A.S., Alexiou, C., Bergemann, C. (2001). *J. Surg. Res.*, **95**, 200–206.

Luksirikul, P., Ballesteros, B., Tobias, G., Moloney, M. G., Green, M.L.H. (2010). *Carbon*, **48**, 1912–1917.

Luo, K., Liu, G., She, W., Wang, Q., Wang, G., He, B., Ai, H., Gong, Q., Song, B., Gu, Z. (2011). *Biomaterials*, **32**, 7951–7960.

Luo, X., Matranga, C., Tan, S., Alba, N., Cui, X.T. (2011). *Biomaterials*, **32**, 6316–6623.

Lutz, J.-F. (2011). *Adv. Mater.*, **23**, 2237–2243.

Luz Garcia-Hernandez, M., Gray, A., Hubby, B., Klinger, O.J., Kast, W.M. (2008). *Cancer Res.*, **68**, 861–869.

Ma, X., Wang, X., Zhou, M., Fei, H. (2013). *Adv Health Mater.*, **2**(12), 1638–1643.

Ma, Y., Wu, S., Wu, T., Chang, Y., Hua, M., Chen, J. (2009) *Biomaterials*, **30**, 3343–3351.

Ma, Z., Ding, T. (2009). *Nanoscale Res. Lett.*, **10**, 1236–1240.

MacDonald, R.C., MacDonald, R.I., Menco, B. Ph. M., Takeshita, K., Subbarao, N.K., Hu, L. (1991). *Biochim. Biophys. Acta*, **1061**, 297–303.

Madero-Visbal, R.A., Alvarado, B.E., Colon, J.F., Baker, C.H., Wason, M.S., Isley, B., Seal, S., Lee, C.M., Das, S., Manon, R. (2012). *Nanomedicine*, **8**, 1223–1231.

Madhusudana Rao, K., Mallikarjuna, B., Krishna Rao, K.S. V, Siraj, S., Chowdoji Rao, K., Subha, M.C.S. (2013). *Colloid Surf. B*, **102**, 891–897.

Maeda, H., Akaike, T., Wu, J., Noguchi, Y., Sakata, Y. (1996). *Immunopharmacology*, **33**, 222–230.

Maeda, H., Takeshita, J., Kanamaru, R., Sato, H., Katoh, J., Sato, H. (1979). *Gann*, **70**, 601–606.

Maeda, H., Takeshita, J., Yamashita, A. (1980). *Eur. J. Cancer*, **16**, 723–731.

Maeda, H., Wu, J., Sawa, T., Matsumura, Y., Hori, K. (2000). *J. Controlled Release*, 65, 271–284.

Magnusson, J.P., Khan, A., Pasparakis, G., Saeed, A.O., Wang, W., Alexander, C. (2008). *J. Am. Chem. Soc.*, **130**, 10852–10853.

Mahmoudi, M., Hosseinkhani, H., Hosseinkhani, M., Boutry, S., Simchi, A., Journeay, W.S., Subramani, K., Laurent, S. (2010). *Chem. Rev.*, **2**, 253–280.

Mahoney, M.J., Saltzman, W.M. (2001). *Nat. Biotechnol.*, **19**, 934–939.

Maitz, M.F., Freudenberg, U., Tsurkan, M.V., Fischer, M., Beyrich, T., Werner, C. (2013). *Nat. Commun.*, **4**, 2168.

Majoros, I.J., Myc, A., Thomas, T., Mehta, C.B., Baker, J.R. (2006). *Biomacromolecules*, **7**, 572–579.

Majoros, I.J., Thomas, T.P., Mehta, C.B., Baker, J.R. (2005). *J. Med. Chem.*, **48**, 5892–5899.

Majoros, I.J., Williams, C.R., Baker, J., James, R. (2008). *Curr. Top. Med. Chem.*, **8**, 1165–1179.

Malik, M.A., Zulu, M.M., O'Brien, P., Wakefield, G. (1998). *J. Mater. Chem.*, **8**, 1885–1888.

Malik, N., Evagorou, E.G., Duncan, R. (1999). *Anti-Cancer Drug*, **10**, 767.

Mallidi, S., Larson, T., Aaron, J., Sokolov, K., Emelianov, S. (2007). *Opt. Express*, **11**, 6583–6588.

Malvindi, M.A., Di Corato, R., Curcio, A., Melisi, D., Rimoli, M.G., Tortiglione, C., Tino, A., George, C., Brunetti, V., Cingolani, R., Pellegrino, T., Ragusa, A. (2011). *Nanoscale*, **12**, 5110–5119.

Mann, A.P., Tanaka, T., Somasunderam, A., Liu, X., Gorenstein, D.G., Ferrari, M. (2011). *Adv. Mater.*, **23**, H278–H282.

Mann, J.F.S., Shakir, E., Carter, K.C., Mullen, A.B., Alexander, J., Ferro, V.A. (2009). *Vaccine*, **27**, 3643–3649.

Manolova, V., Flace, A., Bauer, M., Schwarz, K., Saudan, P., Bachmann, M.F. (2008). *Eur. J. Immunol.*, **38**, 1404–1413.

Manzoor, A.A., Lindner, L.H., Landon, C.D., Park, J.Y., Simnick, A.J., Dreher, M.R., Das, S., Hanna, G., Park, W., Chilkoti, A., Koning, G.A., Ten, H.agen, T.L., Needham, D., Dewhirst, M.W. (2012). *Cancer Res.*, **72**, 5566–5575.

Maquieira, Á., Brun, E.M., Garcés-García, M., Puchades, R. (2012). *Anal. Chem.*, **84**, 9340–9348.

Marcinowski, K.J., Shao, H., Clancy, E.L., Zagorski, M.G. (1998). *J. Am. Chem. Soc.*, **120**, 11082–11091.

Margolis, L.B., Namiot, V.A., Kljukin, L.M. (1983). *Biochim. Biophys. Acta*, **735**, 193–195.

Marin, A., Muniruzzaman, M., Rapoport, N. (2001). *J. Controlled Release*, **71**, 239–249.

Marin, A., Sun, H., Husseini, G.A., Pitt, W.G., Christensen, D.A., Rapoport, N.Y. (2002). *J. Controlled Release*, **84**, 39–47.

Marrache, S., Dhar, S. (2012). *Proc. Natl. Acad. Sci. U S A*, **109**, 16288–16293.

Marradi, M., Chiodo, F., Garcia, I., Penades, S. (2013). *Chem. Soc. Rev.*, **42**, 4728–4745.

Marsch, D. (2010). *Biochim. Biophys. Acta*, **1798**, 688–699.

Marsh, J.N., Hu, G., Scott, M.J., Zhang, H., Goette, M.J., Gaffney, P.J., Caruthers, S.D., Wickline, S.A., Abendschein, D., Lanza, G.M. (2011). *Nanomedicine (London)*, **6**, 605–615.

Martelli, G., Zope, H.R., Brovia, C.apell, M., Kros, A. (2013). *Chem. Commun.*, **49**(85), 9932–9934.

Martin, F.J., Heath, T.D., New, R.R.C. (1990). Covalent attachment of proteins to liposomes, In *Liposomes. A Practical Approach*, Ed. New, R.R.C., IRL Press, Oxford, UK.

Martins, M.B., Corvo, M.L., Marcelino, P., Marinho, H.S., Feio, G., Carvalho, A. (2013a). *Nanomedicine*, **10**(1), 207–214.

Martins, N.C.T., Freire, C.S.R., Neto, C.P., Silvestre, A.J.D., Causio, J., Baldi, G., Sadocco, P., Trindade, T. (2013b). *Colloid Surface A*, 111–119.

Mashat, A., Deng, L., Altawashi, A., Sougrat, R., Wang, G., Khashab, N.M. (2012). *J. Mater. Chem.*, **22**, 11503–11508.

Massart, R. (1980). *CR Hebd. Acad. Sci. C*, **1**, 1–3.

Masuda, T., Akita, H., Niikura, K., Nishio, T., Ukawa, M., Enoto, K., Danev, R., Nagayama, K., Ijiro, K., Harashima, H. (2009). *Biomaterials*, **27**, 4806–4814.

Matsumoto, N.M., Gonzalez-Toro, D.C., Chacko, R.T., Maynard, H.D., Thayumanavan, S. (2013). *Polym. Chem.*, **4**, 2464–2469.

Matsumoto, S., Christie, R.J., Nishiyama, N., Miyata, K., Ishii, A., Oba, M., Koyama, H., Yamasaki, Y., Kataoka, K. (2009). *Biomacromolecules*, **10**, 119–127.

Matsumura, Y., Maeda, H. (1986). *Cancer Res.*, **46**, 6387–6392.

Matsuzaki, K., Noguch, T., Wakabayashi, M., Ikeda, K., Okada, T., Ohashi, Y., Hoshino, M., Naiki, H. (2007). *Biochim. Biophys. Acta*, **1768**, 122–130.

Mattoussi, H., Palui, G., Na, H.B. (2012). *Adv. Drug Delivery Rev.*, **2**, 138–166.

Matyjaszewski, K., Davis, T.P. (2002). *Handbook of Radical Polymerization*, John Wiley and Sons, Hoboken.

May, J.P., Li, S.D. (2013). *Expert Opin. Drug Delivery*, **10**, 511–527.

May, R., Machesky, L. (2001). *J. Cell Sci.*, **114**, 1061–1077.

Mayer, L.D., Hope, M.J., Cullis, P.R. (1986). *Biochim. Biophys. Acta*, **858**, 161–168.

Mayhew, E., Lazo, R., Vail, W.J., King, J., Green, A.M. (1984). *Biochim. Biophys. Acta*, **775**, 169–174.

McCarthy, J.R., Sazonova, I.Y., Erdem, S.S., Hara, T., Thompson, B.D., Patel, P., Botnaru, I., Lin, C.P., Reed, G.L., Weissleder, R., Jaffer, F.A. (2012). *Nanomedicine (London)*, **7**, 1017–1028.

McCarthy, T.D., Karellas, P., Henderson, S.A., Giannis, M., O'Keefe, D.F., Heery, G., Paull, J.R., Matthews, B.R., Holan, G. (2005). *Mol. Pharm.*, **2**, 312–318.

McDaniel, J.R., Bhattacharyya, J., Vargo, K.B., Hassouneh, W., Hammer, D.A., Chilkoti, A. (2013a). *Angew. Chem., Int. Ed. Engl.*, **52**, 1683–1687.

McDaniel, J.R., Dewhirst, M.W., Chilkoti, A. (2013b). *Int. J. Hyperthermia*, **29**, 501–510.

McDaniel, J.R., Macewan, S.R., Dewhirst, M., Chilkoti, A. (2012). *J. Controlled Release*, **159**, 362–367.

McDonald, T.O., Qu, H., Saunders, R., Ulijn, R.V., Saunders, B.R. (2009). *Soft Matter*, **5**, 1728–1734.

McFarland, A.D., Van Duyne, R.P. (2003). *Nano Lett.*, **8**, 1057–1062.

McLaurin, J., Chakrabartty, A. (1996). *J. Biol. Chem.*, **271**, 26482–26489.

McNeeley, K.M., Annapragada, A., Bellamkonda, R.V. (2007). *Nanotechnology*, **18**, 385101.

McNeeley, K.M., Karathanasis, E., Annapraganda, A., Ballahanda, R. (2009). *Biomaterials*, **30**, 3986–3995.

Mecke, A., Majoros, I.J., Patri, A.K., Baker, J.R., Jr., Holl, M.M., Orr, B.G. (2005). *Langmuir*, **21**, 10348–10354.

Mecke, A., Uppuluri, S., Sassanella, T.M., Lee, D.K., Ramamoorthy, A., Baker, J.R., Jr., Orr, B.G., Banaszak Holl, M.M. (2004). *Chem. Phys. Lipids*, **132**, 3–14.

Medina, S.H., El-Sayed, M.E. (2009). *Chem. Rev.*, **109**, 17.

Medintz, I.L., Uyeda, H.T., Goldman, E.R., Mattoussi, H. (2005). *Nat. Mater.*, **6**, 435–446.

Meers, P. (2001). *Adv. Drug Delivery Rev.*, **53**, 265–272.

Meers, P., Neville, M., Malinin, V., Scotto, A.W., Sardaryan, G., Kurumunda, R., Mackinson, C., James, G., Fisher, S., Perkins, W.R. (2008). *J. Antimicrob. Chemother.*, **61**, 859–868.

Ménager, C., Sandre, O., Mangili, J., Cabuil, V. (2004). *Polymer*, **45**, 2475–2481.

Ménard-Moyon, C., Venturelli, E., Fabbro, C., Samorì, C., Da Ros, T., Kostarelos, K., Prato, M., Bianco, A. (2010). *Expert Opin. Drug Discovery*, **5**, 691–707.

Meng, F., Hennink, W.E., Zhong, Z. (2009). *Biomaterials*, **30**, 2180–2198.

Meng, H., Mai, W.X., Zhang, H., Xue, M., Xia, T., Lin, S., Wang, X., Zhao, Y., Ji, Z., Zink, J.I., Nel, A.E. (2013). *ACS Nano*, **7**, 994–1005.

Meng, J., et al. (2009). *Mater. Sci. Eng. C*, **29**, 1467–1479.

Menger, F.M., Chlebowski, M.E., Galloway, A.L., Lu, H., Seredyuk, V.A., Sorrells, J.F., Zhang, H. (2005). *Langmuir*, **21**, 10336–10341.

Menjoge, A.R., Kannan, R.M., Tomalia, D.A. (2010). *Drug Discovery Today*, **15**, 171–185.

Menon, J.U., Jadeja, P., Tambe, P., Vu, K., Yuan, B., Nguyen, K.T. (2013). *Theranostics*, **3**, 152–166.

Menon, S., Thekkayil, R., Varghese, S., Das, S. (2011). *J. Polym. Sci., Part A: Polym. Chem.*, **49**, 5063–5073.

Merbach, A.E., Toth, E. (2001). NMR-MRI, μSR and Mössbauer spectroscopies in molecular magnets, In *The Chemistry of Contrast Agents in Medical Magnetic Resonance Imaging*, John Wiley and Sons, USA.

Metz, N., Theato, P. (2009). *Macromolecules*, **42**, 37–39.

Meure, L.A., Foster, N.R., Deghhani, F. (2008). *AAPS Pharm. Sci. Tech.*, **9**, 798–809.

Meyer, D.E., Chilkoti, A. (2004). *Biomacromolecules*, **5**, 846–851.

Meyer, R.R., Sloan, J., Dunin-Borkowski, R.E., Kirkland, A.I., Novotny, M.C., Bailey, S.R., Hutchison, J.L., Green, M.L. (2000). *Science*, **289**, 1324–1327.

Mh, P.-L., Hickling, W.j JE, Rodriguez, O.c GA, Albanese, C. NM, Van, K.euren, E. FS, Dollahon, N. SS (2013). *ACS Nano*, **7**(10), 9040–9048.

Michaels, A.M., Jiang, J., Brus, L. (2000). *J. Phys. Chem. B*, **50**, 11965–11971.

Michalet, X., Pinaud, F.F., Bentolila, L.A., Tsay, J.M., Doose, S., Li, J.J., Sundaresan, G., Wu, A. M., Gambhir, S.S., Weiss, S. (2005). *Science*, **5709**, 538–544.

Miklya, I. (2010). *Neuropsychopharmacol. Hung*, **12**, 395–403.

Mintzer, M.A., Dane, E.L., O'Toole, G.A., Grinstaff, M.W. (2011). *Mol. Pharm.*, **9**, 342–354.

Mintzer, M.A., Grinstaff, M.W. (2011). *Chem. Soc. Rev.*, **40**, 173–190.

Mintzer, M.A., Simanek, E.E. (2009). *Chem. Rev.*, **109**, 259–302.

Mirkin, C.A., Letsinger, R.L., Mucic, R.C., Storhoff, J.J. (1996). *Nature*, **6592**, 607–609.

Misselwitz, B., Schmitt-Willich, H., Ebert, W., Frenzel, T., Weinmann, H.-J. (2001). *Mang. Reson. Mater. Phy.*, **12**, 128–134.

Mitchell, D.J., Ninham, B.W. (1981). *J. Chem. Soc., Faraday Trans. 2*, **77**, 601–629.

Mitragotri, S., Lahann, J. (2009). *Nat. Mater.*, **8**, 15–23.

Mittal, G., Sahana, D.K., Bhardwaj, V., Ravi Kumar, M.N. (2007). *J. Controlled Release*, **119**, 77–85.

Miyata, T., Asami, N., Jige, M., Uragami, T. (2011). *Polym. Prepr. (Am. Chem. Soc., Div. Polym. Chem.)*, **C**, 581–582.

Miyata, T., Asami, N., Uragami, T. (1999). *Nature*, **399**, 766–769.

Moad, G., Solomon, D.H. (2006). *The Chemistry of Radical Polymerization*, Elsevier Amsterdam.

Mochalin, V.N., Shenderova, O., Ho, D., Gogotsi, Y. (2012). *Nat. Nanotechnol.*, **1**, 11–23.

Modo, M., Hoehn, M., Bulte, J.W.M. (2005). *Mol. Imaging*, **4**, 143–164.

Moghimi, S., Pavey, K., Hunter, A. (2003). *FEBS Lett.*, **547**, 177–182.

Moghimi, S.M. (1999). *Biochim. Biophys. Acta*, **1472**, 399–403.

Moghimi, S.M., Andersen, A.J., Ahmadvand, D., Wibroe, P.P., Andresen, T.L., Hunter, A.C. (2011). *Adv. Drug Delivery Rev.*, **63**, 1000–1007.

Moghimi, S.M., Hunter, A.C., Andresen, T.L. (2012). *Annu. Rev. Pharmacol. Toxicol.*, **52**, 481–503.

Moghimi, S.M., Hunter, A.C., Murray, J.C. (2001). *Pharmacol. Rev.*, **53**, 283–318.

Mohan, P., Rapoport, N. (2010). *Mol. Pharm.*, **7**, 1959–1973.

Mohs, A.M., Lu, Z.-R. (2007). *Expert Opin. Drug Delivery*, **4**, 149–164.

Moingeon, P., de Taisne, C., Almond, J. (2002). *Br. Med. Bull.*, **62**, 29–44.

Moller, R., Fritzsche, W. (2007). *Curr. Pharm. Biotechnol.*, **5**, 274–285.

Monica, P., Monica, B., Cosmin, F., Simion, A. (2012). *Nanotechnology*, **5**, 1–10.

Monsky, W.L., Fukumura, D., Gohongi, T., Ancukiewcz, M., Weich, H.A., Torchilin, V.P., Yuan, F., Jain, R.K. (1999). *Cancer Res.*, *59*, 4129–4135.

Montenegro, L., Trapani, A., Latrofa, A., Puglisi, G. (2012). *J. Nanosci. Nanotechnol.*, **12**, 330–337.

Monthioux, M. (2002). *Carbon*, **40**, 1809–1823.

Montomoli, E., Piccirella, S., Khadang, B., Mennitto, E., Camerini, R., De Rosa, A. (2011). *Expert Rev. Vaccines*, **10**, 1053–1061.

Montserret, R., McLeish, M.J., Böckmann, A., Geourjon, C., Penin, F. (2000). *Biochemistry*, **39**, 8362–8373.

Moon, J.J., Suh, H., Bershteyn, A., Stephan, M.T., Liu, H., Huang, B., Sohail, M., Luo, S., Ho Um, S., Khant, H., Goodwin, J.T., Ramos, J., Chiu, W., Irvine, D.J. (2011). *Nat. Mater.*, **10**, 243–251.

Moore, A., Marecos, E., Bogdanov, A., Jr and Weissleder, R. (2000). *Radiology*, **214**, 568–574.

Mora, J.R., Zielinski, T.L., Nelson, B.P., Getts, R.C. (2008). *BioTechniques*, **44**, 815–819.

Moradi, M., Heuts, J.P.A., van Herk, A.M., The effect of monomer feed composition on the morphology of the products of RAFT mediated vesicle templating polymerizations, to be published.

Morales, M.A., Jain, T.K., Labhasetwar, V., Leslie-Pelecky, D.L. (2005). *J. Appl. Phys.*, **97**, 10Q905–10Q905-3.

Morales-Narváez, E., Montón, H., Formicheva, A., Merkoçi, A. (2012). *Anal. Chem.*, **84**, 6821–6827.

Morel, A.-L., et al. (2008). *ACS Nano*, **2**, 847–856.

Morel, S., Didierlaurent, A., Bourguignon, P., Delhaye, S., Baras, B., Jacob, V., Planty, C., Elouahabi, A., Harvengt, P., Carlsen, H., Kielland, A., Chomez, P., Garçon, N., Van Mechelen, M. (2011). *Vaccine*, **29**, 2461–2473.

Morgan, D.A., Sloan, J., Green, M.L. (2002). *Chem. Commun.*, 2442–2443.

Morgan, M.T., Nakanishi, Y., Kroll, D.J., Griset, A.P., Carnahan, M.A., Wathier, M., Oberlies, N.H., Manikumar, G., Wani, M.C., Grinstaff, M.W. (2006). *Cancer Res.*, **66**, 11913–11921.

Moriggi, L., Cannizzo, C., Dumas, E., Mayer, C.R., Ulianov, A., Helm, L. (2009). *J. Am. Chem. Soc.*, **131**, 10828–10829.

Morimoto, N., Hirano, S., Takahashi, H., Loethen, S., Thompson, D.H., Akiyoshi, K. (2012). *Biomacromolecules*, **14**, 56–63.

Morisada, S., Miyata, N., Iwahori, K. (2002). *J. Microbiol. Meth.*, **2**, 141–148.

Morosini, V., Bastogne, T., Frochot, C., Schneider, R., Francois, A., Guillemin, F., Barberi-Heyob, M. (2011). *Photoch. Photobiol. Sci.*, **5**, 842–851.

Morral-Ruiz, G., Melgar-Lesmes, P., Solans, C., Garcia-Celma, M.J. (2013). *J. Controlled Release*, **171**, 163–171.

Morris, K. (2004). *Lancet Neurol.*, **3**, 202.

Mosca, F., Tritto, E., Muzzi, A., Monaci, E., Bagnoli, F., Iavarone, C., O'Hagan, D., Rappuoli, R., De Gregorio, E. (2008). *Proc. Natl. Acad. Sci.*, **105**, 10501–10506.

Moskovits, M. (2005). *J. Raman Spectrosc.*, **6–7**, 485–496.

Motokawa, R., Morishita, K., Koizumi, S., Nakahira, T., Annaka, M. (2005). *Macromolecules*, **38**, 5748–5760.

Moughton, A.O., Hillmyer, M.A., Lodge, T.P. (2012). *Macromolecules*, **45**, 2–19.

Mourtas, C., Canovi, M., Zona, C., Aurilia, D., Niarakis, A., LaFeria, B., Salmona, M., Nicotra, F., Gobbi, M., Antimisiaris, S.G. (2011). *Biomaterials*, **31**, 1635–1645.

Moyer, H.R., Delman, K.A. (2008). *Int. J. Hyperthermia*, **24**, 251–261.

Muhammad, F., Guo, M., Guo, Y., Qi, W., Qu, F., Sun, F., Zhao, H., Zhu, G. (2011a). *J. Mater. Chem.*, **35**, 13406–13412.

Muhammad, F., Guo, M., Qi, W., Sun, F., Wang, A., Guo, Y., Zhu, G. (2011b). *J. Am. Chem. Soc.*, **23**, 8778–8781.

Mukherjee, P., Bhattacharya, R., Wang, P., Wang, L., Basu, S., Nagy, J.A., Atala, A., Mukhopadhyay, D., Soker, S. (2005). *Clin. Cancer Res.*, **11**, 3530–3534.

Mulik, R.S., Monkkönen, J., Juvonen, R.O., Mahadik, K.R., Paradkar, A.R. (2010). *Mol. Pharm.*, **7**, 815–825.

Muller, R.H., Keck, C.M. (2004). *J. Biotechnol.*, **113**, 151–170.

Munoz-Ubeda, M., Rodriguez-Pulido, A., Nogales, A., Martin-Molina, A., Aicart, E., Junquera, E. (2010). *Biomacromolecules*, **11**, 3332–3340.

Mura, C.V., Becker, M.L., Orellana, A., Wolff, D. (2002). *J. Immunol. Meth.*, **1–2**, 263–271.

Murphy, C.J., Thompson, L.B., Chernak, D.J., Yang, J.A., Sivapalan, S.T., Boulos, S.P., Huang, J., Alkilany, A.M., Sisco, P.N. (2011). *Curr. Opin. Colloid Inter. Sci.*, **2**, 128–134.

Murphy, E.A., Majeti, B.K., Barnes, L.A., Makale, M., Weis, S.M., Lutu-Fuga, K., Wrasidlo, W., Cheresh, D.A. (2008). *Proc. Natl. Acad. Sci. U S A*, **105**, 9343–9348.

Murphy, R.M. (2002). *Annu. Rev. Biomed. Eng.*, **4**, 155–174.

Murphy, S.L., Xu, J., Kochanek, K.D. (2012). *Natl. Vital Stat. Rep.*, **60**, 1–52.

Murray, C.B., Kagan, C.R., Bawendi, M.G. (2000). *Annu. Rev. Mater. Sci.*, **1**, 545–610.

Murray, C.B., Norris, D.J., Bawendi, M.G. (1993). *J. Am. Chem. Soc.*, **19**, 8706–8715.

Murthy, N., Thng, Y.X., Schuck, S., Xu, M.C., Fréchet, J.M.J. (2002). *J. Am. Chem. Soc.*, **124**, 12398–12399.

Murthy, N., Xu, M., Schuck, S., Kunisawa, J., Shastri, N., Fréchet, J.M.J. (2003). *Proc. Natl. Acad. Sci. U S A*, **100**, 4995–5000.

Musacchio, T., Vaze, O., D'souza, G., Torchilin, V.P. (2010). *Bioconjug. Chem.*, **21**, 1530–1536.

Muyal, J.P., Singh, S.K., Fehrenbach, H. (2008). *CRC Cr. Rev. Biotechn.*, **28**, 239–251.

Na, K., Lee, K.H., Lee, D.H., Bae, Y.H. (2006). *Eur. J. Pharm. Sci.*, **27**, 115e22.

Nagahama, R., Matoba, T., Nakano, K., Kim-Mitsuyama, S., Sunagawa, K., Egashira, K. (2012). *Arterioscler. Thromb. Vasc. Biol.*, **32**, 2427–2434.

Nagamitsu, A., Greish, K., Maeda, H. (2009). *Jpn. J. Clin. Oncol.*, **39**, 756–766.

Nagamoto, T., Hattori, Y., Takayama, K., Maitani, Y. (2004). *Pharm. Res.*, **21**, 671–674.

Nagasaki, Y., Yasugi, K., Yamamoto, Y., Harada, A., Kataoka, K. (2001). *Biomacromolecules*, **2**, 1067–1070.

Nagase, H., Woessner, J.F., Jr. (1999). *J. Biol. Chem.*, **274**, 21491–21494.

Nagata, T., Koide, Y. (2010). *J. Biomed. Biotechnol.*, **2010**, 1–11.

Nagatani, N., Shinkai, M., Nagase, Y., Honda, H., Hata, K., Mizuno, H., Ueda, M., Kobayashi, T. (2000). *Biotechnol. Lett.*, **12**, 999–1002.

Nair, B.G., Nagaoka, Y., Morimoto, H., Yoshida, Y., Maekawa, T., Kumar, D.S. (2010). *Nanotechnology*, **21**, 45–51.

Najlah, M., Freeman, S., Attwood, D., D'Emanuele, A. (2006). *Int. J. Pharm.*, **308**, 175–182.

Najlah, M., Freeman, S., Attwood, D., D'Emanuele, A. (2007). *Int. J. Pharm.*, **336**, 183–190.

Nakamura, H., Ito, N., Kotake, F., Mizokami, Y., Matsuoka, T. (2000). *J. Gastroenterol.*, **35**, 849–855.

Nakamura, M., Davila-Zavala, P., Tokuda, H., Takakura, Y., Hashida, M. (1998). *Biochem. Biophys. Res. Commun.*, **245**, 235–239.

Nam, J.M., Stoeva, S.I., Mirkin, C.A. (2004). *J. Am. Chem. Soc.*, **19**, 5932–5933.

Nam, Y., Park, J., Pak, Y.K., Pak, J.J. (2012). *J. Nanosci. Nanotechnol.*, **12**, 5547–5554.

Nance, E.A., Woodworth, G.F., Sailor, K.A., Shih, T.Y., Xu, Q., Swaminathan, G., Xiang, D., Eberhart, C., Hanes, J. (2012). *Sci. Transl. Med.*, **4**, 149ra119.

Nandy, B., Maiti, P.K. (2010). *J. Phys. Chem. B*, **115**, 217–230.

Nandy, B., Santosh, M., Maiti, P.K. (2012). *J. Biosci.*, **37**, 457–474.

Nappini, S., Bombelli, F.B., Bonini, M., Norden, B., Baglioni, P. (2010). *Soft Matter*, **6**, 154–162.

Nathan, C.F., Hibbs Jr, J.B. (1991). *Curr. Opin. Immunol.*, **3**, 65–70.

Navani, N.K., Li, Y. (2006). *Curr. Opin. Chem. Biol.*, **10**, 272–281.

Nayak, S., Lee, H., Chmielewski, J., Lyon, L.A. (2004). *J. Am. Chem. Soc.*, **126**, 10258–10259.

Needham, C.J., Williams, A.K., Chew, S.A., Kasper, F.K., Mikos, A.G. (2012). *Biomacromolecules*, **13**, 1429–1437.

Needham, D., Anyaambhatta, G., Kong, G., Dewhirst, M. (2000). *Cancer Res.*, **60**, 1197–1201.

Needham, D., MacIntosh, T.J., Lasic, D.D. (1992). *Biochim. Biophys. Acta*, **1108**, 40–48.

Neely, A., Perry, C., Varisli, B., Singh, A.K., Arbneshi, T., Senapati, D., Kalluri, J.R., Ray, P.C. (2010). *ACS Nano*, **3**, 2834–2840.

Negishi, Y., Endo-Takahashi, Y., Matsuki, Y., Kato, Y., Takagi, N., Suzuki, R., Maruyama, K., Aramaki, Y. (2012). *Mol. Pharm.*, **9**, 1834–1840.

Negishi, Y., Hamano, N., Tsunoda, Y., Oda, Y., Choijamts, B., Endo-Takahashi, Y., Omata, D., Suzuki, R., Maruyama, K., Nomizu, M., Emoto, M., Aramaki, Y. (2013). *Biomaterials*, **34**, 501–507.

Neimert-Andersson, T., Hällgren, A.-C., Andersson, M., Langebäck, J., Zettergren, L., Nilsen-Nygaard, J., Draget, K.I., van Hage, M., Lindberg, A., Gafvelin, G., Grönlund, H. (2011). *Vaccine*, **29**, 8965–8973.

Nel, A.E., Madler, L., Velegol, D., Xia, T., Hoek, E.M.V., Somasundaran, P., Klaessig, F., Castranova, V., Thompson, M. (2009). *Nat. Mater.*, **8** (7), 543–557.

Nelson, A.L., Dhimolea, E., Reichert, J.M. (2010). *Nat. Rev. Drug Discovery*, **9**, 767–774.

Neradovic, D., Soga, O., Van, N.ostrum, C.F., Hennink, W.E. (2004). *Biomaterials*, **25**, 2409–2418.

Neuberger, T., Schöpf, B., Hofmann, H., Hofmann, M., Von, R.echenberg, B. (2005). *J. Magn. Magn. Mater.*, **293**, 483–496.

Neves, M.C., Martins, M.A., Soares-Santos, P.C.R., Rauwel, P., Ferreira, R.A.S., Monteiro, T., Carlos, L.D., Trindade, T. (2008). *Nanotechnology*, **15**, 1–9.

New, R.R.C. (1990). Characterization of liposomes, In *Liposomes. A Practical Approach*, Ed. New, R.R.C., IRL Press, Oxford, UK.

New, R.R.C. (1990). Preparation of liposomes, In *Liposomes. A Practical Approach*, Ed. New, R.R.C., IRL Press, Oxford, UK.

Nguyen, D., Zondanos, H.S., Farrugia, J.M., Serelis, A.K., Such, C.H., Hawkett, B.S. (2008). *Langmuir*, **24**, 2140–2150.

Nie, S. (2010). *Nanomed.*, **5**, 523–528.

Nie, S., Xing, Y., Kim, G.J., Simons, J.W. (2007). *Annu. Rev. Biomed. Eng.*, **9**, 257–288.

Niederhafner, P., Šebestík, J., Ježek, J. (2005). *J. Pept. Sci.*, **11**, 757–788.

Nieva, J.L., Goni, F.M., Alonso, A. (1989). *Biochemistry*, **28**, 7364–7367.

Niikura, K., Iyo, N., Matsuo, Y., Mitomo, H., Ijiro, K. (2013). *ACS Appl. Mater. Interfaces*, **5**, 3900–3907.

Nikoobakht, B., El-Sayed, M.A. (2003). *Chem. Mater.*, **10**, 1957–1962.

Nilsson, M.R. (2004). *Methods*, **34**, 151–160.

Nisbert, D.R., Crompton, K.E., Horne, M.K. (2008). *J. Biomed. Mater. Res. B Appl. Biomater.*, **87**, 251–263.

Nishikawa, K., Hagisawa, K., Kinoshita, M., Shono, S., Katsuno, S., Doi, M., Yanagawa, R., Suzuki, H., Iwaya, K., Saitoh, D., Sakamoto, T., Seki, S., Takeoka, S., Handa, M. (2012). *J. Thromb. Haemost.*, **10**, 2137–2148.

Niu, J., Azfer, A., Rogers, L.M., Wang, X., Kolattukudy, P.E. (2007) *Cardiovasc. Res.*, **73**, 549–559.

Niu, L., Meng, L., Lu, Q. (2013). *Macromol. Biosci.*, (6), 735–744.

Njoki, P.N., Lim, I.I.S., Mott, D., Park, H.-Y., Khan, B., Mishra, S., Sujakumar, R., Luo, J., Zhong, C.-J. (2007). *J. Phys. Chem. C*, **40**, 14664–14669.

Nolan, C.M., Gelbaum, L.T., Lyon, L.A. (2006). *Biomacromolecules*, **7**, 2918–2922.

Nolan, C.M., Reyes, C.D., Debord, J.D., García, A.J., Lyon, L.A. (2005). *Biomacromolecules*, **6**, 2032–2039.

Nomikou, N., Mchale, A.P. (2010). *Cancer Lett.*, **296**, 133–143.

Novavax (2013). A study to evaluate the immune response and safety of a seasonal virus-like particle influenza vaccine in healthy young adults, NCT01561768, May 8, 2013, Available from www.clinicaltrials.gov.

Novobrantseva, T.I., Borodovsky, A., Wong, J., Klebanov, B., Zafari, M., Yucius, K., Querbes, W., Ge, P., Ruda, V. M., Milstein, S., Speciner, L., Duncan, R., Barros, S., Basha, G., Cullis, P., Akinc, A., Donahoe, J.S., Narayanannair Jayaprakash, K., Jayaraman, M., Bogorad, R.L., Love, K., Whitehead, K., Levins, C., Manoharan, M., Swirski, F.K., Weissleder, R., Langer, R., Anderson, D.G., de Fougerolles, A., Nahrendorf, M., Koteliansky, V. (2012). *Mol. Ther. Nucleic Acids*, **1**, e4.

Nugroho, A.K., Li, G., Grossklaus, A., Danhof, M., Bouwstra, J.A. (2004). *J. Controlled Release*, **96**, 159–167.

Nukolova, N.V., Yang, Z., Kim, J.O., Kabanov, A.V., Bronich, T.K. (2011). *React. Funct. Polym.*, **71**, 315–323.

Nunn, A.V.W., Barnard, M.L., Bhakoo, K., Murray, J., Chilvers, E.J., Bell, J.D. (1996). *FEBS Lett.*, **392**, 295–298.

Nyholm, D., Nilsson Remahl, A.I.M., Dizdar, N., Constatinescu, R., Holmberd, B., Jansson, R., et al. (2005). *Neurology*, **64**, 216–223.

Ochoteco, E., Murillo, N., Rodriguez, J., Pomposo, J.A., Grande, H. (2006). Intrinsically conducting polymers based electrochemical, In *Encyclopaedia of Sensors*, Vol. X, pp. 1–19, Ed. Dickey E.C., American Scientific.

Oda, S., Nagahama, R., Nakano, K., Matoba, T., Kubo, M., Sunagawa, K., Tominaga, R., Egashira, K. (2010). *J. Vasc. Surg.*, **52**, 412–420.

Oerlemans, C., Bult, W., Bos, M., Storm, G., Nijsen, J.F., Hennink, W.E. (2010). *Pharm. Res.*, 27, 2569–2589.

Ogawa, Y., Ogawa, K., Kokufuta, E. (2004). *Langmuir*, 20, 2546–2552.

Ogawara, K., Yoshida, M., Furumoto, K., Takakura, Y., Hashida, M., Higaki, K., Kimura, T. (1999). *J. Drug Targeting*, **7**, 213–221.

Oh, E., Hong, M.Y., Lee, D., Nam, S.H., Yoon, H.C., Kim, H.S. (2005). *J. Am. Chem. Soc.*, **10**, 3270–3271.

Oh, J.K., Lee, D.I., Park, J.M. (2009). *Prog. Polym. Sci.*, **34**(12), 1261–1282.

Oh, J.K., Siegwart, D.J., Lee, H., Sherwood, G., Peteanu, L., Hollinger, J.O., Kataoka, K., Matyjaszewski, K. (2007). *J. Am. Chem. Soc.*, **129**, 5939–5945.

Oh, J.K., Tang, C., Gao, H. (2006). *J. Am. Chem. Soc.*, 5578–5584.

Oh, N.M., Oh, K.T., Baik, H.J., Lee, B.R., Lee, A.H., Youn, Y.S., Lee, E.S. (2010). *Colloids Surf. B Biointerfaces*, **78**, 120–126.

Oh, N.M., Oh, K.T., Youn, Y.S., Lee, D.-K., Cha, K.-H., Lee, D.H., Lee, E.S. (2013). *Colloid Surf. B*, **101**, 298–306.

Oishi, M., Hayama, T., Akiyama, Y., Takae, S., Harada, A., Yamasaki, Y., Nagatsugi, F., Sasaki, S., Nagasaki, Y., Kataoka, K. (2005). *Biomacromolecules*, **6**, 2449–2454.

Oishi, M., Hayashi, H., Iijima, M., Nagasaki, Y. (2007a). *J. Mater. Chem.*, **17**, 3720–3725.

Oishi, M., Hayashi, H., Itaka, K., Kataoka, K., Nagasaki, Y. (2007b). *Colloid Polym. Sci.*, **285**, 1055–1060.

Okada, T., Ikeda, K., Wakabayashi, M., Ogawa, M., Matsuzaki, K. (2008). *J. Mol. Biol.*, **382**, 1066–1074.

Okamura, Y., Eto, K., Maruyama, H., Handa, M., Ikeda, Y., Takeoka, S. (2010). *Nanomedicine*, **6**, 391–396.

Okamura, Y., Takeoka, S., Eto, K., Maekawa, I., Fujie, T., Maruyama, H., Ikeda, Y., Handa, M. (2009). *J. Thromb. Haemost.*, **7**, 470–477.

Okusanya, O.O., Bhavnani, S.M., Hammel, J., Minic, P., Dupont, L.J., Forrest, A., Mulder, G.J., Mackinson, C., Ambrose, P.G., Gupta, R. (2009). *Antimicrob. Agents Chemother.*, **53**, 3847–3854.

Oliveira, H. (2012). *Wiley Interdiscip. Rev. Nanomed. Nanobiotechnol.*, **4**, 525–546.

Ong, W., Yang, Y., Cruciano, A.C., McCarley, R.L. (2008). *J. Am. Chem. Soc.*, **130**, 14739–14744.

Onoue, S., Aoki, Y., Kawabata, Y., Matsui, T., Yamamoto, K., Sato, H., Yamauchi, Y., Yamada, S. (2011). *J. Pharm. Sci.*, **100**, 622–633.

Ooya, T., Lee, J., Park, K. (2004). *Bioconjug. Chem.*, **15**, 1221–1229.

Orive, G., Hernández, R.M., Gascón, A.R., Calafiore, R., Chang, T.M.S., De Vos, P., et al. (2003). *Nat. Med.*, **9**, 104–107.

Ornelas-Megiatto, C., Shah, P.N., Wich, P.R., Cohen, J.L., Tagaev, J.A., Smolen, J.A., Wright, B. D., Panzner, M.J., Youngs, W.J., Frechet, J.M., Cannon, C.L. (2012). *Mol. Pharm.*, **9**, 3012–3022.

Ortega, P., Cobaleda, B.M., Hernández-Ros, J.M., Fuentes-Paniagua, E., Sánchez-Nieves, J., Tarazona, M.P., Copa-Patino, J.L., Soliveri, J., de la Mata, F.J., Gómez, R. (2011). *Org. Biomol. Chem.*, **9**, 5238–5248.

Otake, K., Imura, T., Sakai, H., Abe, M. (2001). *Langmuir*, **17**, 3898–3901.

Otake, K., Shimomura, T., Goto, T., Imura, T., Furuya, F., Yoda, S., Takebayashi, Y., Sakai, H., Abe, M. (2006). *Langmuir*, **22**, 2543–2550.

Otsuka, H., Nagasaki, Y., Kataoka, K. (2003). *Adv. Drug. Delivery Rev.*, **55**, 403–419.

Oussoren, C., Storm, G. (2001). *Adv. Drug Delivery Rev.*, **50**, 143–156.

Ouyang, D., Zhang, H., Herten, D.P., Parekh, H.S., Smith, S.C. (2010a). *J. Phys. Chem. B*, **114**, 9220–9230.

Ouyang, D., Zhang, H., Parekh, H.S., Smith, S.C. (2010b). *J. Phys. Chem. B*, **114**, 9231–9237.

Ouyang, D., Zhang, H., Parekh, H.S., Smith, S.C. (2011). *Biophys. Chem.*, **158**, 126–133.

Overall, C.M., Kleifeld, O. (2006). *Nat. Rev. Cancer*, **6**, 227–239.

Oyewumi, M.O., Kumar, A., Cui, Z. (2010). *Expert Rev. Vaccines*, **9**, 1095–1107.

Paasonen, L., et al. (2007a). *Bioconj. Chem.*, **18**, 2131–2136.

Paasonen, L., et al. (2007b). *J. Controlled Release*, **122**, 86–93.

Pack, D.W., Hoffman, A.S., Pun, S., Stayton, P.S. (2005). *Nat. Rev. Drug Discovery*, **4**, 581–593.

Padegimas, L., Kowalczyk, T.H., Adams, S., Gedeon, C.R., Oette, S.M., Dines, K., Hyatt, S.L., Sesenoglu-Laird, O., Tyr, O., Moen, R.C., Cooper, M.J. (2012). *Mol. Ther.*, **20**, 63–72.

Padera, T.P., Stoll, B.R., Tooredman, J.B., Capen, D., Di, T.omaso, E., Jain, R.K. (2004). *Nature*, **427**, 695.

Padin-Iruegas, M.E., Misao, Y., Davis, M.E., Segers, V.F., Esposito, G., Tokunou, T., Urbanek, K., Hosoda, T., Rota, M., Anversa, P., Leri, A., Lee, R.T., Kajstura, J. (2009). *Circulation*, **120**, 876–887.

Pagliari, F., Mandoli, C., Forte, G., Magnani, E., Pagliari, S., Nardone, G., Licoccia, S., Minieri, M., Di Nardo, P., Traversa, E. (2012). *ACS Nano*, **6**, 3767–3775.

Pai, A.S., Rubinstein, I., Önyüksel, H. (2006). *Peptides*, **27**, 2858–2866.

Paik, T., Gordon, T.R., Prantner, A.M., Yun, H., Murray, C.B. (2013). *ACS Nano*, **7**, 2850–9.

Pan, G.Q., Zhang, Y., Ma, Y., Li, C.X., Zhang, H.Q. (2011). *Angew. Chem., Int. Ed.*, **40**, 11731–11734.

Pan, Y., Du, X., Zhao, F., Xu, B. (2012a). *Chem. Soc. Rev.*, **7**, 2912–2942.

Pan, Y.-J., Chen, Y.-Y., Wang, D.-R., Wei, C., Guo, J., Lu, D.-R., Chu, C.-C., Wang, C.-C. (2012b). *Biomaterials*, **33**, 6570–6579.

Papagiannaros, A., Dimas, K., Papaioannou, G.T., Demetzos, C. (2005). *Int. J. Pharm.*, **302**, 29–38.

Papp, I., Sieben, C., Sisson, A.L., Kostka, J., Böttcher, C., Ludwig, K., Herrmann, A., Haag, R. (2011). *ChemBioChem*, **12**, 887–895.

Parasuraman, D., Serpe, M.J. (2011). *ACS Appl. Mater. Interfaces*, **3**, 2732–2737.

Pardridge, W.M. (2005). *NeuroRx.*, **2**, 129–138.

Parekh, H. (2007). *Curr. Pharm. Design.*, **13**, 2837–2850.

Parham, P. (2000). *The Immune System*, 3rd Ed., Garland, USA.

Parikh, N.I., Gona, P., Larson, M.G., Fox, C.S., Benjamin, E.J., Murabito, J.M., O'Donnell, C.J., Vasan, R.S., Levy, D. (2009). *Circulation*, **119**, 1203–1210.

Park, J., Lim, D.-H., Lim, H.-J., Kwon, T., Choi, J.-s., Jeong, S., Choi, I.-H., Cheon, J. (2011). *Chem. Commun.*, **47**, 4382–4384.

Park, J.H., Von, M.altzahn, G., Xu, M.J., Fogal, V., Kotamraju, V.R., Ruoslahti, E., Bhatia, S.N., Sailor, M.J. (2010). *Proc. Natl. Acad. Sci. U S A*, **107**, 981–986.

Park, J.S., Akiyama, Y., Yamasaki, Y., Kataoka, K. (2007). *Langmuir*, **23**, 138–146.

Park, J.W., Kirpotin, D.B., Hong, K., Shalaby, R., Shao, Y., Nielsen, U.B., Marks, J.D., Papahadjopoulos, D., Benz, C.C. (2001). *J. Controlled Release*, **74**, 95–113.

Parnham, M.J., Wetzig, H. (1993). *Chem. Phys. Lipids*, **64**, 263–274.

Parrott, M.C., Luft, J.C., Byrne, J.D., Fain, J.H., Napier, M.E., Desimone, J.M. (2010). *J. Am. Chem. Soc.*, **132**, 17928–17932.

Pasparakis, G.,Vamvakaki, M. (2011). *Polym. Chem.*, **2**, 1234–1248.

Pasut, G., Scaramuzza, S., Schiavon, O., Mendichi, R., Veronese, F.M. (2005). *J. Bioact. Compat. Pol.*, **20**, 213–230.

Patel, D.A., Henry, J.E., Good, T.A. (2007). *Brain Res.*, **1161**, 95–105.

Pathak, C.P., Thigle, S.M. (2010). Google Patents.

Patil, M.L., Zhang, M., Betigeri, S., Taratula, O., He, H., Minko, T. (2008). *Bioconjug. Chem.*, **19**, 1396–1403.

Patnaik, S., Sharma, A.K., Garg, B.S., Gandhi, R.P., Gupta, K.C. (2007). *Int. J. Pharm.*, **342**, 184–193.

Patri, A.K., Majoros, I.J., Baker, J.R. (2002). *Curr. Opin. Chem. Biol.*, **6**, 466–471.

Patri, A.K., Myc, A., Beals, J., Thomas, T.P., Bander, N.H., Baker, J.R. (2004). *Bioconjug. Chem.*, **15**, 1174–1181.

Paul, W.E. (2008). *Fundamental Immunology*, 6th Ed., Lippincott Williams and Wilkins, USA.

Paulo, C.S., Pires, Das, Neves, R., Ferreira, L.S. (2011). *Nanotechnology*, **22**, 494002.

Pavan, G.M., Kostiainen, M.A., Danani, A. (2011). *R. Soc. Chem. Adv.*, **1**, 1677–1681.

Pavan, G.M., Posocco, P., Tagliabue, A., Maly, M., Malek, A., Danani, A., Ragg, E., Catapano, C.V., Pricl, S. (2010). *Chem.: Eur. J.*, **16**, 7781–7795.

Payne, N.I., Timmins, P., Ambrose, C.V. (1986). *J. Pharm. Sci.*, **75**, 325–329.

Pearce, R.C., Railsback, J.G., Anderson, B.D., Sarac, M.F., McKnight, T.E., Tracy, J.B., Melechko, A.V. (2013). *ACS Appl. Mater. Interfaces*, **5**, 878–882.

Pearse, M.J., Drane, D. (2005). *Adv. Drug Delivery Rev.*, **57**, 465–474.

Pedersen, P.J., Adolph, S.K., Subramanian, A.K., Arouri, A., Andresen, T.L., Mouritsen, O.G., Madsen, R., Madsen, M.W., Peters, G.H., Clausen, M.H. (2010). *J. Med. Chem.*, **53**, 3782–3792.

Peer, D., Karp, J.M., Hong, S., Farokhzad, O.C., Margalit, R., Langer, R. (2007). *Nat. Nanotechnol.*, **2**, 751–760.

Pelka, J., et al. (2009). *Chem. Res. Toxicol.*, **22**, 649–659.

Pelton, R. (2000). *Adv. Colloid Interface Sci.*, **85**(1), 1–33.

Pelton, R.H. (2012). In *Hydrogel Micro and Nanoparticles*, Eds. Lyon, L.A., Serpe, M.J., WILEY-VCH Verlag GMBH, Weinheim, Germany.

Pelton, R.H., Chibante, P. (1986). *Colloids Surf.*, **20**, 247–256.

Peng, J., Zhao, L., Zhu, X., Sun, Y., Feng, W., Gao, Y., Wang, L., Li, F. (2013). *Biomaterials*, **34**, 7905–7912.

Peng, X., Schlamp, M.C., Kadavanich, A.V., Alivisatos, A.P. (1997). *J. Am. Chem. Soc.*, **30**, 7019–7029.

Peng, Z.A., Peng, X.G. (2001). *J. Am. Chem. Soc.*, **1**, 183–184.

Peracchia, M., Harnisch, S., Pinto-Alphandary, H., Gulik, A., Dedieu, J., Desmaele, D., d'Angelo, J., Müller, R., Couvreur, P. (1999). *Biomaterials*, **20**, 1269–1275.

Perche, F., Patel, N.R., Torchilin, V.P. (2012). *J. Controlled Release*, **164**, 95–102.

Pereira, S.O., Barros-Timmons, A., Trindade, T. (2013). *Colloid Polym. Sci.*, **1**, 33–50.

Perera, R.H., Solorio, L., Wu, H., Gangolli, M., Silverman, E., Hernandez, C., Peiris, P.M., Broome, A.M., Exner, A.A. (2013). *Pharm. Res.*, **31**(6), 1407–1417.

Perez, F., Joliot, A., Bloch-Gallego, E., Zahraoui, A., Triller, A., Prochiantz, A. (1992). *J. Cell Sci.*, **102**(Pt 4), 717–722.

Pérez-Anes, A., Spataro, G., Coppel, Y., Moog, C., Blanzat, M., Turrin, C.-O., Caminade, A.-M., Rico-Lattes, I., Majoral, J.-P. (2009). *Org. Biomol. Chem.*, **7**, 3491–3498.

Perez-Juste, J., Pastoriza-Santos, I., Liz-Marzan, L.M., Mulvaney, P. (2005). *Coord. Chem. Rev.*, **17–18**, 1870–1901.

Pertinhez, T.A., Bouchard, M., Smith, R.A.G., Dobson, C.M., Smith, L.J. (2002). *FEBS Lett.*, **529**, 193–197.

Pervaiz, S., Irshad, S. (2012). *Pak. J. Biochem. Mol. Biol.*, **45**, 35–43.

Petros, R.A., Desimone, J.M. (2010). *Nat. Rev. Drug Discovery*, **9**, 615–627.

Pich, A., Berger, S., Ornatsky, O., Baranov, V., Winnik, M.A. (2008). *Colloid Polym. Sci.*, **287**, 269–275.

Pich, A., Bhattacharya, S., Lu, Y., Boyko, V., Adler, H. (2004). *Langmuir*, 10706–10711.

Pich, A., Richtering, W. (2010). *Adv. Polym. Sci.*, **234**, 1–37.

Picone, P., Bondi, M.L., Montana, G., Bruno, A., Pitarresi, G., Giammona, G., Di Carlo, M. (2009). *Free Radic. Res.*, **43**, 1133–1145.

Pierobon, P., Cappello, G. (2012). *Adv. Drug Delivery Rev.*, **2**, 167–178.

Pignataro, B. (2010). *Ideas in Chemistry and Molecular Sciences: Advances in Nanotechnology, Materials and Devices*, 1st Ed., Wiley, Weinheim.

Pikabea, A., Ramos, J., Forcada, J. (2014). *Part. Part. Syst. Charact.*, **31**, 101–109.

Pillai, Z.S., Kamat, P.V. (2003). *J. Phys. Chem. B*, **3**, 945–951.

Pillay, C.S., Elliott, E., Dennison, C. (2002). *Biochem. J.*, **363**, 417–429.

Ping, Y., Hu, Q., Tang, G., Li, J. (2013). *Biomaterials*, **34**, 6482–6494.

Pinheiro, P.C., Fateixa, S., Nogueira, H.I.S., Trindade, T. (2013). *Spectrochim. Acta A*, 36–39.

Pinto, R.J.B., Marques, P.A.A.P., Neto, C.P., Trindade, T., Daina, S., Sadocco, P. (2009). *Acta Biomater.*, **6**, 2279–2289.

Pirko, I., Ciric, B., Gamez, J., Bieber, A.J., Warrington, A.E., Johnson, A.J., Hanson, D.P., Pease, L.R., Macura, S.I., Rodriguez, M. (2004). *FASEB J. Off. Publ. Fed. Am. Soc. Exp. Biol.*, **18**, 1577–1579.

Pirmohamed, T., Dowding, J.M., Singh, S., Wasserman, B., Heckert, E., Karakoti, A.S., King, J.E.S., Seal, S., Self, W.T. (2010). *Chem. Commun.*, **46**, 2736–2738.

Pittella, F., Miyata, K., Maeda, Y., Suma, T., Watanabe, S., Chen, Q., Christie, R.J., Osada, K., Nishiyama, N., Kataoka, K. (2012). *J. Controlled Release*, **161**, 868–874.

Plotkin, S.A., Orenstein, W.A., Offit, P.A. (2013). *Vaccines*, 6th Ed., Elsevier Saunders, Edinburgh.

Podgorski, I., Sloane, B.F. (2003). *Biochem. Soc. Symp.*, 263–276.

Polcyn, P., Jurczak, M., Rajnisz, A., Solecka, J., Urbanczyk-Lipkowska, Z. (2009). *Molecules*, **14**, 3881–3905.

Pollock, S., Antrobus, R., Newton, L., Kampa, B., Rossa, J., Latham, S., Nichita, N.B., Dwek, R.A., Zitzmann, N. (2010). *FASEB J.*, **24**, 1866–1878.

Ponce, A.M., Vujaskovic, Z., Yuan, F., Needham, D., Dewhirst, M.W. (2006). *Int. J. Hyperthermia*, **22**, 205–213.

Pons, M., Foradada, M., Estelrich, J. (1993). *Int. J. Pharm.*, **95**, 51–56.

Porcel, E., Liehn, S., Remita, H., Usami, N., Kobayashi, K., Furusawa, Y., Le Sech, C., Lacombe, S. (2010). *Nanotechnology*, **21**, 85103.

Potapova, I., Mruk, R., Prehl, S., Zentel, R., Basche, T., Mews, A. (2003). *J. Am. Chem. Soc.*, **2**, 320–321.

Potineni, A., Lynn, D.M., Langer, R., Amiji, M. (2003). *J. Controlled Release*, **86**, 223–234.

Poupot, M., Griffe, L., Marchand, P., Maraval, A., Rolland, O., Martinet, L., L'Faqihi-Olive, F.-E., Turrin, C.-O., Caminade, A.-M., Fournié, J.-J. (2006). *FASEB J.*, 20, 2339–2351.

PowderMed (2007). *A Safety and Immunology Study of a DNA Trivalent Influenza Vaccine*, NCT00375206, May 6, 2013, Available from http://clinicaltrials.gov.

PowderMed (2008). *Safety Study of HBV DNA Vaccine to Treat Patients With Chronic Hepatitis B Infection*, NCT00277576, May 6, 2013, Available from http://clinicaltrials.gov.

Powell, J., Martinowitz, U., Windyga, J., Di Minno, G., Hellmann, A., Pabinger, I., Maas Enriquez, M., Schwartz, L., Ingerslev, J., LipLong Study, I. (2012). *Thromb. Haemost.*, **108**, 913–922.

Prabaharan, M., et al. (2009a). *Biomaterials*, **30**, 5757–5766.

Prabaharan, M., Grailer, J.J., Steeber, D.A., Gong, S. (2009b). *Macromol. Biosci.*, **9**, 744–753.

Prabhakar, U., Maeda, H., Jain, R.K., Sevick-Muraca, E.M., Zamboni, W., Farokhzad, O.C., Barry, S.T., Gabizon, A., Grodzinski, P., Blakey, D.C. (2013). *Cancer Res.*, **73**, 2412–2417.

Pradhan, N., Efrima, S. (2003). *J. Am. Chem. Soc.*, **8**, 2050–2051.

Pradhan, P., Giri, J., Rieken, F., Koch, C., Mykhaylyk, O., Doblinger, M., Banerjee, R., Bahadur, D., Plank, C. (2010). *J. Controlled Release*, **142**, 108–121.

Praetorius, N.P., Mandal, T.K. (2007). *Recent Patents Drug Delivery Formul.*, **1**, 37–51.

Prajapati, R.N., Tekade, R.K., Gupta, U., Gajbhiye, V., Jain, N.K. (2009). *Mol. Pharm.*, **6**, 940–950.

Preston, T.J., Abadi, A., Wilson, L., Singh, G. (2001). *Adv. Drug. Delivery Rev.*, **49**, 45–61.

Prevot, M., Dé jugnat, C., Möhwald, H., Sukhorukov, G.B. (2006). *ChemPhysChem*, **7**, 2497–2502.

Price, R., Skyba, D., Kaul, S., Skalak, T. (1998). *Circulation*, **98**, 1264–1267.

Prieur, X., Rőszer, T., Ricote, M. (2010). *Biochim. Biophys. Acta (BBA): Mol. Cell Biol. Lipids*, **1801**, 327–337.

Prost, A.C., Menegaux, F., Langlois, P., Vidal, J.M., Koulibaly, M., Jost, J.L., Duron, J.J., Chigot, J.P., Vayre, P., Aurengo, A., Legrand, J.C., Rosselin, G., Gespach, C. (1998). *Int. J. Oncol.*, **13**, 871–875.

Prow, T.W., Rose, W.A., Wang, N., Reece, L.M., Lvov, Y., Leary, J.F (2005). Biosensor-controlled gene therapy/drug delivery with nanoparticles for nanomedicine. In *Advanced Biomedical and Clinical Diagnostic Systems III*, Eds. Vo-Dinh, T., Grundfest, W.S., Benaron, D.A., Cohn, G.E., Proc. SPIE, 5692, SPIE, Bellingham, WA.

Puri, A., et al. (2008). *J. Liposome Res.*, **18**, 293–307.

Purushotham, S., Ramanujan, R.V. (2010). *Acta Biomater.*, **6**, 502–510.

Pusic, K., Aguilar, Z., McLoughlin, J., Kobuch, S., Xu, H., Tsang, M., Wang, A., Hui, G. (2012). *FASEB J.*, **27**, 1153–1166.

Putnam, D. (2006). *Nat. Mater.*, **5**, 439–451.

Qiao, R., Jia, Q., Hüwel, S., Xia, R., Liu, T., Gao, F., Galla, H. J., Gao M (2012). *ACS Nano*, **6**, 3304–3310.

Qiao, Z.-Y., Zhang, R., Du, F.-S., Liang, D.-H., Li, Z.-C. (2011). *J. Controlled Release*, **152**, 57–66.

Qiu, J., Charleux, B., Matyjaszewski, K. (2001). *Prog. Polym. Sci.*, **26**, 2083–2134.

Qiu, L.Y., Wu, X.L., Jin, Y. (2009). *Pharm. Res.*, **26**, 946–957.

Quadir, M.A., Haag, R. (2012). *J. Controlled Release*, **161**, 484–495.

Qualmann, B., Kessels, M.M., Musiol, H.J., Sierralta, W.D., Jungblut, P.W., Moroder, L. (1996). *Angew. Chem., Int. Ed.*, **35**, 909–911.

Quintana, A., Raczka, E., Piehler, L., Lee, I., Myc, A., Majoros, I., Patri, A.K., Thomas, T., Mule, J., Baker, J.R., Jr. (2002). *Pharm. Res.*, **19**, 1310–1316.

Qureshi, A., Roci, I., Gurbuz, Y., Niazi, J.H. (2012). *Biosens. Bioelectron.*, **34**, 165–170.

Raatschen, H.-J., Simon, G.H., Fu, Y., Sennino, B., Shames, D.M., Wendland, M.F., McDonald, D.M., Brasch, R.C. (2008). *Radiology*, **247**, 391–399.

Radadia, A.D., Stavis, C.J., Carr, R., Zeng, H., King, W.P., Carlisle, J.A., Aksimentiev, A., Hamers, R. J., Bashir, R. (2011). *Adv. Funct. Mater.*, **21**, 1040–1050.

Raemdonck, K., Naeye, B., Buyens, K., Vandenbroucke, R.E., Høgset, A., Demeester, J., Smedt, S.C. De (2009). *Adv. Funct. Mater.*, **19**, 1406–1415.

Rahbek, U.L., Nielsen, A.F., Dong, M., You, Y., Chauchereau, A., Oupicky, D., Besenbacher, F., Kjems, J., Howard, K.A. (2010). *J. Drug Targeting*, **18**, 812–820.

Rajapaksa, T.E., Lo, D.D. (2010). *Curr. Immunol. Rev.*, **6**, 29–37.

Rakovich, A., Savateeva, D., Rakovich, T., Donegan, J., Rakovich, Y., Kelly, V., Lesnyak, V., Eychmüller, A. (2010). *Nanoscale Res. Lett.*, **4**, 753–760.

Ramakrishna, S., Mayer, J., Wintermantel, E., Leong, K.W. (2001). *Compos. Sci. Technol.*, **61**, 1189–1224.

Ramos, J., Forcada, J., Hidalgo-Alvarez, R. (2014). *Chem. Rev.*, **114**, 367–428.

Ramos, J., Imaz, A., Callejas.Fernández, J., Barbosa-Barros, L., Estelrich, J., Quesada-Pérez, M., Forcada, J. (2011). *Soft Matter*, **7**(11), 5067–5082.

Ramos, J., Imaz, A., Forcada, J. (2012). *Polym. Chem.*, **3**(4), 852–856.

Rangachari, V., Reed, D.K., Moore, B.D., Rosenberry, T.L. (2006). *Biochemistry*, **45**, 8639–8648.

Rapoport, N. (1999). *Colloids Surf. B Biointerfaces.*, **16**, 93–111.

Rapoport, N. (2007). *Prog. Polym. Sci.*, **32**, 962–90.

Rapoport, N. (2012). *Int. J. Hyperthermia*, **28**, 374–385.

Rapoport, N., Gao, Z., Kennedy, A. (2007). *J. Natl. Cancer Inst.*, **99**, 1095–1106.

Rapoport, N., Kennedy, A.M., Shea, J.E., Scaife, C.L., Nam, K.H. (2010). *Mol. Pharm.*, **7**, 22–31.

Rapoport, N., Nam, K.H., Gupta, R., Gao, Z., Mohan, P., Payne, A., Todd, N., Liu, X., Kim, T., Shea, J., Scaife, C., Parker, D.L., Jeong, E.K., Kennedy, A.M. (2011). *J. Controlled Release*, **153**, 4–15.

Rapoport, N.Y., Kennedy, A.M., Shea, J.E., Scaife, C.L., Nam, K.H. (2009). *J. Controlled Release*, **138**, 268–276.

Raschke, G., Kowarik, S., Franzl, T., Sonnichsen, C., Klar, T.A., Feldmann, J., Nichtl, A., Kurzinger, K. (2003). *Nano Lett.*, **7**, 935–938.

Rasines, B., Hernández-Ros, J.M., de las Cuevas, N., Copa-Patiño, J.L., Soliveri, J., Muñoz-Fernández, M.A., Gómez, R., de la Mata, F.J. (2009). *Dalton Trans.*, 8704–8713.

Rathore, A., Cleary, M., Naito, Y., Rocco, K., Breuer, C. (2012). *Wiley Interdiscip. Rev. Nanomed. Nanobiotechnol.*, **4**, 257–272.

Ravina, M., Paolicelli, P., Seijo, B., Sanchez, A. (2010). *Mini.: Rev. Med. Chem.*, **10**, 73.

Ray, P.C., Darbha, G.K., Ray, A., Walker, J., Hardy, W. (2007). *Plasmonics*, **4**, 173–183.

Reddy, M.K., Wu, L., Kou, W., Ghorpade, A., Labhasetwar, V. (2008). *Appl. Biochem. Biotechnol.*, **151**, 567–577.

Reddy, S.T., Rehor, A., Schmoekel, H.G., Hubbell, J.A., Swartz, M.A. (2006). *J. Controlled Release*, **112**, 26–34.

Reddy, S.T., van der Vlies, A.J., Simeoni, E., Angeli, V., Randolph, G.J., O'Neil, C.P., Lee, L.K., Swartz, M.A., Hubbell, J.A. (2007). *Nat. Biotechnol.*, **25**, 1159–1164.

Reichert, J.M., Rosensweig, C.J., Faden, L.B., Dewitz, M.C. (2005). *Nat. Biotechnol.*, **23**, 1073–1078.

Rejman, J., Oberle, V., Zuhorn, I.S., Hoekstra, D. (2004). *Biochem. J.*, **377**, 159–169.

Ren, T.-B., Feng, Y., Zhang, Z.-H., Li, L., Li, Y.-Y. (2011). *Soft Matter*, **7**, 2329–2331.

Ren, Y., Pastorin, G. (2008). *Adv. Mater.*, **20**, 2031–2036.

Resch-Genger, U., Grabolle, M., Cavaliere-Jaricot, S., Nitschke, R., Nann, T. (2008). *Nat. Methods*, **9**, 763–775.

Reubi, J.C. (1996). *Ann. N. Y. Acad. Sci.*, **805**, 753–759.

Richard, C., de Chermont, Q. le M., Scherman, D. (2008). *Tumori*, **94**, 264–270.

Righettoni, M., Tricoli, A., Gass, S., Schmid, A., Amann, A., Pratsinis, S.E. (2012). *Anal. Chim. Acta*, **738**, 69–75.

Rijcken, C.J., Soga, O., Hennink, W.E., Van, N.ostrum, C.F. (2007). *J. Controlled Release*, **120**, 131–148.

Rippe, C., Rippe, A., Torffvit, O., Rippe, B. (2007). *Am. J. Physiol. Renal Physiol.*, **293**, F1533–F1538.

Robert, C., Thomas, L., Bondarenko, I., O'Day, S., Weber, J., Garbe, C., Lebbe, C., Baurain, J.-F., Testori, A., Grob, J.-J. (2011). *New Engl. J. Med.*, **364**, 2517–2526.

Roberts, W.G., Palade, G.E. (1997). *Cancer Res.*, **57**, 765–772.

Rocha, S., Thunemann, A.F., Pereira Mdo, C., Coelho, M., Mohwald, H., Brezesinski, G. (2008). *Biophys. Chem.*, **137**, 34–42.

Rockenberger, J., Scher, E.C., Alivisatos, A.P. (1999). *J. Am. Chem. Soc.*, **49**, 11595–11596.

Rodríguez, M.C., Gorospe, A., Linazasoro, G., Ramos, E., Rodriguez, P., Obeso, J.A. (2000). *Neurology*, **55**(Suppl 6), 45–51.

Rogach, A.L., Franzl, T., Klar, T.A., Feldmann, J., Gaponik, N., Lesnyak, V., Shavel, A., Eychmüller, A., Rakovich, Y.P., Donegan, J.F. (2007). *J. Phys. Chem. C*, **40**, 14628–14637.

Rogers, T.H., Babensee, J.E. (2011). *Biomaterials*, **32**, 1270–1279.

Rojo, J., Delgado, R. (2004). *J. Antimicrob. Chemoth.*, **54**, 579–581.

Rolland, O., Turrin, C.-O., Bacquet, G., Poupot, R., Poupot, M., Caminade, A.-M., Majoral, J.-P. (2009a). *Tetrahedron Lett.*, **50**, 2078–2082.

Rolland, O., Turrin, C.-O., Caminade, A.-M., Majoral, J.-P. (2009b). *New J. Chem.*, **33**, 1809–1824.

Romagnani, S. (1999). *Inflamm. Bowel Dis.*, **5**, 285–94.

Rosa Borges, A., Wieczorek, L., Johnson, B., Benesi, A.J., Brown, B.K., Kensinger, R.D., Krebs, F.C., Wigdahl, B., Blumenthal, R., Puri, A. (2010). *Virology*, **408**, 80–88.

Rossi, E.A., Goldenberg, D.M., Cardillo, T.M., Stein, R., Chang, C.H. (2009). *Blood*, **114**, 3864–3871.

Roth, J. (1996). *Histochem. Cell Biol.*, **1**, 1–8.

Rovers, S.A., Hoogenboom, R., Kemmere, M.F., Keurentjes, J.T.F. (2012). *Soft Matter*, **8**, 1623–1627.

Roy, D., Cambre, J.N., Sumerlin, B.S. (2010). *Prog. Polym. Sci.*, **35**, 278–301.

Roy, M.J., Wu, M.S., Barr, L.J., Fuller, J.T., Tussey, L.G., Speller, S., Culp, J., Burkholder, J.K., Swain, W.F., Dixon, R.M., Widera, G., Vessey, R., King, A., Ogg, G., Gallimore, A., Haynes, J.R., Heydenburg Fuller, D. (2000). *Vaccine*, **19**, 764–778.

Roy, M.T., Gallardo, M., Estelrich, J. (1997). *Bioconjugate Chem.*, **8**, 941–945.

Ruiz Ortega, J.A., Linazasoro, G. (2004). Estructura y función de los receptores dopaminérgicos. Mecanismos de transducción, in Receptores Dopaminérgicos: De la Estructura a la Función, 51–88, Ed. Luquin R., Entheos, Madrid.

Ruiz-Hernández, E., López-Noriega, A., Arcos, D., Izquierdo-Barba, I., Terasaki, O., Vallet-Regí, M. (2007). *Chem. Mater.*, **19**, 3455–3463.

Rupp, R., Rosenthal, S.L., Stanberry, L.R. (2007). *Int. J. Nanomed.*, **2**, 561.

Russo, A., Degraff, W., Friedman, N., Mitchell, J.B. (1986). *Cancer Res.*, **46**, 2845–2848.

Ryan, C.J. (2013). Clinical trial design strategies in the post-abiraterone/enzalutamide setting, in 2013 GU Cancers Symposium Proceedings, 26–28, Ed. Carducci, M.A., American Society of Clinical Oncology, Alexandria, VA.

Ryu, J.H., Chacko, R.T., Jiwpanich, S., Bickerton, S., Babu, R.P., Thayumanavan, S. (2010). *J. Am. Chem. Soc.*, **132**, 17227–17235.

Sabaté, R., Barnadas-Rodríguez, R., Callejas-Fernández, J., Hidalgo-Álvarez, R., Estelrich, J. (2008). *Int. J. Pharm.*, **347**, 156–162.

Sabaté, R., Espargaró, A., Barbosa-Barros, L., Ventura, S., Estelrich, J. (2012). *Biochimie*, **270–271**, 13–17.

Sabaté, R., Estelrich, J. (2005a). *J. Phys. Chem. B*, **109**, 11027–11032.

Sabaté, R., Estelrich, J. (2005b). *Langmuir*, **21**, 6944–6949.

Sabaté, R., Gallardo, M., Estelrich, J. (2003). *Biopolymers*, **71**, 190–195.

Sabaté, R., Gallardo, M., Estelrich, J. (2005). *Colloids Surf. A.*, **270–271**, 13–17.

Šafařík, I., Šafaříková, M. (2002). Magnetic nanoparticles and biosciences, In *Nanostructured Materials*, pp. 1–23, Eds. Hofmann, H., Rahman, Z., Schubert, U., Springer, Vienna.

Saha, K., Agasti, S.S., Kim, C., Li, X.N., Rotello, V.M. (2012). *Chem. Rev.*, **5**, 2739–2779.

Saito, G., Swanson, J.A., Lee, K.-D. (2003). *Adv. Drug Delivery Rev.*, **55**, 199-215.

Saito, N., Usui, Y., Aoki, K., Narita, N., Shimizu, M., Hara, K., Ogiwara, N., Nakamura, K., Ishigaki, N., Kato, H., Taruta, S., Endo, M. (2009). *Chem. Soc. Rev.*, **38**, 1897–1903.

Salgueiro, A.M., Daniel-da-Silva, A.L., Fateixa, S., Trindade, T. (2013). *Carbohyd. Polym.*, **1**, 100–109.

Salmaso, S., Caliceti, P., Amendola, V., Meneghetti, M., Magnusson, J.P., Pasparakis, G., Alexander, C. (2009). *J. Mater. Chem.*, **19**, 1608–1615.

Salmon, A.H.J., Toma, I., Sipos, A., Muston, P.R., Harper, S.J., Bates, D.O., Neal, C.R., Peti-Peterdi, J. (2007). *Am. J. Physiol. Renal Physiol.*, **293**, F1777-F1786.

Salvador, A., Igartua, M., Hernández, R.M., Pedraz, J.L. (2011). *J. Drug Delivery*, **2011**, 1–23.

Samorí, C., Ali-Boucetta, H., Sainz, R., Guo, C., Toma, F.M., Fabbro, C., Da Ros, T., Prato, M., Kostarelos, K., Bianco, A. (2010). *Chem. Commun.*, **46**, 1494–1496.

Samuelsen, M., Nygaard, U.C., Løvik, M. (2009). *Scand. J. Immunol.*, **69**, 421-428.

Santhosh, P.B., Ulrih, N.P. (2013). *Cancer Lett.*, **336**, 8–17.

Sanz, V., Tilmacîu, C., Soula, B., Flahaut, E., Coley, H.M., Silva, S.R.P., McFadden, J. (2011). *Carbon*, **49**, 5348–5358.

Saovapakhiran, A., D'Emanuele, A., Attwood, D., Penny, J. (2009). *Bioconjug. Chem.*, **20**, 693–701.

Sapsford, K.E., Algar, W.R., Berti, L., Gemmill, K.B., Casey, B.J., Oh, E., Stewart, M.H., Medintz, I.L. (2013). *Chem. Rev.*, **3**, 1904–2074.

Sapsford, K.E., Berti, L., Medintz, I.L. (2006). *Angew. Chem., Int. Ed.*, **28**, 4562–4588.

Saraiva, A.M., Cardoso, I., Saraiva, M.J., Tauer, K., Pereira, M.C., Coelho, M.A., Möhwald, H., Brezesinski, G. (2010). *Macromol. Biosci.*, **10**, 1152–1163.

Saraogi, G.K., Sharma, B., Joshi, B., Gupta, P., Gupta, U.D., Jain, N.K., Agrawal, G.P. (2011). *J. Drug Targeting*, **19**, 219–227.

Sartore, L., Peroni, I., Ferruti, P., Latini, R., Bernasconi, R. (1997). *J. Biomater. Sci. Polym. Ed.*, **8**, 741–754.

Sato, K., Hosokawa, K., Maeda, M. (2003). *J. Am. Chem. Soc.*, **27**, 8102–8103.

Sato,T., Miyagi, T., Hirano, T., Seno, M. (2004). *Polym. Int.*, **53**, 1503–1511.

Saul, J.M., Annapragada, A.V., Bellamkonda, R.V. (2006). *J. Controlled Release*, **114**, 277–287.

Sauzedde, F., Elaïssari, A., Pichot, C. (1999). *Colloid Polym. Sci.*, **277**, 846–855.

Sawai, J. (2003). *J. Microbiol. Methods*, **54**, 177–182.

Sawant, R., Sawant, R., Gultepe, E., Nagesha, D., Papahadjopoulos-Sternberg, B., Sridhar, S., Torchilin, V. (2009). *J. Nanopart. Res.*, **11**, 1777–1785.

Sawant, R.P., Torchilin, V.P. (2010). *Soft Matter*, **6**, 4026–4044.

Sawant, R.R., Jhaveri, A.M., Koshkaryev, A., Qureshi, F., Torchilin, V.P. (2013). *J. Drug Targeting*, **21**(7), 630–638.

Sawant, R.R., Sriraman, S.K., Navarro, G., Biswas, S., Dalvi, R.A., Torchilin, V.P. (2012). *Biomaterials*, **33**, 3942–3951.

Sawant, R.R., Torchilin, V.P. (2009). *Int. J. Pharm.*, **374**, 114–118.

Say, J., Vreden, C., Reilly, D., Brown, L., Rabeau, J., King, N.C. (2011). *Biophys. Rev.*, **4**, 171–184.

Schafer, F.Q., Buettner, G.R. (2001). *Free Radic. Biol. Med.*, **30**, 1191–1212.

Schapira, A.H.V. (2005). *J. Neurol. Neurosurg. Psychiatry*, **76**, 1472–1478.

Schild, H.G. (1992). *Prog. Polym. Sci.*, **17**, 163–249.

Schild, H.H., Kuhl, C.K., Hübner-Steiner, U., Böhm, I., Speck, U. (2006). *Radiology*, **240**, 56–64.

Schlegel, G., Bohnenberger, J., Potapova, I., Mews, A. (2002). *Phys. Rev. Lett.*, **13**, 1–4.

Schmaljohann, D. (2006). *Adv. Drug. Delivery Rev.*, **58**, 1655–1670.

Schneider, G., Subr, V., Ulbrich, K., Decher, G. (2009). *Nano Lett.*, **2**, 636–642.

Schroeder, A., Honen, R., Turjeman, K., Gabizon, A., Kost, J., Barenholz, Y. (2009). *J. Controlled Release*, **137**, 63–68.

Schroit, A.J., Madsen, J., Nayar, R. (1986). *Chem. Phys. Lipids*, **40**, 373–393.

Schulze, E., Ferrucci, J.T., Jr, Poss, K., Lapointe, L., Bogdanova, A., Weissleder, R. (1995). *Invest. Radiol.*, **30**, 604–610.

Schumers, J.M., Fustin, C.A., Gohy, J.F. (2010). *Macromol. Rapid Commun.*, **31**, 1588–1607.

Scott, A. (2010). *Polymer*, **2**, 522–553.

Šebestík, J., Reiniš, M., Ježek, J. (2012). In *Biomedical Applications of Peptide-, Glyco- and Glycopeptide Dendrimers, and Analogous Dendrimeric Structures*, pp. 141–147, Springer.

Seiffert, S., Oppermann, W., Saalwächter, K. (2007). *Polymer*, **48**, 5599–5611.

Senzer, N., Nemunaitis, J., Nemunaitis, D., Bedell, C., Edelman, G., Barve, M., Nunan, R., Pirollo, K.F., Rait, A., Chang, E.H. (2013). *Mol. Ther.*, **21**, 1096–1103.

Seo, H.J., Kim, J.C. (2013). *J. Nanosci. Nanotechnol.*, **13**, 1727–1732.

Seth, R.-N. (2006). *Yale J. Biol. Med.*, **79**, 123–130.

Shah, N., Steptoe, R.J., Parekh, H.S. (2011). *J. Pept. Sci.*, **17**, 470–478.

Shan, Y., Luo, T., Peng, C., Sheng, R., Cao, A., Cao, X., Shen, M., Guo, R., Tomas, H., Shi, X. (2012). *Biomaterials*, **33**, 3025–3035.

Shang, T., Wang, C.-D., Ren, L., Tian, X.-H., Li, D.-H., Ke, X.-B., Chen, M., Yang, A.-Q. (2013). *Nanoscale Res. Lett.*, **8**, 4.

Shao, H., Jao, S.-C., Ma, K., Zagorski, M.G. (1999). *J. Mol. Biol.*, **285**, 755–773.

Shcharbin, D., Jokiel, M., Klajnert, B., Bryszewska, M. (2006). *Bioelectrochemistry*, **68**, 56–59.

Shcharbin, D., Klajnert, B., Bryszewska, M. (2005). *J. Biomater. Sci. Polym. Ed.*, **16**, 1081–1093.

Shchepinov, M., Udalova, I., Bridgman, A., Southern, E. (1997). *Nucleic Acids Res.*, **25**, 4447–4454.

Sheikholeslami, P., Ewaschuk, C.M., Ahmed, S.U., Greenlay, B.A., Hoare, T. (2012). *Colloid Polym. Sci.*, **290**, 1181–1192.

Shen, S., Ren, J., Chen, J., Lu, X., Deng, C., Jiang, X. (2011). *J. Chromatogr. A*, **1218**, 4619–4626.

Sheng, R., Luo, T., Zhu, Y., Li, H., Sun, J., Chen, S., Sun, W., Cao, A. (2011). *Biomaterials*, **32**, 3507–3519.

Shi, D., Matsusaki, M., Akashi, M. (2011). *J. Controlled Release*, **149**, 182–189.

Shi, J., Yu, X., Wang, L., Liu, Y., Gao, J., Zhang, J., Ma, R., Liu, R., Zhang, Z. (2013). *Biomaterials*, **34**(37), 9666–9677.

Shi, X., Wang, S., Meshinchi, S., Van Antwerp, M.E., Bi, X., Lee, I., Baker, J.R. (2007). *Small*, **3**, 1245–1252.

Shieh, M.J., Peng, C.L., Chiang, W.L., Wang, C.H., Hsu, C.Y., Wang, S.J., Lai, P.S. (2010). *Mol. Pharm.*, **7**, 1244–1253.

Shim, M.S., Bhang, S.H., Yoon, K., Choi, K., Xia, Y. (2012). *Angew. Chem., Int. Ed. Engl.*, **51**, 11899–11903.

Shim, M.S., Kwon, Y.J. (2009). *Bioconjug. Chem.*, **20**, 488–499.

Shim, M.S., Kwon, Y.J. (2012). *Adv. Drug. Delivery Rev.*, **64**, 1046–1059.

Shin, S.R., Jung, S.M., Zalabany, M., Kim, K., Zorlutuna, P., Kim, S.B., Nikkhah, M., Khabiry, M., Azize, M., Kong, J., Wan, K.T., Palacios, T., Dokmeci, M.R., Bae, H., Tang, X.S., Khademhosseini, A. (2013). *ACS Nano*, **7**, 2369–2380.

Shinde, S.B., Fernandes, C.B., Patravale, V.B. (2012). *J. Controlled Release*, **159**, 164–180.

Shinkai, M. (2002). *J. Biosci. Bioeng.*, **6**, 606–613.

Shinkai, M., Suzuki, M., Iijima, S., Kobayashi, T. (1995). *Biotechnol. Appl. Biochem.*, **21**(Pt 2), 125–137.

Shiraishi, K., Kawano, K., Minowa, T., Maitani, Y., Yokoyama, M. (2009). *J. Controlled Release*, **136**, 14–20.

Shkilnyy, A., Souce, M., Dubois, P., Warmont, F., Saboungi, M.-L., Chourpa, I. (2009). *Analyst*, **9**, 1868–1872.

Shoffstall, A.J., Atkins, K.T., Groynom, R.E., Varley, M.E., Everhart, L.M., Lashof-Sullivan, M.M., Martyn-Dow, B., Butler, R.S., Ustin, J.S., Lavik, E.B. (2012). *Biomacromolecules*, **13**, 3850–3857.

Shukla, R., Thomas, T.P., Desai, A.M., Kotlyar, A., Park, S.J., Baker, J.R. (2008). *Nanotechnology*, **19**, 295102.

Shukla, R., Thomas, T.P., Peters, J., Kotyla, A., Myc, A., Baker Jr, J.R. (2005). *Chem. Commun.*, 5739–5741.

Shukla, R., Thomas, T.P., Peters, J.L., Desai, A.M., Kukowska-Latallo, J., Patri, A.K., Kotlyar, A., Baker, J.R. (2006). *Bioconjug. Chem.*, **17**, 1109–1115.

Shukla, S., Wu, G., Chatterjee, M., Yang, W., Sekido, M., Diop, L.A., Müller, R., Sudimack, J.J., Lee, R.J., Barth, R.F. (2003). *Bioconjug. Chem.*, **14**, 158–167.

Shvedova, A.A., Fabisiak, J.P., Kisin, E.R., Murray, A.R., Roberts, J.R., Tyurina, Y.Y., Antonini, J.M., Feng, W.H., Kommineni, C., Reynolds, J., Barchowsky, A., Castranova, V., Kagan, V.E. (2008). *Am. J. Respir. Cell Mol. Biol.*, **38**, 579–590.

Siegemund, T., Paulke, B.R., Schmiedel, H., Bordag, N., Hoffmann A., Harkany, T., Tanila, H., Kacza, J., Härtig, W. (2006). *Int. J. Dev. Neurosci.*, **24**, 195–201.

Silva, A.C., Oliveira, T.R., Mamani, J.B., Malheiros, S.M.F., Malavolta, L., Pavon, L.F., Sibov, T.T., Amaro, E., Jr., Tannus, A., Vidoto, E.L.G., Martins, M.J., Santos, R.S., Gamarra, L.F. (2011). *Int. J. Nanomed.*, 591–603.

Silva, G.A. (2006). *Nat. Rev. Neurosci.*, **7**, 65–74.

Silva, G.A. (2010). *Ann. N. Y. Acad. Sci.*, **1199**, 221–230.

Silva, G.A., Czeisler, C., Niece, K.L., Beniash, E., Harrington, D.A., Kessler, J.A., Stupp, S.I. (2004). *Science*, **303**, 1252–1255.

Simões, S., Moreira, J.N., Fonseca, C., Düzgüneş, N., de Lima, M.C. (2004). *Adv. Drug Delivery Rev.*, **56**, 947–965.

Simonyan, V.V., Johnson, J.K., Kuznetsova, A., Yates, J.T. (2001). *J. Chem. Phys.*, **114**, 4180–4185.

Simşek, S., Eroğlu, H., Kurum, B., Ulubayram, K. (2013). *J. Microencapsul.*, **30**, 10–20.

Singh, M., Shah, T., Khosla, K., Singh, P., Molnar, J., Khosla, S., Arora, R. (2012a). *Ther. Adv. Cardiovasc. Dis.*, **6**, 101–114.

Singh, P., Campidelli, S., Giordani, S., Bonifazi, D., Bianco, A., Prato, M. (2009). *Chem. Soc. Rev.*, **38**, 2214–2230.

Singh, R., Pantarotto, D., Lacerda, L., Pastorin, G., Klumpp, C., Prato, M., Bianco, A., Kostarelos, K. (2006). *Proc. Natl. Acad. Sci. U S A*, **103**, 3357–3362.

Singh, S., Blöhbaum, J., Möller, M., Pich, A. (2012b). *J. Polym. Sci. Part A Polym. Chem.*, **50**, 4288–4299.

Singka, G.S.L., Samah, N.A., Zulfakar, M.H., Yurdasiper, A., Heard, C.M. (2010). *Eur. J. Pharm. Biopharm.*, **76**, 275–281.

Sinha, B., Mukherjee, B., Pattnaik, G. (2013). *Nanomedicine*, **9**, 94–104.

Sipe, J.C., Filippi, M., Martino, G., Furlan, R., Rocca, M.A., Rovaris, M., Bergami, A., Zyroff, J., Scotti, G., Comi, G. (1999). *Magn. Reson. Imaging*, **17**, 1521–1523.

Sipe, J.D., Benson, M.D., Buxbaum, J.N., Ikeda, S., Merlini, G., Saraiva, M.J.M., Westermark, P. (2010). *Amyloid*, **17**, 101–104.

Sitharaman, B., Kissell, K., Hartman, K., Tran, L.A., Baikalov, A., Rusakova, I., Sun, Y., Khant, H.A., Ludtke, S.J., Chiu, W., Laus, S., Tóth, E., Helm, L., Merbach, A.E., Wilson, L.J. (2005). *Chem. Commun.*, 3915–3917.

Siti, M., Janib, Ara, S., Moses, J., MacKay, A. (2010). *Adv. Drug Delivery Rev.*, **62**, 1052–1063.

Sivaev, I.B., Bregadze, V.V. (2009). *Eur. J. Inorg. Chem.*, **2009**, 1433–1450.

Sivakumaran, D., Maitland, D., Hoare, T. (2011). *Biomacromolecules*, **12**, 4112–4120.

Sivakumaran, D., Maitland, D., Oszustowicz, T., Hoare, T. (2013). *J. Colloid Interface Sci.*, **392**, 422–430.

Skaat, H., Chen, R., Grinberg, I., Margel, S. (2012). *Biomacromolecules*, **13**, 2662–2670.

Skaat, H., Margel, S. (2009). *Biochem. Biophys. Res. Commun.*, **386**, 645–649.

Skidan, I., Dholakia, P., Torchilin, V. (2008). *J. Drug Targeting*, **16**, 486–493.

Skirtach, A.G., Dejugnat, C., Braun, D., Susha, A.S., Rogach, A.L., Parak, W.J., Möhwald, H., Sukhorukov, G.B. (2005). *Nano Lett.*, **7**, 1371–1377.

Skirtach, A.G., Muñoz Javier, A., Kreft, O., Köhler, K., Piera Alberola, A., Möhwald, H., Parak, W.J., Sukhorukov, G.B. (2006). *Angew. Chem., Int. Ed.*, **28**, 4612–4617.

Slomkowski, S., Alemán, J.V., Gilbert, R.G., Hess, M., Horie, K., Jones, R.G., Kubisa, P., Meisel, I., Mormann, W., Penczek, S., Stepto, R.F.T. (2011). *Pure Appl. Chem.*, **83**(12), 2229–2259.

Smeets, N.M.B., Hoare, T. (2013). *J. Polym. Sci. Part A Polym. Chem.*, **51**, 3027–3043.

Smith, A., Giunta, B., Bickford, P.C., Fountain, M., Tan, J., Shytle, R.D. (2010). *Int. J. Pharm.*, **389**, 207–212.

Smith, A.M., Nie, S. (2010). *Acc. Chem. Res.*, **2**, 190–200.

Smith, B.W., Monthioux, M., Luzzi, D.E. (1998). *Nature*, **396**, 323–324.

Smith, G., Raghunandan, R., Wu, Y., Liu, Y., Massare, M., Nathan, M., Zhou, B., Lu, H., Boddapati, S., Li, J., Flyer, D., Glenn, G. (2012). *PLoS ONE*, **7**, e50852.

Smith, M.H., South, A.B., Gaulding, J.C., Lyon, L.A. (2010). *Anal. Chem.*, **82**, 523–530.

Smola, M., Vandamme, T., Sokolowski, A. (2008). *Int. J. Nanomed.*, **3**, 1–19.

Snowden, M.J. (1992). *J. Chem. Soc. Chem. Commun.*, 803.

Snowden, M.J., Booty, M. (1993). The temperature controlled uptake and release of molecules and ions using a polymeric microgel. In *Encapsulation and Controlled Release*, Vol. 141, Eds. Karsa, D., Stephenson, R., Royal Society of Chemistry, Cambridge, UK.

Snowden, M.J., Thomas, D., Vincent, B. (1993). *Analyst*, **118**, 1367–1369.

Sobhani, Z., Dinarvand, R., Atyabi, F., Ghahremani, M., Adeli, M. (2011). *Int. J. Nanomed.*, **6**, 705–719.

Soenen, S.J., Hodenius, M., De Cuyper, M. (2009). *Nanomedicine (London)*, **4**, 177–191.

Song, C.W., Lyon, J.C., Luo, Y. (1993). Intra-, extracellular pH in solid tumors, influence on therapeutic response, in Drug Resistance in Oncology, 25–51, Ed. Teicher, B.V., Marcel Dekker, New York.

Song, J., Kai, M., Zhang, W., Zhang, J., Liu, L., Zhang, B., Liu, X., Wang, R. (2011). *Peptides*, **32**, 1934–1941.

Sönnichsen, C., Franzl, T., Wilk, T., Plessen, G.v., Feldmann, J. (2002). *New J. Phys*, **1**, 1–8.

Soppimath, K.S., Aminabhavi, T.M., Kulkarni, A.R., Rudzinski, W.E. (2001). *J. Controlled Release*, **70**, 1–20.

Sorrell, C.D., Serpe, M.J. (2012). *Anal. Bioanal. Chem.*, **402**, 2385–2393.

Sosnowski, S., Gadzinowski, M., Slomkowski, S. (1996). *Macromolecules*, **29**, 4556–4564.

Sotiriou, G.A, Pratsinis, S.E. (2010). *Environ. Sci. Technol.*, **44**, 5649–5654.

Soto, C. (2003). *Nat. Rev. Neurosci.*, **4**, 49–60.

Souza, K.C., Salazar-Alvarez, G., Ardisson, J.D., Macedo, W.A.A., Sousa, E.M.B. (2008). *Nanotechnology*, **19**, 185603.

Spanhel, L. (2006). *J. Sol-Gel Sci. Technol.*, **1**, 7–24.

Speiser, D.E., Liénard, D., Rufer, N., Rubio-Godoy, V., Rimoldi, D., Lejeune, F., Krieg, A.M., Cerottini, J.-C., Romero, P. (2005). *J. Clin. Invest.*, **115**, 739–746.

Speiser, D.E., Schwarz, K., Baumgaertner, P., Manolova, V., Devevre, E., Sterry, W., Walden, P., Zippelius, A., Conzett, K.B., Senti, G. (2010). *J. Immunother.*, **33**, 848–858.

Sperling, R.A., Parak, W.J. (2010). *Philos. Trans. R. Soc. A*, **1915**, 1333–1383.

Spira, J., Plyushch, O., Zozulya, N., Yatuv, R., Dayan, I., Bleicher, A., Robinson, M., Baru, M. (2010). *Haemophilia*, **16**, 910–918.

Spokoyny, A.M., Machan, C.W., Clingerman, D.J., Rosen, M.S., Wiester, M.J., Kennedy, R.D., Stern, C.L., Sarjeant, A.A., Mirkin, C.A. (2011). *Nat. Chem.*, **3**, 590–596.

Springer, C., Benita, S., Sherman, Y., Gursoy, N., Gilhar, D., Avital, A. (2005). *Pediatr. Res.*, **58**, 537–541.

Sprong, E., Leswin, S.K., Lamb, D.J., Ferguson, C.J., Hawkett, B.S., Pham, B.T.T., Nguyen, D., Such, C.H., Serelis, A.K., Gilbert, R.G (2006). *Macromol. Symp.*, **231**, 84–93.

Stanczyk, M., Dziki, A., Morawiec, Z. (2012). *Curr. Med. Chem.*, **19**, 4896–4902.

Stano, A., Nembrini, C., Swartz, M.A., Hubbell, J.A., Simeoni, E. (2012). *Vaccine*, **30**, 7541–7546.

Stella, B., Arpicco, S., Peracchia, M.T., Desmaële, D., Hoebeke, J., Renoir, M., D'Angelo, J., Cattel, L., Couvreur, P. (2000). *J. Pharm. Sci.*, **89**, 1452–1464.

Stevanovic, M., Bracko, I., Milenkovic, M., Filipovic, N., Nunic, J., Filipic, M., Uskokovic, D.P. (2013). *Acta Biomater.*, **10**(1), 151–162.

Stewart-Marshall, J. (1980). *Anal. Biochem.*, **104**, 10–14.

Stocchi, L., Nelson, H. (1998). *Dis. Colon Rectum*, **41**, 232–250.

Stolnik, S., Illum, L., Davis, S.S. (2012). *Adv. Drug. Delivery Rev.*, **64**(Suppl), 290–301.

Storhoff, J.J., Lazarides, A.A., Mucic, R.C., Mirkin, C.A., Letsinger, R.L., Schatz, G.C. (2000). *J. Am. Chem. Soc.*, **19**, 4640–4650.

Storm, G. (1999). *Adv. Drug Delivery Rev.*, **40**, 103–127.

Storm, G., ten Kate, M.T., Working, P.K., Bakker-Woudenberg, I.A. (1998). *Clin. Cancer Res.*, **4**, 111–115.

Strable, E., Bulte, J.W.M., Moskowitz, B., Vivekanandan, K., Allen, M., Douglas, T. (2001). *Chem. Mater.*, **13**, 2201–2209.

Strebhardt, K., Ullrich, A. (2008). *Nat. Rev. Cancer*, **8**, 473–480.

Striebel, H.-M., Birch-Hirschfeld, E., Egerer, R., Földes-Papp, Z., Tilz, G.P., Stelzner, A. (2004). *Exp. Mol. Pathol.*, **77**, 89–97.

Stuart, M.A., Huck, W.T., Genzer, J., Muller, M., Ober, C., Stamm, M., Sukhorukov, G.B., Szleifer, I., Tsukruk, V.V., Urban, M., Winnik, F., Zauscher, S., Luzinov, I., Minko, S. (2010). *Nat. Mater.*, **9**, 101–113.

Su, S., Wang, H., Liu, X., Wu, Y., Nie, G. (2013). *Biomaterials*, **34**, 3523–3533.

Su, Z., Zhu, S., Donkor, A.D., Tzoganakis, C., Honek, J.F. (2011). *J. Am. Chem. Soc.*, **133**, 6874–6877.

Sugihara, S., Blanazs, A., Armes, S.P., Ryan, A.J., Lewis, A.L. (2011). *J. Am. Chem. Soc.*, **133**, 15707–15713.

Sui, M., Liu, W., Shen, Y. (2011). *J. Controlled Release*, **155**, 227–236.

Suk, J.S., Boylan, N.J., Trehan, K., Tang, B.C., Schneider, C.S., Lin, J.M., Boyle, M.P., Zeitlin, P.L., Lai, S.K., Cooper, M.J., Hanes, J. (2011). *Mol. Ther.*, **19**, 1981–1989.

Suk, J.S., Lai, S.K., Boylan, N.J., Dawson, M.R., Boyle, M.P., Hanes, J. (2011). *Nanomedicine (London)*, **6**, 365–375.

Sukhanova, A., Devy, M., Venteo, L., Kaplan, H., Artemyev, M., Oleinikov, V., Klinov, D., Pluot, M., Cohen, J.H.M., Nabiev, I. (2004). *Anal. Biochem.*, **1**, 60–67.

Suma, T., Miyata, K., Anraku, Y., Watanabe, S., Christie, R.J., Takemoto, H., Shioyama, M., Gouda, N., Ishii, T., Nishiyama, N., Kataoka, K. (2012). *ACS Nano*, **6**, 6693–6705.

Sumaru, K., Kameda, M., Kanamori, T., Shinbo, T. (2004). *Macromolecules*, **37**, 4949–4955.

Sumer, B., Gao, J. (2008). *Nanomedicine*, **3**, 137–140.

Sun, H.W., Zhang, L.Y., Zhu, X.J., Kong, C.Y., Zhang, C.L., Yao, S. (2009). *Sci. China Ser. B Chem.*, **52**, 69–75.

Sun, L., Ma, X., Dong, C.-M., Zhu, B., Zhu, X. (2012a). *Biomacromolecules*, **13**, 3581–3591.

Sun, L., Yang, Y., Dong, C.M., Wei, Y. (2011). *Small*, **7**, 401–406.

Sun, L., Zhang, X., Zheng, C., Wu, Z., Xia, X., Li, C. (2012b). *RSC Adv.*, **2**, 9904–9913.

Sun, M., Gao, Y., Guo, C., Cao, F., Song, Z., Xi, Y., Yu, A., Li, A., Zhai, G. (2010). *J. Nanopart. Res.*, **12**, 3111–3122.

Sun, X.L., Stabler, C.L., Cazalis, C.S., Chaikof, E.L. (2006a). *Bioconj. Chem.*, **1**, 52–57.

Sun, Y., Xia, Y. (2003). *Analyst*, **6**, 686–691.

Sun, Y.P., Zhou, B., Lin, Y., Wang, W., Fernando, K.A.S., Pathak, P., Meziani, M.J., Harruff, B.A., Wang, X., Wang, H.F., Luo, P.J.G., Yang, H., Kose, M.E., Chen, B.L., Veca, L.M., Xie, S.Y. (2006b). *J. Am. Chem. Soc.*, **24**, 7756–7757.

Sunoqrot, S., Bae, J.W., Pearson, R.M., Shyu, K., Liu, Y., Kim, D.H., Hong, S. (2012). *Biomacromolecules*, **13**, 1223–1230.

Suresh Reddy, J., Venkateswarlu, V., Koning, G.A. (2006). *J. Drug Targeting*, **14**, 13–19.

Suzuki, R., Namai, E., Oda, Y., Nishiie, N., Otake, S., Koshima, R., Hirata, K., Taira, Y., Utoguchi, N., Negishi, Y., Nakagawa, S., Maruyama, K. (2010). *J. Controlled Release*, **142**, 245–250.

Suzuki, R., Takizawa, T., Negishi, Y., Hagisawa, K., Tanaka, K., Sawamura, K., Utoguchi, N., Nishioka, T., Maruyama, K. (2007a). *J. Controlled Release*, **117**, 130–136.

Suzuki, R., Takizawa, T., Negishi, Y., Utoguchi, N., Maruyama, K. (2007b). *J. Drug Targeting*, **15**, 531–537.

Svenson, S. (2009). *Eur. J. Pharm. Biopharm.*, **71**, 445–462.

Svenson, S., Tomalia, D.A. (2012). *Adv. Drug Delivery Rev.*, **57**(15), 2106–2129.

Svirskis, D., Travas-Sejdic, J., Rodgers, A., Garg, S. (2010). *J. Controlled Release*, **146**, 6–15.

Swami, A., Shi, J., Gadde, S., Votruba, A.R., Kolishetti, N., Farokhzad, O.C. (2012). Nanoparticles for targeted and temporally controlled drug delivery. In *Multifunctional Nanoparticles for Drug Delivery Applications*, pp. 2–29, Eds. Svenson, S., Prud'Homme, R.K., Springer US.

Swaminathan, S.K., Roger, E., Toti, U., Niu, L., Ohlfest, J.R., Panyam, J. (2013). *J. Controlled Release*, **171**(3), 280–287.

Swanson, S.D., Kukowska-Latallo, J.F., Patri, A.K., Chen, C., Ge, S., Cao, Z., Kotlyar, A., East, A.T., Baker, J.R. (2008). *Int. J. Nanomed.*, 3, 201.

Szlachcic, A., Pala, K., Zakrzewska, M., Jakimowicz, P., Wiedlocha, A., Otlewski, J. (2012). *Int. J. Nanomed.*, **7**, 5915–5927.

Szoka, F., Olson, F., Heath, T., Vail, W., Mayhew, E., Papahadjopoulos, D. (1980). *Biochim. Biophys. Acta*, **601**, 559–571.

Szoka, F., Papahadjopoulos, D. (1978). *Proc. Natl. Acad. Sci. U S A*, **75**, 4194–4198.

Tabata, Y., Ikada, Y. (1988). *Biomaterials*, **9**, 356–362.

Taghdisi, S.M., Lavaee, P., Ramezani, M., Abnous, K. (2011). *Eur. J. Pharm. Biopharm.*, **77**, 200–206.

Tai, L.A., et al. (2009). *Nanotechnology*, **20**, 135101.

Takahama, H., Minamino, T., Asanuma, H., Fujita, M., Asai, T., Wakeno, M., Sasaki, H., Kikuchi, H., Hashimoto, K., Oku, N., Asakura, M., Kim, J., Takashima, S., Komamura, K., Sugimachi, M., Mochizuki, N., Kitakaze, M. (2009). *J. Am. Coll. Cardiol.*, **53**, 709–717.

Takakura, Y., Hashida, M. (1995). *Crit. Rev. Oncol. Hematol.*, **18**, 207e31.

Takenobu, T., Takano, T., Shiraishi, M., Murakami, Y., Ata, M., Kataura, H., Achiba, Y., Iwasa, Y. (2003). *Nat. Mater.*, **2**, 683–688.

Takeuchi, H., Kojima, H., Yamamoto, H., Kawashima, Y. (2001). *J. Controlled Release*, **75**, 83–91.

Talanov, V.S., Regino, C.A., Kobayashi, H., Bernardo, M., Choyke, P.L., Brechbiel, M.W. (2006). *Nano Lett.*, **6**, 1459–1463.

Talekar, M., Ganta, S., Amiji, M., Jamieson, S., Kendall, J., Denny, W.A., Garg, S. (2013). *Int. J. Pharm.*, **450**, 278–89.

Talsma, H., Jousma, H., Nicolay, K., Crommelin, D.J. A. (1987). *Int. J. Pharm.*, **37**, 171–173.

Tam, J.M., McConville, J.T., Williams, R.O., 3rd Johnston, K.P. (2008). *J. Pharm. Sci.*, **97**, 4915–4933.

Tam, J.P., Lu, Y.A., Yang, J.L. (2002). *Eur. J. Biochem.*, **269**, 923–932.

Tamura, A., Oishi, M., Nagasaki, Y. (2009). *Biomacromolecules*, **10**, 1818–1827.

Tan, J.P., Tan, M.B., Tam, M.K. (2010). *Local Reg. Anesth.*, **3**, 93–100.

Tan, J.P.K., Goh, C.H., Tam, K.C. (2007). *Eur. J. Pharm. Sci.*, **32**, 340–348.

Tan, W.B., Zhang, Y. (2005). *J. Biomed. Mater. Res. A*, **1**, 56–62.

Tanaka, M., Kataoka, H., Mabuchi, M., Sakuma, S., Takahashi, S., Tujii, R., Akashi, H., Ohi, H., Yano, S., Morita, A., Joh, T. (2011). *Anticancer Res.*, **31**, 763–769.

Tanford, C. (1980). *The Hydrophobic Effect. Formation of Micelles and Biological Membranes*, Wiley, New York.

Tang, X., Gao, L., Fan, X., Liang, X., Zhou, Q. (2009). *Macromol. Chem. Phys.*, **210**, 1556–1562.

Tannock, I.F., Rotin, D. (1989). *Cancer Res.*, **49**, 4373–4384.

Tao, S.L., Desai, T.A. (2003). *Adv. Drug Delivery Rev.*, **55**, 315–328.

Tao, X., Yang, Y.-J., Liu, S., Zheng, Y.-Z., Fu, J., Chen, J.-F. (2013). *Acta Biomater.*, **9**, 6431–6438.

Taratula, O., Garbuzenko, O.B., Kirkpatrick, P., Pandya, I., Savla, R., Pozharov, V.P., Minko, T. (2009). *J. Controlled Release*, **140**, 284–293.

Tartaj, P., Morales, M.D.P, Veintemillas-Verdaguer, S., González-Carreño, T., Serna, C.J. (2003). *J. Phys. D: Appl. Phys.*, **36**, R182–R197.

Tashima, Y., Oe, R., Lee, S., Sugihara, G., Chambers, E.J., Takahashi, M., Yamada, T. (2004). *J. Biol. Chem.*, **279**, 17587–17595.

Tassa, C., Shaw, S.Y., Weissleder, R. (2011). *Acc. Chem. Res.*, **10**, 842–852.

Tatham, A.S., Shewry, P.R. (2000). *Trends Biochem. Sci.*, **25**, 567–571.

Taylor, P., Xu, C., Fletcher, P.D. I., Paunov, V.N. (2003). *Phys. Chem. Chem. Phys.*, **5**, 4918–4922.

Tekade, R.K., Kumar, P.V., Jain, N.K. (2009). *Chem. Rev.*, **109**, 49.

Terada, T., et al. (2006). *J. Controlled Release*, **111**, 333–342.

Thambi, T., Deepagan, V.G., Yoo, C.K., Park, J.H. (2011a). *Polymer*, **52**, 4753–4759.

Thambi, T., Yoon, H.Y., Kim, K., Kwon, I.C., Yoo, C.K., Park, J.H. (2011b). *Bioconjug. Chem.*, **22**, 1924–1931.

Thickett, S.C., Gilbert, R.G. (2007). *Polymer*, **48**, 6965–6991.

Thiele, L., Diederichs, J.E., Reszka, R., Merkle, H.P., Walter, E. (2003). *Biomaterials*, **24**, 1409–1418.

Thiery, J.P., Acloque, H., Huang, R.Y.J., Nieto, M.A. (2009). *Cell*, **139**, 871–890.

Thomas, P.J., Qu, B.-H., Pedersen, P.L. (1995). *TIBS*, **20**, 456–459.

Thomas, T.P., Huang, B., Choi, S.K., Silpe, J.E., Kotlyar, A., Desai, A.M., Zong, H., Gam, J., Joice, M., Baker, J.R., Jr. (2012). *Mol. Pharm.*, **9**, 2669–2676.

Thomas, T.P., Patri, A.K., Myc, A., Myaing, M.T., Ye, J.Y., Norris, T.B., Baker, J.R. (2004). *Biomacromolecules*, **5**, 2269–2274.

Thornton, P.D., Mart, R.J., Ulijn, R.V. (2007). *Adv. Mater.*, **19**, 1252–1256.

Thornton, P.D., Mart, R.J., Webb, S.J., Ulijn, R.V. (2008). *Soft Matter*, **4**, 821.

Thurner, G., Abdelmoez, A., Wallnoefer, E., Rohr, I., Klammsteiner, N., Talasz, H., Kremser, C., Jaschke, W., Debbage, P. (2011). *Int. J. Clin. Pharm. Ther.*, **49**, 65.

Tian, L., Shi, C., Zhu, J. (2007). *Chem. Commun.*, **37**, 3850–3852.

Tian, W.D., Ma, Y.Q. (2013). *Chem. Soc. Rev.*, **42**, 705–727.

Tibbe, A.G.J., de Grooth, B.G., Greve, J., Liberti, P.A., Dolan, G.J., Terstappen, L. (1999). *Nat. Biotechnol.*, **12**, 1210–1213.

Tiernan, J.P., Perry, S.L., Verghese, E.T., West, N.P., Yeluri, S., Jayne, D.G., Hughes, T.A. (2013). *Br. J. Cancer*, **108**, 662–667.

Tisch, U., Schlesinger, I., Ionescu, R., Nassar, M., Axelrod, N., Robertman, D., Tessler, Y., Azar, F., Marmur, A., Aharon-Peretz, J., Haick, H. (2013). *Nanomedicine (London)*, **8**, 43–56.

Tivnan, A., Orr, W.S., Gubala, V., Nooney, R., Williams, D.E., Mcdonagh, C., Prenter, S., Harvey, H., Domingo-Fernandez, R., Bray, I.M., Piskareva, O., Ng, C.Y., Lode, H.N., Davidoff, A.M., Stallings, R.L. (2012). *PLoS ONE*, **7**, e38129.

Tkachenko, A.G., Xie, H., Liu, Y.L., Coleman, D., Ryan, J., Glomm, W.R., Shipton, M.K., Franzen, S., Feldheim, D.L. (2004). *Bioconj. Chem.*, **3**, 482–490.

Tokumasu, F., Dvorak, J. (2003). *J. Microsc.: Oxford*, **3**, 256–261.

Tokumasu, F., Fairhurst, R.M., Ostera, G.R., Brittain, N.J., Hwang, J., Wellems, T.E., Dvorak, J.A. (2005). *J. Cell Sci.*, **5**, 1091–1098.

Tokura, S., Hiroshi, T., Ichiro, A. (1999). Immunological aspects of chitin and chitin derivatives administered to animals, In *Chitin and Chitinases*, Vol. **87**, pp. 279–292, Eds. Jollès, P., Birkhäuser, Basel.

Tokuraku, K., Marquandt, M., Ikezu, T. (2009). *PLoS ONE*, **4**, e8492.

Tomali, D., Swanson, D., Huang, B., Heinzelmann, J., Svenson, S., Reyna, L., Zhuravel, C.A., DeMattei, C. (2006). Inc., PCT Patent WO, 115547.

Tomalia, D.A. (2005). *Prog. Polym. Sci.*, **30**, 294–324.

Tomatsu, I., Peng, K., Kros, A. (2011). *Adv. Drug Delivery Rev.*, **63**, 1257–1266.

Tondra, M., Porter, M., Lipert, R.J. (2000). *J. Vac. Sci. Technol. A*, **4**, 1125–1129.

Tong, R., Hemmati, H.D., Langer, R., Kohane, D.S. (2012). *J. Am. Chem. Soc.*, **134**, 8848–8855.

Toole, B.P. (2009). *Clin. Cancer Res.*, **15**, 7462–7468.

Torchilin, V. (2009). *Eur. J. Pharm. Biopharm.*, **71**, 431–444.

Torchilin, V.P. (1996). *J. Liposome Res.*, **6**, 99–116.

Torchilin, V.P. (2005). *Nat. Rev. Drug Discovery*, **4**, 145–160.

Torchilin, V.P. (2006a). *Adv. Drug. Delivery Rev.*, **58**, 1532–1555.

Torchilin, V.P. (2006b). *Annu. Rev. Biomed. Eng.*, **8**, 343–375.

Torchilin, V.P. (2008). *Biopolymers*, **90**, 604–610.

Torchilin, V.P., Levchenko, T.S., Lukyanov, A.N., Khaw, B.A., Klibanov, A.L., Rammohan, R., Samokhin, G.P., Whiteman, K.R. (2001a). *Biochim. Biophys. Acta*, **1511**, 397–411.

Torchilin, V.P., Levchenko, T.S., Whiteman, K.R., Yaroslavov, A.A., Tsatsakis, A.M., Rizos, A.K., Michailova, E.V., Shtilman, M.I. (2001b). *Biomaterials*, **22**, 3035–3044.

Torchilin, V.P., Lukyanov, A.N., Gao, Z., Papahadjopoulos-Sternberg, B. (2003). *Proc. Natl. Acad. Sci. U S A*, **100**, 6039–6044.

Torchilin, V.P., Trubetskoy, V.S. (1995). *Adv. Drug Delivery Rev.*, **16**, 141–155.

Tóth, É., Pubanz, D., Vauthey, S., Helm, L., Merbach, A.E. (1996). *Chem.: Eur. J.*, **2**, 1607–1615.

Toti, U.S., Moon, S.H., Kim, H.Y., Jun, Y.J., Kim, B.M., Park, Y.M., Jeong, B., Sohn, Y.S. (2007). *J. Controlled Release*, **119**, 34–40.

Trevor, A.J., Katzung, B., Masters, S. (2009). *Basic and Clinical Pharmacology*, 11th ed., McGraw-Hill.

Trindade, T., da Silva, A.L.D. (eds.) (2011). *Nanocomposite Particles for Bio-Applications: Materials and Bio-Interfaces*, Pan Stanford, Singapore.

Trindade, T., O'Brien, P. (1996). *Adv. Mater.*, **2**, 161–163.

Trindade, T., O'Brien, P., Pickett, N.L. (2001). *Chem. Mater.*, **11**, 3843–3858.

Trindade, T., O'Brien, P., Zhang, X.-m. (1997). *Chem. Mater.*, **2**, 523–530.

Trindade, T., Thomas, P.J. (2013). Defining and using very small crystals, In *Comprehensive Inorganic Chemistry II*, 2nd ed., pp. 343–369, Eds. Jan, R., Kenneth, P., Elsevier, Amsterdam.

Tripisciano, B.-P. (2008). *Cancer Res.*, **47**(4), 1036–1039.

Tripisciano, C., Costa, S., Kalenczuk, R.J., Borowiak-Palen, E. (2010). *Eur. Phys. J. B*, **75**, 141–146.

Tripisciano, C., Kraemer, K., Taylor, A., Borowiak-Palen, E. (2009). *Chem. Phys. Lett.*, **478**, 200–205.

Trongsatitkul, T., Budhlall, B.M. (2011). *Langmuir*, **27**, 13468–13480.

Trongsatitkul, T., Budhlall, B.M. (2013). *Polym. Chem.*, **4**, 1502–1516.

Troutman, T.S., Leung, S.J., Romanowski, M. (2009). *Adv. Mater.*, **21**, 2334–2338.

Trubetskoy, V.S., Gazelle, G.S., Wolf, G.L., Torchilin, V.P. (1997). *J. Drug Targeting*, **4**, 381–388.

Tsai, M.J., Huang, Y.B., Wu, P.C., Fu, Y.S., Kao, Y.R., Fang, J.Y., Tsai, Y.H.J. (2011). *J. Pharm. Sci.*, **100**, 547–557.

Tsukakoshi K., Abe K., Sode K., Ikebukuro, K. (2012). *Anal. Chem.*, **84**, 5542–5547.

Tsukie, N., Nakano, K., Matoba, T., Masuda, S., Iwata, E., Miyagawa, M., Zhao, G., Meng, W., Kishimoto, J., Sunagawa, K., Egashira, K. (2013). *J. Atheroscler. Thromb.*, **20**, 32–45.

Tülü, M., Ertürk, A.S. (2012). *A Search for Antibacterial Agents*, 6, 89–106, Intech, Croatia.

Turkevich, J., Stevenson, P.C., Hillier, J. (1951). *Discuss Faraday Soc.*, 55–75.

Uesugi, Y., Kawata, H., Jo, J., Saito, Y., Tabata, Y. (2010). *J. Controlled Release*, **147**, 269–277.

Ugarenko, M., Chan, C.K., Nudelman, A., Rephaeli, A., Cutts, S.M., Phillips, D.R. (2009). *Oncol. Res.*, **17**, 283–299.

Ulbrich, K., Šubr, V., Seymour, L.W., Duncan, R. (1993). *J. Controlled Release*, **24**, 181–190.

Underhill, C. (1992). *J. Cell Sci.*, **103**(Pt 2), 293–298.

Urbán, P., Valle-Delgado, J.J., Moles, E., Marques, J., Díez, C., Fernàndez-Busquets, X. (2012). *Curr. Drug Targeting*, **13**, 1158–1172.

Vaishnavi, T.S., Haridoss, P., Vijayan, C. (2008). *Mater. Lett.*, **10–11**, 1649–1651.

Valizadeh, A., Mikaeili, H., Samiei, M., Farkhani, S.M., Zarghami, N., Kouhi, M., Akbarzadeh, A., Davaran, S. (2012). *Nanoscale Res. Lett.*, **1**, 1–14.

Vallet-Regi, M., Rámila, A., del Real, R.P., Pérez-Pariente, J. (2000). *Chem. Mater.*, **13**, 308–311.

Van Berkel, K.Y., Piekarski, A.M., Kierstead, P.H., Pressly, E.D., Ray, P.C., Hawker, C.J. (2009). *Macromolecules*, **42**, 1425–1427.

van Herk, A.M. (Ed.) (2013). *Chemistry and Technology of Emulsion Polymerization*, Wiley, New Jersey.

Van Swaay, D., deMello, A. (2013). *Lab. Chip*, **13**, 752–767.

Van Vlerken, L.E., Amiji, M.M. (2006). *Expert Opin. Drug Delivery*, **3**, 205–216.

van Zijverden, M., Granum, B. (2000). *Toxicology*, **152**, 69–77.

Vangara, K.K., Liu, J.L., Palakurthi, S. (2013). *Anticancer Res.*, **33**, 2425–2434.

Varga, C.M., Wickham, T.J., Lauffenburger, D.A. (2000). *Biotechnol. Bioeng.*, **70**, 593–605.

Varki, A., Lowe, J.B. (2009). In *Essentials of Glycobiology*, 2nd ed., Eds. Varki, A., Cummings, R.D., Esko, J.D., Freeze, H.H., Stanley, P., Bertozzi, C.R., Hart, G.W., Etzler, M.E., Cold Spring Harbor, New York.

Vasantha, V.A., Jana, S., Parthiban, A., Vancso, J.G. (2014). *Chem. Commun.*, **50**, 46–48.

Vasir, J.K., Labhasetwar, V. (2005). *Technol. Cancer Res. Treat.*, **4**, 363–374.

Vasir, J.K., Labhasetwar, V. (2007). Nanoparticles for gene delivery: formulation characteristics, In *Nanoparticulate Drug Delivery Systems*, 291–304.

Vaupel, P., Kallinowski, F., Okunieff, P. (1989). *Cancer Res.*, **49**, 6449–6465.

Verma, A., Stellacci, F. (2010). *Small*, **6**, 12–21.

Veronese, F.M., Pasut, G. (2005). *Drug Discovery Today*, **10**, 1451–1458.

Vihola, H., Laukkanen, A., Hirvonen, J., Tenhu, H. (2002). *Eur. J. Pharm. Sci.*, **16**, 69–74.

Vihola, H., Laukkanen, A., Valtola, L., Tenhu, H., Hirvonen, J. (2005). *Biomaterials*, **26**, 3055–3064.

Villaraza, A.J.L., Bumb, A., Brechbiel, M.W. (2010). *Chem. Rev.*, **110**, 2921.

Villemejane, J., Mir, L.M. (2009). *Br. J. Pharmacol.*, **157**, 207–219.

Vinogradov, S.V., Bronich, T.K., Kabanov, A.V. (2002). *Adv. Drug Delivery Rev.*, **54**(1), 135–147.

Vitaliano, G.D., Vitaliano, F., Rios, J.D., Renshaw, P.F., Teicher, M.H. (2012). *PLoS ONE*, **7**, e35821.

Vo-Dinh, T. (1998). *Trac-Trend Anal. Chem.*, **8–9**, 557–582.

Volodkin, D.V., Skirtach, A.G., Möhwald, H. (2009). *Angew. Chem., Int. Ed.*, **48**, 1807–1809.

Von Andrian, U.H., Mackay, C.R. (2000). *N. Engl. J. Med.*, **343**, 1020–1034.

Wadajkar, A.S., Menon, J.U., Tsai, Y.S., Gore, C., Dobin, T., Gandee, L., Kangasniemi, K., Takahashi, M., Manandhar, B., Ahn, J.M., Hsieh, J.T., Nguyen, K.T. (2013). *Biomaterials*, **34**, 3618–3625.

Wadghiri, Y.Z., Sigurdsson, E.M., Sadowski, M., Elliot, J.I., Li, Y., Scholtzova, H., Tang, C.Y., Aguinaldo, G., Pappolla, M., Duff, K., Wisniewski, T., Turnbull, D.H. (2003). *Magn. Reson. Med.*, **50**, 293–302.

Wadhwa, R., Lagenaur, C.F., Cui, X.T. (2006). *J. Controlled Release*, **110**, 531–541.

Wagner, A., Vorauer-Uhl, K. (2011). *J. Drug Delivery*, Article 591325 (9 pages).

Wagner, A., Vorauer-Uhl, K., Kreismayr, G., Katinger, H. (2002). *J. Liposome Res.*, **12**, 271–283.

Wang, A.Z., Langer, R., Farokhzad, O.C. (2012a). *Annu. Rev. Med.*, **63**, 185–198.

Wang, B., Navath, R.S., Menjoge, A.R., Balakrishnan, B., Bellair, R., Dai, H., Romero, R., Kannan, S., Kannan, R.M. (2010a). *Int. J. Pharm.*, **395**, 298–308.

Wang, B., Xu, C., Xie, J., Yang, Z., Sun, S. (2008a). *J. Am. Chem. Soc.*, **130**, 14436–14437.

Wang, C., Ding, C., Kong, M., Dong, A., Qian, J., Jiang, D., Shen, Z. (2011a). *Biochem. Biophys. Res. Commun.*, **410**, 537–542.

Wang, C., Liu, D., Wang, Z. (2012b). *Chem. Commun.*, **48**, 8392–8394.

Wang, C., Mallela, J., Garapati, U.S., Ravi, S., Chinnasamy, V., Girard, Y., Howell, M., Mohapatra, S. (2013a). *Nanomed. Nanotechnol., Biol. Med.*, **9**, 903–911.

Wang, D., Liu, T., Yin, J., Liu, S. (2011b). *Macromolecules*, **44**, 2282–2290.

Wang, F., Liu, X. (2009). *Chem. Soc. Rev.*, **38**, 976–989.

Wang, H., et al. (2009). *Neuro. Oncol.*, **11**, 790–802.

Wang, H.F., Jia, H.Z., Zhu, J.Y., Chu, Y.F., Feng, J., Zhang, X.Z., Zhuo, R.X. (2012c). *Macromol. Biosci.*, **12**, 1689–1696.

Wang, H.-Q., Wang, J.-H., Li, Y.-Q., Li, X.-Q., Liu, T.-C., Huang, Z.-L., Zhao, Y.-D. (2007). *J. Colloid Interface Sci.*, **2**, 622–627.

Wang, J., Fang, X., Liang, W. (2012d). *ACS Nano*, **6**, 5018–5030.

Wang, J., Huang, G. (2011). *Drug Delivery*, **18**, 631–638.

Wang, L., Erasquin, U.J., Zhao, M., Ren, L., Zhang, M.Y., Cheng, G.J., Wang, Y., Cai, C. (2011c). *ACS Appl. Mater. Interfaces*, **3**, 2885–2894.

Wang, L., Liu, Y., Li, W., Jiang, X., Ji, Y., Wu, X., Xu, L., Qiu, Y., Zhao, K., Wei, T., Li, Y., Zhao, Y., Chen, C. (2011d). *Nano Lett.*, **11**, 772–780.

Wang, S.L., et al. (2008b). *J. Drug Targeting*, **16**, 233–242.

Wang, T., D'souza, G.G., Bedi, D., Fagbohun, O.A., Potturi, L.P., Papahadjopoulos-Sternberg, B., Petrenko, V.A., Torchilin, V.P. (2010b). *Nanomedicine*, **5**, 563–574.

Wang, T., Yang, S., Petrenko, V.A., Torchilin, V.P. (2010c). *Mol. Pharm.*, **7**, 1149–1158.

Wang, W., Chen, C., Qian, M., Zhao, X.S. (2008c). *Anal. Biochem.*, **2**, 213–219.

Wang, W., Zhou, F., Ge, L., Liu, X., Kong, F. (2012e). *Int. J. Nanomed.*, **7**, 2513–2522.

Wang, X., Qiu, X., Wu, C. (1998). *Macromolecules*, **31**, 2972–2976.

Wang, Y., Guo, R., Cao, X., Shen, M., Shi, X. (2011e). *Biomaterials*, **32**, 3322–3329.

Wang, Y., Li, X., Zhou, Y., Huang, P., Xu, Y. (2010d). *Int. J. Pharm.*, **384**, 148–153.

Wang, Y., Wang, H., Liu, D., Song, S., Wang, X., Zhang, H. (2013b). *Biomaterials*, **34**, 7715–7724.

Wang, Y., Zhang, X., Yu, P., Li, C. (2013c). *Int. J. Pharm.*, **441**, 170–180.

Wang, Z., Wang, L., Sun, J., Han, L., Hong, C. (2013d). *Polym. Chem.*, **4**, 1694–1699.

Wang, Z., Yu, Y., Dai, W., Lu, J., Cui, J., Wu, H., Yuan, L., Zhang, H., Wang, X., Wang, J., Zhang, X., Zhang, Q. (2012f). *Biomaterials*, **33**, 8451–8460.

Wang, Z.X., Kanaras, A.G., Bates, A.D., Cosstick, R., Brust, M. (2004). *J. Mater. Chem.*, **4**, 578–580.

Warner, S. (2004). *Scientist*, **18**(16), 38–39.

Warr, L, Walker, Z. (2012). *Q. J. Nucl. Med. Mol. Imaging*, **56**, 39–54.

Wasiak, T., Ionov, M., Nieznanki, K., Nieznanska, H., Klementieva, O., Granell, M., Cladera, J., Majoral, J.-P., Caminade, A.M., Klajnert, B. (2011). *Mol. Pharm.*, **9**, 458–469.

Watanabe, K., Kaneko, M., Maitani, Y. (2012). *Int. J. Nanomed.*, **7**, 3679–3688.

Webb, S.D., Sherratt, J.A., Fish, R.G. (1999). *J. Theor. Biol.*, **196**, 237–250.

Webber, M.J., Tongers, J., Newcomb, C.J., Marquardt, K.T., Bauersachs, J., Losordo, D.W., Stupp, S.I. (2011). *Proc. Natl. Acad. Sci. U S A*, **108**, 13438–13443.

Wedge, S.R., Duncan, R., Kopeckova, P. (1991). *Br. J. Cancer*, **63**, 546–549.

Wei, H., Zhuo, R.-X., Zhang, X.-Z. (2012). *Prog. Polym. Sci.*, **38**(3–4), 503–535.

Weinmann, H.J., Ebert, W., Misselwitz, B., Schmitt-Willich, H. (2003). *Eur. J. Radiol.*, **1**, 33–44.

Weiss, C.K., Landfester, K. (2010). *Adv. Polym. Sci.*, **233**, 185–236.

Weissleder, R., Tung, C.H., Mahmood, U., Bogdanov, A., Jr. (1999). *Nat. Biotechnol.*, **17**, 375–378.

Weitz, T.M., Town, T. (2012). *Int. J. Alzheimer Dis.*, ID 314185, 11 pages. DOI:10.1155/2012/314185.

Wells, L.A., Brook, M.A., Sheardown, H. (2011). *Macromol. Biosci.*, **11**, 988–998.

Wells, T.N., Scully, P., Proudfoot, A.E., Payton, M.A. (1995). *Biochemistry*, **20**, 7896–7903.

Welsch, N., Ballauff, M., Lu, Y. (2011). *Adv. Polym. Sci.*, **234**, 129–163.

Wen, J., Anderson, S.M., Du, J., Yan, M., Wang, J., Shen, M., Lu, Y., Segura, T. (2011). *Adv. Mater.*, **23**, 4549–4553.

Wen, S., Liu, H., Cai, H., Shen, M., Shi, X. (2013). *Adv. Healthcare Mater.*, DOI:10.1002/adhm.201200389.

Westerman, M.A., Cooper-Blacketer, D., Mariash, A., Kotilinek, L., Kawarabayashi, T., Younkin, L.H., Carlson, G.A., Younkin, S.G., Ashe, K.H. (2002). *J. Neurosci.*, **22**, 1858–1867.

White, A.G., Gray, B.D., Pak, K.Y., Smith, B.D. (2012). *Bioorg. Med. Chem. Lett.*, **22**, 2833–2836.

White, M.A., Johnson, J.A., Koberstein, J.T., Turro, N.J. (2006). *J. Am. Chem. Soc.*, **35**, 11356–11357.

Whiteman, K.R., Subr, V., Ulbrich, K., Torchilin, V.P. (2001). *J. Liposome Res.*, **11**, 153–164.

Wiener, E., Brechbiel, M., Brothers, H., Magin, R., Gansow, O., Tomalia, D., Lauterbur, P. (1994). *Magn. Reson. Med.*, **31**, 1–8.

Wiener, E.C., Konda, S., Shadron, A., Brechbiel, M., Gansow, O. (1997). *Invest. Radiol.*, **32**, 748–754.

Wijagkanalan, W., Kawakami, S., Hashida, M. (2011). *Pharm. Res.*, **28**, 1500–1519.

Wike-Hooley, J.L., Haveman, J., Reinhold, H.S. (1984). *Radiother. Oncol.*, **2**, 343–366.

Wilchek, M., Bayer, E.A. (1988). *Anal. Biochem.*, **1**, 1–32.

Wilchek, M., Bayer, E.A. (1989). *Trends Biochem. Sci.*, **10**, 408–412.

Wiley, B., Sun, Y., Xia, Y. (2007a). *Acc. Chem. Res.*, **10**, 1067–1076.

Wiley, B.J., Chen, Y., McLellan, J.M., Xiong, Y., Li, Z.-Y., Ginger, D., Xia, Y. (2007b). *Nano Lett.*, **4**, 1032–1036.

Wiley, B.J., Im, S.H., Li, Z.-Y., McLellan, J., Siekkinen, A., Xia, Y. (2006). *J. Phys. Chem. B*, **32**, 15666–15675.

Williams, S., Lepene, B., Thatcher, C., Long, T. (2009). *Biomacromolecules*, **10**, 155–161.

Wilson, B., Samanta, M.K., Santhi, K., Kumar, K.P., Paramakrishnan, N., Suresh, B. (2008a). *Brain Res.*, **1200**, 159–168.

Wilson, B., Samanta, M.K., Santhi, K., Kumar, K.P., Paramakrishnan, N., Suresh, B. (2008b). *Eur. J. Pharm. Biopharm.*, **70**, 75–84.

Wilson, N.S., Yang, B., Morelli, A.B., Koernig, S., Yang, A., Loeser, S., Airey, D., Provan, L., Hass, P., Braley, H., Couto, S., Drane, D., Boyle, J., Belz, G.T., Ashkenazi, A., Maraskovsky, E. (2012). *Immunol. Cell Biol.*, **90**, 540–552.

Wilson, W.L., Szajowski, P.F., Brus, L.E. (1993). *Science*, **5137**, 1242–1244.

Winkler, C., Kirik, D., Björklund, A. (2005). *Trends Neurosci.*, **28**, 86–92.

Winsor, P.A. (1948). *Trans. Faraday Soc.*, **44**, 376–398.

Winter, P.M., Caruthers, S.D., Zhang, H., Williams, T.A., Wickline, S.A., Lanza, G.M. (2008). *JACC Cardiovasc. Imaging*, **1**, 624–634.

Wittig, A., Michel, J., Moss, R.L., Stecher-Rasmussen, F., Arlinghaus, H.F., Bendel, P., Mauri, P.L., Altieri, S., Hilger, R., Salvadori, P.A. (2008). *CRC Cr. Rev. Oncol.: Hem.*, **68**, 66–90.

Wohlfart, S., Gelperina, S., Kreuter, J. (2012). *J. Controlled Release*, **161**, 264–273.

Wolinsky, J.B., Grinstaff, M.W. (2008). *Adv. Drug Delivery Rev.*, **60**, 1037–1055.

Won, Y.Y., Sharma, R., Konieczny, S.F. (2009). *J. Controlled Release*, **139**, 88–93.

Wong, A.D., Dewit, M.A., Gillies, E.R. (2012). *Adv. Drug Delivery Rev.*, **64**, 1031–1045.

Wong, K.K.Y., Liu, X. (2010). *MedChemComm*, **2**, 125–131.

Wong, P.T., Schauerte, J.A., Wisser, K.C., Ding, H., Lee, E.L., Steel, D.G., Gafni, A. (2009). *J. Mol. Biol.*, **386**, 81–96.

Woolley, A.T., Cheung, C.L., Hafner, J.H., Lieber, C.M. (2000). *Chem. Biol.*, **7**, R193–204.

Wu, C.H., Cao, C., Kim, J.H., Hsu, C.H., Wanebo, H.J., Bowen, W.D., Xu, J., Marshall, J. (2012a). *Nano Lett.*, **12**, 5475–5480.

Wu, G. (2008). *J. Am. Chem. Soc.*, **130**, 8175–8177.

Wu, H., Tang, L., An, L., Wang, X., Zhang, H., Shi, J., Yang, S. (2012b). *React. Funct. Polym.*, **72**, 329–336.

Wu, J., Akaike, T., Maeda, H. (1998). *Cancer Res.*, **58**, 159–165.

Wu, J.-Y., Liu, S.-Q., Heng, P.W.-S., Yang, Y.-Y. (2005). *J. Controlled Release*, **102**, 361–372.

Wu, L., Man, C., Wang, H., Lu, X., Ma, Q., Cai, Y., Ma, W. (2013). *Pharm. Res.*, **30**, 412–423.

Wu, P., Chen, X., Hu, N., Tam, U.C., Blixt, O., Zettl, A., Bertozzi, C.R. (2008). *Angew. Chem., Ger. Ed.*, **120**, 5100–5103.

Wu, W., Li, R., Bian, X., Zhu, Z., Ding, D., Li, X., Jia, Z., Jiang, X., Hu, Y. (2009). *ACS Nano*, **3**, 2740–2750.

Wu, W., Shen, J., Gai, Z., Hong, K., Banerjee, P., Zhou, S. (2011). *Biomaterials*, **32**, 9876–9887.

Wu, X.Y., Liu, H.J., Liu, J.Q., Haley, K.N., Treadway, J.A., Larson, J.P., Ge, N.F., Peale, F., Bruchez, M.P. (2003). *Nat. Biotechnol.*, **1**, 41–46.

Wu, Y., Hu, H., Hu, J., Liu, S. (2012c). *Macromol. Rapid Commun.*, **33**, 1852–1860.

Wybenga, D.R., Pileggi, V.J., Dirstine, P.H., Di Giorgio, J. (1970). *J. Clin. Chem.*, **16**, 980–984.

Xia, W., Wang, P., Lin, C., Li, Z., Gao, X., Wang, G., Zhao, X. (2012). *J. Controlled Release*, **157**, 427–436.

Xiang, B., Dong, D.W., Shi, N.Q., Gao, W., Yang, Z.Z., Cui, Y., Cao, D.Y., Qi, X.R. (2013). *Biomaterials*, **34**, 6976–6991.

Xiang, L.Z., Yuan, Y., Xing, D., Ou, Z.M., Yang, S.H., Zhou, F.F. (2009). *J. Biomed. Opt.*, **14**, 02100.

Xiang, S.D., Scholzen, A., Minigo, G., David, C., Apostolopoulos, V., Mottram, P.L., Plebanski, M. (2006). *Methods*, **40**, 1–9.

Xiao, J., Qi, L. (2011). *Nanoscale*, **4**, 1383–1396.

Xiao, L., Zhao, D., Chan, W.-H., Choi, M.M. F., Li, H.-W. (2010). *Biomaterials*, **31**, 91–98.

Xiao, Y., Gao, X., Taratula, O., Treado, S., Urbas, A., Holbrook, R.D., Cavicchi, R.E., Avedisian, C.T., Mitra, S., Savla, R., Wagner, P.D., Srivastava, S., He, H. (2009). *BMC Cancer*, **9**, 351.

Xie, J., Chen, K., Huang, J., Lee, S., Wang, J., Gao, J., Li, X., Chen, X. (2010a). *Biomaterials*, **31**, 3016–3022.

Xie, K., Doles, J., Hemann, M.T., Walker, G.C. (2010b). *Proc. Natl. Acad. Sci. U S A*, **107**, 20792–20797.

Xie, Y., Bagby, T.R., Cohen, M., Forrest, M.L. (2009). *Expert Opin. Drug Delivery*, **6**, 785–792.

Xie, Y., He, Y., Irwin, P.L., Jin, T., Shi, X. (2011). *Appl. Environ. Microbiol.*, **77**, 2325–2331.

Xing, M.Y., Guo, W.H., Cheng, Z.P., Zhang, Z.B., Zhu, J., Zhu, X.L. (2011a). *J. Controlled Release*, **152**, E104–E105.

Xing, S., Guan, Y., Zhang, Y. (2011b). *Macromolecules*, **44**, 4479–4486.

Xing, T., Mao, C., Lai, B., Yan, L. (2012). *ACS Appl. Mater. Interfaces*, **4**, 5662–5672.

Xiong, F., Zhu, Z.Y., Xiong, C., Hua, X.Q., Shan, X.H., Zhang, Y., Gu, N. (2012c). *Pharm. Res.*, **29**, 1087–1097.

Xiong, H.-M. (2010). *J. Mater. Chem.*, **21**, 4251–4262.

Xiong, H.-M., Xie, D.-P., Guan, X.-Y., Tan, Y.-J., Xia, Y.-Y. (2007). *J. Mater. Chem.*, **24**, 2490–2496.

Xiong, H.-M., Xu, Y., Ren, Q.-G., Xia, Y.-Y. (2008). *J. Am. Chem. Soc.*, **24**, 7522–7523.

Xiong, M.-H., Bao, Y., Yang, X.-Z., Wang, Y.-C., Sun, B., Wang, J. (2012a). *J. Am. Chem. Soc.*, **134**, 4355–4362.

Xiong, M.-H., Li, Y.-J., Bao, Y., Yang, X.-Z., Hu, B., Wang, J. (2012b). *Adv. Mater.*, **24**, 6175–6180.

Xiong, W., Wang, W., Wang, Y., Zhao, Y., Chen, H., Xu, H., Yang, X. (2011). *Colloid Surf. B*, **84**, 447–453.

Xiong, X.B., Lavasanifar, A. (2011). *ACS Nano*, **5**, 5202–5213.

Xiong, X.Y., Tam, K.C., Gan, L.H. (2005). *J. Controlled Release*, **103**, 73–82.

Xu, C.J., Xu, K.M., Gu, H.W., Zheng, R.K., Liu, H., Zhang, X.X., Guo, Z.H., Xu, B. (2004a). *J. Am. Chem. Soc.*, **32**, 9938–9939.

Xu, C.J., Xu, K.M., Gu, H.W., Zhong, X.F., Guo, Z.H., Zheng, R.K., Zhang, X.X., Xu, B. (2004b). *J. Am. Chem. Soc.*, **11**, 3392–3393.

Xu, D., Yu, J., Liu, Y., Sun, H., Xu, J., Sheng, K., Yao, S., Xu, Y., Lu, H. (2006). *Int. J. Nanosci.*, **5**, 753–756.

Xu, G., Yong, K.T., Roy, I., Mahadan, S.D., Ding, H., Schwartz, S.A., Prasad, P.N. (2008a). *Bioconjug. Chem.*, **19**, 1179–1185.

Xu, H., Regino, C.A., Bernardo, M., Koyama, Y., Kobayashi, H., Choyke, P.L., Brechbiel, M.W. (2007a). *J. Med. Chem.*, **50**, 3185–3193.

Xu, H., Regino, C.A., Koyama, Y., Hama, Y., Gunn, A.J., Bernardo, M., Kobayashi, H., Choyke, P.L., Brechbiel, M.W. (2007b). *Bioconjug. Chem.*, **18**, 1474–1482.

Xu, H.X., Bjerneld, E.J., Kall, M., Borjesson, L. (1999). *Phys. Rev. Lett.*, **21**, 4357–4360.

Xu, R., Pelicano, H., Zhou, Y., Carew, J.S., Feng, L., Bhalla, K.N., Keating, M.J., Huang, P. (2005). *Cancer Res.*, **65**, 613–621.

Xu, R., Wang, Y., Wang, X., Jeong, E.-K., Parker, D.L., Lu, Z.-R. (2007c). *Exp. Biol. Med.*, **232**, 1081–1089.

Xu, S., Xu, Q., Zhou, J., Wang, J., Zhang, N., Zhang, L. (2013a). *J. Nanosci. Nanotechnol.*, **13**, 129–138.

Xu, X., et al. (2013b). *Proc. Natl. Acad. Sci. U S A*, **110**, 18638–18643.

Xu, X.Y., Ray, R., Gu, Y.L., Ploehn, H.J., Gearheart, L., Raker, K., Scrivens, W.A. (2004c). *J. Am. Chem. Soc.*, **40**, 12736–12737.

Xu, Z., Gu, W., Chen, L., Gao, Y., Zhang, Z., Li, Y. (2008b). *Biomacromolecules*, **9**, 3119–3126.

Xuan, J., Boissiere, O., Zhao, Y., Yan, B., Tremblay, L., Lacelle, S., Xia, H. (2012). *Langmuir*, **28**, 16463–16468.

Yallapu, M.M., Jaggi, M., Chauhan, S.C. (2011). *Drug Discovery Today*, **16**(9/10), 457–463.

Yan, B., Boyer, J.C., Branda, N.R., Zhao, Y. (2011). *J. Am. Chem. Soc.*, **133**, 19714–19717.

Yan, F., Li, L., Deng, Z., Jin, Q., Chen, J., Yang, W., Yeh, C.K., Wu, J., Shandas, R., Liu, X., Zheng, H. (2013a). *J. Controlled Release*, **166**, 246–255.

Yan, F., Texter, J. (2006). *Soft Matter*, **2**, 109–118.

Yan, G.-P., Robinson, L., Hogg, P. (2007). *Radiography*, **13**, e5–e19.

Yan, Q., Han, D., Zhao, Y. (2013b). *Polym. Chem.*, **4**, 5026–5037.

Yanagi, K., Iakoubovskii, K., Matsui, H., Matsuzaki, H., Okamoto, H., Miyata, Y., Maniwa, Y., Kazaoui, S., Minami, N., Kataura, H. (2007). *J. Am. Chem. Soc.*, **129**, 4992–4997.

Yanagi, K., Miyata, Y., Kataura, H. (2006). *Adv. Mater.*, **18**, 437–441.

Yang, C.-C., Yang, S.-Y., Chieh, J.-J., Horng, H.-E., Hong, C.-Y., Yang, H.-C., Chen, K.H., Shih, B.Y., Chen, T.-F., Chiu, M.-J. (2011a). *ACS Chem. Neurosci.*, **2**, 500–505.

Yang, D., Hu, J., Fu, S. (2009). *J. Phys. Chem. C*, **18**, 7646–7651.

Yang, H., Lopina, S.T., DiPersio, L.P., Schmidt, S.P. (2008a). *J. Mater. Sci.: Mater. Med.*, **19**, 1991–1997.

Yang, L., Zhang, L., Webster, T.J. (2011b). *Nanomedicine (London)*, **6**, 1231–1244.

Yang, S.-T., Luo, J., Zhou, Q., Wang, H. (2012). *Theranostics*, **2**(3), 271–282.

Yang, W., Ratinac, K.R., Ringer, S.P., Thordarson, P., Gooding, J.J., Braet, F. (2010). *Angew. Chem., Int. Ed. Engl.*, **49**, 2114–2138.

Yang, W., Wu, G., Barth, R.F., Swindall, M.R., Bandyopadhyaya, A.K., Tjarks, W., Tordoff, K., Moeschberger, M., Sferra, T.J., Binns, P.J. (2008b). *Clin. Cancer Res.*, **14**, 883–891.

Yang, X., Zheng, R., Cai, Y., Liao, M., Yuan, W., Liu, Z. (2012a). *Int. J. Nanomed.*, **7**, 2077–2086.

Yang, X.L., Luo, Y.L., Xu, F., Chen, Y.S. (2013a). *Pharm. Res.*, **31**(2), 291–304

Yang, X.-Z., Dou, S., Wang, Y.-C., Long, H.-Y., Xiong, M.-H., Mao, C.-Q., Yao, Y.-D., Wang, J. (2012b). *ACS Nano*, **6**, 4955–4965.

Yang, Y., Velmurugan, B., Liu, X., Xing, B. (2013b). *Small*, **9**, 2937–2944.

Yang, Y.Q., Zhao, B., Li, Z.D., Lin, W.J., Zhang, C.Y., Guo, X.D., Wang, J.F., Zhang, L.J. (2013c). *Acta Biomater.*, **9**(8), 7679–7690.

Yankner, B.A. (1996). *Nat. Med.*, **2**, 850–852.

Yao, M., Stenmark, P., Abou-Hamad, E., Nitze, F., Qin, J., Goze-Bac, C., Wågberg, T. (2011). *Carbon*, **49**, 1159–1166.

Yavlovich, A., Singh, A., Blumenthal, R., Puri, A. (2011). *Biochim. Biophys. Acta*, **1808**, 117–126.

Yavlovich, A., Smith, B., Gupta, K., Blumenthal, R., Puri, A. (2010). *Mol. Membr. Biol.*, **27**, 364–381.

Yavuz, M.S., Cheng, Y., Chen, J., Cobley, C.M., Zhang, Q., Rycenga, M., Xie, J., Kim, C., Song, K.H., Schwartz, A.G., Wang, L.V., Xia, Y. (2009). *Nat. Mater.*, **8**, 935–939.

Yeh, R.W., Sidney, S., Chandra, M., Sorel, M., Selby, J.V., Go, A.S. (2010). *N. Engl. J. Med.*, **362**, 2155–2165.

Yin, J., Hu, H., Wu, Y., Liu, S. (2011). *Polym. Chem.*, **2**, 363–371.

Yin, J.J., Sharma, S., Shumyak, S.P., Wang, Z.X., Zhou, Z.W., Zhang, Y., Guo, P., Li, C.Z., Kanwar, J.R., Yang, T., Mohapatra, S.S., Liu, W., Duan, W., Wang, J.C., Li, Q., Zhang, X., Tan, J., Jia, L., Liang, J., Wei, M.Q., Li, X., Zhou, S.F. (2013a). *PLoS ONE*, **8**, e62289.

Yin, T., Wang, P., Li, J., Zheng, R., Zheng, B., Cheng, D., Li, R., Lai, J., Shuai, X. (2013b). *Biomaterials*, **34**, 4532–4543.

Yinghuai, Z., Cheng Yan, K., Maguire, J.A., Hosmane, N.S. (2007). *Curr. Chem. Biol.*, **1**, 141–149.

Yoo, S.I., Yang, M., Brender, J.R., Subramanian, V., Sun, K., Joo, N.E., Jeong, S.-H., Ramamoorthy, A., Kotov, N. (2011). *Angew. Chem., Int. Ed. Engl.*, **50**, 5110–5115.

Yoon, K.-Y., Hoon Byeon, J., Park, J.-H., Hwang, J. (2007). *Sci. Total Environ.*, **373**, 572–575.

Yoon, S.R., Yang, H.M., Park, C.W., Lim, S., Chung, B.H., Kim, J.D. (2012). *J. Biomed. Mater. Res. A*, **100**, 2027–2033.

Yoong, S.L., Wong, B.S., Zhou, Q.L., Chin, C.F., Li, J., Venkatesan, T., Ho, H.K., Yu, V., Ang, W.H., Pastorin, G. (2014). *Biomaterials*, **35**, 748–759.

Yordanov, A.T., Lodder, A.L., Woller, E.K., Cloninger, M.J., Patronas, N., Milenic, D., Brechbiel, M.W. (2002). *Nano Lett.*, **2**, 595–599.

Yoshida, M., Mata, J., Babensee, J.E. (2007). *J. Biomed. Mater. Res. Part A*, **80A**, 7–12.

Yoshida, M., Watanabe, Y., Sato, M., Maehara, T., Aono, H., Naohara, T., Hirazawa, H., Horiuchi, A., Yukumi, S., Sato, K., Nakagawa, H., Yamamoto, Y., Sugishita, H., Kawachi, K. (2010). *Int. J. Cancer*, **126**, 1955–1965.

You, J., Zhang, P., Hu, F., Du, Y., Yuan, H., Zhu, J., Wang, Z., Zhou, J., Li, C. (2013). *Pharm. Res.*, **31**(3), 554–565.

Younes, A., Gopal, A.K., Smith, S.E., Ansell, S.M., Rosenblatt, J.D., Savage, K.J., Ramchandren, R., Bartlett, N.L., Cheson, B.D., de Vos, S. (2012). *J. Clin. Oncol.*, **30**, 2183–2189.

Yu, A.K., Kudrinskiy, A.A., Olenin, A.Y., Lisichkin, G.V. (2008). *Russ. Chem. Rev.*, **3**, 233–257.

Yu, B., Mao, Y., Bai, L.Y., Herman, S.E., Wang, X., Ramanunni, A., Jin, Y., Mo, X., Cheney, C., Chan, K.K., Jarjoura, D., Marcucci, G., Lee, R.J., Byrd, J.C., Lee, L.J., Muthusamy, N. (2013a). *Blood*, **121**, 136–147.

Yu, C., Hu, Y., Duan, J., Yuan, W., Wang, C., Xu, H., Yang, X.D. (2011a). *PLoS ONE*, **6**, e24077.

Yu, H., Zou, Y., Wang, Y., Huang, X., Huang, G., Sumer, B.D. (2011b). *ACS Nano*, **2011**, 9246–9255.

Yu, J., Xie, X., Zheng, M., Yu, L., Zhang, L., Zhao, J., Jiang, D., Che, X. (2012a). *Int. J. Nanomed.*, **7**, 5079–5090.

Yu, S.S., Lau, C.M., Barham, W.J., Onishko, H.M., Nelson, C.E., Li, H., Smith, C.A., Yull, F.E., Duvall, C.L., Giorgio, T.D. (2013b). *Mol. Pharm.*, **10**, 975–987.

Yu, T., Liu, X., Bolcato-Bellemin, A.L., Wang, Y., Liu, C., Erbacher, P., Qu, F., Rocchi, P., Behr, J.P., Peng, L. (2012b). *Angew. Chem., Ger. Ed.*, **124**, 8606–8612.

Yu, Y.-L., Zhang, M.-J., Xie, R., Ju, X.-J., Wang, J.-Y., Pi, S.-W., Chu, L.-Y. (2012c). *J. Colloid Interface Sci.*, **376**, 97–106.

Yu, Z., Sun, J., Pan, C., Hong, C. (2012d). *Chem. Commun.*, **48**, 5623–5625.

Yu, Z.Q., Yan, J.J., You, Y.Z., Zhou, Q.H. (2012e). *Int. J. Nanomed.*, **7**, 5819–5832.

Yuan, F., Chen, Y., Dellian, M., Safabakhsh, N., Ferrara, N., Jain, R.K. (1996). *Proc. Natl. Acad. Sci. U S A*, **93**, 14765–14770.

Yuan, F., Dellian, M., Fukumura, D., Leunig, M., Berk, D.A., Torchilin, V.P., Jain, R.K. (1995). *Cancer Res.*, **55**, 3752–3756.

Yuan, Q., Zhang, Y., Chen, T., Lu, D., Zhao, Z., Zhang, X., Li, Z., Yan, C.H., Tan, W. (2012). *ACS Nano*, **6**, 6337–6344.

Yue, H., Wei, W., Yue, Z., Lv, P., Wang, L., Ma, G., Su, Z. (2010). *Eur. J. Pharm. Sci.*, **41**, 650–657.

Yue, Z.G., Wei, W., Lv, P.P., Yue, H., Wang, L.Y., Su, Z.G., Ma, G.H. (2011). *Biomacromolecules*, **12**, 2440–2446.

Yun, C.S., Javier, A., Jennings, T., Fisher, M., Hira, S., Peterson, S., Hopkins, B., Reich, N.O., Strouse, G.F. (2005). *J. Am. Chem. Soc.*, **9**, 3115–3119.

Yurek, D.M., Flectcher, A.M., Kowalczyk, T.H., Padegimas, L., Cooper, M.J. (2009). *Cell Transplant.*, **18**, 1183–1196.

Zabadal, M., Klan, P. (2001). *Chem. List.*, **95**, 694–699.

Zahr, A.S., Davis, C.A., Pishko, M.V. (2006). *Langmuir*, **22**, 8178–8185.

Zanoli, L., D'Agata, R., Spoto, G. (2012). *Anal. BioAnal. Chem.*, **5**, 1759–1771.

Zeleňáková, A., Zeleňák, V., Degmová, J., Kováč, J., Sedláčková, K., Abkowitz, M.A., Sitek, J. (2008). *Rev. Adv. Mater. Sci.*, **18**, 501–504.

Zha, L., Banik, B., Alexis, F. (2011). *Soft Matter*, **7**(13), 5908–5916.

Zha, Z., Zhang, S., Deng, Z., Li, Y., Li, C., Dai, Z. (2013). *Chem. Commun. (Camb.)*, **49**, 3455–3457.

Zhang, A., Zhang, Z., Shi, F., Xiao, C., Ding, J., Zhuang, X., He, C., Chen, L., Chen, X. (2013a). *Macromol. Biosci.*, **13**(9), 1249–1258.

Zhang, G., Yu, Y., Chen, X., Han, Y., Di, Y., Yang, B., Xiao, F., Shen, J. (2003a). *J. Colloid Interface Sci.*, **263**, 467–472.

Zhang, H.-J., Xin, Y., Yan, Q., Zhou, L.-L., Peng, L., Yuan, J.-Y. (2012a). *Macromol. Rapid Commun.*, **33**, 1952–1957.

Zhang, H.-J., Xiong, H.-M., Ren, Q.-G., Xia, Y.-Y., Kong, J.-L. (2012b). *J. Mater. Chem.*, **26**, 13159–13165.

Zhang, J., Misra, R.D.K. (2007). *Acta Biomater.*, **3**, 838–850.

Zhang, J.L., Srivastava, R.S., Misra, R.D.K. (2007a). *Langmuir*, **23**, 6342–6351.

Zhang, K., Chang, H.Y., Fu, A.H., Alivisatos, A.P., Yang, H. (2006). *Nano Lett.*, **4**, 843–847.

Zhang, K., Wu, X.Y. (2002). *J. Controlled Release*, **80**, 169–178.

Zhang, L., Gong, F., Zhang, F., Ma, J., Zhang, P., Shen, J. (2013b). *Int. J. Nanomed.*, **8**, 1517–1524.

Zhang, L., Gu, F.X., Chan, J.M., Wang, A.Z., Langer, R.S., Farokhzad, O.C. (2008). *Clin. Pharmacol. Ther.*, **83**, 761–769.

Zhang, L., Guo, R., Yang, M., Jiang, X., Liu, B. (2007b). *Adv. Mater.*, **19**, 2988e92.

Zhang, L., Jiang, Y., Ding, Y., Povey, M., York, D. (2007c). *J. Nanopart. Res.*, **9**, 479–489.

Zhang, M.Z., Yu, R.N., Chen, J., Ma, Z.Y., Zhao, Y.D. (2012c). *Nanotechnology*, **23**, 485104.

Zhang, W., Jiang, J., Qin, C.H., Perez, L.M., Parrish, A.R., Safe, S.H., Simanek, E.E. (2003b). *Supramol. Chem.*, **15**, 607–616.

Zhang, W., Shi, L., Wu, K., An, Y. (2005). *Macromolecules*, **38**, 5743–5747.

Zhang, X., Du, F., Huang, J., Lu, W., Liu, S., Yu, J. (2012d). *Colloids Surf. B Biointerfaces*, **100**, 155–162.

Zhang, X., Meng, L., Lu, Q., Fei, Z., Dyson, P.J. (2009a). *Biomaterials*, **30**, 6041–6047.

Zhang, Y., Calon, F., Zhu, C., Boado, R.J., Pardridge, W.M. (2003c). *Hum. Gene Ther.*, **14**, 1–12.

Zhang, Y., Hong, H., Cai, W. (2011). *Curr. Med. Chem.*, **18**, 4185–4194.

Zhang, Y.L., Yang, Y., Zhao, J.H., Tan, R.Q., Cui, P., Song, W.J. (2009b). *J. Sol-Gel Sci. Technol.*, **2**, 198–203.

Zhang, Z., Liu, Z., Ma, L., Jiang, S., Wang, Y., Yu, H., Yin, Q., Cui, J., Li, Y. (2013c). *Mol. Pharm.*, **10**, 2426–2434.

Zhang, Z.Y., Smith, B.D. (2000). *Bioconjug. Chem.*, **11**, 805–814.

Zhao, E., Zhao, Z., Wang, J., Yang, C., Chen, C., Gao, L., Feng, Q., Hou, W., Gao, M., Zhang, Q. (2012). *Nanoscale*, **16**, 5102–5109.

Zhao, M., Beauregard, D.A., Loizou, L., Davletov, B., Brindle, K.M. (2001). *Nat. Med.*, **7**, 1241–1244.

Zhao, N., Gao, M. (2009). *Adv. Mater.*, **21**, 184–187.

Zhao, N., Roesler, S., Kissel, T. (2011a). *Int. J. Pharm.*, **411**, 197–205.

Zhao, W., Chiuman, W., Lam, J.C.F., McManus, S.A., Chen, W., Cui, Y., Pelton, R., Brook, M.A., Li, Y. (2008). *J. Am. Chem. Soc.*, **11**, 3610–3618.

Zhao, Y. (2007). *Chem. Rec.*, **7**, 286–294.

Zhao, Y., Butler, E.B., Tan, M. (2013). *Cell Death Dis.*, **4**, e532.

Zhao, Y., Fan, X., Liu, D., Wang, Z. (2011b). *Int. J. Pharm.*, **409**, 229–236.

Zheng, D., Duan, C., Zhang, D., Jia, L., Liu, G., Liu, Y., Wang, F., Li, C., Guo, H., Zhang, Q. (2012). *Int. J. Pharm.*, **436**, 379–386.

Zheng, Y., Micic, M., Mello, S.V., Mabrouki, M., Andreopoulos, F.M., Konka, V., Pham, S.M., Leblanc, R.M. (2002). *Macromolecules*, **35**, 5228–5234.

Zhou, J., Wu, W., Caruntu, D., Yu, M.H., Martin, A., Chen, J.F., O'Connor, C.J., Zhou, W.L. (2007). *J. Phys. Chem. C*, **111**, 17473–17477.

Zhou, T., Wu, W., Zhou, S. (2010). *Polymer*, **51**, 3926–3933.

Zhou, T., Xiao, C., Fan, J., Chen, S., Shen, J., Wu, W., Zhou, S. (2013). *Acta Biomater.*, **9**, 4546–4557.

Zhu, H., Wang, X., Li, Y., Wang, Z., Yang, F., Yang, X. (2009). *Chem. Commun.*, **34**, 5118–5120.

Zhu, L., Kate, P., Torchilin, V.P. (2012). *ACS Nano*, **6**, 3491–3498.

Zhu, L., Torchilin, V.P. (2013). *Integr. Biol.*, **5**, 96–107.

Zhu, L., Wang, T., Perche, F., Taigind, A., Torchilin, V.P. (2013). *Proc. Natl. Acad. Sci. U S A*, **110**(42), 17047–17052.

Zhu, S., Hong, M., Tang, G., Qian, L., Lin, J., Jiang, Y., Pei, Y. (2010a). *Biomaterials*, **31**, 1360–1371.

Zhu, S., Hong, M., Zhang, L., Tang, G., Jiang, Y., Pei, Y. (2010b). *Pharmaceut. Res.*, **27**, 161–174.

Zhu, S., Zhou, Z., Zhang, D. (2007). *ChemPhysChem*, **8**, 2478–2483.

Ziegler, C., Eychmüller, A. (2011). *J. Phys. Chem. C*, **11**, 4502–4506.

Zigeuner, R.E., Riesenberg, R., Pohla, H., Hofstetter, A., Oberneder, R. (2003). *J. Urol.*, **2**, 701–705.

Zong, H., Thomas, T.P., Lee, K.-H., Desai, A.M., Li, M.-h., Kotlyar, A., Zhang, Y., Leroueil, P.R., Gam, J.J., Banaszak Holl, M.M. (2012). *Biomacromolecules*, **13**, 982–991.

Zu, B.Y., Pan, G.Q., Guo, X.Z., Zhang, Y., Zhang, H.Q. (2009). *J. Polym. Sci. Polym. Chem. Ed.*, **47**, 3257–3270.

Index